All aspects of space plasmas in the solar system are introduced and explored in this text for senior undergraduate and graduate students. *Introduction to Space Physics* provides a broad, yet selective, treatment of the complex interactions of the ionized gases of the solar-terrestrial environment. The book includes extensive discussions of the sun and solar wind, the magnetized and unmagnetized planets, and the fundamental processes of space plasmas, including shocks, plasma waves, ULF waves, wave–particle interactions, and auroral processes. The text devotes particular attention to space-plasma observations and integrates these with phenomenological and theoretical interpretations.

Highly coordinated chapters, written by experts in their fields, combine to provide a comprehensive introduction to space physics. Based on an advanced undergraduate and graduate course presented in the Department of Earth and Space Sciences at UCLA, the text will be valuable to both students and professionals in the field.

INTRODUCTION TO SPACE PHYSICS

INTRODUCTION TO SPACE PHYSICS

EDITED BY

Margaret G. Kivelson   **Christopher T. Russell**
University of California, Los Angeles

Published by the Press Syndicate of the University of Cambridge
The Pitt Building, Trumpington Street, Cambridge CB2 1RP
40 West 20th Street, New York, NY 10011-4211, USA
10 Stamford Road, Oakleigh, Melbourne 3166, Australia

© Cambridge University Press 1995

First published 1995

Printed in the United States of America

Library of Congress Cataloging-in-Publication Data
Introduction to space physics / edited by Margaret G. Kivelson,
  Christopher T. Russell.
      p. cm.
    ISBN 0-521-45104-3. – ISBN 0-521-45714-9 (pbk.)
    1. Sun. 2. Space plasmas. 3. Planets. I. Kivelson, M. G.
(Margaret Galland), 1928–   . II. Russell, Christopher T.
QB521.I65   1995
523.7 – dc20   94-19084
                                                                CIP

A catalog record for this book is available from the British Library.

ISBN 0-521-45104-3 Hardback
ISBN 0-521-45714-9 Paperback

CONTENTS

List of Contributors   *page* xi
Preface   xiii

1   A BRIEF HISTORY OF SOLAR-TERRESTRIAL PHYSICS
C. T. Russell   1
- 1.1   Ancient Auroral Sightings   1
- 1.2   Early Measurements of the Geomagnetic Field   3
- 1.3   The Emergence of a Scientific Discipline   5
- 1.4   The Ionosphere and Magnetosphere   9
- 1.5   The Solar Wind   13
- 1.6   Magnetospheric Exploration   15
- 1.7   Planetary and Interplanetary Exploration   21
- 1.8   Concluding Remarks   26
- *Additional Reading*   26
- *Problems*   26

2   PHYSICS OF SPACE PLASMAS
M. G. Kivelson   27
- 2.1   Introduction   27
- 2.2   Single-Particle Motion   27
- 2.3   Collections of Particles   34
- 2.4   The Plasma State   38
- 2.5   The Fluid Description of a Plasma   41
- 2.6   Two Applications of the MHD Equations   50
- 2.7   Conclusion   53
- **Appendix 2A:** Some Properties of Nonrelativistic Charged Particles in Magnetic Fields   53
- *Additional Reading*   54
- *Problems*   55

3   THE SUN AND ITS MAGNETOHYDRODYNAMICS
E. R. Priest   58
- 3.1   Introduction   58
- 3.2   The New Sun   62
- 3.3   The Role of the Magnetic Field   65
- 3.4   MHD Equilibria, Waves, and Instabilities   69

3.5	Solar Activity	70
3.6	Prominences	74
3.7	Coronal Heating	79
3.8	Solar Flares	82
3.9	Conclusion	88
	Additional Reading	88
	Problems	89

4 THE SOLAR WIND
A. J. Hundhausen — 91

4.1	Introduction	91
4.2	A Quick Survey of Solar-Wind Properties	92
4.3	The Basic Concept of Solar-Wind Formation in the Solar Corona	96
4.4	The Magnetic Structure of the Corona and Solar Wind	110
4.5	The Major Time-Dependent Disturbances of the Solar Wind	124
	Additional Reading	128
	Problems	128

5 COLLISIONLESS SHOCKS
D. Burgess — 129

5.1	Introduction	129
5.2	Shocks without Collisions	134
5.3	Shock Structure: How Shocking?	145
5.4	Things That Haven't Been Mentioned	155
Appendix 5A:	The de Hoffman–Teller Frame	156
Appendix 5B:	Energetic Particles and Foreshocks	158
Appendix 5C:	Determining the Shock-Normal Direction	161
	Additional Reading	163
	Problems	163

6 SOLAR-WIND INTERACTIONS WITH MAGNETIZED PLANETS
R. J. Walker and C. T. Russell — 164

6.1	Introduction	164
6.2	Planetary Magnetic Fields	164
6.3	Size of the Magnetospheric Cavity	168
6.4	Shape of the Magnetospheric Cavity	172
6.5	Self-Consistent Models	174
6.6	Flow around the Magnetosphere	177
6.7	Concluding Remarks	180
	Additional Reading	181
	Problems	181

7 IONOSPHERES
J. G. Luhmann   183

- 7.1 Introduction   183
- 7.2 Ion Production   184
- 7.3 Ion Loss   192
- 7.4 Determining Ionospheric Density from Production and Loss Rates   193
- 7.5 An Example: The Earth's Ionosphere   196
- 7.6 Other Considerations Relating to Ionospheres   199
- 7.7 Final Notes   202
    - *Additional Reading*   202
    - *Problems*   202

8 PLASMA INTERACTIONS WITH UNMAGNETIZED BODIES
J. G. Luhmann   203

- 8.1 Introduction   203
- 8.2 Plasma Interactions with Moonlike Bodies   203
- 8.3 Plasma Interactions with Bodies with Atmospheres   207
- 8.4 Concluding Remarks   224
    - *Additional Reading*   225
    - *Problems*   225

9 THE MAGNETOPAUSE, MAGNETOTAIL, AND MAGNETIC RECONNECTION
W. J. Hughes   227

- 9.1 Introduction   227
- 9.2 The Magnetopause   228
- 9.3 The Geomagnetic Tail   232
- 9.4 Magnetic Reconnection   236
- 9.5 Reconnection at the Magnetopause   259
- 9.6 Reconnection and the Plasma-Sheet Boundary Layer   276
- 9.7 Is Steady-State Convection Possible in the Tail?   283
- 9.8 Conclusion   284
    - *Additional Reading*   284
    - *Problems*   285

10 MAGNETOSPHERIC CONFIGURATION
R. A. Wolf   288

- 10.1 Introduction   288
- 10.2 Magnetic-Field Configuration of the Earth's Magnetosphere   288
- 10.3 Plasma in the Earth's Middle and Inner Magnetosphere 290

10.4	Electric Fields and Magnetospheric Convection	300
10.5	Adiabatic Invariants and Particle Drifts	304
10.6	Ionosphere–Magnetosphere Coupling	320
10.7	Ionospheric Currents	323
10.8	Magnetic-Field-Aligned Potential Drops	324
10.9	Loss of Magnetospheric Particles into the Earth's Atmosphere	325
10.10	Concluding Comment	327
	Additional Reading	327
	Problems	328

11 PULSATIONS AND MAGNETOHYDRODYNAMIC WAVES

M. G. Kivelson   330

11.1	Introduction	330
11.2	Basic Equations	332
11.3	Equations for Linear Waves	334
11.4	Waves in Cold Plasmas	335
11.5	Waves in Warm Plasmas	340
11.6	Ionospheric Boundary Conditions	343
11.7	MHD Waves in a Dipolar Magnetic Field	345
11.8	Sources of Wave Energy	349
11.9	Instabilities	350
11.10	Waves in Planetary Magnetospheres and Elsewhere	352
	Additional Reading	353
	Problems	353

12 PLASMA WAVES

C. K. Goertz and R. J. Strangeway   356

12.1	Introduction	356
12.2	Waves in a Two-Fluid Plasma	356
12.3	Waves in an Unmagnetized Plasma	360
12.4	Waves in a Magnetized Plasma	375
12.5	Kinetic Theory and Wave Instabilities	392
	Additional Reading	398
	Problems	398

13 MAGNETOSPHERIC DYNAMICS

R. L. McPherron   400

13.1	Introduction	400
13.2	Types of Magnetic Activity	402
13.3	Measures of Magnetic Activity: Geomagnetic Indices	408
13.4	Solar-Wind Control of Geomagnetic Activity	410
13.5	Magnetospheric Control of Geomagnetic Activity	420
13.6	Phenomenological Models of Substorms	430

	13.7 Conclusions	441
	Appendix 13A: Instruments for Measuring Magnetic Fields	443
	Appendix 13B: Standard Indices of Geomagnetic Activity	451
	Additional Reading	457

14 THE AURORA AND THE AURORAL IONOSPHERE

H. C. Carlson, Jr., and A. Egeland — 459

14.1	Introduction	459
14.2	Auroral-Particle Precipitation: The Auroral Spectrum	463
14.3	Auroral Distribution in Space and Time	476
14.4	The Auroral Substorm	486
14.5	The Auroral Ionosphere	489
14.6	Auroral Effects on Radio Waves	493
14.7	Energy Transfer to the Ionosphere	494
14.8	Relation to Boundaries and Physical Processes in the Magnetosphere–Ionosphere–Thermosphere	497
14.9	Stable Sun-Aligned Arc: Energetics and Thermal Balance	498
	Additional Reading	500
	Problems	500

15 THE MAGNETOSPHERES OF THE OUTER PLANETS

C. T. Russell and R. J. Walker — 503

15.1	Introduction	503
15.2	The Variation in the Solar-Wind Properties	504
15.3	Magnetospheric Size	507
15.4	The Role of Reconnection	511
15.5	Interaction of Moons with their Magnetospheres	512
15.6	Radiation Belts	514
15.7	Waves and Instabilities	515
15.8	Radio Emissions	517
15.9	Concluding Remarks	519
	Additional Reading	519
	Problems	519

Appendix 1: Notation, Vector Identities, and Differential Operators	521
Appendix 2: Fundamental Constants and Plasma Parameters of Space Physics	529
Appendix 3: Geophysical Coordinate Transformations	531
References	545
Index	563

CONTRIBUTORS

D. Burgess
Astronomy Unit
Queen Mary and Westfield College
London, U.K.

H. C. Carlson, Jr.
Phillips Laboratory
Geophysics Directorate
Hanscom AFB
Bedford, MA

A. Egeland
Department of Physics
University of Oslo
Blindern, Oslo, Norway

C. K. Goertz (deceased)
Department of Physics and
  Astronomy
University of Iowa
Iowa City, IA

W. J. Hughes
Center for Space Physics
Boston University
Boston, MA

A. J. Hundhausen
High Altitude Observatory
National Center for Atmospheric
  Research
Boulder, CO

M. G. Kivelson
Department of Earth and Space
  Sciences and
Institute of Geophysics and
  Planetary Physics
University of California
Los Angeles, CA

J. G. Luhmann
Institute of Geophysics and
  Planetary Physics
University of California
Los Angeles, CA

R. L. McPherron
Department of Earth and Space
  Sciences and
Institute of Geophysics and
  Planetary Physics
University of California
Los Angeles, CA

E. R. Priest
Mathematical and Computational
  Sciences Department
The University
St. Andrews, Scotland

C. T. Russell
Department of Earth and Space
  Sciences and
Institute of Geophysics and
  Planetary Physics
University of California
Los Angeles, CA

R. J. Strangeway
Institute of Geophysics and
 Planetary Physics
University of California
Los Angeles, CA

R. J. Walker
Institute of Geophysics and
 Planetary Physics
University of California
Los Angeles, CA

R. A. Wolf
Department of Space Physics and
 Astronomy
Rice University
Houston, TX

PREFACE

THE IONIZED GASES of the solar-terrestrial environment interact in very complex and sometimes counterintuitive ways. Our intuition about gases is trained in situations in which collisions are important, but in most of the ionized gases in the solar system the magnetic and electric fields control the motion of the particles, with collisions and gravitational fields being less important. In an introductory text such as this it is difficult to decide where to begin to discuss these interactions. One could start with the simplest systems and then add complexity; one could order the material by spatial location, discussing the sun first and then proceeding to follow the energy flow outward past all the planets; or one could follow a chronological approach, according to the order of discovery. There is much to justify a spatial approach, because the sun is the energy source for most of the plasma we encounter, either through coupling with the solar wind or through photoionization. On the other hand, the chronological approach follows the way scientists originally learned about how the solar terrestrial environment behaves. This approach has the advantage that the earliest concepts were simple and grew gradually in complexity, but it has the disadvantage that some of the early ideas were wrong and that sometimes science progresses in convoluted ways. Thus, this approach can be quite inefficient.

In this book we shall attempt to combine the three approaches. We shall try always to reduce topics to their basics before introducing the complications. The overall ordering of the book will follow the energy flow, starting with the sun, but first, Chapter 1 will provide some historical perspective. The historical approach is interesting, and it allows us a quick overview of the entire field before becoming too involved with the details. We shall begin with ancient observations of the "northern lights," which we now refer to as the aurora borealis, or simply the aurora, and work our way up to the era of space exploration.

Chapter 2 covers the physics of the plasmas we encounter in space. In this chapter we describe some of the most basic physical processes that occur in space plasmas and the equations that govern them. In particular we introduce the magnetohydrodynamic (MHD) approximations that are so useful in describing the solar-terrestrial environment. In Chapter 3, E. R. Priest discusses both the "old" sun and the "new" sun. The new sun has emerged from the old sun of observational solar physics through the application of MHD treatments of solar phenomena.

In Chapter 4, A. J. Hundhausen takes us from the solar corona to the farthest reaches of the solar system, the heliopause, where the solar wind stops. D. Burgess follows this with a discussion of that ubiquitous process in space plasmas – the collisionless shock. Collisionless shocks are caused by processes on the sun, by the interaction of fast streams and slow streams in the solar wind, and by the diversion of the solar wind about the intrinsic and induced magnetospheres of all the planets.

In Chapter 6, R. J. Walker and C. T. Russell describe the interaction of the solar wind with a magnetized planet, both what happens to the planetary magnetic field and how the interaction affects the solar wind. In Chapters 7 and 8, J. G. Luhmann describes how an ionosphere is formed from a planetary atmosphere by the ionizing radiation from the sun and then how such an ionosphere creates a magnetic barrier to the solar-wind flow and an induced magnetosphere that deflects the solar wind, much as does an intrinsic planetary magnetic field.

In Chapter 9, W. J. Hughes examines the processes whereby energy is transferred to the earth's magnetosphere by the solar wind, the storage of that energy in the tail, and the eventual release of that energy into the inner magnetosphere and ionosphere in a magnetospheric substorm. In Chapter 10, R. A. Wolf describes the processes occurring in the inner magnetosphere.

Chapters 11 and 12 cover the wave processes in the magnetosphere. In Chapter 11, M. G. Kivelson discusses the phenomena known collectively as magnetic pulsations, which often involve the oscillation of an entire magnetic-field line. Chapter 12 reviews the waves that occur at higher frequencies, principally interacting with the energetic electrons in the magnetosphere. The majority of the material in this chapter was prepared by C. K. Goertz, who was killed in a most unfortunate incident before completion of this work. We are most grateful to R. J. Strangeway who took over the writing of the chapter.

Chapter 13, by R. L. McPherron, deals with magnetospheric dynamics and geomagnetic activity and the current systems responsible for this behavior.

In Chapter 14, H. C. Carlson, Jr., and A. Egeland cover the auroral ionosphere, where much of the energy transferred to the magnetosphere from the solar wind is ultimately deposited. Chapter 15 covers the magnetospheres of the outer planets. The book closes with a set of useful appendices covering various topics of practical importance, such as vector operations and coordinate transformations.

It is our intention in this book to provide an introduction to space physics for the beginning graduate student. Nevertheless, much of the material is suitable for upper-division undergraduates and has been tested on both undergraduates and graduate students in our Department of Earth and Space Sciences at the University of California, Los Angeles (UCLA).

The assembling of this book began with the convening of a "Rubey Colloquium" in March 1990. The Department of Earth and Space Sciences at UCLA holds such a colloquium annually in honor of the late W. W. Rubey (1898–1974), a career geologist with the U.S. Geological Survey and a professor of geology and geophysics at UCLA. That colloquium brought together the authors of this book, who presented lectures associated with each of the chapters over the course of a week between the winter and spring quarters. Initial drafts of the chapters were distributed at that time. Since then they have been refined and edited in an attempt to produce a more uniform style, to eliminate unnecessary duplication of material, and to fill in some of the gaps in coverage. We are particularly grateful to the Department of Earth and Space Sciences for providing the funding to initiate this project and to the National Aeronautics and Space Administration for sustaining it through their Space Grant University program administered by the California Space Institute. We are also most grateful to A. McKnight, Linda Kim, and Rose Silva, who provided clerical assistance for this project, and to UCLA students M. Ginskey, T. Meseroll, T. Mulligan, and J. Newbury, who helped to proofread the volume.

C. T. Russell
M. G. Kivelson

1 A BRIEF HISTORY OF SOLAR-TERRESTRIAL PHYSICS

C. T. Russell

SOLAR-TERRESTRIAL PHYSICS is principally concerned with the interaction of energetic charged particles with the electric and magnetic fields in space. In the vicinity of the earth, most of these charged particles derive their energy ultimately from the sun or from the interaction of the solar wind with the earth's magnetosphere. These interactions are complex, because the magnetic and electric fields that determine the motion of the particles are affected in turn by the motion of these charged particles. Some solar-terrestrial research is carried out on the surface of the earth with cameras, photometers and spectrometers, and magnetometers and other devices sensitive to the processes occurring high in the upper atmosphere and magnetosphere, but today most of this research is carried out using rockets and satellites that enable measurements to be obtained directly in the regions of strongest interactions. In recent years, these in situ data have resulted in explosive growth in our knowledge and understanding of solar-terrestrial processes. Nevertheless, the field has had a long history of investigation, starting well before the advent of satellites and rockets. We shall briefly review that history in order to provide a context for our later, more physically oriented presentation of the processes occurring in the solar-terrestrial environment.

1.1 ANCIENT AURORAL SIGHTINGS

The emerging field of solar-terrestrial physics began with a growing appreciation of two phenomena: the aurora and, later, the geomagnetic field. Because it can be observed visually, the aurora was the first of these phenomena to be recorded. Most other phenomena of this emerging field awaited the advent of new technology, such as the compass in the case of geomagnetism, before they were discovered. References to the aurora are contained in the ancient literature from both East and West. Several passages in the Old Testament appear to have been inspired by auroral sightings, and Greek literature includes references to phenomena most likely to have been auroral phenomena. For example, Xenophanes, in the sixth century B.C., mentions "moving accumulations

FIG. 1.1. Early drawing of the aurora, 12 January 1570. (Original print in Crawford Library, Royal Observatory, Edinburgh.)

of burning clouds." Chinese literature also describes possible auroral sightings, of which several occurred prior to 2000 B.C.

Because the phenomenon was not understood, much fear and superstition surrounded those early sightings of the aurora. Figure 1.1, inspired by an auroral display in 1570, illustrates the lack of scientific understanding prevalent at that time. The seventeenth century marked the beginning of scientific theories concerning the origin of the lights in the north. Galileo Galilei, for example, proposed that the aurora was caused by air rising out of the earth's shadow to where it could be illuminated by sunlight. He also appears to have coined the term *aurora borealis,* meaning "dawn of the north." Pierre Gassendi, a French mathematician and astronomer, at about the same time, deduced that auroral displays must be occurring at great heights, because they were seen to have the same configuration when observed at places quite remote from one another. His contemporary, René Descartes, seems to have been the originator of the idea that auroras were caused by reflections from ice crystals in the air at high latitudes. From about 1645 to about 1715, both solar activity and auroral sightings declined, although neither were completely lacking.

Edmund Halley, after finally, at the age of 60, having personally observed an auroral display, seems to have been the first to suggest that the auroral phenomenon was ordered by the direction of the earth's magnetic field. In 1731, the French philosopher de Mairan ridiculed the currently popular idea that the aurora was a reflection of polar ice and snow, and he also criticized Halley's theory. He suggested that the aurora was connected to the solar atmosphere, and he suspected a connection between the return of sunspots and the aurora. After that time, studies of geomagnetism and the aurora became more firmly linked.

1.2 EARLY MEASUREMENTS OF THE GEOMAGNETIC FIELD

The earliest indication of the existence of the geomagnetic field was the direction-finding capability of the compass. As compasses were improved, more and more was learned about the geomagnetic field. The earliest reliable evidence of Chinese knowledge that a compass points north or south dates from the eleventh century. The encyclopedist Shon-Kau (A.D. 1030–93) stated that "fortune-tellers rub the point of a needle with the stone of the magnet in order to make it properly indicate the south." In the European literature, the earliest mention of the compass and its application to navigation appeared in two works by Alexander Neekan, a monk of St. Albans (A.D. 1157–217), entitled *De Untensilibus* and *De Rerum*. In the former he described the use of the magnetic needle to indicate north and noted that mariners used that means to find their course when the sky was cloudy. In the second, he described the needle as being placed on a pivot, a second-generation form of the compass. In neither work did he describe the instrument as a novelty; it was in common use at that time. Official records indicate that by the fourteenth century, many sailing ships carried compasses.

The direction of magnetic north and that of geographic or true north differ over most of the globe. The measure of this difference is called the declination. It is not clear when magnetic declination was actually discovered. However, a letter written by Georg Hartmann, vicar of St. Sebald's at Nürnberg, to Duke Albrecht of Prussia in 1544 showed that he had observed the declination of Rome in 1510 to be 6° east, whereas it was 10° at Nürnberg. Also, it is known that between the years 1538 and 1541, João de Castro made 43 determinations of declination during a voyage along the west coast of India and in the Red Sea.

The geomagnetic field is also inclined to the horizontal. To measure this inclination, one must pivot a needle about a horizontal axis. Georg Hartmann's letter also discussed such an observation, but the angle of inclination was incorrect for his point of observation. William Gilbert ascribed the discovery of the magnetic dip or inclination to an Englishman, Robert Norman, who in 1576 published a work with the title *The newe Attractiue containyng a short discourse of the Magnes or Lodestone, and amongst other his vertues, of a newe discouered secret and subtill propertie, concernyng the Declinyng of the Needle, touched there with onder the plaine of the Horizon. Now first found out by* ROBERT NORMAN *Hydrographer. Here onto are annexed certaine necessarie rules for the art of Nauigation, by the same R.N. Imprinted at London by John Kyngston, for Richard Ballard, 1581.*

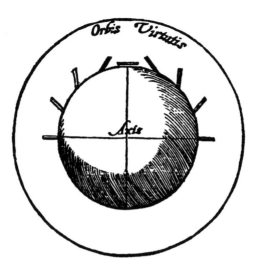

FIG. 1.2. Illustration of magnetic-dipole character of the earth's main magnetic field, as shown in Gilbert's *De Magnete*.

The year 1600 saw the publication of the famous treatise *De Magnete*, by William Gilbert, who in 1601 was appointed chief physician in personal attendance to Queen Elizabeth. This treatise consists of six books containing a total of 115 chapters. The central theme of the book is also the title of Chapter 17, Book 1: "That the globe of the earth is magnetic, a magnet; how in our hands the magnet stone has all the primary forces of the earth, while the earth by the same powers remains constant in a fixed direction in the universe." Figure 1.2 is Gilbert's woodcut showing the distribution of magnetic inclination or dip over the earth, and over a small spherical lodestone, which he called a "terrella." Gilbert believed that the terrestrial magnetic field was constant, but it is not. Henry Gellibrand, professor of astronomy at Gresham College, discovered that magnetic declination changed with time, and he published his discovery in a work entitled *A discourse mathematical on the variation of the magneticall needle. Together with its admirable diminuation lately discovered. London 1635.*

Another early pioneer in the study of geomagnetism was Edmund Halley, who published in 1683 and 1692 two works on the theory of geomagnetism, but needed to test his theory further. King William III put at his disposal the ship *Paramour Pink,* on which Halley made two voyages: in October 1698 to the North Atlantic Ocean, and in September 1700 to the South Atlantic Ocean. Those voyages were the first purely scientific expeditions, and they returned measurements of great value both for practical navigation and for the theory of navigation. Those investigations led to the publication of two geomagnetic charts: "New and Correct Chart showing the Variations of the Compass in the Western and Southern Oceans, as observed in year 1700 by his Majesty's Command by Edm. Halley" and "Sea Chart of the whole world, showing the Variations of the Compass," published in 1701 and 1702, respectively.

1.3 THE EMERGENCE OF A SCIENTIFIC DISCIPLINE

Despite the fact that the sun is the most luminous object we can see, the solar side of solar-terrestrial physics awaited technological change as surely as the study of geomagnetism awaited the development of the compass and its successor the magnetometer. Sunspots, magnetized cool spots in the solar photosphere, are generally too small to be resolved by the naked eye. Thus the study of sunspots did not begin until the invention of the telescope. Galileo Galilei was one of the first to use this new invention to study them. Sunspot studies proceeded slowly, perhaps because very few sunspots occurred during the period called the Maunder minimum, from about 1645 to 1700. The now-familiar 11-yr periodicity in sunspot number illustrated in Figure 1.3 was not discovered until 1851. The sunspot or solar cycle is discussed in greater depth in Chapter 3, which reviews our current understanding of the physics of the sun, in which magnetism plays a significant role that is only gradually being understood.

Perhaps the first discovery in the emerging discipline we now call solar-terrestrial physics was the observation in 1722 by George Graham, a famous London instrument maker, that the compass is always in motion. Graham's discovery was confirmed in 1740 by Anders Celsius in Uppsala, Sweden. His observations were continued by O. Hiorter to a total of over 20,000 observations made on more than 1,000 different days. From those data Hiorter discovered the diurnal variation of the geomagnetic field. Magnetic perturbations vary systematically with local time, which is determined by the longitudinal separation between the meridian of the observer and that containing the sun, which is called the noon meridian. These perturbations are due to the rotation of one's observation station under current systems flowing in the upper atmo-

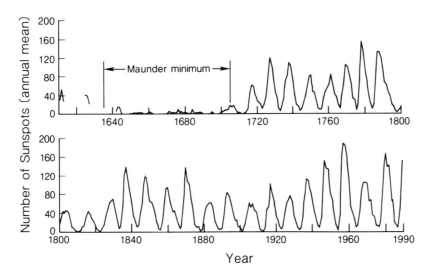

FIG. 1.3. Sunspot cycle since A.D. 1610.

sphere that are fixed with respect to the sun. Perhaps even more important, on April 5, 1741, Hiorter discovered that geomagnetic and auroral activities were correlated. Simultaneous observations in London by Graham confirmed the occurrence of strong geomagnetic activity on that day. In 1770, J. C. Wilcke noted that auroral rays extend upward along the direction of the magnetic field. That was the same year that Captain James Cook first reported the southern counterpart of the aurora borealis, the aurora australis, or "dawn of the south." Twenty years later, the English scientist Henry Cavendish used triangulation to estimate the height of auroras as between 52 and 71 miles. Earlier attempts at triangulation by Halley and Mairan had been much less accurate.

The great advance of the early nineteenth century was the development of a network to make frequent simultaneous observations with widely spaced magnetometers. C. F. Gauss was one of the leaders of that effort and one of the foremost pioneers in the mathematical analysis of the resulting measurements, which allowed contributions to the geomagnetic field from below the surface of the earth to be separated from those contributions arising high in the atmosphere. Meanwhile, Heinrich Schwabe, on the basis of his sunspot measurements taken between 1825 and 1850, deduced that the variation in the number of sunspots was periodic, with a period of about 10 yr. Magnetic observatories had spread to the British colonies by 1839. Edward Sabine was assigned to supervise four of those observatories (Toronto, St. Helena, Cape of Good Hope, and Hobarton). Using the data from those observatories, he was able to show in 1851 that the intensity of geomagnetic disturbances varied in concert with the sunspot cycle. Chapter 13 discusses our modern understanding of these disturbances.

The next discovery to provide a link between the sun and geomagnetic activity was Richard Carrington's sighting of a great flare of white light on the sun, on September 1, 1859. Carrington, who was sketching sunspot groups at the time, was startled by the flare, and by the time he was able to summon someone to witness the event a minute later, he was dismayed to find that it had weakened greatly in intensity. Fortunately, it had been simultaneously noted by another observer some miles away. Furthermore, at the moment of the flare, the Kew Observatory (London) measurements of the magnetic field had been disturbed. Today we realize that that disturbance of the magnetic field was caused by an increase in the electric currents flowing overhead in the earth's ionosphere. Such currents flow in response to electric fields in the ionosphere. The extreme radiation by ultraviolet rays and x-rays from the flare increased the ionization and hence the electrical conductivity of the ionosphere, causing more current to flow in response to the unaltered electric field. Finally, 18 h later, one of the strongest magnetic storms ever recorded broke out. Auroras were seen as far south as Puerto Rico.

To have arrived that quickly, the disturbance would have had to

1.3 THE EMERGENCE OF A SCIENTIFIC DISCIPLINE

travel from the sun at over 2,300 km·s^{-1}. As discussed in Chapter 4, we know today that the sun and the earth are linked by the supersonic solar wind, but 2,300 km·s^{-1} is a high velocity even for the disturbed solar wind. When such disturbances arrive at the earth, the terrestrial field jumps quite abruptly, indicating that the discontinuity in the interplanetary medium that is flowing by the earth is quite thin. The thinness of these disturbance fronts strongly suggests that they are caused by shocks in the interplanetary medium, despite the collisionless nature of the gas there. Usually, collisions are needed to account for the dissipation and heating that occur at a shock. That was the first indication of the existence of collisionless shocks, which since the advent of interplanetary exploration have been found to be ubiquitous in the solar system, as discussed in Chapter 5. Shortly after those observations, in 1861, Balfour Stewart noted the occurrence of pulsations in the earth's magnetic field, with periods of minutes. We now know that the magnetosphere pulsates at a wide variety of periods. These pulsations are discussed in further detail in Chapter 11.

The nineteenth century also brought another simple but important observation about the aurora. Captain John Franklin, the ill-fated English Arctic explorer whose party perished in 1845 attempting to discover the Northwest Passage, noted that auroral frequency did not increase all the way to the pole, according to observations made during his 1819–22 journeys. In 1860, Elias Loomis of Yale was one of the first to plot the zone of maximum auroral occurrence, which roughly corresponds to what today we call the auroral zone. The auroral zone is an oval band around the magnetic pole, roughly 20–25° from the pole.

Precursors to our modern understanding of the aurora began to appear in the late nineteenth century. About 1878, H. Becquerel suggested that particles were shot off from the sun and were guided by the earth's magnetic field to the auroral zone. He believed that sunspots ejected protons. A similar theory was espoused by E. Goldstein. In 1897, the great Norwegian physicist Kristian Birkeland made his first auroral expedition to northern Norway. However, it was not until after his third expedition in 1902–3, during which he obtained extensive data on the magnetic perturbations associated with auroras, that he concluded that large electric currents flowed along magnetic-field lines during aurora. The invention of the vacuum tube led to the understanding that the aurora was in some way similar to the cathode rays in those devices. Soon Sir William Crookes demonstrated that cathode rays were bent by magnetic fields, and shortly thereafter J. J. Thomson showed that cathode rays consisted of the tiny, negatively charged particles we now call electrons. Birkeland adopted those ideas for his auroral theories and attempted to verify his theories with both field observations and laboratory experiments. Specifically, he conducted experiments with a magnetic dipole inside a model earth, which he called a terrella. Figure 1.4 shows Birkeland in his laboratory beside his terrella experiment. Those

FIG. 1.4. Kristian Birkeland (left) in his laboratory with his terrella and with his assistant, O. Devik (right), about 1909. (Photo courtesy of A. Egeland.)

experiments showed that electrons incident on the terrella would produce patterns quite reminiscent of the auroral zone. He believed, as we do today, that those particles came from the sun.

K. Birkeland's work inspired the Norwegian mathematician Carl Størmer, whose subsequent calculations of the motion of charged particles in a dipole magnetic field in turn supported Birkeland. Figure 1.5 shows Størmer and his assistant Olaf (not Kristian) Birkeland. As is evident from this photograph, the advent of the camera was an important advance in the study of the aurora. It was through measurements such as these that Størmer accurately determined the height of the aurora. Figure 1.6 illustrates one of Størmer's charged-particle orbit calculations in a forbidden region to which charged particles from the sun would not have direct access. In such a region, charged particles would spiral around the magnetic field and bounce back and forth along it, reflected by the converging magnetic-field geometry. Størmer's contributions became much more relevant and appreciated after the discovery of the earth's radiation belts, whose particle motions resemble those of Figure 1.6. Birkeland's work was not appreciated until even later. A more detailed discussion of the trajectories of charged particles in the earth's magnetic field can be found in Chapter 10, and more about the aurora can be found in Chapter 14.

All that work proceeded despite Lord Kelvin's 1882 argument that he had provided absolutely conclusive evidence against the supposition that terrestrial magnetic storms were due to magnetic action in the sun

FIG. 1.5. Auroral physicists C. F. Størmer, standing, and Olaf Birkeland, seated, in northern Norway, ca. 1910. (Photo courtesy of A. Egeland.)

or to any kind of dynamic action taking place within the sun. Lord Kelvin also claimed "that the supposed connection between magnetic storms and sunspots is unreal, and the seeming agreement between periods has been a mere coincidence." More telling was the criticism of A. Schuster that a beam of electrons from the sun could not hold together against their mutual electrostatic repulsion.

1.4 THE IONOSPHERE AND MAGNETOSPHERE

The electrically conducting region above about 100 km altitude that we now call the ionosphere may rightly be claimed to have been discovered by Balfour Stewart. In his 1882 *Encyclopaedia Britannica* article entitled "Terrestrial Magnetism" he concluded that the upper atmosphere was the most probable location of the electric currents that produce the solar-controlled variation in the magnetic field measured at the surface of the earth. He noted that "we know from our study of aurora that there are such currents in these regions – continuous near the poles and occasional in lower latitudes." He proposed that the primary causes of the daily variations in the intensity of the surface magnetic field were "convective currents established by the sun's heating influence in the upper regions of the atmosphere." These currents "are to be regarded as conductors moving across lines of magnetic force and are thus the vehicle of electric currents which act upon the magnetic field." Those statements are very close to modern atmospheric-dynamo theory. However, it was left to A. Schuster to put the dynamo theory into quantitative form.

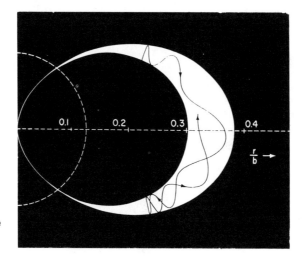

FIG. 1.6. Trajectory of an energetic charged particle in the "forbidden" zone of a dipole magnetic field, as drawn by Størmer. (From Rossi and Olbert, 1970.)

The turn of the century brought another new invention that was used to probe the solar-terrestrial environment: the radio transmitter and receiver. In 1902, A. E. Kennelly and O. Heaviside independently postulated the existence of a highly electrically conducting ionosphere to explain G. Marconi's transatlantic radio transmissions. Verification of the existence of the ionosphere did not come until much later, in 1925, when E. V. Appleton and M. A. F. Barnett in the United Kingdom, and shortly thereafter G. Breit and M. A. Tuve in America, established the existence and altitude of the Kennelly-Heaviside layer, as it was known then. The original method of Breit and Tuve, using short pulses of radio energy at vertical incidence and timing the arrival of a reflected signal in order to infer the altitude of the electrically reflecting layer, is still used today for sounding the ionosphere. In drawing diagrams of the electromagnetic waves reflected by the ionosphere, Appleton used the letter E for the electric vector of the downcoming wave. When he found reflections from a higher layer, he used the letter F for the electric vector of those reflected waves, and when he occasionally got reflections from a lower layer, he naturally used the letter D. When it came time to name these layers, he chose the same letters, leaving the letters A, B, and C for possible later discoveries that never came. So now the ionospheric layers are called the D, E, and F layers, as illustrated in Figure 1.7. We now know that all planets with atmospheres have electrically conducting ionospheres like that of the earth. Chapter 7 discusses how these are formed.

At about that same time, progress was being made in understanding the auroral glows. Spectroscopy, together with photography, permitted first the determination of the wavelength and then the identity of the excited molecule that was radiating. There were initial successes, beginning with Lars Vegard's work in Norway relating auroral emissions to emission bands from known atmospheric gases such as nitrogen. How-

1.4 THE IONOSPHERE AND MAGNETOSPHERE

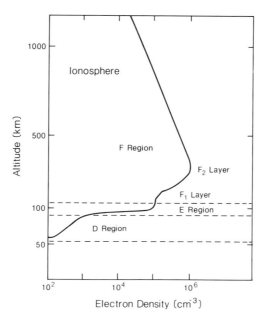

FIG. 1.7. Electron density of the earth's ionosphere as a function of altitude.

ever, identification of the yellow-green line at 557.7 nm was elusive. Finally, H. Babcock's precise measurements in 1923 allowed John McLennan to identify it as a metastable transition of atomic oxygen. At atmospheric pressures close to that at the surface of the earth, collisions between molecules de-excite the molecules before they have a chance to radiate if they happen to become excited into one of the metastable states. However, at the altitude of the aurora, collisions are so rare that the time between collisions is longer than the lifetimes of the metastable states, and the excitation energy of the state can be released by radiation. A similar line in the auroral spectrum is the 630.0-nm red line of atomic oxygen. This metastable transition has a lifetime of 110 s and can radiate only above some 250 km. Those discoveries led to the realization that the varied colors of the aurora were simply related to height. In low-altitude auroras, below 100 km, where collisions quench even the oxygen green line, the blue and red nitrogen bands predominate. From 100 to 250 km, the oxygen green line is strongest. Above 250 km, the red line is most important.

Although most of the auroral forms are associated with electrons, some aurora are due to precipitating protons. The first observations of the proton aurora were made in 1939. Measurements of the Doppler shifts of the proton emissions permitted estimates of the energy of the precipitating particles from the ground. Chapter 14 contains more detailed discussion of the aurora and auroral ionosphere.

With the concept of the ionosphere firmly established, scientists began to wonder about the upper extension of the ionosphere, linked magnetically to the earth, which region we today call the magnetosphere. In 1918, Sydney Chapman postulated a singly-charged beam from the sun as the cause of worldwide magnetic disturbances. That was a revival of an old idea that had previously been criticized by Schuster. Chapman was soon challenged by Frederick Lindemann, who pointed out that mutual electrostatic repulsion would destroy such a stream. Lindemann, instead, suggested that the stream of charged particles contained particles of both signs in equal numbers. We would now call such a stream a "plasma." That proposal was a breakthrough, and it permitted Chapman and his co-workers, in a series of papers beginning in 1930, to lay the foundations for our modern understanding of the interaction of the solar wind with the magnetosphere.

In the rarefied conditions of outer space, where collisions between particles are infrequent, the ion-electron gas, or plasma, is highly electrically conducting. Thus, Chapman and Ferraro proposed that as the plasma from the sun approached the earth, the earth would effectively see a mirror magnetic-dipole moment advancing on the earth, as illustrated in Figure 1.8. The net result of that advancing mirror field would be to compress the terrestrial field. Eventually, as sketched in Figure 1.9, the plasma would surround the earth on all sides, and a cavity would be carved out of the solar plasma by the terrestrial magnetic field. That is very similar to our modern concept of the geomagnetic cavity, which is discussed in greater detail in Chapter 6.

After the compression of the magnetosphere, which is detected by ground-based magnetometers as a sharp increase in the magnetic field, the magnetosphere becomes inflated. Chapman and Ferraro correctly interpreted that subsequent decrease in the magnetic field at the surface of the earth as the appearance of energetic plasma deep inside the magnetosphere, forming a ring of current around the earth in the near-equatorial regions. The development of this ring current in what we now call a geomagnetic storm is discussed at greater length in Chapters 10 and 13.

At the same time that the ionosphere was being discovered by virtue of its effects on man-made radio signals, natural radio emissions were also being explored, and the magneto-ionic theory developed for the man-made signals was being applied to those natural emissions. The first report of those electromagnetic signals in the audio-frequency range was an observation of what have become known as "whistlers," coming from a 22-km telephone line in Austria in 1886. Whistlers are short bursts of audio-frequency radio noise of continuously decreasing pitch. In 1894, British telephone operators heard "tweeks," possibly whistlers generated by lightning, and a "dawn chorus" generated deep in the magnetosphere during a display of aurora borealis. Little work was done on those observations because of the lack of suitable analysis equipment at the time. During World War I, equipment installed to eavesdrop on enemy telephone conversations picked up whistling sounds. Soldiers at the front would say, "You could hear the grenades fly." H. Barkhausen reported on those observations in 1919 and suggested that they were

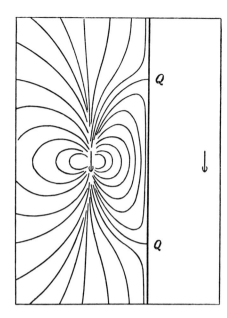
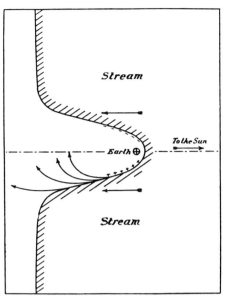

(*Left*) **FIG. 1.8.**
Compression of a dipole field by an advancing infinite, superconducting slab. The magnetic field is due to the original dipole plus an image dipole an equal distance behind the front, as shown by the right-hand arrow. (From Chapman and Bartels, 1940.)

(*Right*) **FIG. 1.9.**
Expected evolution of the front of superconducting plasma as it passes the earth. This model was proposed by Chapman and Ferraro in the 1930s to explain the phenomena of the geomagnetic storm. (From Chapman and Bartels, 1940.)

correlated with meteorological influences. However, he could not duplicate the phenomenon in laboratory experiments. In 1925, T. L. Eckersley also described that phenomenon and ascribed it to the dispersion of an electrical impulse in a medium loaded with free ions. Eventually, after much work and several incorrect explanations, in 1935 Eckersley concluded that the distinctive swooping sound of whistlers was due to the dispersion of a burst of electromagnetic noise traveling through the ionosphere. Very little work was done on whistlers until the early 1950s, at which time L. R. O. Storey, with a homemade spectrum analyzer, conducted a thorough study of whistlers. He found that whistlers are caused by lightning flashes, whose electromagnetic energy then echoes back and forth along field lines in the upper ionosphere, as illustrated in Figure 1.10. A major implication of those findings was that the electron density in the outer ionosphere, which is now called the plasmasphere, was unexpectedly high. Storey also found other types of audio-frequency, or very low frequency (VLF), emissions that are not associated with lightning and are now known to be generated within the magnetospheric plasma. Chapter 12 discusses the generation and propagation of these waves.

1.5 THE SOLAR WIND

If the auroras were caused by electrons, and if those electrons came from the sun, as was commonly believed among solar-terrestrial researchers in the first half of the twentieth century, then those electrons would have to travel in the company of an equal number of ions, or else the beam would disrupt. That idea can be considered the first model of

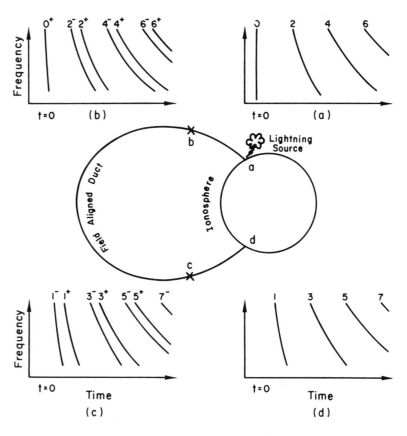

FIG. 1.10. Dispersion of whistler-mode waves generated by lightning, as seen at four different locations. The different velocities of propagation as functions of wave frequency (dispersion) cause the wave arrival to be delayed by a different amount at each frequency. The delay depends on the distance traveled and the properties of the plasma traversed by the wave. (From Russell, 1972.)

the streaming ionized plasma of interplanetary space that we call the solar wind. It was an essential element of the geomagnetic-storm model of Chapman and Ferraro, but in their model the solar wind was intermittent. It flowed only at active times. However, in 1943, C. Hoffmeister noted that a comet tail was not strictly radial, but lagged behind the comet's radial direction by about 5°. In 1951, L. Biermann correctly interpreted that lag in terms of an interaction between the comet tail and a solar wind. That wind was said to flow at about 450 km · s^{-1} at all times and in all directions from the sun, although he assumed that the electron density was about 600 cm^{-3}, two orders of magnitude too high. Several years later, in 1957, Hannes Alfvén postulated that the solar wind was magnetized and that the solar-wind flow draped that magnetic field over the comet, forming a long magnetic tail downstream in the antisolar direction, as illustrated in Figure 1.11. The cometary ions were confined by that tail in a narrow ribbon between the two tail "lobes." In 1958, E. W. Parker provided the theoretical underpinning for such a flow of magnetized plasma, and in 1962 he showed that in order to be consistent with the geomagnetic records, the electron density of the solar wind should seldom exceed 30 cm^{-3}. Confirmation was not long in coming. That was the dawn of the space age, and soon observations were being returned by both Soviet and American space probes that

1.6 MAGNETOSPHERIC EXPLORATION

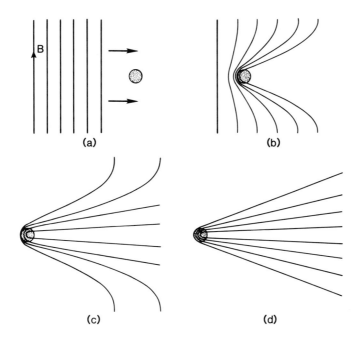

FIG. 1.11. Original model of the formation of a type I (plasma) cometary tail, according to H. Alfvén. In this model the solar-wind magnetic field is draped over the comet by the motion of the plasma from left to right. (From Alfvén, 1957.)

clearly confirmed the existence of the solar wind and its entrained magnetic field, measured its properties, and demonstrated its pivotal role in controlling geomagnetic activity and the aurora. Chapter 4 discusses the solar wind in greater detail.

1.6 MAGNETOSPHERIC EXPLORATION

Rockets provided the opportunity to begin to explore the magnetosphere. In the early and middle 1950s, James Van Allen and his colleagues launched a series of rocket flights into the Arctic and Antarctic ionosphere, reaching heights up to 110 km. Those flights detected either energetic electrons or the bremsstrahlung radiation from such electrons. The year 1957 marked the beginning of an International Geophysical Year (IGY), an 18-month period of worldwide geophysical studies. It also marked the launch of *Sputnik 1*. The attendant space race began a period of explosive growth in our knowledge of the terrestrial magnetosphere and its interaction with the solar wind. In 1958, *Explorer 1* carried a Geiger counter that enabled Van Allen to discover the trapped radiation belts. Instrumentation developed by Konstantin I. Gringauz for the Soviet *Luna* probes provided the first measurements of the solar wind, and instrumentation developed by Conway Snyder and Marcia Neugebauer for *Mariner 2* in 1962 provided the first detailed study of this plasma.

Battery-powered *Explorer 10,* launched in 1961, was the first spacecraft to provide measurements across the magnetopause, the boundary

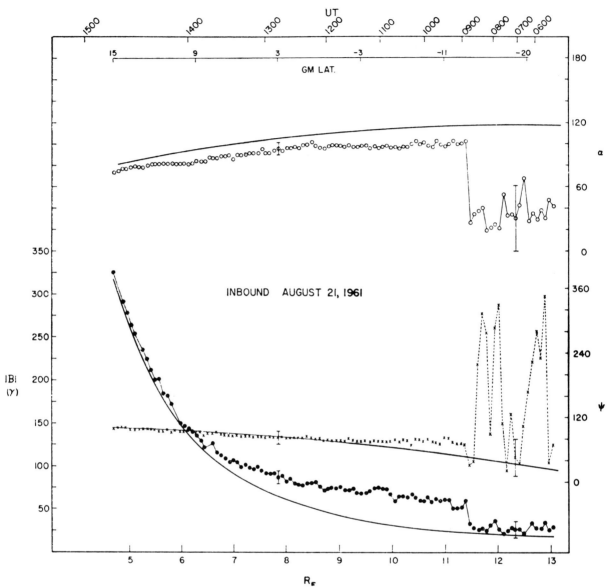

FIG. 1.12. Measurements of magnetic-field strength in the outer magnetosphere and through the magnetopause by the *Explorer 12* spacecraft. Two angles and the field magnitude are shown. Smooth lines are dipole values. (From Cahill and Patel, 1967.)

between the flowing solar wind and the earth's magnetic field, but the first detailed examination of that boundary awaited measurements with a solar-powered spacecraft, *Explorer 12,* which provided 4 months of data and coverage from the noon meridian to the dawn meridian. Figure 1.12 shows magnetic measurements obtained by *Explorer 12* during a traversal of the outer magnetosphere, through the magnetopause, and out into the magnetosheath. It was clear from the data provided by the many spacecraft that were launched into the solar wind during those

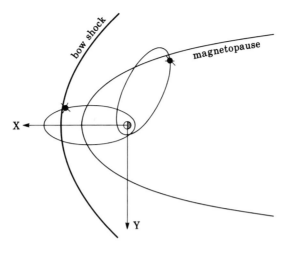

FIG. 1.13. Mapping of the magnetopause and bow shock as the earth goes around the sun. During the course of a year, the magnetopause and bow-shock surfaces maintain their orientation about the earth–sun line, while the spacecraft orbit is fixed in inertial space. Thus the orbits appear to sweep through these boundaries in the course of a year.

early years that the solar wind undergoes an abrupt transition prior to reaching the magnetopause. This transition is a shock produced as the high-speed solar wind encounters the earth as an obstacle in its flow. The existence of a shock in a collisionless plasma surprised many physicists. In the intervening years it has become clear that the electric and magnetic fields in the plasma can alter the motion of the particles in a manner similar to ordinary collisions. These changes provide the dissipation needed to form a shock. The physics of this process is discussed in Chapter 5. The shock allows the solar wind, which flows faster than the speed of compressional waves in a plasma, to be slowed, heated, and deflected around the planet.

It was not until the launch of the first Orbiting Geophysical Observatory (OGO) in 1964 that scientists obtained time-resolution measurements of sufficient accuracy to study the bow shock. The *OGO 1, 3,* and *5* spacecraft in highly eccentric orbits mapped the locations of both boundaries as the earth orbited the sun and the orbits precessed relative to the magnetosphere, as shown in Figure 1.13. Those measurements and data from other spacecraft launched in the 1960s, such as the Interplanetary Monitoring Platform (IMP) spacecraft and the VELA spacecraft, revealed that the structure of the bow shock was very sensitive to the conditions in the plasma, the ratio of the speed of the solar wind to the speed of compressional waves (Mach number), and the ratio of the thermal pressure to the magnetic pressure (beta). It was also found to be sensitive to the direction of the interplanetary magnetic field. When the magnetic field is almost aligned with the direction of propagation of the shock, the shock normal, the shock is said to be quasi-parallel. When the field is more nearly perpendicular to the normal, it is referred to as being quasi-perpendicular. Ion beams are found upstream of the quasi-parallel shock, as illustrated in Figure 1.14, and these ion beams interact with the incoming solar-wind plasma to produce copious large-amplitude waves, called upstream waves, shown in Figure 1.15.

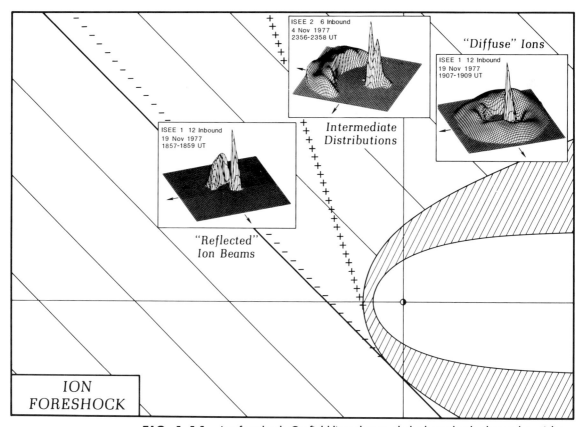

FIG. 1.14. Ion foreshock. On field lines that touch the bow shock, charged particles can spiral upstream along the magnetic field. The solar-wind electric field causes these particles to drift antisunward. The fastest particles (electrons) are affected the least, and the slowest particles the most, by this drift. Representative distribution functions of ions are shown for various locations. The narrow peak represents the unperturbed solar-wind beam. The broader distributions represent the back-streaming ions. (From Russell and Hoppe, 1983.)

The shock is important because it modifies the properties of the solar-wind flow before the flow interacts with the earth's magnetic field, but the processes acting at the magnetopause are the ones finally responsible for determining how much energy the magnetosphere receives from the solar-wind flow. One can imagine a very inviscid interaction in which the solar wind is completely diverted by the magnetosphere and there is very little drag and hence little momentum transfer across the boundary. In fact, this situation does occur when the interplanetary magnetic field is northward, but when the interplanetary field is southward, the momentum transfer from the solar wind increases markedly.

The most important clue that the nature of the processes at the magnetopause changes with variations in the properties of the solar wind is the observation that geomagnetic activity is controlled by the north–south component of the interplanetary magnetic field. The availability of

1.6 MAGNETOSPHERIC EXPLORATION 19

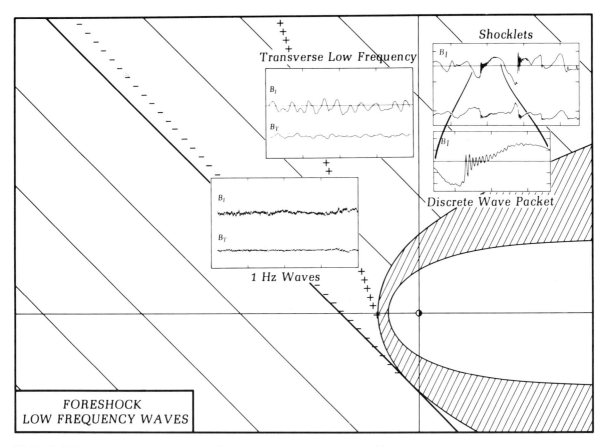

FIG. 1.15. Back-streaming ions and electrons can stimulate a variety of low-frequency waves (with periods of seconds to many tens of seconds). Waves representative of different regions of the near-earth solar wind are illustrated here. (From Russell and Hoppe, 1983.)

extended measurements in the solar wind, especially from *Explorer 33* and *35,* allowed researchers such as Roger Arnoldy and Joan Hirshberg to study this control. The mechanism by which the interplanetary magnetic field exerts this control is known as "reconnection," which was postulated by James Dungey in 1961. As illustrated in Figure 1.16, interplanetary and planetary magnetic fields become linked. As a result, magnetic flux is transported from the dayside of the magnetosphere to the nightside. This magnetic flux builds up in the tail until reconnection occurs there too and returns the magnetic flux to the magnetosphere proper. Spacecraft such as *OGO 5,* launched in 1968, showed the erosion of the dayside magnetosphere and the corresponding activity in the magnetotail. This process, which also leads to activation of the aurora, is called a "substorm." As implied by the name substorm, often there are occasions of more major activity covering the entire magnetosphere, which are called geomagnetic storms. The most popular model of substorms is called the near-earth neutral-point model, where the neutral

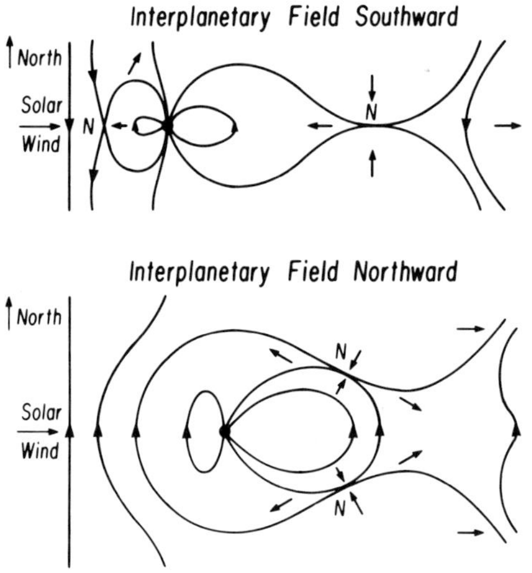

FIG. 1.16. Topology of the magnetosphere for northward and southward interplanetary fields, according to J. W. Dungey in the early 1960s. In the steady state, the plasma flows as indicated by the short arrows. (From Dungey, 1963a.)

point, a location where the magnetic field vanishes, is the site of the reconnection. It was not until the late 1970s, after the launch of the dual co-orbiting satellites *ISEE 1* and *2,* that the reconnection mechanism gained general acceptance. Those satellites returned plasma data of sufficiently high resolution to show the accelerated flows at the magnetopause and in the magnetotail. Chapter 9 provides further details about these processes. However, even today there is debate as to where reconnection occurs and how important it is relative to other processes. Furthermore, it has been found that three-dimensional structures are present, and their explication will require not just two spacecraft, but a whole cluster. Thus there is much to be done in future magnetospheric exploration.

Laboratory plasma measurements have also been useful in understanding the magnetosphere. Figure 1.17 shows a wire model of the magnetosphere developed by Igor Podgorny and his colleagues at the Space Research Institute in Moscow, based on laboratory experiments undertaken in the 1960s, illustrating the development of cusp-shaped openings in the field pattern on the dayside and a long tail at night. Figure 1.18 shows a three-dimensional sketch of the magnetosphere, representing the structure that has been inferred from spacecraft observations. It is deep within this magnetosphere that the radiation-belt particles bounce and drift, as illustrated in Figure 1.19. As noted earlier, the converging magnetic-field lines of the dipole magnetic field stop the forward motion of the particles and accelerate them back toward the equatorial regions. While gyrating and bouncing, these particles also drift, because their gyroradii are greater when they gyrate into weaker fields than when they are in the stronger fields on the inner part of their trajectories. The dipole magnetic field of the earth can confine particles over a wide range of energies, the more energetic of which Van Allen encountered when he and his colleagues discovered the radiation belts.

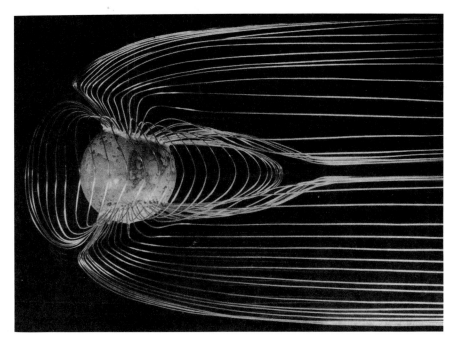

FIG. 1.17. Three-dimensional wire model of the magnetosphere based on the laboratory experimental data of Podgorny and colleagues. (From Podgorny, 1976.)

Figure 1.20 shows the intensities of very energetic protons and electrons in the inner magnetosphere. These particles enter the radiation belts through a variety of means, including radial diffusion from more distant regions, with accompanying acceleration and the decay of neutrons from the sputtering of the atmosphere by cosmic rays. Whereas energetic protons form a single belt, as illustrated in the top panel of Figure 1.20, electrons, as illustrated in the bottom panel, form two belts separated by a region called the slot. This slot in the flux of electrons of fixed energy is formed when naturally occurring electromagnetic waves interact with the gyromotion of the electrons, causing them to spiral into the atmosphere, where they are lost through collisions. The inner electron belt is quite stable and the outer belt quite variable. The radiation belts and the motions of charged particles are discussed further in Chapter 10.

1.7 PLANETARY AND INTERPLANETARY EXPLORATION

The earth is but one test bed for the physical processes occurring in space plasmas. There is much occurring at the other planets and in the interplanetary plasma itself. The solar-wind properties evolve with heliocentric distance, and on the way through the solar system the solar wind encounters a variety of obstacles to its flow, including magnetized bodies, unmagnetized bodies, some with atmospheres and others without, and eventually the heliopause, where the solar wind and the interstellar wind meet.

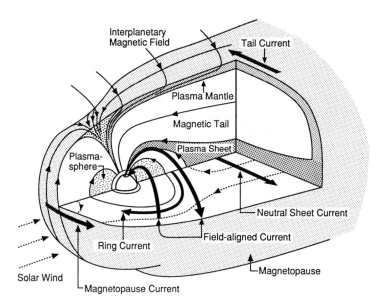

FIG. 1.18. Three-dimensional cutaway view of the magnetosphere showing currents, fields, and plasma regions.

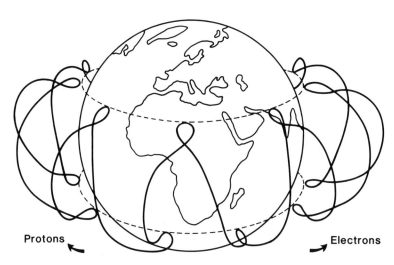

FIG. 1.19. Longitudinal drift of energetic charged particles in the earth's dipolar magnetic field. Electrons drift eastward in the direction of the earth's rotation, and protons westward.

The earliest deep-space probes were the *Mariner 2, 4,* and *5* spacecraft that in the early 1960s went to Venus, Mars, and back to Venus again. Those missions showed that both Venus and Mars were quite different from the earth in their interactions with the solar wind, because neither planet had any significant magnetic moment. However, it was not until after the *Venera 9* and *10* orbiters in 1975, the *Pioneer Venus* orbiter in 1978, and the *Phobos* mission to Mars in 1989 that the details of the solar-wind interactions with these planets became fully understood. At these two planets, the extreme ultraviolet radiation from the sun ionizes the upper atmosphere. It also creates a hot neutral atmosphere that extends into the solar wind. As shown in Figure 1.21, the ionospheric pressure, consisting of both thermal and magnetic components, balances the dynamic pressure of the solar-wind flow. The neutral

1.7 PLANETARY AND INTERPLANETARY EXPLORATION

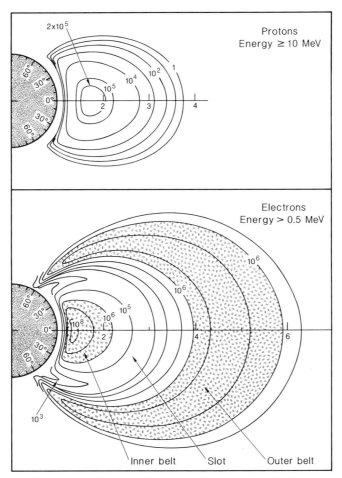

FIG. 1.20. Earth's radiation belts. The top panel shows the contours of the omnidirectional flux (particles per square centimeter per second) of protons with energies greater than 10 MeV. The bottom panel shows the contours of the omnidirectional flux of electrons with energies greater than 0.5 MeV.

atmosphere that extends into the solar wind becomes ionized and adds to the solar-wind flow, further decelerating it.

The slowing down of the magnetized flow around the planet leads to the draping of magnetic-field lines over the obstacle and the formation of a long tail. In this respect, the solar-wind interaction with Venus and Mars resembles that with a comet. This interaction was probed by the International Cometary Explorer (ICE) spacecraft at comet Giacobini-Zinner in 1985 and by the *VEGA 1* and *2, Giotto, Suisei,* and *Sakigake* spacecraft at the comet Halley in 1986. Chapter 8 describes in greater detail the solar-wind interaction with such unmagnetized bodies.

Mariner 10, with a gravitational assist from Venus, made three passes by Mercury in 1974 and 1975. As illustrated in Figure 1.22, *Mariner 10* found a minimagnetosphere very much like that of the earth. Mercury, however, has almost no atmosphere, so that the ionospheric current systems that we believe to be so important for the terrestrial magnetosphere must be absent from Mercury. Thus we expect Mercury's magnetosphere to be quite different in some respects from that of the earth. There has been little investigation of even the meager data from the

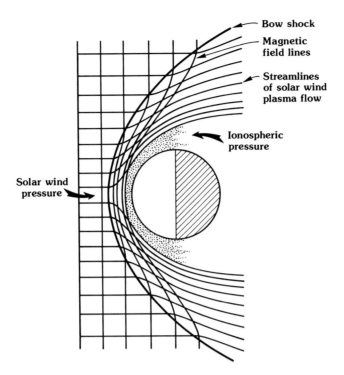

FIG. 1.21. Solar-wind interaction with an unmagnetized planet. The ionospheric pressure stands off the solar-wind flow, so that the streamlines going from left to right flow around the planet. The magnetic field, which is shown here perpendicular to the flow in the solar wind, is bent around the obstacle by this interaction (From Luhmann, 1986.)

Mariner 10 mission, and currently there are no plans to return to Mercury. Thus the Mercury magnetosphere will remain mysterious for many years to come.

In 1972 and 1973, the first spacecraft to the outer solar system, *Pioneer 10* and *11,* were launched, reaching Jupiter in December 1973 and 1974, with *Pioneer 11* going on to Saturn in 1979. Now *Pioneer 10* and *11* are heading out of the solar system, with *Pioneer 10* going downwind relative to the interstellar medium, and *Pioneer 11* upwind. *Voyager 1* and *2* were launched in 1977 and reached Jupiter in 1979 and Saturn in 1980 and 1981. *Voyager 2* then went on to successful encounters with Uranus in 1986 and Neptune in 1989. Both *Voyager 1* and *2* are now heading upwind toward the heliopause.

Those missions revealed well-developed magnetospheres at all the outer planets, each with a bow shock, magnetopause, and magnetotail. The magnetosphere of Jupiter distinguishes itself because its rapid rotation, coupled with a strong plasma source at the moon Io, causes the magnetosphere to be distorted into a disklike geometry. Jupiter is also a source of intense radio waves. Saturn has a simpler magnetosphere, with no strong mass source. Its magnetic field is almost perfectly aligned with its rotation axis.

Uranus and Neptune have unusual magnetospheres because of the unusual orientations of their planetary magnetic fields. Both fields are very complex, and when each is fitted with a dipole moment, the best-fit

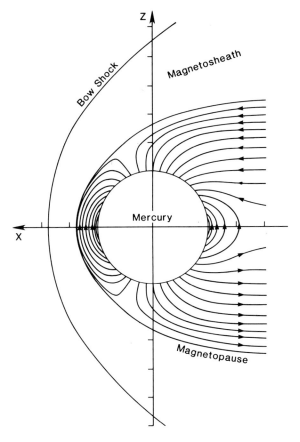

FIG. 1.22. Magnetic-field configuration and noon–midnight cross section of Mercury's magnetosphere. (From Russell et al., 1988.)

dipole is at a large angle to the rotation axis, and the dipole moment is offset from the center of the planet.

Uranus's spin axis is nearly in its orbital plane and is currently nearly pole-on to the sun. Because its magnetic axis is at such a large angle to its rotation axis, its magnetosphere undergoes large oscillations during the course of a day. However, currently the angle between the magnetic moment and the solar-wind flow is not much larger than the maximum angles possible at the earth. Neptune has a more customary rotation axis, roughly perpendicular to its orbit around the sun, but its planetary magnetic field is even more complex than that of Uranus. In both magnetospheres, the radiation belts are much more benign than those of the earth. For Uranus and Neptune we shall have to rely on the findings provided by *Voyager 2* for the foreseeable future, but plans are being made for a Saturn orbiter, called *Cassini,* to arrive in 2004, and a Jovian orbiter, *Galileo,* has been launched and is on its way to Jupiter, to arrive in 1995. Chapter 15 describes the phenomena occurring in the magnetospheres of these outer planets.

1.8. CONCLUDING REMARKS

The field of solar-terrestrial physics has advanced greatly since the earliest recorded sightings of the aurora. We now have physical models for almost all the observed phenomena. Sometimes several competing models exist. The discipline has evolved from one of remote sensing to in situ observations, theory, and computer modeling. In fact, it is now perhaps more appropriate to refer to the field as "space physics," as one of the major journals in the field does, rather than solar-terrestrial physics.

ADDITIONAL READING

Brekke, A., and A. Egeland. 1983. *The Northern Light*. Berlin: Springer-Verlag.

Chapman, S., and J. Bartels. 1940. *Geomagnetism*. Oxford University Press.

Eather, R. H. 1980. *Majestic Lights*. Washington, DC: American Geophysical Union.

Gilbert W. 1893. *De Magnete,* trans. P. Fleury Mottelay. Reprinted 1958, New York: Dover.

Helliwell, R. A. 1965. *Whistlers and Related Ionospheric Phenomena*. Stanford University Press.

PROBLEMS

1.1. Discuss the role of new technology in the development of solar-terrestrial physics.

1.2. At its typical velocity of 440 km $\cdot$ s^{-1}, how long does it take the solar wind to arrive at Mercury, Earth, Jupiter, and Pluto?

2 PHYSICS OF SPACE PLASMAS

M. G. Kivelson

2.1 INTRODUCTION

PLASMAS IN SPACE are extremely tenuous gases of ionized particles in which, on average, there is no net charge. In this book we shall describe many different plasmas in which the densities are much lower than that of an extremely good laboratory vacuum. Because there are very few close encounters between particles in a very low density gas, we need only consider the responses of the charged particles to the force fields in which they move. Calculating the force fields can be complicated, because in the presence of Coulomb interactions a specific particle feels the effects of even very remote particles. Fortunately, the other particles need be considered only in an average sense, and so only the "collective interactions" of the particles are important. Because the collective interactions impose some long-range order, and because in a magnetized plasma the magnetic field creates important links between remote spatial regions, it is also possible to discuss space plasmas in terms of fluid physics. The plasma fluid is a conductor, and that determines its responses in important ways. In this chapter we shall present the important theoretical concepts for both the particle and fluid descriptions of space plasmas. The ideas presented in this chapter will be used to interpret observations. The meaning of the theory undoubtedly will become clearer when it is applied in following chapters. Underlying the material to follow are a few fundamental concepts. Most of the important results follow from conservation laws, including the conservation of mass, energy, and charge. It is also essential to include the relationships among forces and the time variation of momentum (basically, one of Newton's laws), as well as the relations of electromagnetic theory.

2.2 SINGLE-PARTICLE MOTION

A plasma is an electrically neutral gas composed predominantly of charged particles; the roles of electrical and magnetic forces are critical to understanding the behavior of the plasma. This makes it apparent that the equations of importance must contain electromagnetic forces. If an electric field **E** and a magnetic induction **B** act on a particle with charge q and velocity **v**, the particle experiences a force $\mathbf{F}_L$, called the Lorentz force:

$$\mathbf{F}_L = q\mathbf{E} + q\mathbf{v} \times \mathbf{B} \qquad \text{(Lorentz-force law)} \qquad (2.1)$$

which is expressed here in a set of units called the Système International (SI) or the International System. SI units express mass, length, and time in kilograms, meters, and seconds, respectively, E in volts per meter (V · m^{-1}), D in coulombs per square meter (C · m^{-2}), charge in coulombs (C), B in tesla (T), H in ampere-turns per meter (At · m^{-1}), and current in amperes (A). Another commonly used system of units is called the Gaussian system. These units express mass, length, and time in grams, centimeters, and seconds, respectively, E in statvolts per centimeter, D in statcoulombs per square centimeter, charge in statcoulombs, B in gauss, H in oersteds, and current in statamperes.

Table 2.1 shows Maxwell's equations in full form, introducing the displacement field **D** and the magnetic intensity **H**. These satisfy the so-called constitutive relations: $\mathbf{D} = \epsilon \mathbf{E}$ and $\mathbf{H} = \mathbf{B}/\mu$. In space-physics applications, these field variables, which are convenient for dealing with dense matter, are rarely used. In space plasmas, set $\mathbf{D} = \epsilon_0 \mathbf{E}$ and $\mathbf{H} = \mathbf{B}/\mu_0$ in SI units, and $\mathbf{D} = \mathbf{E}$ and $\mathbf{H} = \mathbf{B}$ in Gaussian units. Here $\epsilon_0 = 8.854 \times 10^{-12}$ farad per meter (F · m^{-1}) is the permittivity of free space and $\mu_0 = 4\pi \cdot 10^{-7}$ henry per meter (H · m^{-1}) is the permeability of free space. These equations will be discussed later in this chapter.

It is often useful to express the fields **E** and **B** in terms of potential functions: φ, the scalar potential, and **A**, the vector potential. In SI units, the fields are found from

$$\mathbf{B} = \nabla \times \mathbf{A} \quad \text{and} \quad \mathbf{E} = -\nabla \phi - \frac{\partial \mathbf{A}}{\partial t}$$

The velocity has been assumed to be sufficiently small that relativistic effects need not be taken into account. Although **B** is properly the magnetic induction, we shall normally use the more common term "magnetic field" in referring to it. From Newton's laws, for a particle of mass m, the rate of change of momentum ($m\mathbf{v}$) is given by

$$m\frac{d\mathbf{v}}{dt} = q\mathbf{E} + q\mathbf{v} \times \mathbf{B} + \mathbf{F}_g \qquad (2.2)$$

where $\mathbf{F}_g$ represents nonelectromagnetic forces such as gravitational forces that may be present. Often we shall deal with systems for which one can show that the nonelectromagnetic forces are so small that they can be ignored. In parts of this chapter, we shall make that assumption; that is, we shall set $\mathbf{F}_g = 0$. However, elsewhere the gravitational forces will be important, as, for example, in discussions of the solar corona and of planetary ionospheres. Equation (2.1) shows that the magnetic

TABLE 2.1. Maxwell's Equations in Different Systems of Units

SI Units	Gaussian Units	
$\nabla \cdot \mathbf{D} = \rho_q$	$\nabla \cdot \mathbf{D} = 4\pi\rho_q$	Poisson's equation
$\dfrac{\partial \mathbf{B}}{\partial t} = -\nabla \times \mathbf{E}$	$\dfrac{1}{c}\dfrac{\partial \mathbf{B}}{\partial t} = -\nabla \times \mathbf{E}$	Faraday's law
$\nabla \times \mathbf{H} = \mathbf{j} + \dfrac{\partial \mathbf{D}}{\partial t}$	$\nabla \times \mathbf{H} = \dfrac{4\pi}{c}\mathbf{j} + \dfrac{1}{c}\dfrac{\partial \mathbf{D}}{\partial t}$	Ampère's law
$\nabla \cdot \mathbf{B} = 0$	$\nabla \cdot \mathbf{B} = 0$	Divergenceless $\mathbf{B}$
$\mathbf{F} = q[\mathbf{E} + \mathbf{u} \times \mathbf{B}]$	$\mathbf{F} = q[\mathbf{E} + \dfrac{1}{c}\mathbf{u} \times \mathbf{B}]$	Lorentz-force law

field acts to change the motion of a charged particle only in directions perpendicular to that motion. In a uniform magnetic field with $\mathbf{E} = 0$, a charged particle moves in a circle. To see this, assume that $\mathbf{B}$ is along z and write the x- and y-components of equation (2.2) as

$$m\dot{v}_x = qv_y B; \quad m\dot{v}_y = -qv_x B \tag{2.3}$$

By substitution,

$$\ddot{v}_j = -(qB/m)^2 v_j = -\Omega_c^2 v_j \quad \text{or} \quad \ddot{x}_j = -\Omega_c^2 x_j \tag{2.4}$$

for $j = x, y$, which does indeed imply circular motion (in the left-hand sense if $q > 0$, and in the right-hand sense if $q < 0$), with an angular frequency

$$\Omega_c = qB/m \tag{2.5}$$

which is called the cyclotron frequency or the gyrofrequency. Appendix 2A contains the equations that allow us to evaluate the related frequency $f_c = \Omega_c/2\pi$ (units are cycles per second or hertz) for electrons and ions for a given magnitude of B. The radius of the circle is determined by the magnitude of the particle's velocity perpendicular to the magnetic field (designated as $v_\perp$ and called the perpendicular velocity) and the magnitude of the magnetic field and is given by

$$\rho_c = \dfrac{v_\perp}{\Omega_c} = \dfrac{mv_\perp}{qB} \tag{2.6}$$

This quantity is called the cyclotron radius, Larmor radius, or gyroradius. A listing of the values of the gyroradius as a function of energy for protons and electrons in the dipole equator at different distances from the earth is given in Table 10.1. In that table, and elsewhere in this book, the particle energy is given in electron volts (eV). This is an energy unit that represents the energy that a particle carrying a charge e gains or loses in falling through a potential drop of 1 volt (V): 1 eV $= 1.6022 \times 10^{-19}$ joules (J). Typical energies of interest for discussions of space plasmas range from a few electron volts to kiloelectron volts (1 keV = 10^3 eV) to megaelectron volts (1 MeV = 10^6 eV). Cosmic-ray energies may reach gigaelectron volts (1 GeV = 10^9 eV) or higher.

The circular motion in the uniform magnetic field does not change the particle's kinetic energy, $\frac{1}{2}mv^2$, as can be proved by dropping nonelectromagnetic forces and setting $\mathbf{E}=0$ in equation (2.2) and taking the scalar product with the vector velocity to obtain

$$m\frac{d\mathbf{v}}{dt}\cdot\mathbf{v} = \frac{d(\frac{1}{2}mv^2)}{dt} = q\mathbf{v}\cdot(\mathbf{v}\times\mathbf{B}) = 0 \qquad (2.7)$$

This shows that no work is done, because there is no motion in the direction of the force.

If the electric field does not vanish, or if there are gradients in the magnetic field, the particle motion will be modified. Consider first what happens if the electric field is nonvanishing. In space plasmas, even if there is an electric field, it is quite often possible to suppose that its component parallel to the magnetic field vanishes (i.e., $\mathbf{E}\cdot\mathbf{B} = E_\parallel B = 0$). This is because equation (2.2) with $\mathbf{F}_g = 0$ tells us that only the electric field can exert force in the direction along $\mathbf{B}$. The force moves positively charged particles in the direction of $\mathbf{E}_\parallel$ and negatively charged particles in the opposite direction. If enough particles are present, the relative displacement can produce charge-separation fields that will grow until they cancel out $\mathbf{E}_\parallel$, after which no further forces will act along the magnetic-field direction.

The response of the charged particles to a perpendicular component of $\mathbf{E}$ is quite different. Because this aspect of particle motion is very important in accounting for the special properties of a plasma, it is important that we try to understand it, not only mathematically but also in an intuitive manner. Figure 2.1 shows schematically how positively and negatively charged particles move in a uniform magnetic field if an electric field perpendicular to the magnetic field is also present. To understand the motion, think first about how the particle would move if only the magnetic field were present. It would gyrate in a circle, and

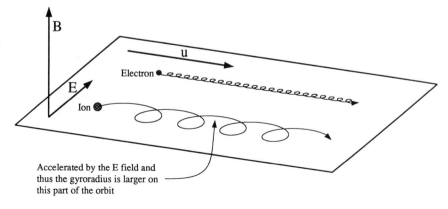

FIG. 2.1. Schematic showing the motions of ions (charge e) and electrons (charge $-e$) in a uniform magnetic field $\mathbf{B}$ in the presence of an electric field $\mathbf{E}$ perpendicular to $\mathbf{B}$. The diagram represents motion in a plane perpendicular to the magnetic field. For both signs of the charge, the motion along the magnetic field is at constant velocity, unaffected by the presence of the fields.

the direction of motion around the circle would depend on the sign of the particle's charge. The radius of the circle, ρ_L, would vary with the particle's mass and would therefore be much larger for an ion than for an electron if their velocities were the same. The electrical force accelerates the particle during part of each orbit and decelerates it during the remaining part of the orbit. The result is that the orbit is a distorted circle, with a larger-than-average radius of curvature during half of the orbit and a smaller-than-average radius of curvature during the remaining half of the orbit. The diagram shows that a net displacement in a direction perpendicular to **E** and independent of the sign of the charge results. If the motion is followed over several gyration orbits, the particle drift velocity, $\mathbf{u}_E$, can be determined, and it satisfies the equation

$$\mathbf{u}_E = \mathbf{E} \times \mathbf{B}/B^2 \tag{2.8}$$

which says that the drift velocity is perpendicular to both the electric and magnetic fields. This is sometimes referred to as an "*E*-cross-*B* drift," but its magnitude is inversely proportional to the magnitude of **B**. The *E*-cross-*B* drift does not introduce currents into the plasma, as $\mathbf{u}_E$ is independent of both q and m. The electrons and the ions will move in different directions if there are collisions. Such differential motions of charges constitute a current, a point that will be particularly important in understanding the properties of the ionosphere. One of the exercises at the end of this chapter asks the reader to calculate how the average motion of the charged particles will change if they occasionally collide with neutrals, stop, and then accelerate again.

Any force that is capable of accelerating and decelerating particles as they gyrate about the magnetic field will result in drifts perpendicular in both the field and the force. The general expression for the drift velocity produced by an arbitrary force **F** is

$$\mathbf{u}_F = \mathbf{F} \times \mathbf{B}/qB^2 \tag{2.9}$$

If the force is charge-independent, the drift direction will depend on the sign of the charge on the particle, and therefore currents perpendicular to the magnetic field $\mathbf{j}_\perp$ will develop.

In the foregoing discussion, the magnetic field is assumed to be uniform. This means that its strength and direction will not change anywhere in the system. Although sometimes this may be a rather good approximation (e.g., the earth's magnetic field hardly changes over laboratory dimensions), more often that is not the case. For example, the earth's magnetic field is approximated to lowest order as a dipole field (see Chapter 6) that changes in direction and magnitude both along and across the field. This means that descriptions of particle motions in changing magnetic fields are needed.

A gradient in the field strength in the direction perpendicular to **B** will produce a drift velocity. If a particle gyrates in a field whose strength changes from one side of its gyration orbit to the other, the instantaneous radius of curvature of the orbit, ρ_L, will become alternately smaller and larger. Averaged over several gyrations, the particle will drift, this time in a direction perpendicular to both the magnetic field and the direction in which the strength of the field changes. The drift velocity, which we shall call $\mathbf{u}_g$ (for gradient drift velocity), is given by

$$\mathbf{u}_g = \tfrac{1}{2} m v_\perp^2 \mathbf{B} \times \nabla B / q B^3 \tag{2.10}$$

Equation (2.10) shows that the drift velocities produced by a gradient in the magnetic field will depend on the sign of the particle's charge, and so this drift will cause electric currents to flow across the magnetic field.

The curvature of magnetic-field lines will introduce additional drifts, because as the particles move along the field direction, they will experience a centrifugal acceleration. The instantaneous radius of the gyration orbit will increase away from the center of curvature of the field line, and the particle will drift in a direction perpendicular to $\mathbf{B}$. Defining the radius of curvature R_c as

$$\frac{\hat{\mathbf{n}}}{R_c} = -(\hat{\mathbf{b}} \cdot \nabla)\hat{\mathbf{b}} \tag{2.11}$$

where $\hat{\mathbf{b}} = \mathbf{B}/B$, and $\hat{\mathbf{n}}$ is a unit vector perpendicular to $\mathbf{B}$ that points away from the center of curvature, one can express the curvature drift velocity $\mathbf{u}_c$ as

$$\mathbf{u}_c = \frac{m v_\parallel^2 \mathbf{B} \times (\hat{\mathbf{b}} \cdot \nabla)\hat{\mathbf{b}}}{qB^2} = -\frac{m v_\parallel^2 \mathbf{B} \times \hat{\mathbf{n}}}{R_c \, qB^2} \tag{2.12}$$

In the latter expression, the centrifugal force produced by motion along the field appears explicitly. Thus, this drift velocity can be seen to be a special case of the expression obtained for drift produced by an arbitrary force $\mathbf{F}$.

Particle motion in prescribed electric and magnetic fields is governed by equations (2.1) and (2.2). However, the fields themselves must satisfy various conditions. Time-varying magnetic fields and spatially varying electric fields are related by the requirement

$$\frac{\partial \mathbf{B}}{\partial t} = -\nabla \times \mathbf{E} \quad \text{(Faraday's law)} \tag{2.13}$$

This is the equation that expresses the fact that a changing magnetic field drives an electromotive force (emf) through a coiled wire. When applied to charged particles, it implies that if a particle experiences time-varying magnetic fields, it will simultaneously be subjected to electrical forces that will change its energy.

A remarkable feature of the motion of charged particles in collisionless plasmas (i.e., plasmas in which a particle can move for many gyroradii without being significantly influenced by the effects of close encounters with other particles) is that even though the energy changes, there is a quantity that will remain constant if the field changes slowly enough. By "slowly enough" we mean that the field changes encountered by the particle within a single gyration orbit will be small compared with the initial field. If this condition is satisfied, then the particle's "magnetic moment"

$$\mu = \frac{\tfrac{1}{2}mv_\perp^2}{B} \tag{2.14}$$

will remain constant. Note that if μ remains constant as the particle moves across the field into regions of different field magnitudes, some acceleration is required.

The quantity μ is also called the *first adiabatic invariant*. Here, "adiabatic" refers to the requirement that μ may not remain invariant or unchanged unless the parameters of the system, such as its field strength and direction, change slowly.

It may not be obvious why μ is called a magnetic moment, as that name usually refers to a property of a current loop defined as the current flowing through the loop times the area of the loop. It is possible to show that equation (2.14) is of precisely that form. This means that we can use the equation for the force on a dipole magnetic moment

$$\mathbf{F} = \boldsymbol{\mu} \cdot \nabla B = -\mu \frac{dB}{dz} \tag{2.15}$$

(where $\boldsymbol{\mu} = \mu \hat{\mathbf{b}}$ and $\hat{\mathbf{b}}$ is a unit vector along the magnetic field) to find how gradients in field strength along the field direction will affect particle motion. The force is always exerted along the field and away from the direction of increasing field. This means that $v_\parallel$ decreases to 0 at some maximum field strength and then changes sign. We can work out how strong the field is at that point by remembering two things. First, we have assumed that $E_\parallel = 0$, and we know that the magnetic field does not accelerate the particle; so motion along the field does not change the particle's energy. Therefore, as the parallel velocity ($v_\parallel = \mathbf{v} \cdot \hat{\mathbf{b}}$) decreases, the perpendicular velocity must increase, but it cannot become greater than the total velocity. In addition, μ is conserved as the particle moves along the field. Together these ideas imply that the particle turns around when the field strength satisfies $B = \tfrac{1}{2}mv^2\mu$. Because the particle's motion along the field reverses at this point, we say that the particle experiences a *mirror force,* and we call the place where it turns around the *mirror point*. The type of motion that results is illustrated in Figure 2.2. There is also a second adiabatic invariant, this one associated with the cyclical bounce motion between mirror points. Chapter 10 provides further discussion of the second adiabatic invariant. There is a third adiabatic invariant as well, associated with particle drift in closed orbits in a magnetic field such as the earth's dipole field. It will be defined later in this chapter.

2.3 COLLECTIONS OF PARTICLES

Up to this point, we have described the motions of individual particles. Plasmas consist of collections of particles, some with positive and some with negative charges. Let us next consider how to deal with collections of particles, particularly large numbers of particles that have different velocities. The approach is quite sensible. We describe the properties of a large number of particles by saying how many of them there are per unit volume of a six-dimensional space that is called *phase space*. The phase-space density $f(\mathbf{r}, \mathbf{v}, t)$ is also called the single-particle distribution function. Referring to this quantity as a density may sound strange, but it provides a count of the number of particles per unit volume of ordinary or configuration space that also fall into a particular unit volume in a velocity space whose axes are labeled v_x, v_y, and v_z. "Volume" in phase space needs to be specified. Often, interest focuses on a differential element of phase-space volume, which can be denoted as $d\mathbf{v}\, d\mathbf{r} = dv_x\, dv_y\, dv_z\, dx\, dy\, dz$ at phase-space position $(\mathbf{v}, \mathbf{r})$, where the number of particles in the differential phase-space volume

$$= f(\mathbf{r}, \mathbf{v}, t)\, d\mathbf{v}\, d\mathbf{r}$$

The following discussion makes a distinction between the terms *volume* (which refers to configuration space only) and *phase-space volume* (which refers to a six-dimensional space).

If there are particles of different masses or charges, it may be necessary to keep track of each type (designated by a subscript) separately. The symbol that is used for this density is $f_s(\mathbf{r}, \mathbf{v}, t)$, the phase-space density for species s particles. In order to determine how many particles of type s there are per unit volume of ordinary space, one must count all the particles in the phase-space volume, regardless of their velocities; that is, we integrate over all possible velocities

$$n_s(\mathbf{r}, t) = \int d\mathbf{v}\, f_s(\mathbf{r}, \mathbf{v}, t) \qquad (2.16)$$

Here, $n_s(\mathbf{r}, t)$ is the number density (number per unit volume) of type s particles, and it is related to the mass density ρ_s by $\rho_s(\mathbf{r}, \mathbf{v}, t)/m_s$. The number density is the zeroth-order moment of the distribution. Other quantities of interest are higher-order moments of the distribution obtained by integrating powers of the velocity and closely related quantities over the phase-space distribution function. For example, it is often important to determine the average velocity $\mathbf{u}_s$ of a collection of particles. The first moment of the distribution

$$\mathbf{u}_s(\mathbf{r}, t) = \int d\mathbf{v}\, \mathbf{v} f_s(\mathbf{r}, \mathbf{v}, t) \Big/ \int d\mathbf{v}\, f_s(\mathbf{r}, \mathbf{v}, t) \qquad (2.17)$$

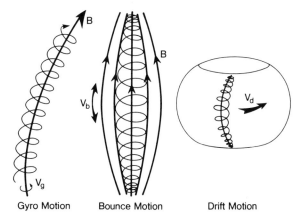

FIG. 2.2. Schematic of particle motion in a magnetic field. On time scales of gyroperiods, the particle spirals about the magnetic field (see illustration of gyromotion at the left). If there is a field-aligned gradient of the field strength, the component of velocity parallel to the field decreases as the particle moves into regions of increasing field magnitude, although the total velocity is conserved. Eventually, the parallel velocity reverses. The reflection of the parallel motion is called "magnetic mirroring." Motion between mirror points is called bounce motion (central panel). As the particle bounces, it also drifts about the source of the field, tracing out a drift shell (see the right-hand illustration).

If this flow or "bulk" velocity is different for ions and electrons, then electric currents will flow in the system. Notice the notation that has been used to distinguish between the velocity of an individual particle, **v**, and the average velocity of a collection of particles, $\mathbf{u}_s$. Often the velocities of the individual particles are orders of magnitude larger than the flow velocity, but if the velocities are random, they contribute little to the average velocity.

To obtain the average kinetic energy per particle of type s in a unit volume of the plasma, we add together the energies of all the particles by integrating over velocities and divide by the total number of particles in the volume. Evidently there is kinetic energy associated with the flow velocity $\mathbf{u}_s$, but more interesting is the kinetic energy of random motion relative to the average. The equation for that part of the kinetic energy is

$$\left\langle \tfrac{1}{2}m_s(\mathbf{v}-\mathbf{u}_s)^2 \right\rangle = \int d\mathbf{v}\, \tfrac{1}{2}m_s(\mathbf{v}-\mathbf{u}_s)^2 f_s(\mathbf{r},\mathbf{v},t) \Big/ \int d\mathbf{v}\, f_s(\mathbf{r},\mathbf{v},t) \qquad (2.18)$$

The angle brackets are used to designate the average in the volume.

The hydrostatic partial pressure of species s particles is related to their average random energy by

$$p_s/n_s = (2/N) \left\langle \tfrac{1}{2}m_s(\mathbf{v}-\mathbf{u}_s)^2 \right\rangle \qquad (2.19)$$

where N is the number of independent components of velocity, normally three. In SI units, pressure is measured in pascals (Pa), with 1 Pa equal to one newton per square meter (1 N · m^{-2}).

For systems in equilibrium, the phase-space distribution is the Maxwellian distribution, the equilibrium distribution function for particles of type s flowing with velocity $\mathbf{u}_s$. It is given by

$$f_s(\mathbf{r},\mathbf{v}) = A_s \exp\left[-\frac{\tfrac{1}{2}m_s(\mathbf{v}-\mathbf{u}_s)^2}{kT_s}\right] \qquad \text{(Maxwellian distribution)} \quad (2.20a)$$

Here, k is the Boltzmann constant, 1.3807×10^{-23} joules per degree Kelvin (J · K^{-1}), and A_s is a constant that is related to the number density such that equation (2.16) holds: $A_s = n_s(m/2\pi kT)^{\frac{3}{2}}$.

In some situations, the direction of the particle velocity is not of interest. Various expressions for the distribution function can then become relevant. For example, from equation (2.20a), for the case $\mathbf{u}_s = 0$, we can obtain

$$\int_{\text{directions}} f_s(\mathbf{r}, \mathbf{v}, t) \, d\Omega_v v^2 \, dv = [4\pi f_s(\mathbf{r}, |v|, t)v^2]dv = g(\mathbf{r}, v) \, dv \quad (2.20b)$$

which expresses the number (per unit volume) of particles with speeds between v and $v + dv$. Here, $g(\mathbf{r}, v) \, dv$ is the number of particles per unit speed with speeds between v and $v + dv$. This quantity increases quadratically with v for small v and falls off exponentially with v for large v. The plots in Figure 2.3 show that the variation of a Maxwellian distribution as a function of velocity in a particular direction, say v_x, differs from its variation as a function of the velocity magnitude v.

For monatomic particles, the temperature, properly defined only for an equilibrium particle distribution, is related to the average energy of random motion by equation (2.18), which yields

$$\langle \tfrac{1}{2}m_s(\mathbf{v} - \mathbf{u}_s)^2 \rangle = NkT_s/2 \quad (2.21)$$

Here, once again, N indicates the number of spatial "degrees of freedom" of the particle distribution. In plasma-physics applications, N is normally 3, corresponding to the three independent directions of velocity space over which the integration in equation (2.18) is carried out. This is true even if the particles are moving in a magnetic field, because the particle velocity is fully specified either by giving $v_x - u_{sx}$, $v_y - u_{sy}$, and $v_z - u_{sz}$ or by giving $v_\parallel$, $|\mathbf{v}_\perp - \mathbf{u}_s|$, and ϕ_v, which describes the direction of the perpendicular velocity relative to the center of gyromotion. From (2.19) and (2.21), we obtain the ideal-gas law, $p_s = n_s kT_s$.

Equation (2.20a) shows that in a Maxwellian distribution, the proportion of particles with large random velocities $|\mathbf{v} - \mathbf{u}_s|$ increases with T_s. This point can be conveniently expressed by defining the most probable thermal speed of the plasma, v_{Ts}:

$$v_{Ts} = (2kT_s/m_s)^{\frac{1}{2}} \quad \text{(most probable thermal speed)} \quad (2.22)$$

and noting that it increases with temperature.

Particle distributions do not always take the Maxwellian form, nor is it always possible to represent the actual distributions analytically. Nonetheless, there are other analytical forms that sometimes are good approximations to the observed distributions. If, for example, there is a notable difference in the distributions of random velocities parallel and perpendicular to the magnetic field, it may be useful to represent the distribution as a product of Maxwellians with different parallel and perpendicular temperatures:

2.3 COLLECTIONS OF PARTICLES

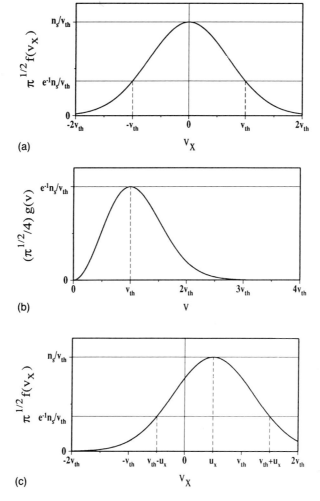

FIG. 2.3. Maxwellian distributions: In (a) and (b) the plasma is at rest. Part (a) shows the dependence of a three-dimensional Maxwellian distribution on one component of the velocity, here taken as v_x. The dependence on the other components of velocity has been removed by integration. Part (b) shows the dependence of $g(\mathbf{r}, v)$ on the speed, v; $v_{th} = (2kT_s/m_s)^{1/2}$ is the thermal speed. In (c) the plasma is flowing toward positive x at a velocity u_x. Notice that in (a) and (c), the position of the maximum is unaffected by changes of temperature. As the temperature increases, v_{th} increases, and the height of the peak of the distribution drops, whereas the width of the distribution increases. In (b), the position of the maximum shifts to the right as temperature increases. Its height increases.

$$f_s(\mathbf{r}, \mathbf{v}) = A'_s \exp\left[-\frac{\frac{1}{2}m_s(v_\parallel - u_{\parallel s})^2}{kT_{\parallel s}}\right] \exp\left[-\frac{\frac{1}{2}m_s(\mathbf{v}_\perp - \mathbf{u}_{\perp s})^2}{kT_{\perp s}}\right] \quad (2.23)$$

(bi-Maxwellian distribution)

Here, $A'_s = A_s T_s^{3/2}/T_{\perp s} T_{\parallel s}^{1/2}$. For this distribution, we can specify the average random energy of motion separately for motion parallel and perpendicular to the field:

$$\langle \tfrac{1}{2}m_s(v_\parallel - u_{\parallel s})^2 \rangle = \tfrac{1}{2}kT_{\parallel s} \quad \text{and} \quad \langle \tfrac{1}{2}m_s(\mathbf{v}_\perp - \mathbf{u}_{\perp s})^2 \rangle = kT_{\perp s}$$

This form reflects the fact that there is one degree of freedom for motion along the magnetic field and two degrees of freedom for motion across it. The definitions of parallel and perpendicular pressures follow from these relations and the obvious generalization of equation (2.19).

Sometimes the Maxwellian form is followed quite closely up to some energy, but the number of particles at high energies follows a power-law decrease rather than the exponential decrease of the Maxwellian distribution. The kappa distribution was devised to represent this situation in a convenient analytical form. For a single species, it is given by

$$f_s(\mathbf{r}, \mathbf{v}) = A_{\kappa s}\left[1 + \frac{\frac{1}{2}m_s(\mathbf{v} - \mathbf{u}_s)^2}{\kappa E_{T_s}}\right]^{-\kappa - 1} \qquad \text{(kappa distribution)} \qquad (2.24)$$

Here there are two parameters κ and E_{T_s} characterizing the distribution, instead of the single temperature parameter of the Maxwellian distribution. E_{T_s} is closely related to the temperature, and κ characterizes the departure from the Maxwellian form. This rather cumbersome formula has some highly useful characteristics. At high energies ($E \gg \kappa E_{T_s}$) it falls off more slowly than does a Maxwellian, that is, like a power law in kinetic energy with a spectral index κ. In the limit $\kappa \to \infty$, it becomes a Maxwellian with temperature $kT = E_{T_s}$. For the kappa distribution or for other (often more realistic) non-Maxwellian distributions, equation (2.21) does not hold.

Although the phase-space distribution function is central to a theoretical interpretation of the behavior of collections of particles, it is measured only indirectly. Measurements provide the differential directional flux of particles within a range of solid angles $d\Omega$ and within an energy band of width dW about W, where W is the kinetic energy, $\frac{1}{2}m_s v^2$. Representing this flux as $\partial^2 J / \partial \Omega\, \partial W$, we find

$$f(\mathbf{r}, \mathbf{v}) = [m^2/2W]\, \partial^2 J / \partial \Omega\, \partial W \qquad (2.25)$$

2.4 THE PLASMA STATE

The definitions presented in the preceding section are applicable to collections of noninteracting particles quite generally. Here we shall consider features that distinguish a plasma from other noninteracting gases. We remarked earlier that plasmas are ionized gases assumed to be electrically neutral. We also pointed out that long-range electromagnetic forces are important, but that the density should be low so that near-neighbor interactions can be disregarded. Let us make these ideas more quantitative by considering, for example, the spatial scale over which electrical neutrality is required.

The electrostatic potential of an isolated particle (assumed to be an ion of charge q in this example) is $\phi = q/4\pi\epsilon_0 r$. However, if the ion is embedded in a gas containing ions and electrons, the electrons will be attracted to it, and their distribution will alter the potential, reducing it at large distances. (Ions are repelled, but their greater inertia means that their responses to changes will be small and sluggish.) The shielded potential takes the form

$$\phi = qe^{-r/\lambda_D}/(4\pi\epsilon_0 r) \tag{2.26}$$

where λ_D is called the Debye length. In an electron-proton plasma,

$$\lambda_D = (\epsilon_0 kT/ne^2)^{\frac{1}{2}} \quad \text{(Debye length)} \tag{2.27}$$

where n is the electron-number density. Notice that the potential becomes very small when r/λ_D becomes large, and thus the Debye length gives an estimate of the spatial scale over which a plasma ion influences its surroundings.

The Debye length is helpful in understanding how a spacecraft affects the space plasmas that surround it. In a collisionless plasma, a spacecraft can develop a net negative charge, because for equal ion and electron temperatures, the electron flux is $(m_i/m_e)^{\frac{1}{2}}$ larger than the ion flux, and the spacecraft potential becomes negative. Solar radiation can liberate photoelectrons from the surface of the spacecraft, often producing a sufficiently large negative current that the spacecraft potential will become positive. Any net charge will perturb the plasma in the immediate vicinity of the spacecraft, a region referred to as a "plasma sheath." The scale size of the perturbed region will be λ_D. At distances large compared with λ_D, the plasma will be completely unaffected by the presence of the spacecraft.

It is useful to consider a sphere of radius λ_D centered on a plasma ion. This is referred to as the Debye sphere. The number of particles within the Debye sphere, $N_D = 4\pi n \lambda_D^3/3$, is proportional to $T^{\frac{3}{2}}/n^{\frac{1}{2}}$ and should be large for the expected shielding to occur. We shall generally assume that the number is large and that the expected shielding occurs; then statistical treatments in terms of phase-space distribution functions will be valid. Correspondingly, $g = N_D^{-1}$ is small. Plasma physicists may refer to either N_D or g as the plasma parameter, but they do not agree on which to choose. They all agree, however, that for a quasi-neutral charged-particle gas to behave as an ideal plasma, N_D must be much greater than 1, which implies that $g \ll 1$. This means that some combination of high temperature and low density is required. As well, the density of neutral particles must be low enough that the characteristic plasma frequencies will not be small compared with the collision frequency, so that damping of the plasma perturbations will occur very slowly.

Such combinations of parameters occur naturally in many parts of the universe. Much of the region from the surface of the sun to the planetary ionospheres is highly if not fully ionized. The solar wind, the particle radiation belts surrounding magnetized planets, the auroral ionosphere, and lightning discharges are all examples of naturally occurring plasmas. At the earth, plasmas are found in the familiar glow of various gas-discharge light sources, as well as in the laboratory, especially in devices for the study of controlled thermonuclear fusion. These systems are of great interest to physicists, but laboratory studies of their behaviors often are complicated by the interactions of the plasmas with the con-

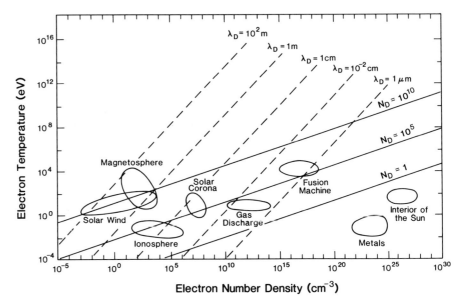

FIG. 2.4. Parameters of various plasmas plotted versus electron temperature and number density. Values of the Debye length λ_D and the number of particles in the Debye sphere N_D, are also plotted.

tainers that confine them and by the relatively high densities of neutral particles that they contain. Naturally occurring plasmas often are relatively insensitive to conditions at their boundaries, and they may contain very low densities of neutral particles. Thus, studies of space plasmas are of great interest to laboratory-plasma physicists and to astrophysicists. The range of parameters over which plasmas can exist is illustrated in Figure 2.4 and Table 2.2. The plasmas in the magnetosphere and the solar wind are seen to be nearly ideal plasmas. Spacecraft measurements of such plasmas can be made without significantly affecting the system being studied, because the scale of a spacecraft typically is small compared with the scale of a space plasma.

In discussing the behavior of charged particles in magnetic fields, we were led to definitions of a characteristic length scale, the gyroradius, and a characteristic frequency, the gyrofrequency. We have also identified a characteristic length, λ_D, that is independent of the magnetic field. There is also an important frequency that is independent of the magnetic field. It is called the plasma frequency and is given as an angular frequency by

$$\omega_{ps} = (n_s e^2/\epsilon_0 m_s)^{\frac{1}{2}} \quad \text{(plasma frequency)} \tag{2.28}$$

This is the natural frequency of plasma oscillations resulting from charge-density perturbations. Its significance in the treatment of plasmas will become more apparent when plasma waves are discussed in Chapter 12.

It is implicit in much of the foregoing discussion that a plasma contains only ions and electrons, but such a condition is not strictly necessary for an ionized gas to behave as an ideal plasma. All that is necessary is that the density of electrically neutral particles be low enough that

TABLE 2.2. Properties of Typical Plasmas

Plasma Type	Density (m^{-3})	Temperature (eV)	Debye Length (m)	N_D
Interstellar	10^6	10^{-1}	1	10^6
Solar wind	10^7	10	10	10^{10}
Solar corona	10^{12}	10^2	10^{-1}	10^9
Solar atmosphere	10^{20}	1	10^{-6}	10^2
Magnetosphere	10^7	10^3	10^2	10^{13}
Ionosphere	10^{12}	10^{-1}	10^{-3}	10^4
Gas discharge	10^{20}	1	10^{-6}	10^2
Fusion machine	10^{22}	10^5	10^{-5}	10^7

collisions between charged and neutral particles will occur only infrequently. In particular, it is necessary that the collision frequency be substantially lower than the lowest of the natural frequencies of importance (i.e., Ω_{cs} and ω_{ps}).

2.5 THE FLUID DESCRIPTION OF A PLASMA

We began this chapter with a discussion of individual particle motions, and we have just seen how to describe the properties of a collection of many particles in terms of averages over small spatial volumes (i.e., quantities like density, temperature, and pressure). Fortunately, the average properties are governed by the basic conservation laws for mass, momentum, and energy in a fluid. If the effects of electromagnetic fields can be ignored, the relevant fluid equations are the equations of hydrodynamics, but because electric and magnetic fields and currents are always important in plasmas, we have to include their effects. This means that we must introduce the equations of *magnetohydrodynamics* (MHD). The name may sound a bit ominous, but the equations are not unfamiliar. They incorporate familiar mechanical laws, but they also take account of electromagnetic properties. In this section we again treat the plasma as an electrically neutral fluid, and we assume two species, s, representing ions and electrons.

We start with the equation that states what happens to the number density of the fluid as it moves from one place to another; it says that the number of particles can change only if there are particle sources (S_s) or losses (L_s):

$$\frac{\partial n_s}{\partial t} + \nabla \cdot n_s \mathbf{u}_s = S_s - L_s \quad \text{(continuity equation)} \quad (2.29a)$$

Here n_s, $\mathbf{u}_s$, S_s, and L_s may be functions of position and time. $S_s - L_s$ gives the net rate at which particles of type s are added per unit volume. Plasma ions and electrons may be added (i.e., S_s may be nonvanishing), such as by ionization of neutrals; charge exchange can replace one type of ion by another under appropriate circumstances. Loss processes for particles s include recombination of ions and electrons to form neutrals, charge exchange, and so forth. In many applications of interest, the source and loss rates are small enough to allow us to set the right-hand side of equation (2.29a) equal to zero. In ionospheric applications, this may not be a good approximation.

If S_s and L_s vanish, equation (2.29a) implies mass conservation. To prove this, multiply equation (2.29a) by m_s, integrate over a fixed volume, and identify $\int d\mathbf{x}\, \rho_s = M_s$ as the total mass of species s in the volume. Here, $\rho_s = m_s n_s$. Then

$$\frac{\partial M_s}{\partial t} + \int_{\text{volume}} \nabla \cdot (\rho_s \mathbf{u}_s)\, d\mathbf{r} = \frac{\partial M_s}{\partial t} + \int_{\text{surface}} d\mathbf{S} \cdot \rho_s \mathbf{u}_s = 0 \qquad (2.30)$$

where $d\mathbf{S}$ is an element of the surface bounding the volume of interest (taken positive outward), and the integral is to be taken over the entire bounding surface of the volume. As $\rho_s \mathbf{u}_s$ gives the mass flux (i.e., the rate per unit area at which a mass of species s crosses a surface perpendicular to the flow velocity), the surface integral represents the rate at which a type s mass flows into or out of the volume. Temporal variations of the mass within the volume result only if there is a net inward or outward flow of mass through the boundaries of the region of interest. If there is a source of mass within the volume, equation (2.30) must be modified. In the fluid, as for individual particles, momentum and force are related. The fluid form of this relation for a single species is

$$\left(\frac{\partial (\rho_s \mathbf{u}_s)}{\partial t} + \nabla \cdot (\rho_s \mathbf{u}_s \mathbf{u}_s) \right) = -\nabla p_s + \rho_{qs} \mathbf{E} + \mathbf{j}_s \times \mathbf{B} + \rho_s \mathbf{F}_g / m_s \qquad (2.31a)$$

or

$$\rho_s \left(\frac{\partial \mathbf{u}_s}{\partial t} + \mathbf{u}_s \cdot \nabla \mathbf{u}_s \right) + m_s \mathbf{u}_s (S_s - L_s) = \qquad (2.31b)$$

$$-\nabla p_s + \rho_{qs} \mathbf{E} + \mathbf{j}_s \times \mathbf{B} + \rho_s \mathbf{F}_g / m_s \qquad \text{(momentum equation)}$$

where $\rho_{qs} = q_s n_s$ is the charge density and $\mathbf{j}_s = q_s n_s \mathbf{u}_s$ is the current density (i.e., the charge per second that flows across a unit area perpendicular to the flow direction). The last term in the equations represents the density of nonelectrical forces like gravity, and it is analogous to the force $\mathbf{F}_g$ that was introduced in equation (2.2). The preceding two forms are equivalent. We can prove this by expanding the derivatives on the left side of (2.31a). There will then be four terms, of which two can be expressed in terms of $S_s - L_s$ from equation (2.29a). The term in brackets in (2.31b) is sometimes called the convective derivative of $\mathbf{u}_s$. It gives the rate of change of $\mathbf{u}_s$ resulting both from moving into a different spatial region and from explicit time variations of the system.

The expression (2.31a) is particularly useful for interpreting the equations. First consider what happens if the fluid is not flowing. Then $\mathbf{u}_s = 0$, but its time derivative need not vanish, and the left side of equation (2.31a) represents the net change of the momentum density of a fluid element. On the right side of the equation are three terms that represent the density of the forces acting on the fluid element. The first term is the pressure force. Uniform pressure alone does not change fluid momentum, but if the pressure on one side of a fluid element differs from that on the other (i.e., there is a spatial gradient), then the fluid moves toward the direction of lower pressure. The second and third terms represent the forces exerted by the electric and magnetic fields, respectively. As in the analogous single-particle equation [see equation (2.2)], the magnetic field exerts no force if the current is field-aligned. Equation (2.31a) contains a term (second on the left side) that enters only if the fluid is flowing. This term accounts for the fact that the flowing plasma transports momentum density as it moves from one place to another; a net flux of momentum density through the surface bounding the volume will change the momentum density of the fluid within.

Let us now drop the source and loss terms and assume only protons and electrons to be present in the fluid. By multiplying the continuity equations (2.29a) for each species by m_s and adding them together, we find

$$\frac{\partial \rho}{\partial t} + \nabla \cdot \rho \mathbf{u} = 0 \tag{2.29b}$$

in which we have assumed $n_p = n_e$. Here ρ is the total mass density, and $\mathbf{u}$ is the center of mass velocity. Next, we add together the momentum equations for ions and electrons to obtain

$$\left(\frac{\partial (\rho \mathbf{u})}{\partial t} + \nabla \cdot \rho \mathbf{u} \mathbf{u} \right) = -\nabla p + \mathbf{j} \times \mathbf{B} + \rho \mathbf{F}_g / m_p \tag{2.32a}$$

or

$$\rho \left(\frac{\partial \mathbf{u}}{\partial t} + \mathbf{u} \cdot \nabla \mathbf{u} \right) = -\nabla p + \mathbf{j} \times \mathbf{B} + \rho \mathbf{F}_g / m_p \tag{2.32b}$$

Here we have assumed that $m_e \ll m_p$. Notice that no electric-field force is present in equations (2.32a) and (2.32b), because we have assumed charge neutrality. Because the plasma has many of the properties of a gas of noninteracting particles, we can relate the pressure that appears in equations (2.31a) and 2.31b) to the plasma temperature using the ideal-gas law:

$$p = n_p k T_p + n_e k T_e \quad \text{(ideal-gas law)} \tag{2.33a}$$

If $n_e = n_p = n$, and if the electron and ion species have the same temperature, then

$$p = 2nkT \tag{2.33b}$$

2.5.1 Maxwell's Equations

The physical quantities important in a magnetized plasma cannot be specified arbitrarily. If finite-velocity particles are introduced into a preexisting magnetic field, their velocities will change, because the magnetic field will exert a force on them. The forces will be different for electrons and ions; so currents will develop. The currents will modify the magnetic field, which in turn will modify the particle motion. Thus, solving the set of equations is somewhat complicated. Any solution must be a self-consistent solution. Sometimes, if we are lucky, the changes in the field resulting from the currents in the plasma will be sufficiently small compared with the field of external sources that we can approximate the field as known, but more often we have to solve for everything at once. This means that we must also consider the equations that describe how current and charge density affect the magnetic and electric fields. These are the equations of electromagnetism, the Maxwell equations. In this section we write them in their general form. Later we shall introduce the approximate forms that are used in formulating MHD.

The electric field obeys Poisson's equation:

$$\nabla \cdot \mathbf{E} = \rho_q / \epsilon_0 \quad \text{(Poisson's equation)} \tag{2.34}$$

where ρ_q is the net charge density. The electric field also enters into Faraday's law, introduced in equation (2.13):

$$\frac{\partial \mathbf{B}}{\partial t} = -\nabla \times \mathbf{E} \quad \text{(Faraday's law)} \tag{2.35}$$

To this must be added Ampère's law, which relates the magnetic field to the net current $\mathbf{j}$

$$\nabla \times \mathbf{B} = \mu_0 \left(\mathbf{j} + \epsilon_0 \frac{\partial \mathbf{E}}{\partial t} \right) \quad \text{(Ampère's law)} \tag{2.36a}$$

and the requirement that $\mathbf{B}$ be divergenceless:

$$\nabla \cdot \mathbf{B} = 0 \tag{2.37}$$

There are several ways to understand what equation (2.37) implies. One interpretation is based on the concept of magnetic flux, Φ [measured in webers (Wb), with 1 Wb = 1 tesla per square meter (1 T · m^{-2})], defined by

$$\Phi = \int \mathbf{B} \cdot d\mathbf{S} \tag{2.38}$$

where the integral is taken over a surface. The divergenceless requirement implies that if the integral is taken over a surface that completely encloses a spatial volume, no net magnetic flux will cross the surface. Flux can cross the surface provided that if it emerges through one part of the surface, just as much flux comes back at some other part of the surface.

The concept of a magnetic-field line provides a different interpretation of equation (2.37). A field line is a curve that is everywhere parallel to the local magnetic-field direction. The requirement that **B** be divergenceless, equivalent to the statement that the magnetic flux crossing a closed surface vanishes, can also be shown to require that no matter how much they wander around space, all magnetic-field lines must eventually close on themselves.

From the divergence of equation (2.36a) and time derivative of equation (2.34) we find

$$\frac{\partial \rho_q}{\partial t} + \nabla \cdot \mathbf{j} = 0 \quad \text{(current continuity)} \tag{2.39}$$

To complete the set of equations, a relation between current density **j** and the fields is needed. Ohm's law is such a relation, but its form depends critically on the nature of the problem being considered. Let us therefore delay its introduction until we have modified the Maxwell equations to obtain the low-frequency limit that is used in MHD.

Maxwell's equations change form slightly if expressed in units other than SI. A system of units often preferred by theorists is the Gaussian system. Table 2.1 gives Maxwell's equations in both SI units and Gaussian units. In the latter system, some new factors of 4π and c, the velocity of light in a vacuum, appear in the equations, but neither the permittivity of free space nor the permeability of free space is present.

2.5.2 Maxwell's Equations for MHD

The fluid relations introduced in equations (2.29)–(2.33) describe properties of the plasma in a locally averaged form. For example, in order to measure the density, one must focus on a spatial region that is small compared with the size of the entire system, but still large enough to contain a statistically significant number of particles. Furthermore, that number must change slowly enough to allow a measurement. Thus, in formulating a fluid description of the plasma, we implicitly assume that the time scales of interest are long compared with the times of microscopic particle motions (i.e., the periods of cyclotron motion and the inverse plasma frequency) and that the spatial scales of interest are large with respect to the Debye length and the thermal gyroradius. Under such circumstances, it is appropriate to use approximate forms of Maxwell's equations. In particular, we can discard the term in equation (2.36a) proportional to the electric field, using the following argument:

$$|\mathbf{\nabla}\times\mathbf{B}|\approx\frac{B}{L}, \qquad \mu_0\epsilon_0\left|\frac{\partial\mathbf{E}}{\partial t}\right|\approx\mu_0\epsilon_0\frac{E}{\tau}=\frac{E}{c^2\tau}$$

where L and τ are characteristic MHD length and time scales, respectively, and $c=(\mu_0\epsilon_0)^{1/2}$ is the velocity of light. For the slow changes over long distances and the nonrelativistic flows required in MHD, the inequality

$$\frac{\mu_0\epsilon_0\left|\frac{\partial\mathbf{E}}{\partial t}\right|}{|\mathbf{\nabla}\times\mathbf{B}|}\approx\frac{E}{Bc}\frac{L}{c\tau}\approx\frac{v}{c}\frac{L}{c\tau}\ll 1$$

is very strongly satisfied, and the second term on the right-hand side of equation (2.36a) can be dropped to give

$$\mathbf{\nabla}\times\mathbf{B}=\mu_0\mathbf{j} \quad \text{(Ampère's law in the MHD limit)} \tag{2.36b}$$

This form of Ampère's law may be more familiar expressed as an integral:

$$\oint_C \mathbf{B}\cdot d\mathbf{s}=\mu_0\int_S \mathbf{j}\cdot d\mathbf{S} \tag{2.40}$$

where S is a surface and C is the curve bounding the surface. Often this form of the equation is applied to such problems as calculating the magnetic field produced by a current flowing through a long thin wire, but it is more generally applicable whether or not the current path is confined to a specified spatial volume. If the current is not so confined, however, the differential form given in equation (2.36b) is more useful. Similar arguments on the relative sizes of the terms can be used in the equation of current continuity, where the term proportional to the time derivative of the charge density is negligibly small because the system is close to charge neutrality. The current-continuity equation takes the very useful form

$$\mathbf{\nabla}\cdot\mathbf{j}=0 \quad \text{(divergenceless current density)} \tag{2.41}$$

This equation implies that all currents in MHD systems must close on themselves. There are neither sources of charge nor sinks. Equation (2.37), $\mathbf{\nabla}\cdot\mathbf{B}=0$, remains unchanged in the MHD approximation, as does Faraday's law:

$$\frac{\partial\mathbf{B}}{\partial t}=-\mathbf{\nabla}\times\mathbf{E} \quad \text{(Faraday's law)} \tag{2.35}$$

Poisson's equation cannot be simplified, because both sides are equally small, but as the charge density does not enter into any of the other equations, this equation is not needed.

At this point, let us count the number of unknowns and the number of equations. Assuming, as noted, that we need not solve for the negligibly small charge density, there are 14 unknowns: **E**, **B**, **j**, **u**, ρ, and p. A vector has three components, and so it corresponds to three unknowns. The relevant equations are (2.29b), (2.32a) or (2.32b), (2.35), (2.36b), and (2.37). Counting (2.32), (2.35), and (2.36b) as three equations apiece, and the scalar equations as one apiece, it may seem that we have 11 in all. In fact, that is not the case, because equation (2.37) should be thought of as a boundary condition in time. This can be proved by taking the divergence of both sides of Faraday's law, equation (2.35). As the divergence of a curl vanishes, we find $\partial \nabla \cdot \mathbf{B}/\partial t = 0$. Thus, if $\nabla \cdot \mathbf{B} = 0$ at some initial time, then it is always zero. This means that we have only 10 independent equations, not enough equations to determine all the unknowns.

Energy conservation has not yet been invoked. The continuity equation for energy takes the form

$$\frac{\partial}{\partial t}(\tfrac{1}{2}\rho u^2 + U) + \nabla \cdot [(\tfrac{1}{2}\rho u^2 + U)\mathbf{u} + p\mathbf{u} + \mathbf{q}] = \mathbf{j} \cdot \mathbf{E} + \rho \mathbf{u} \cdot \mathbf{F}_g/m \tag{2.42}$$

This equation introduces the heat flux, **q**, and the internal-energy density of the monatomic plasma, U, where $U = nNkT/2$ [see equation (2.21)] or, equivalently, $U = Np/2$. Once again, N is the number of degrees of freedom. Sometimes the divergence of $(U+p)\mathbf{u}$ on the left-hand side of equation (2.42) is expressed in terms of the heat function or the enthalpy density, h, where $h = U + p$.

Notice that the dependence on U does not add a new variable. The heat flux, on the other hand, adds to the set of equations a vector that is not a function of the original set of unknowns. This means that we must either use an approximation to express it in terms of the original set or add further equations governing **q**.

Many treatments avoid introducing energy conservation explicitly. Instead, they obtain an additional equation from the assumption that there is no change in the entropy of a fluid element as it moves through the system. This means that the pressure and the density are related by

$$p\rho^\gamma = \text{constant} \quad \text{or} \quad \frac{\partial p}{\partial t} + \mathbf{u} \cdot \nabla p = c_s^2 \left(\frac{\partial \rho}{\partial t} + \mathbf{u} \cdot \nabla \rho\right) \tag{2.43}$$

where c_s is the speed of sound, defined by

$$c_s^2 = \gamma p/\rho \tag{2.44}$$

and γ is the ratio of the specific heat at constant pressure to the specific heat at constant volume, often referred to as the polytropic index. If the fluid is in thermodynamic equilibrium and the relation (2.21) is valid, then $\gamma = (N+2)/N$, or, in a three-dimensional system, $\gamma = \tfrac{5}{3}$. Measurements in space plasmas often yield values of γ that differ from $\tfrac{5}{3}$, which indicates that the idealized approximations used in obtaining an expression for γ are not always valid. We have noted that Ohm's law gives a relation between current and field, the three additional equations that we still require. To derive this law, we go back to (2.31b), the momentum equation for the individual species. In obtaining equation (2.32), we added these equations. Now let us multiply them by q_s/m_s and subtract to find

$$\mathbf{j} = \sigma \left\{ (\mathbf{E} + \mathbf{u} \times \mathbf{B}) + \frac{1}{ne} \nabla p_e - \frac{1}{ne} \mathbf{j} \times \mathbf{B} - \frac{m_e}{ne^2} \left[\frac{\partial \mathbf{j}}{\partial t} + \nabla \cdot (\mathbf{j}\mathbf{u}) \right] \right\} \quad (2.45a)$$

(Ohm's law)

where

$$\mathbf{j} = \sum_s q_s n_s \mathbf{u}_s$$

The electrical conductivity σ will be discussed further in the chapter on ionospheres. In magnetized plasmas, it usually must be described as a tensor, a point to which we shall return later. Again it has been assumed that the electron mass is negligibly small.

Often the last terms on the right-hand side of equation (2.45a) can be dropped, giving

$$\mathbf{j} = \sigma (\mathbf{E} + \mathbf{u} \times \mathbf{B}) \quad (2.45b)$$

If the plasma is collisionless, the conductivity σ may be so large that equation (2.45b) can be satisfied only if

$$\mathbf{E} + \mathbf{u} \times \mathbf{B} = 0 \quad (2.46)$$

Normally, it is equation (2.46) that we shall use with (2.42) to complete the set of MHD equations.

It may not be immediately clear that equation (2.46) is of great importance for understanding plasma flow. First, consider a system in which the plasma is at rest. Then with $\mathbf{u} = 0$, the equation reduces to $\mathbf{E} = 0$; that is, in the rest state of the plasma, the electric field vanishes. That should make sense. Without collisions to impede their flow, charges can move in response to a perturbing electric field until they induce an electric field that cancels out the field that perturbed them. This is possible only if the perturbing electric field is present over a time interval adequate for the charges to respond fully. Thus, we can see that equation (2.46) is not a good approximation when rapid time variations are important, and Chapter 12 provides examples of such cases.

If $\mathbf{u} \neq 0$, equation (2.46) requires that there must be an electric field present, and, conversely, plasma must flow whenever there is an electric field. These ideas will have many applications in the following chapters. When equation (2.46) is valid, one can demonstrate that in MHD fluid flow, the magnetic flux can be *frozen in* to the fluid. To explain this property of a very high conductivity MHD fluid, we need to consider a surface S in the fluid not parallel to lines of magnetic force. We can then calculate Φ, the total magnetic flux crossing that surface, by using

2.5 THE FLUID DESCRIPTION OF A PLASMA

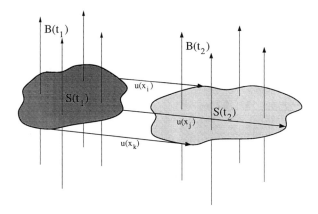

FIG. 2.5. Schematic illustration of frozen-in flux. The fluid that lies on surface $S(t_1)$ threaded by a magnetic field with a component B_{n1} normal to the surface at time t_1 flows through the system and lies on surface $S(t_2)$ threaded by a field with normal component B_{n2} at time t_2. The frozen-in flux condition requires that $B_{n2} = B_{n1}S(t_1)/S(t_2)$.

equation (2.38). The frozen-in-flux conditions tells us that if we follow the fluid initially on the surface S as it moves through the system, the flux through the surface will remain constant even as the surface changes its location and its shape. We can think of the field lines that penetrate the surface as being held in place so tightly that as the surface stretches or shrinks, the field lines will move apart or move closer together. The field accordingly will be weakened or strengthened, but the combination of changed **B** and changed S leaves Φ unchanged. This effect is illustrated schematically in Figure 2.5, where the initial surface is labeled $S(t_1)$, and the surface at a later time is labeled $S(t_2)$. If we follow the field lines that cut the surface at any time along their length, we can map out a spatial volume that is referred to as a *flux tube*. The integral $\int \mathbf{B} \cdot d\mathbf{S}$ is the same for any surface that cuts through the flux tube. Furthermore, the frozen-flux condition implies that all particles initially on a flux tube will remain linked along a single flux tube as they convect through space.

The concept of frozen-in flux allows us to discuss the magnetic-field lines as if they move through the system. Actually, it is the plasma that is on the magnetic-field lines that moves, but the frozen-in flux can be thought of as moving with the plasma. Although the concept is extremely useful, and will be applied in later chapters, it is correct only if all the assumptions leading to equation (2.46) are satisfied, and so it must be used with care.

The concept of magnetic flux also enters into identifying another invariant of particle motion in a magnetic field. As a particle bounces and drifts in a magnetic field that varies slowly along its orbit (e.g., think about a slightly distorted dipole field), it traces out a three-dimensional shell called its *drift shell*, as illustrated in Figure 2.2. The flux that is enclosed by this shell is the same for any surface that intersects it along a closed curve. This is true whether the surface lies near the equator or at higher latitudes. If the field changes slowly, that is, on time scales long compared with the time required for the particle to drift fully around the field, the flux enclosed by the drift shell will remain unchanged. It is this magnetic flux enclosed by the drift shell of a particle that is defined as its third adiabatic invariant.

2.6 TWO APPLICATIONS OF THE MHD EQUATIONS

The MHD equations discussed in this chapter can be applied to many different types of problems; elsewhere in this book there are examples. Here we shall introduce two fundamental applications very briefly.

2.6.1 Magnetic Pressure and Tension

On the right side of the momentum equation (2.32), the magnetic force enters in the form $\mathbf{j} \times \mathbf{B}$. Using Ampère's law (2.36b), we find

$$\mathbf{j} \times \mathbf{B} = \frac{1}{\mu_0} (\nabla \times \mathbf{B}) \times \mathbf{B} = -\nabla B^2/2\mu_0 + (\mathbf{B} \cdot \nabla) \mathbf{B}/\mu_0 \qquad (2.47)$$

The form of (2.47) should now be compared with (2.32). The negative gradient of $B^2/2\mu_0$ can be grouped with the negative gradient of the particle pressure p to see that this part of the magnetic force contributes a magnetic pressure

$$p_B = B^2/2\mu_0 \qquad (2.48)$$

The ratio of the plasma pressure to the magnetic pressure is conventionally represented by the symbol β, where

$$\beta = \frac{p}{B^2/2\mu_0} \qquad (2.49)$$

A plasma is called "cold" if $\beta \ll 1$; for $\beta \gtrsim 1$, the plasma is called "warm," and plasma currents may become important. The additional term in (2.47), $(\mathbf{B} \cdot \nabla)\mathbf{B}/\mu_0$, can be decomposed into two components. One component is field-aligned, $\hat{\mathbf{b}} \mathbf{B} \cdot \nabla \mathbf{B}/\mu_0 = \hat{\mathbf{b}}\hat{\mathbf{b}} \cdot \nabla B^2/2\mu_0$, and cancels the field-aligned component of the magnetic-pressure gradient, so that only the perpendicular components of ∇p_B exert force on the plasma. The second component is antiparallel to the radius of curvature of field lines.

$$(B^2/\mu_0)\hat{\mathbf{b}} \cdot \nabla \hat{\mathbf{b}} = -\frac{\hat{\mathbf{n}} B^2/\mu_0}{R_c}$$

where $\hat{\mathbf{n}}$ is the outward normal. This component of force is present only for curved field lines and is analogous to the perpendicular force exerted by tension in a curved string. In both cases the perpendicular force acts to reduce the curvature. The analogy has led to the use of the term *magnetic tension* to describe this component of force, but we prefer to refer to it as the *curvature force*.

2.6.2 MHD Waves, Characteristic Velocities, and Shocks

Many of the remaining chapters in this book also rely on a familiarity with the fundamental modes of excitation of the system, the natural wave modes of the magnetized fluid. In Chapter 11, the MHD wave solutions are discussed in some detail. Here we shall merely introduce some wave nomenclature and give a sketchy introduction to wave properties to provide background for the chapters that precede the wave chapters. Waves arise from perturbations of a system. A familiar example is a sound wave in a neutral gas. In the dynamics of a neutral gas, electromagnetic fields do not play a role. The underlying perturbation is a pressure change, and sound waves carry pressure perturbations. In a magnetized plasma, dynamics are controlled not only by particle pressure but also by the electromagnetic field. Thus, it is easily understood that waves related to perturbations of both particle pressure and fields are important.

A convenient way to find the response of a system to perturbations is to assume that it starts in equilibrium and that the perturbations are small changes. Mathematically this means that we can linearize the equations governing the system, assuming that we are interested only in the behavior of small quantities. For the MHD equations, we find that the perturbations are governed by wave equations that admit solutions proportional to sines and cosines (or complex exponentials) that depend on $\omega t - \mathbf{k} \cdot \mathbf{x}$. Such solutions are plane waves propagating parallel to k with angular frequency ω and wavelength $\lambda = 2\pi/k$. Phase fronts of these waves advance with a velocity

$$\mathbf{v}_{ph} = \hat{\mathbf{k}}(\omega/k) \tag{2.50a}$$

and this is called the phase velocity. Wave energy is carried at the group velocity

$$\mathbf{v}_g = \nabla_k \omega = \hat{\mathbf{e}}_x \frac{\partial \omega}{\partial k_x} + \hat{\mathbf{e}}_y \frac{\partial \omega}{\partial k_y} + \hat{\mathbf{e}}_z \frac{\partial \omega}{\partial k_z} \tag{2.50b}$$

The equations that determine how ω depends on $\mathbf{k}$ are called the dispersion relations, and they follow from the linearized equations.

In a warm MHD plasma, three different wave solutions are obtained. Two of the wave modes carry changes of plasma and magnetic pressure and changes of plasma density. They are called compressional waves. The third wave mode does not change the plasma pressure or density. Instead, it only causes field lines to bend. This type of wave is called a shear Alfvén wave. The shear Alfvén wave propagates with $v_{ph} = v_A \cos \theta$, where v_A is called the Alfvén velocity and satisfies

$$v_A = B/(\mu_0 \rho)^{\frac{1}{2}} \tag{2.51}$$

and θ is the angle between $\mathbf{k}$ and $\mathbf{B}$. This means that the wave cannot propagate perpendicular to the background field $\mathbf{B}$. The group velocity of the wave is aligned with **B**; that is, the wave energy is carried along the direction of the background field. For **k** oblique to **B**, the perturbation magnetic-field direction is perpendicular to both **B** and **k**. These important properties of Alfvén waves will be used frequently in the following chapters.

The velocity v_A is a fundamental parameter of MHD plasmas. If we use the concept of magnetic tension discussed earlier, it is possible to describe the Alfvén speed as the velocity of a wave along a string. The Alfvén wave does not change the plasma density, plasma pressure, or field magnitude; it is characterized by oscillating perturbations of the transverse (to **B**) magnetic field, the electric field, the plasma velocity, and the current density.

In compressional waves, all of the plasma parameters can vary, including the field-aligned component of the magnetic field, $b_\parallel$ (which changes the field magnitude and the magnetic pressure), and the plasma pressure. The perturbation magnetic field in the compressional waves lies in the plane of **B** and **k**. This property is very important in distinguishing wave modes.

One of the compressional waves, called the *fast-mode wave,* can propagate in any direction and can transport energy in any direction. It carries perturbations in which the particle pressure and density vary in phase with perturbations of $b_\parallel$. This means that the magnetic pressure and particle pressure increase and decrease at the same time. The phase velocity of the fast mode is larger than or equal to v_A, but its exact value depends on the direction of propagation. For propagation perpendicular to **B**, the phase speed is $v_F = (v_A^2 + c_s^2)^{\frac{1}{2}}$, and for propagation along **B** the phase speed is the larger of v_A and c_s. Here, c_s is the sound speed defined in (2.44).

The other compressional wave is called the *slow-mode wave*. It carries perturbations in which the particle pressure and density vary out of phase with perturbations of $b_\parallel$ and of the magnetic pressure. Its phase velocity, v_S, is less than or equal to v_A, largest in the direction of **B**, and it does not propagate perpendicular to **B**. Note that the phase speeds of the three waves satisfy

$$v_F \geq v_A \geq v_S \tag{2.52}$$

and sometimes the shear Alfvén wave is referred to as the intermediate wave.

Just as it is possible for very large pressure perturbations in a neutral gas to steepen and turn into shock waves, the compressional waves of an MHD plasma can, under appropriate conditions, steepen to form shocks. (The intermediate wave does not steepen to form a shock.) The changes of the particle pressure and field pressure across the shock can be in the same sense (both go up behind the shock) or in opposite senses (particle pressure increases, but field pressure decreases, behind the shock). In the first case, the shock variations follow the rules for fast-mode waves, and the shock is referred to as a *fast shock*. In the second case, the shock variations follow the rules for slow-mode waves, and the shock is referred to as a *slow shock*. Just as in the small-amplitude compressional waves, the change of the field in a (compressional) shock, whether fast or slow, lies in the plane of the propagation vector (normal to the shock front) and the unperturbed magnetic field. This means that the shock normal and the magnetic fields on the two sides of the shock all lie in a plane. This is referred to as the *coplanarity theorem*.

2.7 CONCLUSION

The equations and mathematics in this chapter will become more familiar and relevant as we proceed through the material in this book. In the following chapters, the ideas and equations herein will be used to interpret the data and phenomena of space plasmas. Most of the concepts introduced here will be used repeatedly, and the applications should help to clarify the significance of the background material presented here. It may prove useful to refer back to this chapter when learning about specific processes that take place in space plasmas.

Appendix 2A   SOME PROPERTIES OF NONRELATIVISTIC CHARGED PARTICLES IN MAGNETIC FIELDS

IN THE NONRELATIVISTIC LIMIT approximate relations between particle velocity and kinetic energy are

$$v_p(\text{km} \cdot \text{s}^{-1}) = 440 \sqrt{W_p(\text{keV})}$$
$$v_e(\text{km} \cdot \text{s}^{-1}) = 18{,}800 \sqrt{W_e(\text{keV})}$$

The nonrelativistic approximation is valid for electrons of less than ≈ 25 keV and ions of less than ≈ 50 MeV.

Figure 2A.1 shows how the gyroradius changes with particle energy W and B. Plotted are the

proton gyroradius (in km) = $4{,}600 \sin \alpha \; W^{\frac{1}{2}}(\text{keV})/B(\text{nT})$

electron gyroradius (in km) = $107 \sin \alpha \; W^{\frac{1}{2}}(\text{keV})/B(\text{nT})$

where α is the local pitch angle of the particle. In the figure, it is assumed that $\alpha = 90°$.

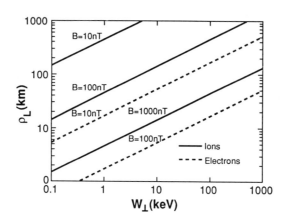

FIG. 2A.1. Larmor radii of ions and electrons as functions of perpendicular energy for selected magnetic-field strengths.

In a dipole field, the equatorial pitch angle α_{eq} and the mirror-point latitude λ_m are related by

$$\sin^2\alpha_{eq} = \cos^6\lambda_m/(4-3\cos^2\lambda_m)^{\frac{1}{2}}$$

The loss cone is bounded at the equator by α_{lc}, whose approximate value at LR_E is

$$\sin\alpha_{lc} = [L^3(4-3/L)^{\frac{1}{2}}]^{-\frac{1}{2}}$$

The full bounce period $T_B(\alpha_{eq})$ within which a bouncing particle returns to its starting point moving in the initial direction is (in seconds)

$$T_B(\alpha_{eq}) = \frac{4LR_E}{v}T(\alpha_{eq})$$

where $T(\alpha_{eq})$ is a numerical multiple that ranges from 0.74 for $\alpha_{eq}=90°$ to 1.38 for $\alpha_{eq}=0°$. For protons (electrons), this becomes

$$T_{Bp}(\alpha_{eq}) = (58L/\sqrt{W_p})T(\alpha_{eq})$$
$$T_{Be}(\alpha_{eq}) = (1.36L/\sqrt{W_e})T(\alpha_{eq})$$

Bounce-averaged azimuthal drift velocities of singly charged particles in a dipole field are

$$v_D(R_E/\text{hour}) = 8.5L^2W(\text{MeV})g(\alpha_{eq})$$

and the corresponding drift period is

$$T_D(\text{min}) = (44/LW(\text{MeV}))g(\alpha_{eq})$$

where

$$g(\alpha_{eq}) = 0.7 + 0.3\sin(\alpha_{eq})$$

ADDITIONAL READING

QUALITATIVE DESCRIPTIONS

Akasofu, S.-I., and Y. Kamide (eds.). 1987. *The Solar Wind and the Earth.* Dordrecht: Reidel.

Friedman, H. 1986. *Sun and Earth*. San Francisco: Freeman (see especially chap. 5).

QUANTITATIVE DISCUSSIONS

Boyd, T. J. M., and J. J. Sanderson. 1969. *Plasma Dynamics*. New York: Barnes & Noble.
Chen, F. F. 1974. *Introduction to Plasma Physics and Controlled Fusion. Vol. 1: Plasma Physics*. New York: Plenum Press.
Dendy, R. O. 1990. *Plasma Dynamics*. Oxford University Press.
Nicholson, D. R. 1983. *Introduction to Plasma Theory*. New York: Wiley.
Nishida, A. 1978. *Geomagnetic Diagnosis of the Magnetosphere*. Berlin: Springer-Verlag.
Priest, E. R. (ed.). 1985. *Solar System Magnetic Fields*. Dordrecht: Reidel (see especially chap. 2, introduction to Alfvén and other MHD waves, and chap. 8, on shocks).

PROBLEMS

2.1. The Lorentz force on a charged particle is

$$\mathbf{F}_L = q\mathbf{E} + q\mathbf{v} \times \mathbf{B}$$

In the absence of an electric field, show that a charged particle's motion can be resolved into two components: one constant, along the magnetic field, and one periodic, perpendicular to the magnetic field. Show that the gyroradius ρ_c is given by

$$\rho_c = \frac{mv_\perp}{qB}$$

and that the gyrofrequency is given by

$$\Omega_c = \frac{qB}{m}$$

Calculate the gyrofrequency (in hertz) of a proton in a 100-nT field, of an electron in a 1,000-nT magnetic field, of a singly ionized oxygen atom in a 50,000-nT field. At roughly what distances from the earth would these gyrofrequencies be found in the equatorial plane? What is the gyroradius of a proton moving transverse to a 100-nT field at 2×10^5 m · s^{-1}? How does this distance compare with the distance (near the equator) over which the earth's dipole field changes by a factor of 2 near the region where $B = 100$ nT? Would you expect the proton to conserve its first adiabatic invariant?

2.2. For each of the following fields, sketch the particle trajectories separately for electrons and protons. Define your coordinate system. Illustrate clearly the direction of the magnetic and electric fields and your coordinate axes. Sketch trajectories in the planes that best illustrate the motions of the charged particles. Describe the form of motion normal to the plane. Explain what you are assuming about a particle's thermal energy, in particular its magnitude relative to the electric-drift speed.

(a) Assume a static uniform magnetic field oriented along the *x*-axis, with no electric field. The particles have initial velocity $v_x = 0$ and $v_z = v_0$.

(b) Assume a static uniform magnetic field oriented along the z-axis, with no electric field. Charged particles are initially moving with nonvanishing v_x and v_z.

(c) Assume a static uniform magnetic field oriented along the y-axis, with a static electric field along z. Charged particles are initially at rest.

(d) Assume a static uniform magnetic field **B** oriented along the y-axis, with a static electric field **E** along z. Charged particles are initially moving in the x-direction with velocity $v_0 = v_x + v_\perp$, where

$$v_x = \frac{|\mathbf{E} \times \mathbf{B}|}{B^2} \quad \text{and} \quad v_\perp > v_x$$

(e) Assume a magnetic field along the z-direction increasing in strength with increasing z. Charged particles initially have velocities v_x and v_z, with $v_x \ll v_z$.

(f) Which of the situations (a)–(e) will give rise to currents in plasmas with equal numbers of positive and negative charges?

(g) If in part(d) the electric field is 10 mV · m^{-1} and the magnetic field is 100 nT, how fast will particles drift?

2.3. Consider the situation presented in part (d) of the preceding problem, assuming equal ion and electron temperatures, with a uniform magnetic field $\mathbf{B} = B_\mathbf{y}$ and static electric field $\mathbf{E} = E_\mathbf{z}$. Suppose that slowly moving neutral atoms are also present (ion thermal speed much greater than neutral velocity, ~0). Sketch qualitatively the drift paths as in Problem 2.2, and calculate ion- and electron-drift velocities (speed and direction) assuming that

(a) collisions occur, on average, once every ion gyroperiod,

(b) collisions occur, on average, twice every gyroperiod,

(c) collisions occur once every five electron gyroperiods.

2.4. A gas in thermal equilibrium has particles at all velocities. The most probable distribution for the particles is that of a Maxwellian

$$f(v) = A \exp(-\tfrac{1}{2} m v^2 / kT)$$

in one dimension, where A is a normalization constant. This distribution is graphically depicted in Figure 2.3a. The width of the distribution is characterized by a constant T that we call temperature. The number density of particles n is given by integrating over all velocities. In three dimensions the relation is

$$n = \int_{-\infty}^{+\infty} f(v) \, d^3v$$

(a) Derive the constant A in the distribution function for a one-dimensional Maxwellian and for a three-dimensional Maxwellian.

(b) If the average value of a function $g(v)$ is

$$\langle g(v) \rangle = \frac{\int_{-\infty}^{+\infty} g(v) f(v) \, d^3v}{\int_{-\infty}^{+\infty} f(v) \, d^3v}$$

derive the average (velocity)^2 of a one-dimensional Maxwellian of temperature T and its average kinetic energy. The square root of this (velocity)^2 is sometimes called the thermal velocity, but it differs from the most probable speed in the distribution [equation (2.22)].

2.5. A particle of mass m and charge e, initially at rest at the origin, is subject to constant fields $\mathbf{E}=E_y$ and $\mathbf{B}=B_z$. Show that it moves on the cycloid

$$x = \frac{E}{\Omega B}(\Omega t - \sin \Omega t)$$

$$y = \frac{E}{\Omega B}(1 - \cos \Omega t)$$

in the plane $z=0$, where $\Omega = eB/m$. The motion is periodic with period Ω. What is the wavelength of the motion?

2.6. The magnetic-field strength in the earth's magnetic equatorial plane is given by

$$B = B_0 (R_E/r)^3$$

where $B_0 = 0.3$ G, R_E is an earth radius, and r is the geocentric distance. Derive an expression for the drift period (the time it takes a particle to drift around the earth) of a particle on the equatorial plane with a pitch angle of 90° and energy W. (Hint: Think carefully about what this implies). Evaluate this period for both a proton and an electron of 1 keV energy at a distance of $5R_E$ from the center of the earth. Compare the answer to

(a) the drift induced by the force of gravity on the same particles,
(b) the orbital period of an uncharged particle at the same position.

3 THE SUN AND ITS MAGNETOHYDRODYNAMICS

E. R. Priest

3.1 INTRODUCTION

FOR ASTRONOMERS, the sun is a fairly ordinary star of spectral type G2V and with a magnitude of 4.8. However, its proximity to the earth makes it of most immediate interest to us and also the most accessible for study. The vital statistics of the sun are as follows:

Age = 4.5×10^9 yr
Mass = 1.99×10^{30} kg
Radius = 696,000 km (696 Mm)
Mean density = 1.4×10^3 kg·m^{-3} (1.4 g·cm^{-3})
Mean distance from earth (1 AU) = 150×10^6 km (215 solar radii)
Surface gravity = 274 m·s^{-2}
Escape velocity at surface = 618 km·s^{-1}
Radiation emitted (luminosity) = 3.86×10^{26} W (3.86×10^{33} erg·s^{-1})
Equatorial rotation period = 26 days
Mass loss rate = 10^9 kg·s^{-1}
Effective blackbody temperature = 5,785K
Inclination of sun's equator to plane of earth's orbit = 7°
Composition: approximately 90% H, 10% He, 0.1% other elements (C, N, O, ...)

In order to put some of these figures into perspective, we may note that the sun's mass is 330,000 times that of the earth, and its radius is 109 times larger. The mean density of the earth is only about four times that of the sun, and the atmospheric density at the surface of the earth is 1 kg·m^{-3}, which makes the earth's sea-level atmospheric pressure five times the sun's surface pressure. One astronomical unit (AU) is only 215 times the radius of the sun, and it takes sunlight 8 min to reach the earth. The sun's emitted radiation amounts to 1 kW·m^{-2} at the surface of the earth. The surface gravity is 27 times greater at the sun, and the sun's rotation produces an equatorial velocity of 2 km·s^{-1}.

The sun is a massive ball of gas held together and compressed under its own gravitational attraction. It consists principally of hydrogen (90%) and helium (10%). These gases are mostly ionized because of the very high temperature in the sun. Elements such as C, N, and O compose

TABLE 3.1. Solar-system, Coronal, and Solar-Wind Compositions

Element[a]	Solar System[b]	Corona[c]	Solar Wind[d]
H	1,350	—	1,900
He	108	72	75
C	0.60	0.41	0.43
N	0.12	0.12	0.15
O	1.00	1.00	1.00
Ne	0.14	0.14	0.17
Na	0.003	0.012	—
Mg	0.053	0.192	—
Al	0.004	0.147	—
Si	0.050	0.176	0.22
S	0.026	0.043	—
Ar	0.005	0.004	0.004
Ca	0.003	0.014	—
Fe	0.045	0.223	0.190
Ni	0.002	0.008	—

[a]Abundances referenced to that of oxygen. Only elements with abundances greater than 10^{-3} are listed.
[b]Values from Anders and Ebihara (1982).
[c]Values derived from measurements of solar-energetic particles (Crook et al., 1984; Breneman and Stone, 1985).
[d]Values from Bochsler and Geiss (1989).

about 0.1% of the mass of the sun and are present in roughly the same proportions as on the earth. Table 3.1 lists our current best estimates of the composition of the solar system, the solar corona, and the solar wind.

The sun's atmosphere consists of three layers. The lowest is the *photosphere*, a thin "skin" only 500 km thick that emits most of the sun's light and that has a density of 10^{23} m^{-3}. Above it lies the rarer and more transparent *chromosphere*, with a density of 10^{17} m^{-3}, and the *corona*, with a density typically of 10^{15} m^{-3} near the sun, extending out to the earth's orbit (where the density is 10^{7} m^{-3}) and beyond. The temperature falls to a minimum value of 4,200K in the photosphere (by way of comparison, the temperature of red-hot iron is 1,400K, and the white-hot filament of an electric lightbulb is at 3,900K). The effective temperature of the sun is 5,785K. The spectral radiance of such a blackbody is shown in Figure 3.1. At active times, there may be additional radiation at longer and shorter wavelengths, as indicated. The integrated radiation R from a blackbody of temperature T is given by the Stefan-Boltzmann law:

$$R = \sigma T^4$$

where σ has the numerical value 5.67×10^{-8} W · m^{-2} · K^{-4}.

Progressing outward into the sun's atmosphere, the temperature rises

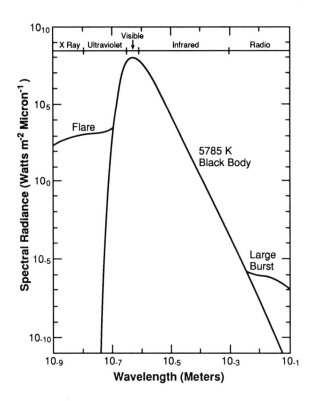

FIG. 3.1. Spectral radiance of the surface of the sun as a function of wavelength.

slowly through the chromosphere, and then, at an altitude of 2 or 3 Mm, it suddenly increases to 2×10^6K or so in the corona. This extremely high coronal temperature was discovered only in 1940. But that high temperature is somewhat deceptive, for if one's hand were immersed in a bottle of coronal gas, its temperature would rise by only a fraction of a degree, because the low density of the corona means that it contains only a minute amount of heat!

The traditional view of the sun was that, apart from its sunspots, the magnetic field was completely unimportant and had a weak uniform value of about 1 gauss (G). The atmosphere was thought to be heated by sound waves, with the excess pressure of the corona driving a spherically symmetric expansion known as the solar wind. The interior was completely hidden from view, but theoretical models suggested that in the core there existed a successful fusion reactor operating at a temperature of 1.5×10^7K and generating energy through fusion of 5 million tons of hydrogen per second to form helium. Table 3.2 lists the fusion reactions believed to create this helium. One by-product of this reaction is a flux of neutrinos, which are expected to be produced at a rate of six solar neutrino units (SNU). Figure 3.2 shows the traditional view of the structure of the solar interior. The interior temperature decreases with radius so rapidly that the outer one-quarter is unstable convectively, to give a turbulent *convection zone*. Radiation generated in the core leaks out extremely slowly. The interior is very opaque, like a thick fog, and so the radiation undergoes many deflections. If it came straight out, it

TABLE 3.2. Fusion Reactions in the Interior of the Sun

$$2(^1H + {}^1H) \rightarrow 2(^2H + e^+ + e^- + \nu)$$
$$2(e^+ + e^-) \rightarrow 2\gamma$$
$$2(^2H + {}^1H) \rightarrow 2(^3He + \gamma)$$
$$^3He + {}^3He \rightarrow {}^4He + 2^1H + \gamma$$

$$4^1H \rightarrow {}^4He + 5\gamma + 2\nu \quad \text{(overall)}$$

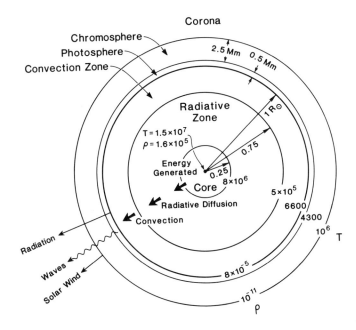

FIG. 3.2. Overall structure of the solar interior (including the core, the radiative zone, and the convection zone) and the solar atmosphere (i.e., the photosphere, chromosphere, and corona).

would reach the surface in only 2 s, but there are so many scatterings that it takes 10 million yr! This region between the core and the convection zone is known as the *radiative zone*.

The appearance of the sun varies at different levels. The photosphere represents the top of the convection zone and so is covered uniformly with a granular pattern outlining the convection cells. A white-light picture shows up dark spots – the sunspots – in two bands, one north of the equator and the other to the south. However, by observing the sun at different wavelengths, pictures of the atmosphere at different altitudes can be obtained. For example, a so-called Hα filter reveals the chromosphere (Figure 3.3), with a great deal of structure. The areas around sunspots are brighter than normal and are known as *active regions*. Occasionally, such a region may brighten very rapidly to give a *solar flare*. Also, there are thin dark structures known as *filaments* or *prominences*. At times of solar eclipse, one can glimpse the corona for a few minutes (Figure 3.4), revealing many beautiful structures.

3.2 THE NEW SUN

Many features of this traditional view have been completely transformed in the past few years. The revolution in solar physics has been caused partly by theoretical advances, partly by high-resolution observations from the ground, and partly by x-ray observations from space using satellites such as *Skylab* and the *Solar Maximum Mission* (*SMM*). We now realize that the plasma atmosphere of the sun is highly structured and dynamic, with most of what we see being caused by the magnetic field, and this interaction between the plasma and magnetic field is, for the most part, described by magnetohydrodynamics.

Detections of electron neutrinos at the earth indicate a flux of only 2 SNU. Perhaps this is caused by the core temperature being lower than expected due to mixing, or perhaps because electron neutrinos each have a mass that makes two-thirds of them change into μ- and τ-neutrinos before they reach the earth.

A major discovery is that the sun is oscillating globally, with an amplitude of only a few centimeters per second. More than a thousand different normal modes of vibration have been detected. The sun is ringing like a bell, and its modes are being used to probe the solar interior and deduce its structure, just as seismology determines the interior structure of the earth. The oscillations are sound waves trapped between the surface and a depth that depends on the order of the spherical harmonic. Around the world, networks of instruments are

FIG. 3.3. Chromosphere in Hα, showing active regions (bright) and filaments or prominences (thin dark ribbons on the disk). (Courtesy B. Schmieder, Meudon Observatory.)

FIG. 3.4. Corona in white light during an eclipse, indicating (1) a prominence, (2) a streamer, and (3) a coronal hole. (Courtesy High Altitude Observatory.)

being built to observe the sun continuously for several years to better resolve these signals. Preliminary results have shown the location of the base of the convection zone, and they suggest that the core is rotating rigidly, but the convection zone is rotating differentially, with the faster rotation at the equator. This differential rotation shows up at the surface, where the equator rotates with a period of 26 days, whereas regions near the poles take 37 days to rotate. Also, the frequencies appear to change with the solar cycle by 0.1%.

Convective instability occurs when the temperature gradient is large enough and a blob is displaced upward from equilibrium; its density will be lower than the ambient density, and so a buoyancy force will push it up still farther. At the photosphere, the resulting granulation covers the solar surface, each granule typically having a size of 500 km and a lifetime of 5 min. However, there is also a larger pattern, known as *supergranulation,* with a scale of 30 Mm, outflows of 500 m · s^{-1}, and a lifetime of 1 day. At the edges of supergranule cells we find that the magnetic field is concentrated to values of 1 kG. Such *intense-flux tubes* occur especially at the junctions between three cells.

Our understanding of the corona, too, has changed enormously. Observations of the corona in white light from space for much longer periods than are allowed by an eclipse have been made with coronagraphs. As illustrated in Figure 3.5, huge erupting bubbles known as *coronal mass ejections* have been discovered propagating ahead of erupting prominences. Furthermore, as shown in Figure 3.6, soft x-ray telescopes have revealed the corona direct, without the need for

FIG. 3.5. Composite picture, from three instruments, of a coronal mass ejection — a huge bubble of plasma and magnetic flux erupting from the sun. The limb of the sun is imaged by a prominence monitor, the inner corona by a *k*-coronameter, and the outer corona by the Solar Maximum Mission coronagraph. (Courtesy A. Hundhausen, High Altitude Observatory.)

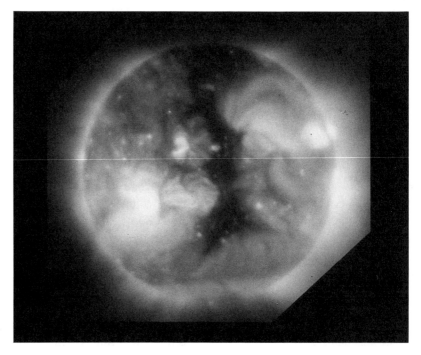

FIG. 3.6. Corona, in soft x-rays. (Courtesy D. Webb, AS&E.)

occultations produced by the moon or by disks in coronagraphs. Soft x-ray pictures show an intriguing new world, with myriad *coronal loops* and *bright points,* where magnetic fields are interacting. Dark regions, known as *coronal holes,* are areas where the magnetic field is open and through which the solar wind is streaming outward. Regions where the magnetic field is closed, on the other hand, are able to contain dense plasma. The basic properties of the sun are described in greater detail in many books, such as those by Noyes (1982) and Wentzel (1989).

3.3 THE ROLE OF THE MAGNETIC FIELD

The sun's magnetic field has several effects on the plasma. Some are passive: It may channel particles, plasma, and heat and may thermally insulate one part of the plasma from a neighboring part. But some are active: It may exert a force on the plasma and thus create structure or accelerate the plasma; it may store energy for a while and then suddenly release it; it may support waves or drive instabilities. The interaction between a plasma and a magnetic field can be modeled according to the principles of magnetohydrodynamics (MHD), in which the plasma can be treated as a continuous medium.

The equations of MHD unify the equations of slow electromagnetism and fluid mechanics. Maxwell's equations comprise Ampère's law (2.36), the divergence of the magnetic field (2.37), Faraday's law (2.35), and Poisson's equation (2.34). These are accompanied by Ohm's law (2.45), the continuity equation (2.29), the momentum equation (2.31), and the ideal-gas law (2.33). In addition, an energy equation is required in order to determine the temperature (T) and thus close the system. In MHD, the displacement current, $\epsilon_0 \, \partial E/\partial t$, is neglected, which is valid when the plasma speed u is much slower than the speed of light. The divergence of the electric field may be used, if necessary, to determine the charge density (ρ_c).

The resulting equations may seem at first to be rather complicated, but they reduce to two main equations, one for the plasma velocity **u** and one for the magnetic field **B**. Ampère's law becomes

$$\mathbf{j} = \nabla \times \frac{\mathbf{B}}{\mu_0} \tag{3.1}$$

which determines the current density **j** once **B** is known. In order of magnitude,

$$\mathbf{j} \sim \frac{\mathbf{B}}{\mu_0 L}$$

where L is the scale length for magnetic variations. Ohm's law is

$$\mathbf{E} = -\mathbf{u} \times \mathbf{B} + \frac{\mathbf{j}}{\sigma} \tag{3.2}$$

which determines the electric field in terms of **u** and **B**. Taking the curl and using Faraday's law will give the first of our two equations, namely, the *induction equation:*

$$\frac{\partial \mathbf{B}}{\partial t} = \nabla \times (\mathbf{u} \times \mathbf{B}) + \eta \nabla^2 \mathbf{B} \tag{3.3}$$

where $\eta = 1/(\mu_0 \sigma)$ is the *magnetic diffusivity* and here is assumed uniform.

The ratio of the first term to the second term on the right of (3.3) is the *magnetic Reynolds number:*

$$R_m = \frac{uL}{\eta} = \mu_0 \sigma u L \tag{3.4}$$

which for most solar phenomena on global length scales (~1 Mm, say) is enormous ($10^6 - 10^{12}$). Here, L is a characteristic scale length for changes of the field and the flow. Thus the magnetic field is frozen to the plasma, and the electric field does not drive the current but is simply $\mathbf{E} = -\mathbf{u} \times \mathbf{B}$. The exception is in intense current concentrations or sheets, where L is so small that $R_m \leq 1$. If the first term on the left of (3.3) is negligible, we have a simple diffusion equation:

$$\frac{\partial \mathbf{B}}{\partial t} = \eta \nabla^2 \mathbf{B} \tag{3.5}$$

which implies that irregularities diffuse away on a time scale

$$\tau_d = \frac{L^2}{\eta} \tag{3.6}$$

known as the *diffusion time,* and with a (diffusion) speed

$$v_d = \frac{\eta}{L} \tag{3.7}$$

where τ_d is the time scale for magnetic-energy conversion into heat by ohmic dissipation and is normally very long. For example, a length scale L of 10^7 m and a temperature of 10^6K give a diffusion time of 10^{14} s! Thus it is only in regions of intense magnetic gradient (and therefore current) that L is small enough to produce time scales of interest in, for example, coronal heating or solar flares. Such scales may be present in shock waves or equilibria with current sheets or reconnecting configurations.

In a one-dimensional magnetic field $B(x, t)\hat{y}$, (3.5) becomes

$$\frac{\partial B}{\partial t} = \eta \frac{\partial^2 B}{\partial x^2} \tag{3.8}$$

which has the same form as the heat-conduction equation that describes the flow of heat from hot to cool regions and the smoothing out of a

3.3 THE ROLE OF THE MAGNETIC FIELD

temperature gradient. Thus, in a similar way, magnetic-field gradients diffuse away in time. For example, if we start with a current sheet with

$$B = \begin{cases} B_0, & x > 0 \\ -B_0, & x < 0 \end{cases}$$

the solution of (3.8) is

$$B = B_0 \mathrm{erf}\left(\frac{x}{(4\eta t)^{\frac{1}{2}}}\right)$$

where

$$\mathrm{erf}(\xi) = \frac{2}{\pi^{\frac{1}{2}}} \int_0^{\xi} e^{-u^2} \, du$$

The total flux

$$\int_{-\infty}^{\infty} B \, dx$$

remains constant, while the magnetic energy

$$\int_{-\infty}^{\infty} \frac{B^2}{(2\mu_0)} dx$$

falls in time; it is converted ohmically into heat (j^2/σ) where $\mu_0 j = \partial B/\partial x$.

The second main equation of MHD is the momentum equation:

$$\rho \frac{d\mathbf{u}}{dt} = -\nabla p + \mathbf{j} \times \mathbf{B} + \rho \mathbf{g} \tag{3.9}$$

On the right-hand side, the first two terms represent the effects of thermal pressure and of magnetic pressure and curvature. When the plasma beta [see equation (2.49)] is small, the magnetic forces usually dominate the thermal-pressure forces. This occurs, for example, in active regions. Equating the left-hand side to the magnetic force in order of magnitude gives a speed of

$$u = \frac{B}{(\mu_0 \rho)^{\frac{1}{2}}} \equiv v_A$$

which is the *Alfvén speed* and is the typical speed to which magnetic forces can accelerate plasma. Equating the sizes of the first and third terms on the right (with $p = R\rho T$) gives a length scale of

$$L = \frac{RT}{g} \equiv H$$

which is known as the *scale height* for the falloff of the pressure with height. For example, it is about 500 km in the chromosphere, and 50,000 km in the corona. It explains why the pressure decreases with height so rapidly in the photosphere, and much more slowly in the corona. For a simple one-dimensional atmosphere with $p=p(z)$, such a hydrostatic balance gives

$$-\frac{dp}{dz}-\rho g=0$$

where $\rho=p/(RT)$, and so, if the temperature is locally uniform, the solution is

$$p=p_0 e^{-z/H}$$

which exhibits the exponential pressure falloff explicitly.

The magnetic force can be decomposed by writing

$$\mathbf{j}\times\mathbf{B}=(\boldsymbol{\nabla}\times\mathbf{B})\times\frac{\mathbf{B}}{\mu_0}=-\boldsymbol{\nabla}\left(\frac{B^2}{2\mu_0}\right)+\frac{(\mathbf{B}\cdot\boldsymbol{\nabla})\mathbf{B}}{\mu_0} \qquad (3.10)$$

in which the first term on the right represents the effect of a magnetic-pressure force acting from regions of high to low magnetic pressure $[B^2/(2\mu_0)]$. This has the same form as the normal plasma-pressure gradient $\boldsymbol{\nabla}p$, in which, for example, a pressure $p(x)$ that is increasing with x produces a pressure force $-dp/dx$ in the negative x-direction. The second term in (3.10) represents the effect of a magnetic-tension effect that gives a force when the field lines are curved and thus tends to shorten them. By putting $\mathbf{B}=B\hat{\mathbf{b}}$, where $\hat{\mathbf{b}}$ is a unit vector along the field, it can be written

$$(\mathbf{B}\cdot\boldsymbol{\nabla})\frac{\mathbf{B}}{\mu_0}=B\frac{d}{ds}\left(\frac{B\hat{\mathbf{b}}}{\mu_0}\right)=\frac{B^2}{\mu_0}\frac{d\hat{\mathbf{b}}}{ds}+\frac{B}{\mu_0}\frac{dB}{ds}\hat{\mathbf{b}}$$

in which the first term on the right is

$$-\frac{B^2\hat{\mathbf{n}}}{\mu_0 R}$$

in terms of the unit vector $\hat{\mathbf{n}}$ along the principal normal and the radius of curvature R. As noted in Chapter 2, the second term on the right cancels the gradient in pressure along the magnetic field in (3.10), so that there is (obviously) no $\mathbf{j}\times\mathbf{B}$ force along the magnetic field.

The two equations (3.3) and (3.9) are supplemented by the continuity equation

$$\frac{d\rho}{dt}+\rho\boldsymbol{\nabla}\cdot\mathbf{u}=0$$

and an energy equation

$$\frac{\rho^\gamma}{\gamma-1}\frac{d}{dt}\left(\frac{p}{\rho^\gamma}\right)=-\boldsymbol{\nabla}\cdot(\kappa\boldsymbol{\nabla}T)-\rho^2 Q(T)+\frac{j^2}{\sigma}$$

which describes how the entropy of a moving element of plasma changes because of three effects on the right-hand side: the conduction of heat, which tends to equalize temperatures along the magnetic field; the optically thin radiation, with a temperature dependence $Q(T)$; and ohmic heating. κ is the thermal conductivity. For temperatures between 10^5K and a few times 10^6K, Q decreases with temperature, and this tends to drive a *radiative instability,* because if the temperature falls, the radiation will increase, and so the plasma will tend to cool further.

Further details of the MHD equations and their use in modeling solar and solar-system phenomena can be found elsewhere (Priest, 1982, 1985).

3.4 MHD EQUILIBRIA, WAVES, AND INSTABILITIES

Equilibria of sunspots, prominences, coronal loops, and other solar structures are described by the force balance

$$\mathbf{j} \times \mathbf{B} - \nabla p + \rho \mathbf{g} = 0 \tag{3.11}$$

Along the magnetic field there is no contribution from the magnetic force, and so we have a hydrostatic balance between pressure gradients and gravity. In places such as active regions, where the magnetic field dominates, (3.11) reduces to the disarmingly simple form

$$\mathbf{j} \times \mathbf{B} = 0 \tag{3.12}$$

and the fields are said to be *force-free,* where

$$\mathbf{j} = \nabla \times \frac{\mathbf{B}}{\mu}$$

and

$$\nabla \cdot \mathbf{B} = 0$$

Thus the electric current is parallel to the magnetic field, and so

$$\nabla \times \mathbf{B} = \alpha \mathbf{B} \tag{3.13}$$

where α is a scalar function of position. Taking the divergence of (3.13) gives

$$\mathbf{B} \cdot \nabla \alpha = 0$$

so that α is constant along a field line. If α is uniform, the curl of (3.13) yields

$$(\nabla^2 + \alpha^2)\mathbf{B} = 0 \tag{3.14}$$

Solutions to this are known as linear or constant-α fields and are well understood. The particular case $\alpha = 0$ gives potential fields with zero current. Of all the fields in a finite volume with a given value for the normal component on the boundary, the field that has the smallest magnetic energy is the potential field.

Very little has been done to study fields with nonuniform α. For example, two-dimensional force-free arcades have fields that can be written as

$$(B_x, B_y, B_z) = \left(B_x(A), \frac{\partial A}{\partial z}, -\frac{\partial A}{\partial y}\right)$$

in terms of a flux function ($A(y, z)$) that satisfies

$$\nabla^2 A = -\frac{d}{dA}\left(\frac{1}{2}B_x^2\right) \tag{3.15}$$

Here, z is vertical, and x lies in the solar surface along the axis of the arcade. The difficulty is that we would like to impose not $B_x(A)$ but rather the displacement

$$\Delta x(y) = B_x \int \frac{dy}{B_y}$$

of the footpoints of coronal field lines in the solar surface ($z=0$, say), together with the normal component (B_z) there, where the integral is evaluated along field lines. Zwingmann (1986) solved the problem numerically and found a continuous sequence of equilibria as the shear was increased. By contrast, if a plasma pressure is imposed as well, then a catastrophic loss of equilibrium can occur.

There are many different kinds of MHD instabilities, as described by Bateman (1978). They include the following: interchange modes, in which field lines are wrapped around plasma in a concave manner; Rayleigh-Taylor instability, in which plasma is supported by a field against gravity, which may create structure in prominences; sausage and kink modes of a flux tube; Kelvin-Helmholtz instability, in which plasma flows over a magnetic surface; resistive modes of a sheared magnetic field, which drive reconnection; convective instability when a temperature gradient is too large, which can concentrate flux tubes in the photosphere; radiative instability, which creates cool loops and prominences up in the corona; and magnetic buoyancy instability of a magnetic field that decreases with height, which causes flux tubes to rise through the convection zone. In each case, the question of nonlinear development and saturation is important.

3.5 SOLAR ACTIVITY

There are many important topics in the field of solar MHD, and we can touch on only a few of them here. For example, dynamo theory deals with the way the global magnetic field is generated in a cyclical manner. Magnetoconvection theory studies the way turbulent convective motions concentrate magnetic flux. Coronal-loop theory deals with their equilibrium structure and stability, together with the waves and flows that are supported. Solar-wind theory concerns the way outflow is channeled along open field lines and around closed field regions. In this section we shall comment briefly on sunspots, whose structure, waves, and instabilities have been studied in detail. In the following sections we shall address the topics of coronal heating, prominences, and solar flares.

The existence of sunspots has been known since at least the fourth century B.C. They can be as large as 20 Mm in diameter, each consisting of a central dark umbra at a temperature of 4,100K and a field strength up to 0.3 tesla (T), surrounded by a penumbra of light and dark radial filaments. Figure 3.7 illustrates a typical sunspot. The field is almost vertical in the umbra, and more horizontal in the penumbra, and magnetostatic models of its structure have been constructed. There is a radial Evershed outflow in the penumbra of 6 km · s^{-1}.

Some sunspots are unipolar, some bipolar, and others more complex. They can last for up to 100 days or so, and they occur in two zones on either side of the equator. As discussed in Chapter 1 and illustrated in Figure 1.3, the number of spots varies, with an 11-yr period, but there were very few during the Maunder minimum (1645–1715), when the earth's climate was cooler than normal. As illustrated in Figure 3.8, the sunspot zones start at high latitudes and move toward the equator as the cycle progresses. Sunspot pairs exhibit polarity rules, such that the

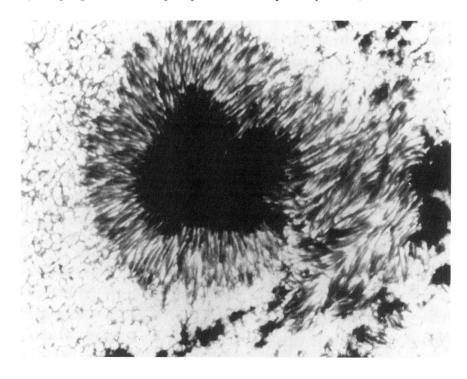

FIG. 3.7. Sunspot at high resolution in the photosphere, showing the umbra and penumbra. (Courtesy R. Muller, Pic du Midi.)

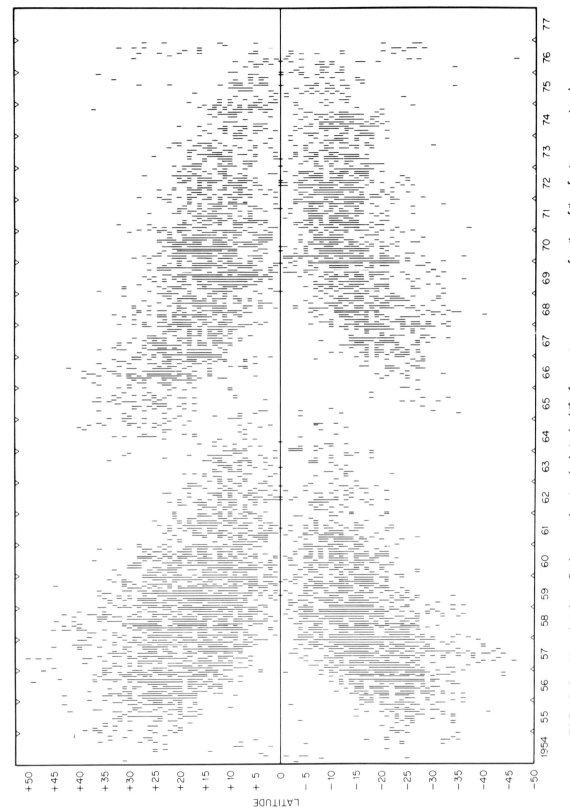

FIG. 3.8. Maunder's butterfly diagram showing the latitude drift of sunspot occurrence as a function of time for two sunspot cycles.

TABLE 3.3. The Properties of Solar Cycles 1–22

Cycle	Start[a] (date)	Solar Maximum (date)	End[b] (date)	Maximum[c] Sunspot Number	Length (yr)	Rise (yr)	Fall (yr)
1	3/17/55	6/17/61	5/17/66	86.5	11.25	6.25	5.00
2	6/17/66	9/17/69	5/17/75	115.8	9.00	3.25	5.75
3	6/17/75	5/17/78	8/17/84	158.5	9.25	2.92	6.33
4	9/17/84	2/17/88	4/17/98	141.2	13.67	3.42	10.25
5	5/17/98	2/18/05	7/18/10	49.2	12.25	6.75	5.50
6	8/18/10	4/18/16	4/18/23	48.7	12.75	5.67	7.08
7	5/18/23	11/18/29	10/18/33	71.7	10.50	6.50	4.00
8	11/18/33	3/18/37	6/18/43	146.9	9.67	3.33	6.33
9	7/18/43	2/18/48	11/18/55	131.6	12.42	4.58	7.83
10	12/18/55	2/18/60	2/18/67	97.9	11.25	4.17	7.08
11	3/18/67	8/18/70	11/18/78	140.5	11.75	3.42	8.33
12	12/18/78	12/18/83	2/18/90	74.6	11.25	5.00	6.25
13	3/18/90	1/18/94	12/19/01	87.9	11.83	3.83	8.00
14	1/19/02	2/19/06	7/19/13	64.2	11.58	4.08	7.50
15	8/19/13	8/19/17	7/19/23	105.4	10.00	4.00	6.00
16	8/19/23	4/19/28	8/19/33	78.1	10.08	4.67	5.42
17	9/19/33	4/19/37	1/19/44	119.2	10.42	3.58	6.83
18	2/19/44	5/19/47	3/19/54	151.8	10.17	3.25	6.92
19	4/19/54	3/19/58	9/19/64	201.3	10.50	3.92	6.58
20	10/19/64	11/19/68	5/19/76	110.6	11.67	4.08	7.58
21	6/19/76	12/19/79	8/19/86	164.5	10.25	3.50	6.75
22	9/19/86	7/19/89		158.1	2.83		
Avg.				113.8	11.02	4.29	6.73

[a] Solar minimum.
[b] Month before solar minimum.
[c] Smoothed with 13-month running average.

leading spot of a pair in one hemisphere tends to show the same polarity throughout a cycle, while the leading spots in the other hemisphere have the opposite polarity. At the start of a new cycle, the polarity of new spots changes. Table 3.3 lists some of the properties of sunspot cycles 1 to 22.

A sunspot is dark because it is cooler than the surrounding photosphere. Cooling occurs locally, because the magnetic field inhibits convection and thus allows the spot temperature to become lower. In the normal photosphere convection mixes the surface and the hotter subsurface layers and thus makes the surface hotter than it would otherwise be. A magnetic-flux tube below the surface tends to rise by the process of *magnetic buoyancy*. Lateral total (plasma plus magnetic) pressure balance between the flux tube and its surrounding field-free region (denoted by subscript zero) implies

$$p + \frac{B^2}{2\mu} = p_0$$

and so

$$p < p_0$$

If the temperature difference is not too great, this in turn implies

$$\rho < \rho_0$$

so that the tube is less dense than its surroundings and experiences an upward buoyancy force. When the tube rises and breaks through the solar surface, it can then create a pair of sunspots of opposite polarity, as often observed.

It has long been thought that a dynamo mechanism is operating throughout the convection zone, creating the field that oscillates every 11 yr. One ingredient is the effect of differential rotation, which stretches out an initially poloidal field and creates a toroidal field with the sense required to explain the sunspot polarity laws. The second is that as an element of toroidal flux rises, it tends to be twisted by Coriolis forces in such a sense as to create new poloidal flux of the opposite polarity. The effect of many turbulent eddies may be modeled by an equation of the form

$$\frac{\partial \mathbf{B}}{\partial t} = \text{curl}(\alpha \mathbf{B}) + \eta \nabla^2 \mathbf{B} \tag{3.16}$$

where the first term on the right is the so-called α effect. These ideas have been applied to other stars, accretion disks, and galaxies, but the current feeling is that in the sun the dynamo mechanism is much more likely to be operating near the base of the convection zone, rather than throughout it.

3.6 PROMINENCES

Prominences appear as thin dark filaments on the disk in Hα pictures, but in reality they are huge vertical sheets of plasma a hundred times cooler and denser than the surrounding corona. Figures 3.9 and 3.10 show prominences on the limb and in projection on the disk, respectively. Densities typically are $10^{16}-10^{17}$ m^{-3}, and temperatures are 5,000–8,000K; the dimensions are 200 Mm long, 50 Mm high, and 6 Mm wide. They remain stable for months and are supported against gravity by a magnetic field of strength 0.5–1 mT (5–10 G) and inclined at a small angle (15°) to the prominence axis. There is much fine-scale structure in the form of thin threads of width 300 km, although their cause is unknown. Plage (or active-region) prominences are smaller and lower than

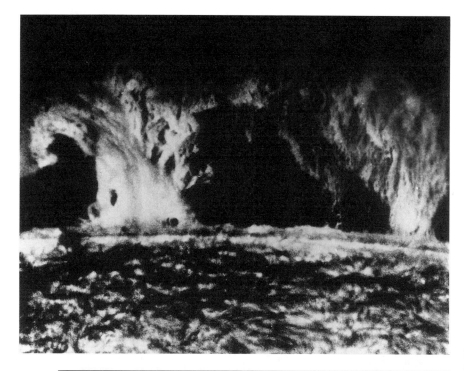

FIG. 3.9. A prominence in Hα at the limb, showing fibril structure and feet that reach down to the surface. (Courtesy Big Bear Solar Observatory.)

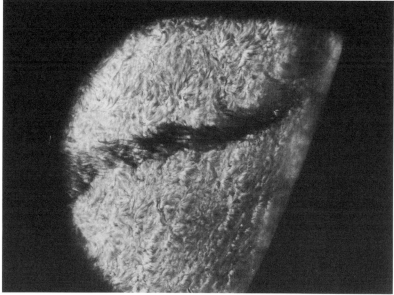

FIG. 3.10. Prominence on the disk at high resolution. (Courtesy O. Engvold.)

their large quiescent cousins, with densities greater than 10^{17} m^{-3} and fields of 2–20 mT (20–200 G).

Prominences lie above a reversal in the line-of-sight photospheric magnetic field, but the direction in which the field passes through the prominence may be normal (the same as one would expect from a simple arcade above the photospheric polarity) or inverse (in the opposite

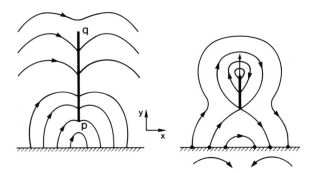

FIG. 3.11. Magnetic-field lines in a vertical section across models of prominences with normal (left) and inverse (right) polarity.

direction), as exemplified in the models of Kippenhahn and Schlüter (1957) and Kuperus and Raadu (1974), respectively, and illustrated in Figure 3.11. Leroy (1989) found that one-third of his sample of prominences were of normal polarity, and two-thirds were of inverse polarity. There are also complex flow patterns in prominences, namely, upflows of 3 km·s^{-1} in H-α on the disk and upflows on either side of a prominence of 6–10 km·s^{-1} at 10^5K.

Prominences probably form because of radiative instability, which can be demonstrated most simply as follows: Consider a uniform hot equilibrium in an arcade of density ρ_0 and temperature T_0 between mechanical heating ($h_0\rho_0$) and radiation simply proportional to density squared

$$0 = h_0\rho_0 - Q_0\rho^2{}_0 = 0$$

and perturb this at constant pressure

$$\rho c_p \frac{\partial T}{\partial t} = h_0\rho - Q_0\rho^2 + \kappa_\| \frac{\partial^2 T}{\partial s^2}$$

where the latter term represents thermal conduction in a direction s along the magnetic field. Writing the perturbed temperature as

$$T = T_0 + T_1 \exp\left(\omega t + \frac{2\pi i s}{L}\right)$$

then gives the growth rate as

$$\omega = \frac{Q_0\rho_0}{c_p T_0} - \frac{\kappa_\| 4\pi^2}{\rho_0 L^2} \tag{3.17}$$

When the loop length L is small, ω is negative, and conduction damps away the perturbation, but when L exceeds a critical value, radiative instability occurs. The classic model for support of a prominence is due to Kippenhahn and Schlüter (1957), who assume a uniform temperature (T) and horizontal field (B_x), but a vertical field [$B_z(x)$] and pressure [$p(x)$] that vary with horizontal position (x). The horizontal and vertical components of force balance are then

$$p + \frac{B^2}{2\mu} = \text{constant}$$

and

$$\rho g = \frac{dB_z}{dx}\frac{B_x}{\mu}$$

so that the magnetic field both supports and compresses the plasma. The solution of these two equations, with $\rho = p/(RT)$ is

$$B_z = B_0 \tanh\frac{x}{\ell}, \qquad p = \frac{B_0^2}{2\mu}\operatorname{sech}^2\frac{x}{\ell} \qquad (3.18)$$

where the prominence width ℓ is

$$\ell = \frac{B_x H}{B_0}$$

and the vertical field tends to $\pm B_0$ as x approaches $\pm\infty$. This solution has been extended to include temperature variations (Milne, Priest, and Roberts, 1979) and to allow variations with height (Ballester and Priest, 1987).

A simple model for a potential field outside a prominence can be constructed using complex-variable theory, with $z = x + iy$ and $B_y + iB_x$ an analytic function of z (Malherbe and Priest, 1983). Thus

$$B_y + iB_x = -\frac{B_0[(p^2+z^2)(q^2+z^2)]^{\frac{1}{2}}}{z(z+ih)^2} - \frac{B_1}{z}$$

gives a prominence of normal polarity, with the prominence represented by a cut in the complex plane from $z = ip$ to $z = iq$. Similarly, the form

$$B_y + iB_x = \frac{B_0}{z}[(p^2+z^2)(q^2+z^2)]^{\frac{1}{2}} + B_1(z-ip)$$

gives a model of inverse polarity. In the latter case, the observed steady upflows can be explained as a response to slow footpoint motions toward the polarity-inversion line if the prominence lies along the boundary of a giant convection cell.

More recently, a twisted-flux-tube model for prominences (Figure 3.12) has been proposed (Priest, Hood, and Anzer, 1989) that agrees much better with observations than do the classic models. The basic geometry is a large-scale flux tube that is slowly twisted up either by Coriolis forces or by flux cancellation (Martin, 1986; Van Ballegooijen and Martens, 1989). Eventually, a dip with upward curvature near the summit is formed, and at that point the prominence begins to form by radiative instability. As the twist or flux cancellation continues, the prominence grows in length, until the twist or length becomes too great.

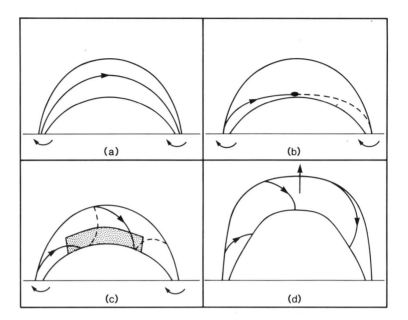

FIG. 3.12. Twisted-flux-tube model of prominences due to Priest et al. (1989) showing (a) the initial large flux tube, (b) the formation of a dip by twisting, (c) the lengthening of the prominence as the twist continues, and (d) the eruption when the twist is too great.

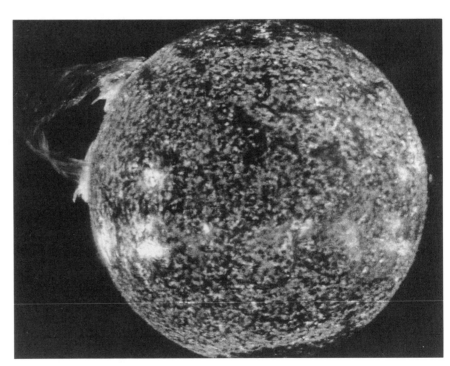

FIG. 3.13. Erupting prominence. (Courtesy Naval Research Laboratory.)

The prominence becomes unstable and erupts, as shown in Figure 3.13. At that point in its life it undergoes a metamorphosis and reveals its true geometry for the first time as a flux tube. More details of prominences can be found in the books by Poland (1986), Priest (1989), and Tandberg-Hanssen (1990).

3.7 CORONAL HEATING

Observations of the corona using soft x-rays have changed our view of the corona and have provided the stimulus for modeling possible heating mechanisms. X-ray bright points typically endure 8 h and consist of several small loops. Closed field regions may somehow act as a source for the slow solar wind, whereas the fast solar wind comes from coronal holes, where the density and temperature are 5×10^{11} m^{-3} and 1.6×10^6 K at a height of one solar radius. There are many kinds of loops, including interconnecting loops that join different active regions and have lengths (L) of 20–700 Mm, a temperature (T) of 2×10^6 K, and a density (n) of 0.7×10^{15} m^{-3}. Quiet-region loops have $L = 20$–700 Mm, $T = 1.5$–2.1×10^6 K, and $n = 0.2$–1.0×10^{15} m^{-3}, whereas active-region loops have $L = 10$–100 Mm, $T = 2.2$–2.8×10^6 K, and $n = 0.5$–5.0×10^{15} m^{-3}. Active regions, in particular, are highly dynamic, with continual activity and a wide range of flows.

The heat required to balance radiation, conduction, and flows varies considerably. In a coronal hole it is 600 W · m^{-2} (mainly for the flow) and the transition-region pressure is 0.007 N · m^{-2}; in quiet regions it is 300 W · m^{-2} (mainly for conduction and radiation), and the transition-region pressure is 0.02 N · m^{-2}; in active regions the heating is 5,000 W · m^{-2}, and the pressure is 0.2 N · m^{-2}. The observed acoustic flux for wave periods between 10 and 300 s is 10^5 W · m^{-2} at the photosphere, but only 10 W · m^{-2} at the transition region.

Thus, it is widely believed that the mechanism must be magnetic, especially because the highly magnetic regions are observed to be brighter. Also, the Poynting flux $|\mathbf{E} \times \mathbf{H}| = uB^2/\mu$ is 10^4 W · m^{-2} for a flow speed of 0.1 km · s^{-1}, say, and a field strength of 10 mT. It should be noted that coronal holes, coronal loops, and x-ray bright points may well have different heating mechanisms. Also, the fact that the corona is highly inhomogeneous and dynamic could be crucial for heating.

In closed magnetic regions, the response of the coronal field to motions of its footpoints depends on the time scale (τ_f) of such motions. If the motions are more rapid than the time (L/v_A), then the coronal field will try to evolve through a series of equilibria. The main question here is: Are the equilibria smooth, or do they contain regions of very high magnetic gradient (current sheets) where energy may be released by magnetic reconnection?

In a uniform medium, we have seen how different types of waves can be produced – Alfvén waves, slow or fast magnetoacoustic waves. In a nonuniform medium, several new features arise. A flux tube may exhibit either kinklike or sausagelike modes, and outside the tube the disturbance may be either evanescent (decaying away exponentially with distance) or propagating (carrying energy away from the tube). Also, loops may respond to photospheric motions by resonating at discrete frequencies (Hollweg, 1984):

$$\omega = \frac{n v_A}{2L}$$

where n is an integer, giving typical periods of $100/n$ s for a loop length L of 10^5 km and an Alfvén speed of 2,000 km·s^{-1}. These are the particular frequencies that a loop would pick out from a broadband excitation, and so one would expect to observe discrete frequencies in the corona, with spatial peaks located a half, a third, a quarter, and so on, along a loop. Shorter loops would show higher frequencies.

One way of dissipating Alfvén waves is by phase mixing (Heyvaerts and Priest, 1983). Consider a unidirectional field $B_0(x)\hat{z}$ with a nonuniform Alfvén speed $v_A(x)$. A perturbation $u_y \hat{y}$ satisfies the wave equation

$$\frac{\partial^2 u_y}{\partial t^2} = v_A^2(x) \frac{\partial^2 u_y}{\partial z^2}$$

and so, for example, standing waves with an imposed wave number k_z have solutions of the form

$$u_y \sim e^{i(\omega(x)t - k_z z)}$$

so that each surface oscillates with its own frequency

$$\omega(x) = k_z v_A(x) \tag{3.19}$$

However, as time proceeds, the phases become mixed and generate steep gradients in the x-direction, because

$$\frac{\partial u_y}{\partial x} = \frac{d\omega}{dx} t u_y$$

After only a few wave periods, the gradients are steep enough for ohmic and viscous dissipation to damp the wave. A similar process occurs for propagating waves, which exhibit phase mixing in space rather than time.

Another dissipation mechanism is resonant absorption, for which global modes having the other polarization

$$u_x = u(x) e^{i(\omega t - kz)}$$

satisfy

$$\frac{d}{dx}\left(\rho_0(\omega^2 - k^2 v_A^2) \frac{du}{dx} \right) - k^2 \rho_0(\omega^2 - k^2 v_A^2) u = 0$$

This equation possesses a singularity for given ω and k at a location where $v_A(x) = \omega/k$, and we find that the energy accumulates and eventually dissipates at such a resonant surface (Poedts and Goossens, 1987).

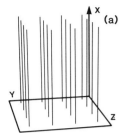 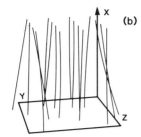 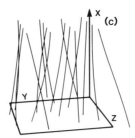 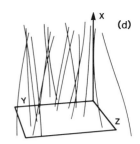

FIG. 3.14. Sequential snapshots of the numerical experiment of Mikic and associates on the three-dimensional structure formed when the ends of initially straight magnetic-field lines are braided about one another. (From Mikic et al., 1989.)

The alternative to wave heating is dissipation in current sheets. Parker (1972) first suggested that the coronal magnetic field evolves because of slow photospheric motions into equilibria that contain current sheets. Heyvaerts and Priest (1984) and Browning and Priest (1985) have extended the idea by proposing that the corona is heated by resistive turbulence as the coronal field evolves and relaxes to its lowest energy state. They set up the technical apparatus to estimate the evolution and heat input. Previously, people had considered an evolution of the coronal field through nonlinear force-free states subject to the constraint that the field line connections be imposed. Instead, Heyvaerts and Priest (1984) generalized an idea of Taylor (1974) for laboratory machines that was in turn based on previous astrophysical ideas of Woltjer (1958). They proposed that the coronal field evolves subject to a new constraint and that continual small-scale reconnections reduce the energy to that of a linear force-free field. The magnetic helicity is

$$K = \int_V \mathbf{A} \cdot \mathbf{B} \, dV \tag{3.20}$$

where $\mathbf{B} = \nabla \times \mathbf{A}$, and the new constraint determines the time evolution of K according to

$$\frac{dK}{dt} = \int_S (\mathbf{A} \cdot \mathbf{u}) \mathbf{B} \cdot d\mathbf{S} \tag{3.21}$$

as motions (**u**) on the surface inject or extract magnetic helicity from the volume. This equation determines the global uniform value of α in the force-free field. Mikic, Schnack, and Van Hoven (1989) have recently constructed an impressive three-dimensional ideal MHD numerical experiment (Figure 3.14) showing how current concentrations are built up in an initially uniform field by random footpoint motions that braid the field lines. They find that the field passes through a series of smooth equilibria, with a transfer to small scales as the current density grows exponentially in filamentary structures.

3.8 SOLAR FLARES

The overall picture for a large solar flare (Figure 3.15) is that during the *preflare phase* an active-region prominence and its overlying arcade rise slowly due to some kind of weak eruptive instability or nonequilibrium. At the flare *onset*, the field lines that have been stretched out start to break and reconnect, which releases energy impulsively and causes the prominence suddenly to erupt much more rapidly. During the *main phase*, reconnection continues and creates hot x-ray loops and Hα ribbons at their footpoints as the field continues to close down. The increase in altitude of the reconnection point causes the locations of the hot loops to rise and those of the chromospheric ribbons to move apart.

Solar-flare MHD is concerned with two main issues, namely, the nature of the initial eruptive process and the details of the energy release. The initial instability has been analyzed by considering either a magnetic-flux tube or a magnetic arcade, and in each case the anchoring or line tying of the photospheric footpoints in the solar surface provides a strong stabilizing feature that allows magnetic energy to be built up in a stable manner by twisting or shearing motions. Hood and Priest (1979) and Einaudi and Van Hoven (1983) found that a straight flux tube may become unstable when the twist is too great. Also, Browning and Priest (1984) considered a curved flux tube in equilibrium under a balance

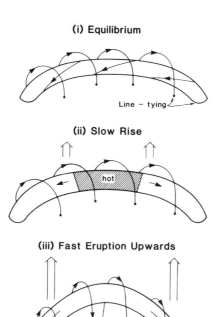

FIG. 3.15. Development of a flare, showing (i) the initial equilibrium magnetic structure, (ii) the slow rise of the structure during the preflare phase, and (iii) the rapid rise at the initiation of the flare in response to reconnection in the stretched-out field.

between magnetic buoyancy and magnetic tension. They found a loss of equilibrium when the footpoint separation or twist was too large. Simple force-free coronal arcades in which all the field lines are tied to the photosphere appear to be stable (Hood and Priest, 1980), but arcades with magnetic islands that may support a prominence, as in the twisted-flux-tube model (see Section 3.6), can become unstable if the twist or the height of the prominence is too great.

A recent analysis by Demoulin and Priest (1988) models a prominence as a sheet of mass and current supported in a linear force-free two-dimensional field with field components

$$(B_x, B_y, B_z) = \left(B_x(A), \frac{\partial A}{\partial z}, -\frac{\partial A}{\partial y}\right)$$

where the flux function A is given by

$$\nabla^2 A + \alpha^2 A = \delta(y)\delta(z-h)$$

when the prominence is located at height $z=h$ above the photospheric surface ($z=0$). The solution of this can be written as the sum of a complementary function and a particular integral (a Green's function), and the normal field component B_z is imposed at the photosphere. The condition for prominence equilibrium

$$IB_y = mg$$

at the prominence of mass m then determines the way the prominence current I varies with its height (h). The startling feature is that, when the prominence reaches a critical height, there is a catastrophe with no neighboring equilibrium, and it erupts.

Mikic, Barnes, and Schnack (1988) have modeled the global eruption of a coronal arcade numerically (Figure 3.16). They have a periodic set of arcades and impose a shearing motion of amplitude $0.01v_A$ at the base. With 100×256 mesh points, their magnetic Reynolds number is 10^4. The arcade evolves through a series of equilibria and then at some point reconnects and forms a plasmoid, which is ejected out of the top of the numerical box.

Some interesting advances have just taken place in the basic theory for steady nonlinear reconnection. Sweet (1958) and Parker (1957) first considered this process in an order-of-magnitude manner. As illustrated in Figure 3.17, oppositely directed field lines are carried into a current sheet of length L and width ℓ by a flow u_i. In the sheet, the magnetic field is no longer frozen to the plasma and may slip through it and reconnect, eventually being expelled from the ends of the sheet at the Alfvén speed (v_A). Thus, continuity of mass flux into and out of the sheet gives

$$Lu_i = \ell v_A$$

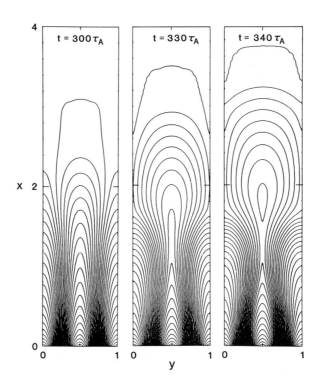

FIG. 3.16. Numerical experiment on global eruption, showing the magnetic-field lines at three times. (From Mikic et al., 1988.)

while the condition for a steady state that the magnetic field is carried in at speed u_i at the same speed as it is trying to diffuse gives

$$u_i = \frac{\eta}{\ell}$$

Elimination of ℓ implies

$$u_i^2 = \eta \frac{v_A}{L}$$

or, in dimensionless notation,

$$M_i = \frac{1}{R_{mi}^{\frac{1}{2}}} \tag{3.22}$$

where $M = u/v_A$ is the Alfvén Mach number and $R_m = Lv_A/\eta$ is the magnetic Reynolds number in terms of the Alfvén speed. Unfortunately, this gives a reconnection rate much too slow for a flare.

Petschek, however, suggested that the current sheet or diffusion region could be very much smaller, and the reconnection therefore much faster, if slow-mode shock waves stand in the flow and propagate from the ends of the current sheet. The flow u_e and field B_e at large distances can then be quite different from their values (u_i and B_i) at the input to the diffusion region. Petschek treated the inflow region as being a small perturbation to a uniform field (B_e) and flow (u_e). As the plasma gets

3.8 SOLAR FLARES

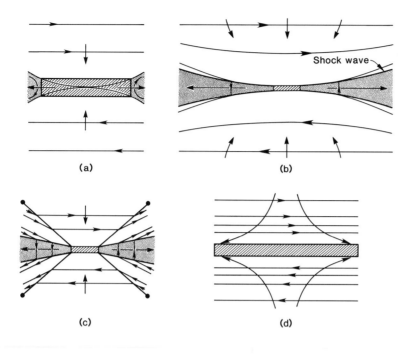

FIG. 3.17. Magnetic-field lines (arrowheads in middles of lines) and plasma-velocity directions (arrowheads on ends of lines) for the classic energy-conversion mechanisms of (a) Sweet-Parker, (b) Petschek, (c) Sonnerup, and (d) Sonnerup and Priest, showing the effect of reconnection on the initially antiparallel magnetic-field geometry and the resulting flows.

closer to the diffusion region, the field lines curve in order to allow the slow shocks to exist, and the field strength decreases from B_e to

$$B_i = B_e - \frac{4B_N}{\pi} \log_e \frac{L_e}{L}$$

where the normal field at the shocks B_N is just $v_e\sqrt{\mu\rho}$. By putting $B_i = \frac{1}{2}B_e$, Petschek then estimates the maximum possible reconnection rate as

$$M_e^* = \frac{\pi}{8 \log R_{me}} \tag{3.23}$$

which is roughly 0.01, much larger than the Sweet-Parker rate (3.7) when $R_m \ll 1$.

Vasyliunas (1975) clarified the physics and distinguished between (1) an inflow that is a diffuse *fast-mode expansion*, for which the magnetic-field strength decreases, the plasma pressure decreases, and the flow converges as it comes in (as in Petschek's model), and (2) a diffuse *slow-mode expansion*, for which the field increases, the pressure decreases, and the flow diverges.

Forbes and I wanted to understand this distinction mathematically and were puzzled that the findings in numerical experiments often were quite different from Petschek's picture. We decided to solve steady, two-dimensional, incompressible MHD equations in a finite region and try and find the reconnection rate (u_e) measured in terms of the Alfvén

FIG. 3.18. Notation for magnetic reconnection, showing magnetic-field lines (arrowheads in middles of lines) and plasma-velocity directions (arrowheads on ends of lines) for one half of the configuration. Plasma flux and magnetic flux flow in through the top boundary (with speed u_e and field strength B_e) and out through the side boundaries, on which the boundary conditions are indicated. A diffusion region exists near the origin of length L, and plasma flux and magnetic flux enter it with speed u_i and field strength B_i.

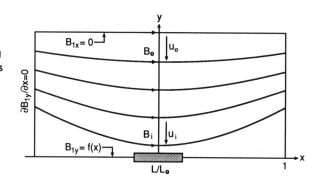

speed (Priest and Forbes, 1986). The geometry of this solution is shown in Figure 3.18. The equations are

$$\rho(\mathbf{u} \cdot \nabla)\mathbf{u} = -\nabla p + \mathbf{j} \times \mathbf{B}$$
$$\mathbf{E} + \mathbf{u} \times \mathbf{B} = 0$$
$$\nabla \cdot \mathbf{u} = \nabla \cdot \mathbf{B} = 0$$

where $\mathbf{j} = \text{curl } \mathbf{B}/\mu$. Expanding about a uniform field $B_0 \hat{\mathbf{x}}$ by putting

$$\mathbf{B} = \mathbf{B}_0 + \mathbf{B}_1 + \cdots, \quad \mathbf{u} = \mathbf{u}_1 + \mathbf{u}_2 + \cdots$$

gives, at first order,

$$j_1 B_0 = \frac{dp_1}{dy}$$

or

$$\nabla^2 A_1 = -\frac{\mu}{B_0} \frac{dp_1}{dy} \tag{3.24}$$

where

$$\mathbf{B}_1 = \left(\frac{\partial A_1}{\partial y}, -\frac{\partial A_1}{\partial x} \right)$$

Separable solutions of (3.24) that make $B_{1x} = 0$ on the inflow ($y = 1$), $\partial B_{1y}/\partial x = 0$ on the side boundaries ($x = \pm 1$), and $B_{1y} = 0$ on the symmetry axis ($x = 0$) are

$$B_{1x} = \sum_0^\infty a_n \{b - \cos[(n+\tfrac{1}{2})\pi x] \sinh[(n+\tfrac{1}{2})\pi(1-y)]\}$$

$$B_{1y} = \sum_0^\infty a_n \sin[(n+\tfrac{1}{2})\pi x] \cosh[(n+\tfrac{1}{2})\pi(1-y)]$$

where the coefficients a_n are determined to be

$$a_n = \frac{4 B_N \sin[(n+\tfrac{1}{2})\pi L/L_e]}{(L/L_e)(n+\tfrac{1}{2})^2 \pi^2 \cosh[(n+\tfrac{1}{2})\pi]}$$

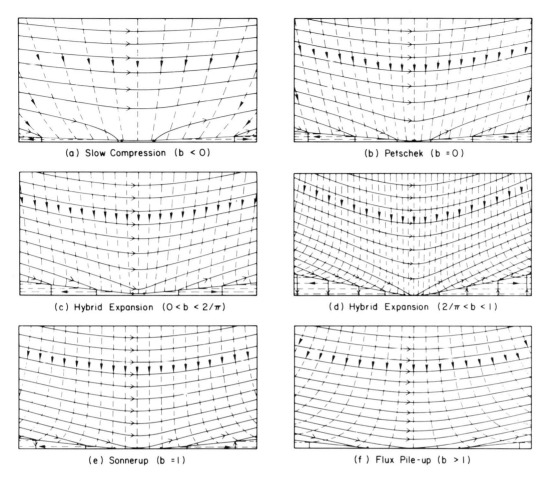

FIG. 3.19. Magnetic-field lines (solid) and streamlines (dashed) for several regions of the unified theory of almost-uniform steady reconnection, indicated by different values of the parameter b. (From Priest and Forbes, 1986.)

by the condition that on $y = 0$,

$$B_{1y} = \begin{cases} 2B_N x L_e/L, & 0 \leq x \leq L/L_e \\ 2B_N, & x \geq L/L_e \end{cases}$$

As illustrated in Figure 3.19, the solutions depend on the parameter b, which in turn depends on the nature of the inflow: At the point $(1, 1)$ the horizontal flow component is

$$u_x(1,1) \sim b - \frac{2}{\pi}$$

When $b = 0$, we recover Petschek's mechanism. When $b < 0$, there is a strong inflow, with the character of a slow compression and slower reconnection than in the Petschek mechanism. As b increases, the inflow direction changes, and the maximum reconnection rate increases. In other words, the regime of reconnection depends on the nature of the boundary conditions on the inflow. In particular, values of $b > 1$ give regimes of flux pileup, with long central current sheets, diverging inflow, and a reconnection rate faster than that of Petschek.

3.9 CONCLUSION

As we have seen, the sun is intrinsically an object of great fascination, with a rich variety of MHD phenomena that are not yet well understood, but are being modeled by analytical and numerical computations. There are two theoretical possibilities for heating the solar corona, namely, magnetic dissipation either in Alfvén waves or in current sheets. Prominences are created by radiative instability and are supported in large flux tubes. Solar flares are due to an eruptive MHD instability or catastrophe, followed by reconnection as the magnetic field closes down.

Many of these basic processes on the sun occur elsewhere in the solar system and indeed in other astronomical objects. So, in the future we hope that there will be more cross-fertilization between solar physicists and magnetosphericists and that together we may understand and appreciate the beautiful universe in which we live.

Finally, the solar community is delighted at the decision of the European Space Agency and the National Aeronautics and Space Administration to go ahead with the Solar and Heliospheric Observatory (SOHO) space mission. SOHO will sit at the Lagrangian L1 point for several years, with several sets of instruments that should produce new surprises. In particular, they aim to determine the internal structure of the sun by detecting the global oscillations with much greater precision, to make observations of the heating of the corona, to determine the structure of prominences and other atmospheric features, and to study the way the solar wind is accelerated. We wait eagerly for these new developments.

ADDITIONAL READING

Bateman, G. 1978. *MHD Instabilities*. Cambridge, MA: MIT Press.
Noyes, R. W. 1982. *The Sun, Our Star*. Cambridge, MA: Harvard University Press.
Poland, A. (ed.). 1986. *Coronal and Prominence Plasmas,* NASA CP-2422. Washington, DC: NASA.
Priest, E. R. 1982. *Solar MHD*. Dordrecht: Reidel.
Priest, E. R. 1985. *Solar System Magnetic Fields*. Dordrecht: Reidel.
Priest, E. R. 1989. *Dynamics and Structure of Quiescent Solar Prominences*. Dordrecht: Kluwer.
Priest, E. R., and A. W. Hood. 1991. *Advances in Solar System MHD*. Cambridge University Press.
Tandberg-Hanssen, E. 1990. *Dynamics of Solar Prominences,* ed. E. Tandberg-Hanssen. IAU Colloquium No. 117. Hvar.
Wentzel, D. G. 1989. *The Restless Sun*. Washington, DC: Smithsonian Institution Press.

PROBLEMS

3.1 The solar radius is 6.960×10^5 km; the sun weighs 1.989×10^{30} kg; it emits 3.9×10^{26} J·s^{-1} and rotates with a 25.34-day period. Calculate the acceleration due to gravity at the surface of the sun, the escape velocity from the solar surface, and the moment of inertia of the sun, assuming uniform density with depth. (Note that a neutral particle of mass m that leaves the surface of the sun in a radial direction moving at the *escape velocity* will reach infinity with zero kinetic energy.) If all the energy emitted comes from fusion in the core, how much mass is burned (converted into energy) per second on the sun? What is the energy flux from the sun at 1 AU (149.6×10^6 km)? What is the apparent rotational period of the sun as viewed from the earth?

3.2 Find the equation of a field line and sketch the magnetic-field lines corresponding to $B_x = y$, $B_y = x$, making sure that their relative spacing indicates the field strength.

3.3 Verify that equation (3.8) has a solution of the form $B = f(t)\exp(-x^2/(4\eta t))$, and find $f(t)$. Sketch B for several times, given that $B(0, t_0) = B_0$, and find the total magnetic flux.

3.4 Show that equation (3.8) possesses solutions of the form $B(x)$, with $x = t^n$, only when $n = -\frac{1}{2}$. Solve the equation for $B(x)$, with $n = -\frac{1}{2}$, subject to the conditions that $B \to B_0$ when $x \to -\infty$ and $B \to B_1$ when $x \to \infty$.

3.5 Consider the effect of a flow $\mathbf{u} = u_0(-x, y)$ on a field B_y that is initially $B_0 \exp(-x^2/a^2)$. Use the method of characteristics to find $B_y(x, t)$ in terms of B_0, u_0, and a, and sketch the field lines.

3.6 Calculate the magnetic-pressure and -tension forces for the field of Problem 3.2.

3.7 Find the equation of a field line, and sketch the field lines, for the field

$$B_x = B_0 \cos kx \, e^{-kz}$$
$$B_z = -B_0 \sin kx \, e^{-kz}$$

for $|x| < \pi/(2k)$, $z > 0$. Verify that it has zero current. This is a useful model for a coronal arcade.

3.8 Find a solution for a force-free arcade by seeking solutions to (3.14) and (3.13) that are separable in x and z and have forms similar to those of Problem 3.7.

3.9 Find the equation of a field line, and sketch the field lines, for

$$B_x = B_0, \quad B_y = 2B_0 x$$

What is the magnetic force at a point $(1, 0)$? Does pressure or tension dominate?

3.10 What is the magnetic field $B_\phi(r)$ produced by a cylinder of radius a and uniform current density $\mathbf{J}$ along the z-direction? Find the resulting pressure $p(r)$ if the plasma is in equilibrium, and sketch both B_0 and p.

3.11 Prove Parker's theorem that the mean square axial field B_z of a flux tube that is confined by a constant uniform pressure p_e and is in equilibrium is unaffected by twisting.

3.12 The field $B_z(r, t)$ of a cylindrically symmetric flux tube is of the form $B_0\exp(-r^2/4\eta t_0)$ at time t_0. Subsequently it diffuses according to the equation

$$\frac{\partial B_z}{\partial t} = \frac{\eta}{r}\frac{\partial}{\partial r}\left(r\frac{\partial B_z}{\partial r}\right)$$

By seeking a solution of the form $f(t)\exp(-r^2/4\eta t)$, find $B_z(r, t)$ and sketch it as a function of r for several times. Find the magnetic flux, and show that it is conserved. Show that the rate of change of magnetic energy is negative, and comment on the result.

3.13 Find the solution for the magnetic field and pressure in a prominence similar to (3.18) for which the temperature is a given function of x.

3.14 For a cylindrically symmetric magnetic-flux tube of length L in equilibrium under a balance between magnetic and pressure forces, find the pressure and azimuthal field as functions of r if $B_z = B_0(1 + r^2/L^2)$ and the twist Φ (through which a given field line is twisted in going from one end of the tube to the other) is uniform.

3.15 A unidirectional field $B_y(x)$ vanishes at $x = 0$ and is oppositely directed in the two regions $x < 0$ and $x > 0$. Plasma that is of uniform density and is frozen to the magnetic field moves steadily under the action of an electric field $E_z(x, y)$ with velocity components

$$u_x = -u_0 x/a, \qquad u_y = u_0 y/a$$

where u_0 and a are constant. Show that E is uniform, and solve Ohm's law for B_y in the two limits $x \gg l$ and $x \ll l$, where $l^2 = a\eta/u_0$. Make a rough sketch of $B_y(x)$, and comment on its behavior. Also, use the steady equation of motion to find the plasma pressure $p(x, y)$. This solution models the annihilation of a magnetic field by a stagnation-point flow.

3.16 Demonstrate the instability of an x-point field

$$B_x = B_0 y/l, \qquad B_y = B_0 x/l$$

by seeking solutions to the time-dependent equations of motion (with no pressure gradient), induction (with no diffusion), and continuity of the form $B_x = B_0(1 + ae^{\omega t})y/l$, $B_y = B_0(1 + be^{\omega t})x/l$, $u_x = ce^{\omega t}x/l$, $u_y = de^{\omega t}y/l$, $\rho = \rho_0(1 + fe^{\omega t})$. Find the values of the constants a, b, c, d, f, and ω, and suggest what happens ultimately.

4 THE SOLAR WIND

A. J. Hundhausen

4.1 INTRODUCTION

THE SOLAR WIND IS a flow of ionized solar plasma and a remnant of the solar magnetic field that pervades interplanetary space. It is a result of the huge difference in gas pressure between the solar corona and interstellar space. This pressure difference drives the plasma outward, despite the restraining influence of solar gravity. The existence of a solar wind was surmised in the 1950s on the basis of evidence that small variations in the earth's magnetic field (geomagnetic activity) were produced by observable phenomena on the sun (solar activity), as well as from theoretical models for the equilibrium state of the solar corona. It was first observed directly and definitively by space probes in the mid-1960s.

Measurements taken by spacecraft-borne instruments since that time (and until recently) have yielded a detailed description of the solar wind across an area from inside the orbit of Mercury to well beyond the orbit of Saturn. Solar-wind observations and theory have been the subjects of several books (Parker, 1963; Brandt, 1970; Hundhausen, 1972) and a continuing series of conferences summarizing new research (e.g., Pizzo et al., 1987). Why has this distant and tenuous plasma been the subject of sustained interest in the scientific community? The answer to this question probably stems from two important aspects of solar-wind research.

The first of these concerns the role of the solar wind in the interdisciplinary subject known as *solar-terrestrial relations*. As we shall see later in this chapter, the solar wind is significantly influenced by solar activity (or, in physical terms, by changes in the solar magnetic field) and transmits that influence to planets, comets, dust particles, and cosmic rays that are immersed in the wind. Some of these effects will be described more extensively in later chapters. The origin of the solar influence through interaction of the solar magnetic field with the expanding coronal plasma has become a major topic in present-day solar-wind research and will be heavily emphasized in this chapter.

The second important aspect of solar-wind research that helps to explain the sustained interest in the subject concerns the physical processes that occur in its formation and expansion from the hot solar corona to the cool and far more tenuous regions of the outer solar system. This expansion takes the magnetized plasma through huge variations in its properties; for example, collisions among ions or electrons in the expanding plasma are frequent in the corona, but extremely rare in interplanetary space. Thus the physics of this plasma system can be examined under a wide variety of conditions, some of which are difficult to attain in terrestrial laboratories or in the immediate vicinity of the earth. Yet the solar wind is accessible to space probes, and its properties can be measured and its physical processes studied at a level of detail impossible for most astrophysical plasmas.

This chapter begins with a summary of the basic characteristics of the solar wind as it forms in the solar corona and as it sweeps outward past the orbit of the earth. It then introduces our basic understanding of how the solar wind manages to form and escape from the gravitationally bound atmosphere of the sun in an idealized case, that of a steady, radial, spherically symmetric flow from the sun. Finally, it describes a number of important additions to and modifications of that basic picture as the simplifying assumptions behind it are relaxed. In this progress from idealization toward reality, we shall touch upon several of the most interesting and significant questions that occupy the attention of those who study the solar wind today, more than 25 yr after its discovery.

4.2 A QUICK SURVEY OF SOLAR-WIND PROPERTIES

The most extensive and most detailed observations of the solar wind have been made from spacecraft near the orbit of the earth. Some of the physical properties of the plasma and magnetic field at this distance from the sun (1.5×10^{13} cm, or one *astronomical unit* from the sun's center) are summarized in Table 4.1. The solar wind that blows past the earth is hot, tenuous, and fast-moving by terrestrial standards. It consists largely of ionized hydrogen (or of protons and electrons in nearly equal numbers), with a small (5 percent by number) admixture of ionized helium and still fewer ions of heavier elements. Embedded in this plasma

TABLE 4.1. Observed Properties of the Solar Wind near the Orbit of the Earth (1 AU)

Proton density	6.6 cm^{-3}
Electron density	7.1 cm^{-3}
He^{2+} density	0.25 cm^{-3}
Flow speed (nearly radial)	450 km·s^{-1}
Proton temperature	1.2×10^5 K
Electron temperature	1.4×10^5 K
Magnetic field (induction)	7×10^{-9} tesla (T)

TABLE 4.2. Solar-Wind Flux Densities and Fluxes near the Orbit of the Earth

	Flux Density	Flux Through Sphere at 1 AU
Protons	3.0×10^8 cm$^{-2} \cdot$s^{-1}	8.4×10^{35} s^{-1}
Mass	5.8×10^{-16} g$\cdot$cm$^{-2} \cdot$s^{-1}	1.6×10^{12} g$\cdot$s^{-1}
Radial momentum	2.6×10^{-9} pascal (Pa)	7.3×10^{14} newton (N)
Kinetic energy	0.6 erg$\cdot$cm$^{-2} \cdot$s^{-1}	1.7×10^{27} erg$\cdot$s^{-1}
Thermal energy	0.02 erg$\cdot$cm$^{-2} \cdot$s^{-1}	0.05×10^{27} erg$\cdot$s^{-1}
Magnetic energy	0.01 erg$\cdot$cm$^{-2} \cdot$s^{-1}	0.025×10^{27} erg$\cdot$s^{-1}
Radial magnetic flux	5×10^{-9} T	1.4×10^{15} weber (Wb)

is a weak magnetic field oriented in a direction nearly parallel to the ecliptic plane (the plane of the earth's orbit around the sun), but at approximately 45° to a line from the sun to the observer at 1 astronomical unit (AU). It is often useful to describe the solar wind in terms of the fluxes or flux densities of quantities that are conserved in a plasma flow. Table 4.2 gives average values for the flux densities of particles (and mass), momentum in the radial direction, and energy carried by the solar wind past the orbit of the earth. The radial-momentum flux is also often called the dynamic pressure, because of its role in confining the magnetospheric magnetic field. Most of the momentum is carried by the protons; most of the energy is in the form of the kinetic energy of the same particles. If these averages of measurements made near the ecliptic plane are typical of conditions over an entire sun-centered sphere with a radius of 1 AU, we can obtain the total fluxes of these quantities by multiplying by the area of that sphere, 2.82×10^{27} cm^2. Thus the solar wind carries mass away from the sun at a rate of 1.6×10^{12} g$\cdot$s^{-1}, and energy at a rate of 1.8×10^{27} erg$\cdot$s^{-1}. At the former rate, the solar wind will have a negligible effect on the 2×10^{33}-g mass of the sun over its 5-billion-yr ($\sim 10^{17}$ s) lifetime. Similarly, the solar-wind energy flux is less than 10^{-6} of the 4×10^{33} erg$\cdot$s^{-1} radiated from the sun. The present-day solar wind is virtually negligible in the overall mass and energy balance of the sun. Table 4.3 shows a few additional properties of the solar wind near 1 AU that can be computed from the measurements in Table 4.1.

The pressure in an ionized gas with equal proton and electron densities n is

$$p_{\text{gas}} = nk(T_p + T_e)$$

where k is the Boltzmann constant, and T_p and T_e are the proton and electron temperatures. Thus

$$p_{\text{gas}} = 3 \times 10^{-10} \text{ dyn} \cdot \text{cm}^{-2}$$
$$= 30 \text{ pico pascals (pPa)}$$

TABLE 4.3. Some Derived Properties of the Solar Wind near the Orbit of the Earth

Gas pressure	30 pPa
Sound speed	60 km·s^{-1}
Magnetic pressure	19 pPa
Alfvén speed	40 km·s^{-1}
Proton gyroradius	80 km
Proton–proton collision time	4×10^6 s
Electron–electron collision time	3×10^5 s
Time for wind to flow from corona to 1 AU	$\sim$4 days $= 3.5 \times 10^5$ s

Sound waves in an ionized gas with pressure p_{gas} and mass density $\rho = n(m_p + m_e)$, where m_p and m_e are the proton and electron masses, travel at a speed

$$c_s = \left\{\frac{\gamma p}{\rho}\right\}^{\frac{1}{2}} = \left\{\frac{\gamma k}{m_p + m_e}(T_p + T_e)\right\}^{\frac{1}{2}}$$

where γ is the ratio of specific heats at constant pressure and constant volume, and c_s is the speed of sound. Using $\gamma = \frac{5}{3}$ for an ionized hydrogen gas and temperatures from Table 4.1, we find a sound speed at 1 AU of

$$c_s \approx 60 \text{ km} \cdot \text{s}^{-1}$$

Thus the typical speed of the solar-wind flow of 400 km·s^{-1} is almost an order of magnitude greater than the sound speed at 1 AU; the solar-wind flow is highly supersonic. Finally, the presence of a magnetic field in an ionized gas can lead to hydromagnetic effects, as outlined in Chapter 2. In many ways, the magnetic field can be thought of as exerting a pressure

$$p_{\text{mag}} = \frac{B^2}{2\mu_0}$$

Using the average field strength in Table 4.1, we find a magnetic pressure near 1 AU of

$$p_{\text{mag}} \approx 1.5 \times 10^{-10} \text{ dyn} \cdot \text{cm}^{-2}$$
$$\approx 15 \text{ pPa}$$

This value is comparable to the gas pressure, indicating that magnetic effects will be about as important as pressure effects in the solar-wind plasma. In particular, Alfvén waves, small-amplitude waves driven by magnetic fluctuations (see Chapters 2 and 11) in the solar wind, will move at a speed comparable to the sound speed.

In attempting to understand the origin of the solar wind, it is necessary to know a few properties of the solar corona. The temperature of the sun falls from $\sim 15 \times 10^6$K in its interior to $\sim$5,000K at its visible

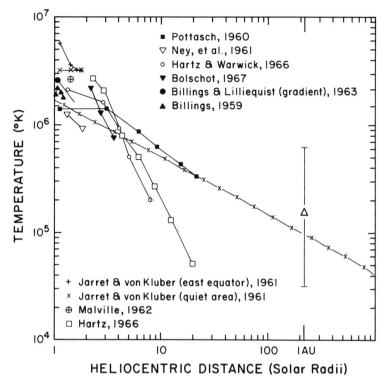

FIG. 4.1. Coronal temperature measurements at various heliocentric distances. (Adapted from Newkirk, 1967.)

surface, where the atmosphere becomes sufficiently thin that light can escape. Surprisingly, at still greater heights in the atmosphere, the temperature rises again, back over the 10^6K level, and then falls off very slowly with increasing height in the corona. It is this slowly varying temperature of about 1.5×10^6K that is the distinguishing characteristic of the corona and, as we shall see shortly, the physical reason for the formation of a solar wind. Figure 4.1 illustrates this remarkable property of the outermost visible (at least at solar eclipses) region of the sun. Figure 4.2 shows the variation of density with height in the corona. The number density of ions (again mainly hydrogen) at the base of the region is $\sim 10^8$ cm^{-3}, small compared with that in the lower regions of the sun, but seven orders of magnitude higher than that in the solar wind at 1 AU. The gas pressure in the corona is then easily calculated to be

$$p_{\text{gas}} \approx 4 \times 10^{-2} \text{ dyn} \cdot \text{cm}^{-2}$$
$$\approx 4 \text{ mPa}$$

Although we still have no good observations of outward-flow speeds in the corona, the absence of large Doppler shifts in coronal emission lines is evidence that flow speeds are small, especially in comparison with the sound speed of ~ 160 km $\cdot$ s^{-1} that applies at a 1.5×10^6K temperature. Finally, models that extend the magnitude of the solar magnetic field observed at its visible surface into the corona indicate an average field strength of a few thousand tesla at the base of the corona. Such a field would be sufficiently strong that the magnetic pressure in this region

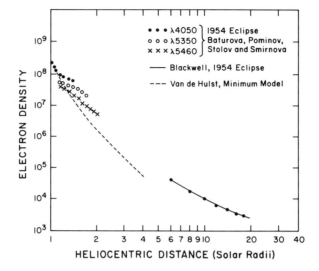

FIG. 4.2. Equatorial electron density for 1954 corona. Number density of electrons (cm^{-3}) inferred by different observers from data obtained at a single eclipse of the sun. (Adapted from Billings, 1966.)

($p_{\text{mag}} \approx 10$ mPa) would be several times greater than the gas pressure. We would thus expect magnetic effects to dominate this region where the solar wind originates. For example, Alfvén waves should travel several times faster than ordinary sound waves in the corona.

Comparison of the coronal and 1-AU values for density and temperature given earlier leads to an obvious conclusion: that these quantities vary with distance from the sun. We shall examine these and other such radial variations in solar-wind properties shortly. Further, the values introduced here are average values or typical values. Even at a given distance from the sun, such as 1 AU, solar-wind properties can vary widely on many different time scales. In particular, the flow speed of the solar wind can be as low as ~200 km · s^{-1} or higher than 1,000 km · s^{-1}. In Section 4.3 we shall examine the basic theory of a uniform, steady coronal expansion in hope of explaining the average solar wind. In doing so, we must not lose sight of the variability of the wind. We shall, in fact, later in this chapter, return to a description of the origin of that variability in the interaction of the magnetic field and the expanding coronal plasma.

4.3 THE BASIC CONCEPT OF SOLAR-WIND FORMATION IN THE SOLAR CORONA

We shall deal here with three facets of what might be termed the classic theory of the solar wind. The first facet involves a fluid model for the equilibrium state of the corona and leads to the basic concept of a continuous supersonic flow of plasma from the corona into interplanetary space. The second involves a description of the configuration of solar magnetic-field lines that are frozen to the expanding plasma and hence are drawn out into interplanetary space by the solar wind. Finally, we shall bring this section to a shocking conclusion by considering how the solar wind in the distant reaches of the solar system manages to merge with the interstellar medium.

4.3.1 The Fluid Theory of Solar-Wind Formation

Our basic understanding of solar-wind formation comes from a theoretical model for the equilibrium state of the hot coronal plasma in the gravitational field of the sun. That theory will be illustrated here starting from the equations describing the conservation of mass (or the continuity equation) and momentum in a fluid, as set forth in Chapter 2:

$$\frac{\partial \rho}{\partial t} + \nabla \cdot \rho \mathbf{u} = 0 \quad \text{(mass)}$$

$$\rho \frac{\partial \mathbf{u}}{\partial t} + \rho \mathbf{u} \cdot \nabla \mathbf{u} = -\nabla p + \mathbf{j} \times \mathbf{B} + \rho \mathbf{F}_g \quad \text{(momentum)}$$

For a plasma flow that is *steady* or independent of time, all time derivatives are zero, and we need deal only with simpler equations:

$$\nabla \cdot \rho \mathbf{u} = 0 \tag{4.1}$$

$$\rho \mathbf{u} \cdot \nabla \mathbf{u} = -\nabla p + \mathbf{j} \times \mathbf{B} + \rho \mathbf{F}_g \tag{4.2}$$

We shall further simplify the solution of these equations by making additional idealizing assumptions. In particular, we shall eliminate any geometric complications by assuming that the system is spherically symmetric, or that all physical properties are functions of the heliocentric distance, r, only. Further, the flow velocity will be assumed to be strictly radial from the sun. It is then natural to write the conservation equations (4.1) and (4.2) in a spherical polar coordinate system with its origin at the sun's center. If $\hat{\mathbf{e}}_r$ is a unit vector that points radially from the sun, we can write the flow speed as

$$\mathbf{u} = u(r)\hat{\mathbf{e}}_r$$

and express the sun's gravitational field as

$$\mathbf{F}_g = -\frac{GM_\odot}{r^2}\hat{\mathbf{e}}_r$$

where G is the gravitational constant, and $M_\odot$ is the mass of the sun. The pressure gradient is

$$\nabla p = \frac{dp}{dr}\hat{\mathbf{e}}_r$$

and the divergence terms on the left-hand sides of equations (4.1) and (4.2) are

$$\nabla \cdot \rho \mathbf{u} = \frac{1}{r^2}\frac{d}{dr}\rho u r^2$$

and

$$\rho \mathbf{u} \cdot \nabla \mathbf{u} = \rho u \frac{du}{dr} \hat{\mathbf{e}}_r$$

If we defer consideration of magnetic effects by neglecting the magnetic force $\mathbf{j} \times \mathbf{B}$, equations (4.1) and (4.2) can finally be written in the form we shall be dealing with in our theoretical model:

$$\frac{1}{r^2} \frac{d}{dr} \rho u r^2 = 0 \qquad (4.3)$$

$$\rho u \frac{du}{dr} = -\frac{dp}{dr} - \rho \frac{GM_\odot}{r^2} \qquad (4.4)$$

This system of equations has a very simple solution that until the late 1950s was regarded as a valid description of the equilibrium state of the solar corona. This solution stems from the further assumption that the corona is in static equilibrium, or that $u(r) = 0$ everywhere. Equation (4.3) is then automatically satisfied, and (4.4), the momentum equation, becomes

$$-\frac{dp}{dr} - \rho \frac{GM_\odot}{r^2} = 0 \qquad (4.5)$$

which is nothing but a statement of the balance between the pressure gradient and gravitational forces in a static atmosphere. If we take coronal protons and electrons to have the same temperature T, the ideal-gas law becomes

$$p = nk(T_e + T_i) = 2nkT$$

Solving for the mass density $\rho = n(m_e + m_p) = nm$, where m is the sum of the proton and electron masses, yields

$$\rho = m \cdot \frac{p}{2kT}$$

Substitution into (4.5), assuming that the temperature T is constant, finally gives the equation for pressure in a *static, isothermal* atmosphere:

$$\frac{1}{p} \frac{dp}{dr} = -\frac{GM_\odot m}{2kT} \frac{1}{r^2}$$

The solution to this differential equation for $p(r)$ is

$$\ln p = \frac{GM_\odot m}{2kT} \frac{1}{r} + K$$

where K is an arbitrary constant. Setting $p(r) = p_0$ at the base of the corona, $r = R$, gives the value of the constant

$$K = \ln p_0 - \frac{GM_\odot m}{2kT}\frac{1}{R}$$

and allows us to write the solution as

$$\ln \frac{p}{p_0} = \frac{GM_\odot m}{2kT}\left(\frac{1}{r} - \frac{1}{R}\right)$$

or

$$p(r) = p_0 \exp\left\{\frac{GM_\odot m}{2kT}\left(\frac{1}{r} - \frac{1}{R}\right)\right\} \quad (4.6)$$

For $r > R$, $1/r - 1/R$ is negative, and $p < p_0$; that is, solution (4.6) shows what we would expect, that $p(r)$ decreases with increasing distance from the sun.

Equation (4.6) is a generalization of a familiar formula for the exponential decay of pressure in a static, isothermal atmosphere. This formula is usually given for an atmosphere that is so "shallow" that the variation in the gravitational field can be neglected over the range of r that is of interest. The more familiar form can be recovered from (4.6) by measuring r from the base of the corona, in terms of the height

$$h = r - R$$

and considering only $h \ll R$. Then

$$\frac{1}{r} = \frac{1}{R+h} = \frac{1}{R(1+h/R)} \approx \frac{1}{R}\left(1 - \frac{h}{R}\right)$$

In this approximation, equation (4.6) becomes

$$p = p_0 \exp\left\{\frac{GM_\odot m}{2kT}\left(\frac{1}{R} - \frac{h}{R^2} - \frac{1}{R}\right)\right\} = p_0 \exp\left\{-\frac{GM_\odot m}{2kTR^2}\cdot h\right\} = p_0 e^{-h/\lambda} \quad (4.7)$$

where $\lambda = 2kTR^2/GM_\odot m = 2kT/mg$ is the *scale height* given in terms of the acceleration due to gravity g at the base of the atmosphere.

The deficiency of equation (4.6) as a model for the equilibrium state of the corona stems from the variation in gravity over the great heliocentric distance spanned by the corona. If we let $r \to \infty$, the pressure given by (4.6) does not continue to decay exponentially as in (4.7); rather, it approaches the value

$$p_\infty = p_0 \exp\left\{-\frac{GM_\odot m}{2kTR}\right\}$$

For a coronal temperature of 10^6K, this is only about e^{-8} or 3×10^{-4} of the high pressure at the base of the corona. It is many orders of magnitude higher than the pressure thought to exist in the interstellar medium

($p_{\text{int}} \approx 10^{-13} - 10^{-14}$ Pa), and thus could not represent an equilibrium state between the corona and that distant medium.

It was this problem that motivated E. N. Parker in the 1950s to reexamine the equilibrium states implied by equations (4.3) and (4.4) by considering solutions with nonzero flow speeds. The first of these equations is satisfied if

$$\rho u r^2 = C \quad \text{(a constant)}$$

The meaning of this expression becomes clear if it is multiplied by 4π to give

$$4\pi r^2 \cdot \rho u = I \quad \text{(a constant)} \tag{4.8}$$

Because ρu is the rate at which mass is carried through a unit area on a sun-centered sphere, and $4\pi r^2$ is a surface area of such a sphere of radius r, then I is just the mass flux (in grams per second) through the entire sphere. For a time-independent radial flow, conservation of matter implies what this expression explicitly tells us: that the total flux through all sun-centered spheres must be the same.

The momentum equation (4.4) involves the same two fluid properties, ρ and u, as our first integral (4.8), plus the pressure p. Under the same isothermal assumption made in our analysis of a static equilibrium,

$$p = 2nkT$$

can be differentiated and substituted into (4.4) to give

$$\rho u \frac{du}{dr} = -2kT \frac{dn}{dr} - \rho \frac{GM_\odot}{r^2}$$

or, using $\rho = nm$ and dividing through by ρ,

$$u \frac{du}{dr} = -\frac{2kT}{m} \frac{1}{n} \frac{dn}{dr} - \frac{GM_\odot}{r^2} \tag{4.9}$$

This equation can be written with the speed $u(r)$ as the only remaining fluid property by writing (4.8) as

$$4\pi r^2 m n u = I$$

solving for

$$n = \frac{I}{4\pi m} \frac{1}{ur^2}$$

and differentiating to obtain

$$\frac{dn}{dr} = \frac{I}{4\pi m} \left\{ -\frac{1}{u} \cdot \frac{2}{r^3} - \frac{1}{r^2} \frac{1}{u^2} \frac{du}{dr} \right\} \tag{4.10}$$

Then

4.3 BASIC CONCEPT OF SOLAR-WIND FORMATION

$$\frac{1}{n}\frac{dn}{dr} = \frac{4\pi m}{I} \cdot ur^2 \cdot \frac{I}{4\pi m}\left\{-\frac{2}{ur^3} - \frac{1}{r^2 u^2}\frac{du}{dr}\right\} = -\frac{2}{r} - \frac{1}{u}\frac{du}{dr}$$

Substitution into (4.9) then reduces the momentum equation to

$$u\frac{du}{dr} = \frac{4kT}{mr} + \frac{2kT}{m}\frac{1}{u}\frac{du}{dr} - \frac{GM_\odot}{r^2} \tag{4.11}$$

This is a differential equation for $u(r)$ and its derivative du/dr in an *expanding, isothermal* atmosphere.

Analysis of equation (4.11) is facilitated by moving all terms involving $u(r)$ to the left-hand side to give

$$u\frac{du}{dr} - \frac{2kT}{m}\frac{1}{u}\frac{du}{dr} = \frac{4kT}{mr} - \frac{GM_\odot}{r^2}$$

or

$$\left(u^2 - \frac{2kT}{m}\right)\frac{1}{u}\frac{du}{dr} = \frac{4kT}{mr} - \frac{GM_\odot}{r^2} \tag{4.12}$$

This form of the momentum equation was recognized by Parker (1958) as revealing the existence of a solar wind. For any realistic coronal temperature T, the second term on the right-hand side of (4.12), GM/r^2, is larger than the first term, $4kT/mr$, at the base of the corona. This is basically a statement that despite its high temperature, the corona is gravitationally bound. Thus the right-hand side of (4.12) is negative near the base of the corona. However, GM/r^2 falls off more rapidly with r than does $4kT/mr$, so that the right-hand side grows with increasing r, passes through zero at the radius

$$r_c = \frac{GM_\odot m}{4kT}$$

and becomes positive at still larger radii. This behavior of the right-hand side of equation (4.12) has a profound influence on the nature of its solutions. If $4kT/mr - GM_\odot/r^2 = 0$ at the special or *critical radius*, r_c, the left-hand side of the equation must also be zero there. This can be achieved in two ways. For most solutions of (4.12),

$$\left.\frac{du}{dr}\right|_{r_c} = 0$$

That is, these solutions have maxima or minima at r_c. Let us now restrict our attention to solutions that start with small values of $u(r)$ near the base of the corona, as we expect from the small Doppler shifts detected in line emission from that region. Then $u^2 - 2kT/m < 0$ for r near R; because the right-hand side of equation (4.12) is also negative there, du/dr must be positive. Thus, $u(r)$ increases with r in these solutions until $r = r_c$ attains its maximum value there, and then decreases (with

$u^2 - 2kT/m$ still negative) for $r > r_c$ because the right-hand side is now positive. For these solutions, $u^2 < 2kT/m$ for all values of r.

However, there are special solutions of (4.12) for which the left-hand side is zero at $r = r_c$, because $u^2 - 2kT/m = 0$ there. In particular, there is a unique solution that starts from a small value of $u(r)$ near the base of the corona, with $u^2 - 2kT/m < 0$ and $du/dr > 0$ for $R < r < r_c$, but which passes through $r = r_c$, with $u^2 = 2kT/m$ and $du/dr|_{r_c}$ still positive. Because $u^2 - 2kT/m$ must then become positive for $r > r_c$, the change in sign on the right-hand side of the equation can be accommodated with du/dr still positive; $u(r)$ can continue to increase for all values of r. As noted in Section 4.2, the sound speed in the corona is given by

$$c_s^2 = \frac{\gamma p}{\rho} = \gamma \frac{2nkT}{nm} = \gamma \frac{2kT}{m}$$

All other solutions (that start with small u near the base of the corona) attain a maximum value of $u(r)$ less than $(2kT/m)^{1/2}$, or less than the sound speed at $r = r_c$. All of these solutions have small speeds at large r and share the problem of the static equilibrium model developed earlier; the pressure approaches an unacceptably large value as $r \to \infty$. The special solution, however, has a flow speed that increases with r, exceeding $(2\gamma kT/m)^{1/2}$, or actually becoming supersonic, as it is observed to be in the solar wind. Because, from

$$n(r) = \frac{I}{4\pi m} \frac{1}{r^2 u(r)}$$

$n(r) \to 0$ as $r \to \infty$, the pressure $p = 2nkT$ must also approach a zero value. Hence, this solution can represent an equilibrium state connecting the high-pressure corona with the very low pressure interstellar medium; it is, however, not a static equilibrium, but one with a steady flow or expansion of coronal plasma away from the sun. This change in behavior as $r \to \infty$ is a direct result of the flow in breaking down the dependence of $n(r)$ on the scale height in the gravitational field and producing the basically *geometric* dependence on r^{-2} through the continuity equation. It was on this basis, before any direct interplanetary observations were available, that Parker proposed the continuous, supersonic expansion of the corona to form a solar wind.

It is not difficult to write down actual solutions to equation (4.12). The reader can easily verify by differentiation that the expression

$$\frac{1}{2}u^2 - \frac{2kT}{m}\ln u = \frac{4kT}{m}\ln r + \frac{GM_\odot}{r} + K'$$

where K' is again an arbitrary constant, is a solution of (4.12). Imposition of the requirement that $u^2 = 2kT/m$ where $r = r_c = GM_\odot m/4kT$ determines the constant K' and gives the explicit form of the special *solar-wind solution*

$$u^2 - \frac{2kT}{m} - \frac{2kT}{m} \ln \frac{mu^2}{2kT} = 8\frac{kT}{m} \ln\left(\frac{r}{r_c}\right) + 2GM_\odot\left(\frac{1}{r} - \frac{1}{r_c}\right) \quad (4.13)$$

Figure 4.3 shows this solution $u(r)$ for a range of values of the coronal temperature T. In the interplanetary region beyond a heliocentric distance of approximately 10 solar radii, these solutions give solar-wind speeds of hundreds of kilometers per second.

A hardheaded reader might question the validity of the model derived earlier, on the basis of the numerous simplifying assumptions made in its development. In the remainder of this chapter we shall consider the relaxation of several of these simplifications; what is important here is that the basic nature of the steady-flow solutions is not changed in the far more complex models that have been developed in the past 30 yr of solar-wind research. For example, relaxation of the isothermal assumption to a specified temperature function $T(r)$ still leads to a *wind* solution as long as $T(r)$ falls off with r less rapidly than $1/r$ [so that the right-hand side of equation (4.12) still changes sign]. In this formulation, the flow speed $u(r)$ increases rapidly near the sun, as in the isothermal model, but approaches a nearly constant value beyond a heliocentric distance of approximately 10 solar radii. A still more realistic examination of the effects of the varying coronal temperature on the characteristics of the coronal expansion can be based on the introduction of an energy-conservation equation that incorporates a reasonable set of physical assumptions as to the mechanisms that transport energy in the corona and solar wind; this approach then requires the far more difficult solution of the coupled mass, momentum, and energy equations. It can be shown that solutions of the special *solar-wind type* exist when energy is carried by the flow and by thermal conduction from a million-degree coronal base. However, it seems unlikely that such a "conductive" model can describe a solar-wind flow consistent both with the observed density and temperature of the corona and with the average properties of the solar wind at 1 AU. Other mechanisms, such as the energy dissipated by Alfvén waves propagating out from the sun (as well as the

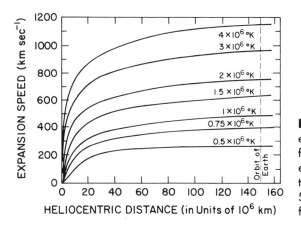

FIG. 4.3. Radial-expansion speed $u(r)$ derived from isothermal coronal-expansion models with coronal temperatures ranging from 5×10^5K to 4×10^6K (Adapted from Parker, 1958.)

force exerted by the pressure of these waves), have been extensively studied as possible factors crucial to the actual formation of the solar wind.

4.3.2 The Spatial Configuration of a Magnetic Field Frozen into the Solar Wind

The classic picture of the interplanetary magnetic field in the presence of an outwardly moving solar wind is a simple application of the concept of a frozen-on magnetic field described in Chapter 2. Referring to the development there, if a magnetic-flux tube exists in a steady plasma flow, the fluid contained in the flux tube at a given location must also be contained within the tube at any other location. If we introduce the concepts of flow tubes and streamlines, in analogy to magnetic-flux tubes and field lines, these plasma flow and magnetic structures are then identical. Physically, we can think of a flow that drags the frozen-in field with it, forming a magnetic structure consistent with the plasma flow.

If we apply this concept to a model of a spherically symmetric, radially expanding solar wind, we might first expect that the resulting interplanetary magnetic field would be extremely simple. If a flow tube were defined to have an infinitesimal area dA_0 on a sphere at the base of our model corona ($r = R$), the uniform radial outflow would give a tube whose cross section at any radius r would be a simple mapping of the original dA_0, but distended because of the spherical geometry of the expansion, as sketched in Figure 4.4. At any r, dA would then be equal to $(r/R)^2 \, dA_0$. Magnetic-field lines passing through dA_0 would also spread out in the radial direction and would define a magnetic-flux tube with the same geometry and the same changing area dA. Conservation of

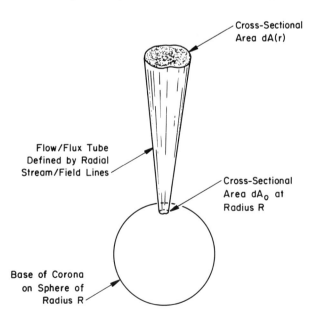

FIG. 4.4. Geometry of a flow (or flux) tube defined by radial streamlines (or magnetic-field lines).

magnetic flux within the tube (as implied by $\nabla \cdot \mathbf{B} = 0$) then would require that if B_0 were the (radial) field at the base of the flux tube, and $B(r)$ were the field at any radius r,

$$B(r)\left(\frac{r}{R}\right)^2 dA_0 = B_0 \, dA_0$$

or

$$B(r) = B_0 \left(\frac{R}{r}\right)^2$$

That is, the intensity of the purely radial magnetic field would fall off as $1/r^2$.

Unfortunately, there is an essential complication that makes this simple picture inappropriate for the solar wind – namely, that the solar atmosphere rotates about an axis that is nearly perpendicular to the ecliptic plane (the plane of the earth's orbit). The rotation rate in this atmosphere varies with location and, on the average, with latitude in a heliographic coordinate system where the solar-rotation axis is used to define the positions in space of the solar poles. Near the solar equator (or in the region from which the solar wind would reach space probes near the ecliptic plane in our radial-expansion model), an average rotation period is 25.4 days. Thus the solar corona and any fixed source, such as we used to define the area element dA_0 in the preceding paragraph, are rotating at an angular rate of

$$\omega = \frac{2\pi \text{ rad}}{25.4 \text{ days}} = 2.7 \times 10^{-6} \text{ rad} \cdot \text{s}^{-1}$$

The effect of this rotation on our description of solar-wind flow tubes/magnetic-flux tubes is sketched in Figure 4.5. As successive parcels of fluid move outward from a fixed solar source at the base of the flux tube, the source moves beneath it because of rotation. Thus the actual trace of fluid parcels emitted from the fixed source takes the shape of a spiral. Magnetic-field lines frozen into the expanding plasma and their fixed source at the base of the corona must take this same form, or be drawn into a spiral configuration. This effect is analogous to the streams of water spiraling out from a rotating water sprinkler, or the grooves in a phonograph record that rotate as the phonograph needle moves in a direction nearly radial to the center of the disk.

The shape of these spiral field lines can be expressed mathematically if we transform ourselves into the frame of reference rotating with the sun, so that the source of the plasma and field lines remains fixed. The plasma still moves outward in the radial direction, with the velocity component $U_r = u(r)$, as in the true stationary frame. However, it now has an apparent component in the direction of solar longitude ϕ that is entirely due to the transformation between coordinate systems,

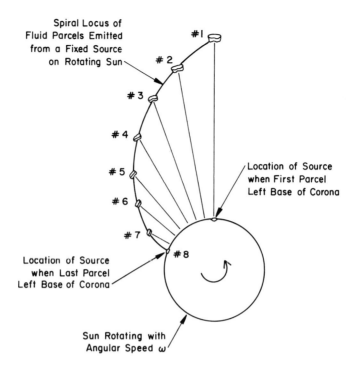

FIG. 4.5. Loci of a succession of fluid parcels (eight of them in this sketch) emitted at constant speed from a source fixed on the rotating sun.

$$U_\varphi = -\omega r$$

Then the magnetic field stretched out along the path of plasma flowing from the fixed source in this coordinate system has components related by

$$\frac{B_\varphi}{B_r} = \frac{U_\varphi}{U_r} = \frac{-\omega r}{u(r)} \tag{4.14}$$

This gives a differential equation for the field lines near the solar equator (solar latitude = 0) as

$$\frac{r \, d\varphi}{dr} = \frac{-\omega r}{u(r)}$$

If the radial-expansion speed is constant, as in the solar wind well out in interplanetary space, this equation becomes

$$\frac{dr}{d\varphi} = -\frac{u}{\omega}$$

and has the obvious solution

$$r = -\frac{u}{\omega}\varphi + K''$$

Specification of the location of the source of a field line at longitude φ_0 at $r = R$ then yields

$$r - R = -\frac{u}{\omega}(\varphi - \varphi_0)$$

This is in fact a geometric figure of Grecian antiquity and respectability known as the spiral of Archimedes.

The field intensity in this magnetic geometry can be derived by drawing a thin flux tube bounded by spiral field lines and again invoking the conservation of magnetic flux. The flux into the tube at its base is simply $B_0 \, dA_0$, as in our previous treatment without rotation. If we draw a cross-sectional area within the tube, then on a spherical surface at radius r its area is again just proportional to r^2, and because the same flux must pass through this surface (none passes through the sides of the flux tube, because they are defined to be parallel to field lines), we have a relation for the radial field component $B_r(r)$,

$$B_r(r) \frac{r^2}{R^2} dA_0 = B_0 \, dA_0$$

or

$$B_r(r) = B_0 \left(\frac{R^2}{r^2} \right)$$

The longitude (or azimuthal) component of the field can then be derived from (4.14):

$$B_\varphi(r) = -\frac{\omega r}{u} B_r = -B_0 \frac{\omega R}{u} \frac{R}{r}$$

Transformation back to the stationary frame of reference does not change these magnetic-field components (for nonrelativistic speeds), but does imply an electric field in the stationary frame where the velocity is not parallel to the magnetic field. Figure 4.6 shows this spiral geometry for a constant solar-wind speed of 400 km · s^{-1}. The field lines carried through interplanetary space become more tightly wound as heliocentric distance increases. At the orbit of the earth, $\omega r = 405$ km · s^{-1}, so that the average radial and longitudinal components of the magnetic field are nearly equal. The angle between the field and a line drawn from the sun to an observer at 1 AU should be close to 45°, as is observed.

4.3.3 The Termination of the Solar Wind

We have, so far, developed a model of a continuously expanding solar wind and frozen-in magnetic field that avoids the problem encountered in static models for the equilibrium of a corona with an extended, high temperature. In solar-wind models based on the isothermal assumption or on more realistic assumptions, where the temperature is allowed to decrease with heliocentric distance r, the gas pressure approaches zero

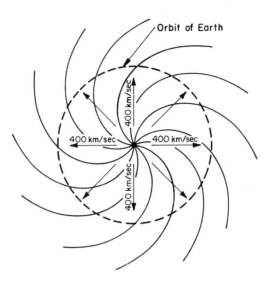

FIG. 4.6. Spiral interplanetary magnetic-field lines frozen into a radial solar-wind expansion at 400 km·s^{-1}. (Adapted from Parker, 1963.)

as $r \to \infty$. The magnetic field in these models becomes more and more azimuthal as $r \to \infty$, so that the magnetic pressure

$$\frac{B^2}{2\mu_0} \approx \frac{B_\varphi^2}{2\mu_0} = \frac{B_0^2}{2\mu_0}\left(\frac{\omega R}{u}\right)^2 \frac{R^2}{r^2}$$

also approaches zero as $r \to \infty$. This behavior has been achieved by having the outward pressure gradient drive a flow of material and break down the simple connection between the pressure and an increasing scale height that led to large pressures as $r \to \infty$ in the static model. In fact, the momentum-flux density in the solar wind, ρu^2, a fluid property that has many of the aspects of a pressure, also approaches zero as $r \to \infty$, because of the nearly $1/r^2$ dependence in ρ implied by the continuity equation.

We are thus led to quite the opposite problem if we consider the need for the solar wind to merge with an interstellar background that has a pressure of $\sim 10^{-13}$ Pa. At some large heliocentric distance r, the momentum-flux density in the solar wind will fall below the interstellar pressure, and one would expect the solar wind to slow down as it reacts to this external-pressure force. Slowing the distant solar wind is, however, not an easy task. The isothermal model described in detail in Section 4.3.1 predicts that at large heliocentric distances, $u(r)$ should continue to slowly increase; for $u \gg (2kT/m)^{\frac{1}{2}}$ and $r \gg r_c$, equation (4.13) predicts that

$$u^2 \approx 8\frac{kT}{m} \ln \frac{r}{r_c}$$

or

$$u(r) \approx 2\left(\frac{2kT}{m}\right)\left(\ln \frac{r}{r_c}\right)^{\frac{1}{2}}$$

Because T is constant, the ratio of flow and sound speeds increases as $(\ln r)^{\frac{1}{2}}$, and the solar wind becomes more supersonic as it flows through the outer solar system. Models that allow the temperature to decrease with heliocentric distance predict that $u(r)$ will approach a finite limiting value as $r \to \infty$. But because $T(r)$ is decreasing, the ratio of flow and sound speeds will again increase. Thus, no internal pressure or magnetic force can have a significant effect in slowing or turning the solar-wind flow to effect a merger with the interstellar plasma or magnetic field. Further, the influence of the latter cannot be felt within the supersonic (and super-Alfvénic) region of solar-wind flow, as no sound or Alfvén waves can propagate toward the sun to warn the solar wind that an obstacle to this flow lies ahead!

One way out of this dilemma is to invoke an effect that occurs in many supersonic flows: the formation of a shock wave that abruptly slows the solar wind to a subsonic speed. Beyond this shock, the solar plasma can "feel" the presence of the interstellar medium by more conventional means and gradually adjust itself to dynamic and pressure equilibrium. If the pressure of the interstellar medium p_{int} is the same in all directions, the shock wave should stand near the heliocentric distance, where the radial-momentum flux in the wind equals the interstellar pressure, or

$$\rho \mathbf{u} \cdot \mathbf{u} = p_{int}$$

For the value of the radial-momentum flux at 1 AU of 2.6×10^{-8} dyn·cm^{-2} given in Table 4.2, an interstellar pressure of 10^{-12} dyn·cm^{-2}, and a $1/r^2$ variation in solar-wind density implied by the continuity integral (4.8) for a constant wind speed, equality is reached for

$$\left(\frac{r_{earth}}{r_{shock}}\right)^2 \approx \frac{10^{-12}}{2.6 \times 10^{-8}}$$

or

$$\frac{r_{shock}}{r_{earth}} \approx (2.6 \times 10^4)^{\frac{1}{2}} \approx 160$$

$$r_{shock} = 160 \text{ AU}$$

This estimate places the shocking conclusion to the supersonic flow of solar wind well beyond any known planet. At such large heliocentric distances, the ratio of the speed of the solar wind flowing into the shock and the sound speed in that wind is large; the shock that terminates the solar-wind flow will be a very strong shock.

More sophisticated models of the shock location yield smaller values of r_{shock}, but generally place it in the range 5–150 AU. For example, the pressure just beyond a strong termination shock is about 0.88 $\rho \mathbf{u} \cdot \mathbf{u}$ (see

Chapter 6). If it is this pressure that comes to equilibrium with p_{int}, the shock distance will be $(0.88)^{\frac{1}{2}}$ of the distance estimated earlier. Questions as to the existence of shock waves in the extremely tenuous and collision-free outer solar system, the possibility of interactions with cosmic rays rather than interstellar plasma, and the effects of an interstellar magnetic field and any motion of the sun through the interstellar gas all enter into a realistic theory of the ultimate deceleration of the solar-wind flow. The dynamics of the region in which this decelerating flow finally adjusts to its interstellar surrounding at some surface known as the *heliopause* are still more influenced by these complications. For a recent review of this topic, the reader is referred to Holzer (1989).

4.3.4 Summary

The idealized models describing the formation of the solar wind through a supersonic expansion of the solar corona, the spatial configuration of the interplanetary magnetic field "frozen into" that expansion, and the ultimate deceleration of the solar wind in the outer solar system were all based on numerous simplifying assumptions. It is worth reiterating that whereas the relaxation of some of these assumptions has been a major source of employment for solar-wind theorists for more than two decades, the basic classic picture of the solar wind has been changed largely in its details. The outward acceleration of corona plasma by a pressure difference between the corona and the interstellar medium is still the basis for current solar-wind models; the questions that remain concern the different pressures (and energies) that must be included to produce quantitative agreement with observations. The spiral magnetic-field structure, the nearly constant flow speed, and the inverse-square falloff in density are so fundamental that observations by space probes in the outer solar system are searched for deviations from these basic patterns.

The remainder of this chapter will deal with two important elaborations of the classic picture. The first of these will bring us to a long-postponed consideration of the influence of the solar magnetic field on the expansion of the corona plasma. The second will consider the existence of time variations in the solar wind and our understanding of their generation and propagation outward through interplanetary space.

4.4 THE MAGNETIC STRUCTURE OF THE CORONA AND SOLAR WIND

The interaction of the solar magnetic field with the expanding coronal plasma occurs through the magnetic force, expressed as $\mathbf{j} \times \mathbf{B}$ in the momentum equation for steady flow,

$$\rho \mathbf{u} \cdot \nabla \mathbf{u} = -\nabla p + \mathbf{j} \times \mathbf{B} + \rho \mathbf{F}_g \qquad (4.15)$$

In the "magnetohydrodynamic approximation," where the time derivative of the electric field is neglected, Ampère's law

$$\mu_0 \mathbf{j} = \nabla \times \mathbf{B}$$

can be used to eliminate **j** from (4.15). The resulting system of equations is closed by the equation $\nabla \cdot \mathbf{B} = 0$ and Ohm's law (as in Chapter 2)

$$\mathbf{j} = \sigma (\mathbf{E} + \mathbf{u} \times \mathbf{B}) \qquad (4.16)$$

In the corona and solar wind, the electrical conductivity is so high that Ohm's law is well approximated by the simpler form expressing the frozen-in nature of the magnetic field:

$$\mathbf{E} + \mathbf{u} \times \mathbf{B} = 0$$

Finding solutions for this magnetohydrodynamic (MHD) system of equations is much more complicated than solving the non-magnetic-fluid equations. The first level of complication is obvious. The momentum equation (4.15) contains an additional term and involves an additional physical quantity as a dependent variable, namely, the vector **B**. Further, the additional equations $\nabla \cdot \mathbf{B} = 0$ and Ohm's law must be solved simultaneously with the mass and momentum equations. A still more severe level of complication, however, stems from the basic nature of the magnetic force on a fluid element. As is evident from the expression $\mathbf{j} \times \mathbf{B}$ for that force, it is always perpendicular to the magnetic-field vector **B**. Some of the geometric simplifications used to facilitate solution of the fluid equations are thus inapplicable to an MHD problem. For example, if any electric currents form in our spiral magnetic-field geometry of Figure 4.6 (as they would if the field strength were nonuniform at the base of the corona or if the radial-expansion speed were not constant), the magnetic force will have a component in the longitude or azimuthal direction. It will thus produce a longitude or azimuthal flow, forcing us to drop the assumption that **u** is purely radial and deal with the true vector nature of the momentum equation. Further, because the simplest possible magnetic field has a dipole configuration (magnetic monopoles do not exist in classic electrodynamics), it would be unrealistic to expect an MHD flow describing the entire corona to be spherically symmetric. An assumption like $\mathbf{u} = \mathbf{u}(r)$ in Section 4.3 can no longer be justified.

4.4.1 The MHD Equations in Field-aligned Coordinates

These complications can be mitigated to some extent by the use of a field-line-oriented coordinate system for steady-flow problems. At any location, let ds be an element of the distance s measured along a field line. For a frozen-in magnetic field, the plasma flows parallel to the field, so that both **B** and **u** are parallel to $d\mathbf{s}$. The mass-conservation equation can then be written

$$\frac{d}{ds}[\rho u A(s)] = 0 \tag{4.17}$$

which has the solution

$$\rho(s)u(s)A(s) = I \quad \text{(a constant)}$$

This differs from the integral (4.8) in our earlier example for radial, spherically symmetric flow only in that $A(s)$ is a more general expression for the flux or flow-tube area than was $4\pi r^2$ in the earlier case. The component of the momentum equation in the direction of $\hat{b}$ (or the parallel component) is then

$$\rho u \frac{du}{ds} = -\frac{dp}{ds} - \rho \frac{GM_\odot}{r^2} \cos \xi \tag{4.18}$$

where ξ is the angle between the field line/streamline and the radial direction at any location. This is a simple generalization of the earlier equation (4.4) involving the proper component of the gravitational force in the direction of $\hat{b}$. It does not contain any magnetic-force term, because that force is perpendicular to the field direction used to define ds.

The momentum equation is, of course, a vector equation and is only partially expressed by (4.18). The remaining components can be deduced by taking the cross product of the vector equation with **B** itself. This transverse-momentum equation in component form for the general three-dimensional case introduces complications too messy to be written out here. Fortunately, the meaning of these equations can be found by dealing with a two-dimensional geometry where the magnetic-field lines lie in a plane and do not depend on the spatial coordinate normal to that plane. Then the current given by $\mu_0 \mathbf{j} = \nabla \times \mathbf{B}$ will always be perpendicular to that plane, and the magnetic force $\mathbf{j} \times \mathbf{B}$ will be back in the field-line plane, but perpendicular to **B** itself. The momentum equation in the direction of the magnetic force can be manipulated into the form

$$jB = F_{\text{mag}} = -\frac{\partial p}{\partial z} - \rho \frac{GM_\odot}{r^2} \sin \xi - \frac{\rho u^2}{R_c} \tag{4.19}$$

where z is the distance measured in the direction of $\mathbf{j} \times \mathbf{B}$, R_c is the radius of curvature of the field line, and ξ is the angle between that direction and the component of the gravitational field projected into the plane defined by the field lines. This is simply a statement of the balance of the magnetic force with the components of the pressure-gradient, gravitational, and centrifugal forces in the direction of $\mathbf{j} \times \mathbf{B}$; the change in the momentum implied by the term $\rho \mathbf{u} \cdot \nabla \mathbf{u}$ and due to the flow of plasma has been reduced to the centrifugal force in this direction perpendicular to the flow.

Although we shall not specifically develop an energy equation here, it can also be simplified by expression in these field-line/flow-line coordinates. This simplification occurs because all energy carried by the plasma flow moves along flow lines. Further, in a plasma with a magnetic field sufficiently strong that the times for collisions among particles are much longer than the gyroperiod, diffusive processes such as heat conduction are severely inhibited across magnetic-field lines and again carry energy largely along the magnetic field.

Several interesting conclusions can be drawn from the form of the equations for mass (4.17) and parallel momentum conservation (4.18). First, they differ only slightly from the similar equations (4.3) and (4.4) in Section 4.3 for a radially expanding, spherically symmetric flow; the only significant difference is in the more general expression $A(s)$ for the cross-sectional area of a flow tube. Thus the solutions along a flow tube will be very similar to those in Section 4.3. This can be most simply illustrated for the case of a system in static equilibrium, for which the momentum equation reduces to the force balance

$$0 = -\frac{dp}{ds} - \rho \frac{GM_\odot}{r^2} \cos \xi$$

$$\frac{dp}{dr} = -\rho \frac{GM_\odot}{r^2}$$

just as in our earlier model, and has exactly the same solution (4.6)! A static equilibrium atmosphere has the same variation in pressure with vertical height or heliocentric distance along a given flux tube as does the nonmagnetic atmosphere. The magnetic force enters into the equilibrium only in the transverse direction, as expressed by equation (4.19) with $u=0$, or by

$$jB = -\frac{\partial p}{\partial z} - \rho \frac{GM_\odot}{r^2} \sin \xi$$

A similar conclusion can be drawn when the flow speed u is not zero and yields a simple explanation for the general success of the models like those from Section 4.3 in giving a viable physical picture of the solar wind despite neglect of magnetic effects. Even in a complicated coronal magnetic geometry, the solar wind expands outward along "open" magnetic-flux tubes. The problem of steady flow along these tubes is complicated only by the factor $A(s)$ in the continuity equation [and the appearance of the derivative of $A(s)$ when the density is eliminated from the parallel momentum equation]. The magnetic effect on the basic outward expansion appears only through that rather subtle geometric factor.

This is not to say that solution of the full set of field-line-oriented equations is a trivial matter; it has been accomplished for magnetic geometries more complex than radial field lines only by numerical techniques. Nonetheless, our second conclusion to be drawn from the form of these equations follows from the separation into a parallel flow problem with no direct magnetic effects and a transverse-force balance that involves the magnetic force but includes the flow of plasma only through the centrifugal-force term. This pattern suggests an iterative approach to the solution of the full set of coupled equations. Starting from an assumed magnetic geometry (consistent, of course, with $\nabla \cdot \mathbf{B} = 0$), the equilibrium of the plasma can be found by solving the parallel equations (4.17) and (4.18). Then, with ρ, $\mathbf{u}$, and p and $\mathbf{B}$ known functions throughout space, equation (4.18) can be used to calculate a current density $\mathbf{j}$ that will give a transverse equilibrium at all locations. A new field geometry, hopefully close to that originally assumed, can then be calculated from $\mathbf{j}$, and the process can be repeated until it converges in the sense that changes in the solution after succeeding steps become small. This was, in fact, the technique used to obtain the first numerical solutions to this problem, as will be described later.

4.4.2 A Model for Coronal Magnetic Structure

Consider now the MHD expansion of an isothermal corona with the simplest of all realistic magnetic fields, a dipole, imposed as a boundary condition at the base of the corona. If the strength of that field were small, we would expect the corona to expand outward, dragging the frozen-in magnetic field with it, as in the fluid model of Section 4.3. If the field were very strong, we would expect the magnetic configuration to be like that of a dipole field in a vacuum (or with no significant electric currents in the corona), with field components in the radial and latitudinal directions of

$$B_r = 2B_0 \sin \theta \left(\frac{R}{r}\right)^3$$

$$B_\theta = B_0 \cos \theta \left(\frac{R}{r}\right)^3$$

where θ is the colatitude in a spherical polar coordinate system whose symmetry axis (z-axis) is aligned with the dipole axis at the coronal base. B_0 is half the field strength at the base of the corona along the dipole axis, or, at $r = R$, $\theta = 0$. (See Chapter 6 for a more detailed discussion of dipole magnetic fields.) All field lines in this magnetic geometry are closed in the sense that each line leaves the base of the corona, extends outward in a plane of constant longitude (in our dipole-aligned coordinate system) to a maximum height at the dipole equator $\theta = \pi/2$, and then returns to the base of the corona on the other side of that

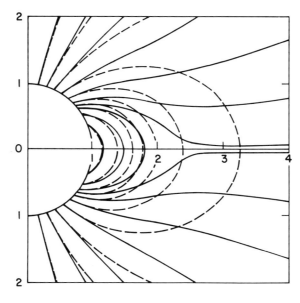

FIG. 4.7. Magnetic-field lines deduced from the isothermal, MHD coronal-expansion model of Pneuman and Kopp (1971) for a dipole field at the base of the corona. The dashed lines are field lines for the pure dipole field.

equator. For the real corona, where the magnetic field is sufficiently strong that the magnetic pressure is a few times larger than the gas pressure, we expect a mixture of these extreme behaviors. Most of the magnetic-field lines that pass through the base of the corona should remain closed, holding the coronal plasma that resides on them in static equilibrium along each field line, as described earlier. A minority of the field lines passing through the base of the corona should be open, in the sense that they are drawn outward by expanding plasma and do not return to the base of the corona at some other location. On those open field lines, solution of equations (4.17) and (4.18) will give a model of solar-wind formation.

This problem was first solved by Pneuman and Kopp (1971) for an isothermal corona, using an iterative approach similar to that outlined earlier. Figure 4.7 shows their solution for the magnetic-field lines in a plane of constant longitude, along with dashed lines showing pure dipole field lines starting from the same footprints at the base of the corona. Closed field lines overlie the "dipole equator," distorted outward from their pure dipole configuration; once again, they contain plasma in static equilibrium, in response to the pressure gradient and gravitational forces. The field lines that extend out from "high-latitude" regions are open, with the coronal plasma on them expanding outward to form a solar wind. These open field lines spread in latitude to cover all space beyond a heliocentric distance of approximately two solar radii; the belt of closed field lines above the equator (remember that this solution is axially symmetric about the vertical dipole axis in Figure 4.7) extends only to a finite height just over two solar radii. Special attention should be given to the two open field lines in Figure 4.7 that are rooted at $\theta \approx 45°$ on opposite sides of the dipole equator, skim along the top of the closed field region to nearly meet at low latitudes, and then extend outward nearly parallel to the equator. Because these field lines come from opposite ends of the dipole, they must represent magnetic fields with different senses or polarities. Thus, the sense or polarity of the distant magnetic field must change sign abruptly at the equator. This implies a thin region of high current density $\mathbf{j} = (1/\mu_0) \nabla \times \mathbf{B}$ at the equator, with $\mathbf{j}$ normal to the plane of Figure 4.7. This current circulates around the dipole, in its equatorial plane and in the same direction as the current in that dipole. It is thus an *interplanetary current sheet* that separates the fields and plasma flows that originated from opposite ends of the dipole field.

Although the Pneuman and Kopp model deals with the simplest of possible real solar fields, it suggests the magnetic configuration for a corona with a more complicated solar field at its lower boundary. Closed magnetic structures should form over those locations where the vertical component of the field at the base of the corona changes its sign, or above what are loosely called neutral lines in the solar magnetic field. Open magnetic structures should form over regions where the vertical component of the field at the base of the corona is of the same sense or sign over a large area, or above what are loosely called unipolar regions in the solar magnetic field. The closed coronal structures should be of limited radial extent, reaching to approximately two solar radii for the largest possible features (in a dipole field), but to smaller heights for structures with smaller lateral extents at the base of the corona. The open coronal structures should spread laterally with increasing height to fill all space above closed regions with outward-flowing solar wind, as well as open magnetic-field lines. Open field structures with opposite magnetic polarities should come into close proximity above the closed magnetic regions and thus form current sheets that extend out into the solar wind.

These generalizations are consistent with intuitive expectations concerning the interaction of the solar magnetic field with the coronal plasma and are confirmed by solutions to the MHD equations that involve fields more complex than a dipole, as have been obtained in the 20 yr since the work of Pneuman and Kopp. More important, the expected magnetic structures seem to be clearly visible in photographs of the solar corona obtained at solar eclipses. For example, the top of Figure 4.8 is a photograph taken during an eclipse in November 1966 from a site in Peru. The bright features at the right and lower left of the sun have the moundlike appearance we would expect for closed magnetic structures; the fine striations within these mounds also suggest closed flux tubes in a plasma where the field is strong enough to prevent cross-field diffusion. The dark feature near the bottom of the photograph spreads laterally with increasing height, as we would expect for an open magnetic structure; fine striations within this region again add to the suggestion that magnetic-flux tubes extend and flare outward. The mod-

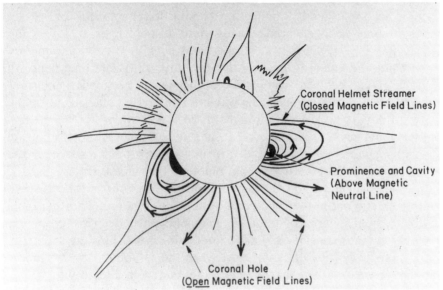

FIG. 4.8. Photograph of the corona obtained at the 1966 solar eclipse and a sketch of the magnetic-field structures believed to exist within the observed coronal features.

erately bright "spikes" extending above the bright mounds are where we would expect to see the current sheets separating open regions of opposite magnetic polarity. This magnetic interpretation of the observed coronal features is sketched in the bottom part of Figure 4.8. The bright, moundlike features with spikes extending above them are traditionally called *coronal helmet streamers*. Many helmet streamers can be shown to lie above neutral lines in the observed solar magnetic field. The example at the bottom of Figure 4.8 encloses a darker cavity that in turn encloses a very bright central feature identified as a solar prominence.

Prominences are known to form above magnetic neutral lines. The dark feature at the bottom of the photograph is an example of a *coronal hole*. Coronal holes are found to lie above unipolar regions in the observed solar magnetic field. They may be more familiar as dark regions in x-ray pictures of the solar disk; such pictures show the cross-sectional area of the hole near the base of the corona, whereas eclipse photographs reveal the vertical structure of coronal holes.

4.4.3 Extension of This Model into Interplanetary Space

The effects of solar rotation are ignored in the preceding model of the coronal magnetic structure. This neglect is entirely reasonable in the corona, where the transformation (see Section 4.3.2) from a stationary coordinate system to one rotating with the sun involves an apparent azimuthal velocity component ωr of 2 km · s^{-1} at $r = 1$ solar radius or 20 km · s^{-1} at 10 solar radii. Comparison with the expansion speeds in an isothermal expansion model ($\sim$10 km · s^{-1} near 1 solar radius to several hundreds of kilometers per second at 10 solar radii) indicates that the deviation from radial flow (in the rotating frame of reference) in this region is small. Extension of this model into interplanetary space can then be accomplished by connecting it to the spiral geometry of Section 4.3.2 at some heliocentric distance where all field lines are open and nearly radial. The spiral-field model involved no essential assumption concerning the value of B_0 at the base of any given field line; hence, a variable B_0 can be used. The pattern of magnetic directions or polarities implied by Figure 4.7, for example, would be carried unchanged by the outward-flowing solar wind throughout the solar system.

We would also expect the current sheet separating regions of opposite magnetic polarity to extend outward into interplanetary space. If the solar magnetic field were a simple dipole, as assumed earlier, and the dipole were aligned with the solar-rotation axis, this current sheet would simply extend out in the solar equatorial plane, separating northern and southern hemispheres of opposite magnetic polarity. This geometric picture becomes considerably more interesting if we allow the magnetic-dipole axis to be tilted at some angle α to the rotation axis. Our model dipolar corona would then look like the sketch in Figure 4.9, with a planar current sheet tilted out of the equatorial plane. Extension into the solar wind would then distort this plane into a spiral surface looking something like a snail's shell. This surface would continue to separate the interplanetary magnetic field into two regions of opposite magnetic polarity. For more complicated solar magnetic fields, the current sheet would be more complicated in the corona and more convoluted when drawn out into a spiral pattern by the solar wind, but it would continue to play the same role in the polarity structure of the interplanetary magnetic field.

This remarkably simple model of the interplanetary magnetic field gives a straightforward explanation for one of the basic properties of the

4.4 MAGNETIC STRUCTURE OF THE SOLAR WIND

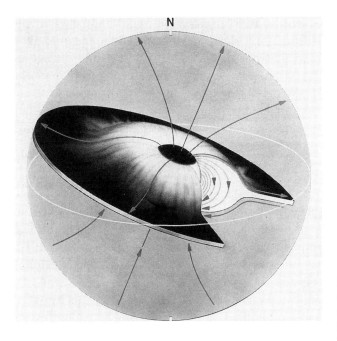

FIG. 4.9. Sketch of the corona and the interplanetary current sheet if the solar dipole field is tilted with respect to the solar-rotation axis (in the vertical direction). (From Hundhausen, 1977.)

interplanetary field discovered in the early days of solar-wind observations. Our initial description of the interplanetary field at 1 AU (Table 4.1) gave its strength and direction, but not its polarity. Solar-wind observations accumulated over long time intervals in the mid-1960s indicated that the polarity of the field was organized in a very simple pattern. The field would point predominantly outward from or inward toward the sun (along the average direction of the field lines at ~45° to the radial direction) for about a week at a time and then change in a relatively short time to the opposite polarity. This pattern was found to repeat, with only minor changes, with a period of 27 days. Because 27 days is the rotation period of the sun as viewed by an observer moving around the sun with the earth, this recurrence tendency suggested explanation in terms of sources fixed on the sun and swept past the earth by solar rotation. For a tilted-dipole corona, rotation of the interplanetary magnetic pattern would place a stationary observer (near the solar equator) alternately above and below the extended current sheet, or alternately in two regions of opposite magnetic polarity, during each 27-day solar-rotation period. A solar magnetic field more complicated than the dipole would lead to a polarity pattern with more frequent crossings of the neutral line and thus shorter intervals of unchanging polarity, but with the same basic 27-day period for repetition of the entire pattern. We have thus found an explanation for the observed "sector" pattern in the interplanetary magnetic field. This pattern has been chronicled in more than 25 yr of direct interplanetary observations and inferred for still earlier times from terrestrial data. Both two-sector and four-sector patterns (and more complicated situations) are clearly present at different epochs. The two-sector pattern is, of course, consistent with our model of coronal structure based on the simplest of assumptions concerning the solar magnetic field: that it is a dipole. The four-sector pattern, as originally discovered in the mid-1960s, implies a slightly more complicated solar magnetic field imposed on the corona, with a significant quadrupole component. What is remarkable here is that these simple field configurations seem to apply to the real corona and solar wind much of the time.

These same simple magnetic configurations are also evident in the appearance of the solar corona. In particular, there have been periods of several years during each of the past two solar activity cycles (see Chapter 3) for which sufficiently good coronal observations are available to show the presence of a nearly continuous belt of helmet streamers about the sun, as predicted by the dipole model. These periods began a year or two after each maximum in solar activity and lasted through the following activity minimum. Near each activity minimum, the belt of helmet streamers was close to the solar equator, implying a dominant solar magnetic dipole that was closely aligned with the solar-rotation axis. Several years before minimum, the belt of helmet streamers was seen to be tilted out of the equatorial plane; the changing appearance of the corona revealed the wobble of these features seen at the limbs of the sun as these tilted features rotated with the sun. Tilt angles as large as $\alpha = 30°$ are consistent with the observations.

The observations described in the preceding paragraph also hint that the magnetic structure of the corona evolves slowly during the solar activity cycle. This is not surprising, as the activity cycle is basically a pattern of cyclic changes in the solar magnetic field. One well-known aspect of those magnetic changes involves the destruction of the large-scale dipole magnetic field of the sun in the epoch when activity rises from minimum to maximum, or the ascending phase of a cycle. By the time of maximum activity, any existing rotation-aligned dipole has disappeared, leaving the large-scale magnetic field in a rather disorganized state. At these times the corona appears to be similarly disorganized; coronal helmet streamers and coronal holes can be found at virtually any solar latitudes, as seen in the eclipse image of Figure 4.10, taken near the 1980 maximum in solar activity. However, a new dipole field with an orientation or polarity opposite that of the old dipole develops and grows in strength as activity decreases on the way to the next activity minimum. It is during this descending phase of the activity cycle that the simple dipolelike configurations appear, with large values of the tilt angle α. As activity decreases, the tilt angle α becomes smaller, until alignment of the dipole and rotation axes (or current sheet and equatorial plane) is approached near the minimum in activity. Figure 4.11 shows an eclipse photo from 1988, just after the most recent minimum. The dipole nature of the outer corona is evident; there are huge coronal holes near both poles of the sun and bright helmet streamers near the solar equator on both sides of the sun. There are also coronal features indicating more complicated magnetic fields near the base of

4.4 MAGNETIC STRUCTURE OF THE SOLAR WIND

FIG. 4.10. Photograph of the corona obtained at the 1980 solar eclipse. (Courtesy High Altitude Observatory.)

FIG. 4.11. Photograph of the corona obtained at the 1988 solar eclipse. (Courtesy R. R. Fisher, High Altitude Observatory.)

the corona. These features (due to the growth of new cycle activity at solar latitudes of about ±45°) usually do not extend sufficiently far out in the corona to disrupt its simple dipole character. This photograph illustrates an important limitation of our simple description of the corona. The real solar magnetic field is extremely complicated, with many components other than a dipole (or quadrupole). The influences of these more complicated fields fall off rapidly with height and thus have little influence on the magnetic structure of the outer corona. It is in this sense that we can describe the corona as dipolelike (or quadrupolelike).

4.4.4 Modulation of the Solar-Wind Speed Related to the Magnetic Structure

Solar-wind observations from the 1960s revealed another important pattern of solar-wind variations that repeated with the 27-day apparent

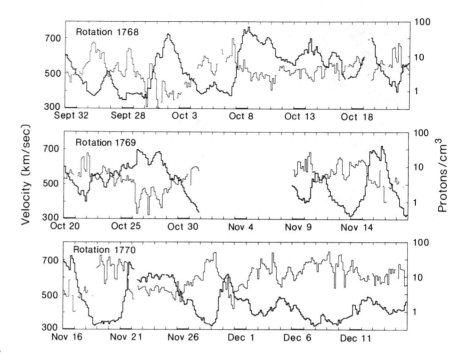

FIG. 4.12. Solar-wind speeds and densities observed with an instrument on the *Mariner 2* space probe in 1962. The display of these quantities as functions of time has been broken into three panels.

rotation period of the sun. Large systematic changes in the solar-wind speed were found to occur in connection with passage of the magnetic sector or polarity features described in Section 4.4.3. The solar-wind speed measured near 1 AU was usually low, 300–400 km · s^{-1}, at sector boundaries (where the magnetic polarity changed sign), rose to high values (up to 700 km · s^{-1}) partway through a sector, and then declined more slowly to the lower level as the next sector boundary was observed. This pattern of recurring high-speed streams is illustrated in Figure 4.12 with observations made from the *Mariner 2* space probe in 1962. During those times when the solar-wind speed was rising from its low (sector-boundary) level, unusually high densities were observed.

These recurrent variations in solar-wind properties can be explained by a spatial dependence of the coronal expansion speed organized by the coronal magnetic field. Namely, it is consistent with an expansion speed that varies with angular displacement from the current sheet in the outer corona. For the dipole coronal field of Section 4.4.2, the solar-wind speed would be low (300–400 km · s^{-1}) along the current sheet and would rise with the absolute value of dipolar magnetic colatitude θ to high values of 700–800 km · s^{-1} for $\theta < 70°$ and $\theta > 110°$. The temporal variations in solar-wind speed seen by a stationary interplanetary observer then result from this spatial pattern sweeping by as it rotates with the sun.

The origin of this spatial variation or modulation in the coronal expansion speed remains unclear. In the dipolar isothermal model of Section 4.4.2, the area $A(s)$ of the flow/flux tubes is a function of colatitude θ.

Because $A(s)$ does enter into the mass and parallel momentum equations (4.17) and (4.18), it could lead to a systematic dependence of expansion speed on θ. Indeed, the expansion speeds derived by Pneuman and Kopp were higher near $\theta = 0$ (the polar regions) than near $\theta = \pi/2$ (the dipolar equator) in the outer corona. However, a latitude variation in coronal temperature over a reasonable range could produce a stronger latitude dependence in the speed $u(r, \theta)$. More realistic treatments of the coronal temperature or of the energy equation, or introduction of wave pressure or energy into the model, would lead to still more ways to induce a latitude variation in the solar-wind speed. Until the importance of these possible effects is sorted out and coronal conditions are better measured, it is unlikely that this situation will be clarified.

Any systematic spatial dependence of the solar-wind speed on solar longitude, as suggested earlier for a coronal current sheet tilted or warped away from the solar equatorial plane, introduces a new complication into our picture of spiral streamlines/field lines in interplanetary space. The shape of the rotation-induced spirals described in Section 4.3.3 depends on the solar-wind speed; for a fast solar-wind flow, the spirals are less tightly wound than for a slow solar-wind flow. Thus, a nonuniform speed would lead to streamlines that would intersect one another. The pattern of speeds within each stream or magnetic sector should then change with increasing distance from the sun as fast-moving solar wind overtakes the slower wind ahead of it and runs away from the slower wind behind it. Thus the profile of high-speed streams, the variation with longitude or time at a given distance from the sun, should evolve toward a rapid-rise, slow-decay shape, as seen in Figure 4.12. The plasma on the rising portions or fronts of the high-speed solar-wind streams should then be compressed, producing the high densities also evident in these observations. The basic geometry of a steepening stream of high-speed solar wind is illustrated in Figure 4.13.

This process of stream steepening explains some of the characteristics of solar-wind speed and density variations seen in interplanetary space near 1 AU. It also leads to a dilemma similar to that encountered in our brief description of the interaction of the solar wind with the interstellar gas in the outer solar system. The convergence of flows near the front of a stream will compress the plasma and produce a ridge of high pressure that will act to stop the steepening. But because the solar wind becomes more and more supersonic as it moves out through the solar system, the pressure-gradient forces at the edges of these high-pressure zones cannot resist the steepening. As before, this dilemma can be solved by the formation of shock fronts bounding the compressed regions; the nonadiabatic conversion of flow energy into thermal energy or pressure by these shocks permits the ultimate cessation of the stream-steepening process. Detailed models predict shock formation somewhere beyond the orbit of the earth and existence of the shocks well out through the solar system. Indeed, most solar-wind streams have been

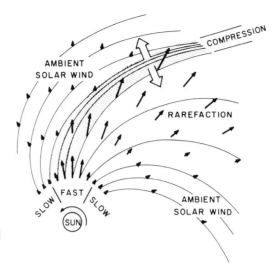

FIG. 4.13. Geometry of the interaction between fast solar wind (on less tightly wound spiral streamlines) and ambient solar wind (on more tightly wound spiral streamlines). The plasma is compressed where streamlines converge. (From Pizzo, 1985.)

found to contain shock fronts when observed by space probes beyond 3 or 4 AU, and the shock-bounded compressed shells, roughly aligned with our now familiar interplanetary spirals, are conspicuous features of the solar wind in the outer solar system.

4.5 THE MAJOR TIME-DEPENDENT DISTURBANCES OF THE SOLAR WIND

In addition to the recurrent variations in solar-wind properties described earlier (and understood in terms of spatial structures that rotate with the sun), there are sporadic disturbances that involve true time variations. The most striking of these are interplanetary shock waves that propagate outward from the sun. An interplanetary observer detects the passage of such a shock wave by abrupt changes in the plasma speed, density, temperature, and magnetic-field strength, with unusual values of these parameters persisting on a time scale of a day or more after these initial changes. An example of the observed passage of a shock wave is shown in Figure 4.14. The existence of this phenomenon was suggested, before examples were actually observed in the solar wind, on the basis of the abrupt onset of some geomagnetic disturbances. Those suggestions led to the realization that shock waves could form in a magnetized plasma with a thickness much smaller than the collision length, or that *collisionless shocks* might exist.

A shock front moving outward through the solar wind actually overtakes the slower-moving plasma ahead of it, accelerating and heating the material that it sweeps up. The shock thus transfers momentum and energy to a widening region of solar-wind plasma. Unless such momentum and energy are continually replenished, the shock must be decelerated as it moves outward through the solar system. Theoretical models

4.5 TIME-DEPENDENT DISTURBANCES OF SOLAR WIND

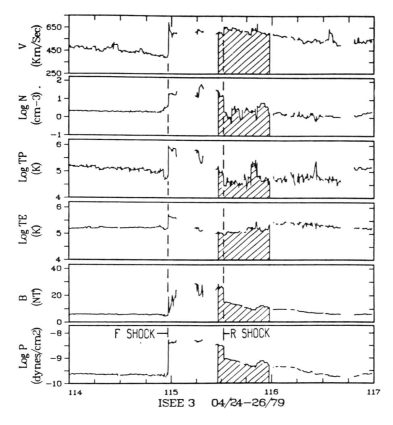

FIG. 4.14. Changes in solar-wind plasma parameters (speed V, density N, proton and electron temperatures T_P and T_E, magnetic-field intensity B, and plasma pressure P) related to the passage of an interplanetary shock wave observed by *ISEE 3*.

of the propagation and deceleration of shock waves in the solar wind require solution of the fluid or MHD equations for a true time-dependent system; the conservation laws used in any model must include the time derivatives that have been ignored in this presentation. Solution of the resulting partial differential equations represents a giant step in difficulty beyond anything described in this chapter. Solutions are usually obtained by extreme geometric and physical simplifications or by sophisticated numerical integrations. Most of the results obtained from these solutions tend to be of special or limited applicability and have not led to much general understanding of shock propagation in the solar wind. It is interesting to note, nonetheless, that the observability of the shock waves that form through the steepening of high-speed solar-wind streams (as described in Section 4.4.4) and of the decelerating shock waves that propagate outward from the sun makes the solar wind a unique laboratory for the study of this phenomenon. It offers the opportunity to witness the birth and vigorous youth of shock waves in the former case, and the old age and decay of shock waves in the latter.

One of the most interesting questions concerning the true time-dependent interplanetary shocks is that of their solar origins. It has been traditional to see the genesis of these shock waves in the phenomenon known as the solar flare. A flare is a sudden brightening of a small region of the sun seen in x-rays, emission lines, and, on rare occasions, the

FIG. 4.15. Coronal mass ejection observed on April 14, 1980, with a coronagraph on the *SMM* spacecraft. The two bright, concentric loops at the upper left are seen to move outward in successive images.

continuum emission from the surface of the sun. This sudden emission of radiation is the result of abrupt and extreme heating of the material at the flare site. Despite the obvious attractiveness of relating the rapid outflow of plasma behind interplanetary shocks to the explosive flare phenomenon, it has proved so difficult to establish a definite connection between flares and interplanetary shocks that some workers in the field (the present author included) have come to doubt the existence of any strong relationship.

An important development in seeking the origin of the major time-dependent disturbances of the solar wind has been the availability of coronal observations from space. Space-borne coronal instruments can take pictures of the source of the solar wind with a time resolution of minutes and with spatial resolution impossible to achieve from the ground except during the rare and fleeting opportunities offered by eclipses. These observations have led to the discovery of major temporal disturbances of the corona known as mass ejections. Figure 4.15 shows an example of the phenomenon. The bright looplike feature in the figure can be seen to move outward through the corona on a time series of images. Figure 4.16 shows an example of the formation of a mass ejection through the disruption of a coronal helmet streamer; the bright material within the mass-ejection loop is a solar prominence that originally marked the neutral line beneath the helmet streamer, much as seen in Figure 4.8. We have previously described coronal helmet streamers as regions of closed magnetic fields wherein the coronal plasma is contained in static equilibrium. Observations like those in Figure 4.16 leave little doubt that mass ejections can involve the violent disruption of that tranquil equilibrium. The study of this phenomenon indicates that many

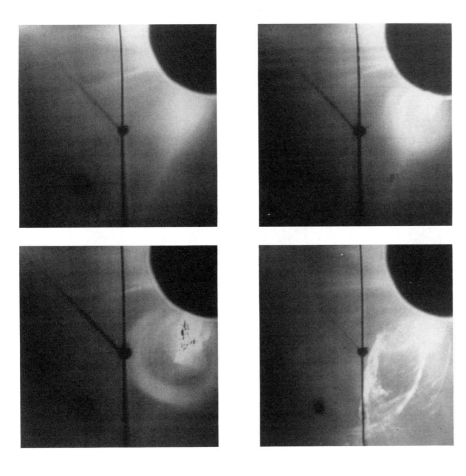

FIG. 4.16. Formation of a coronal mass ejection seen in a time sequence of images obtained on August 18, 1980. The mass ejection forms through the distortion of a coronal helmet streamer as a solar prominence erupts beneath it.

mass ejections follow the pattern suggested by the example of Figure 4.16. They involve the opening of magnetic-field structures that had been closed for many days or even weeks before the ejection, and they add to the solar wind both coronal and chromospheric material (the prominence) that was previously contained by closed magnetic fields.

An aspect of these observations that some have found intriguing (and some others have found dismaying) is another indication of a poor relationship to solar flares. Although flares do occur at about the time and place where some mass ejections appear to form, there are many spectacular ejections that seem to involve no significant enhancement in x-rays or the line emissions common to flares. Most of these events do involve the eruption of material from the solar chromosphere, often in the form of large prominences (as in the examples of Figures 4.15 and 4.16). They are thus related to visible magnetic structures that are not hidden on the far side of the sun. We are thereby led to the interesting possibility that mass ejections are driven by changes in the very large scale magnetic fields that occur within coronal structures, such as helmet streamers, and have only a coincidental relationship to the smaller-scale magnetic fields that seem to be involved in solar flares.

ADDITIONAL READING

The theoretical and observational background for understanding the solar wind in greater breadth and depth can be found in three monographs on the subject:

Parker, E. N. 1963. *Interplanetary Dynamical Processes.* New York: Wiley-Interscience.
Brandt, J. C. 1970. *Introduction to the Solar Wind.* San Francisco: Freeman.
Hundhausen, A. J. 1972. *Coronal Expansion and Solar Wind.* Berlin: Springer-Verlag.

Progress in solar-wind research has been chronicled (along with changing fads and fashions) in the record of a series of meetings on the solar wind published in the following volumes (listed in chronological order):

Mackin, R., and M. Neugebauer (eds.). 1966. *The Solar Wind.* Elmsford, NY: Pergamon.
Sonett, C. P., P. J. Coleman, Jr., and J. M. Wilcox (eds.). 1972. *Solar Wind,* NASA publication SP-308. Washington, DC: NASA.
Russell, C. T. (ed.). 1974. *Solar Wind Three.* Los Angeles: Institute of Geophysics and Planetary Physics, UCLA.
Rosenbauer, H. (ed.). 1981. *Solar Wind Four,* Report no. MPAE-W-100-81-31. Garching: Max-Planck Institut für Aeronomie (und für extraterrestrische Physik).
Neugebauer, M. (ed.). 1983. *Solar Wind Five,* NASA conference publication 2280. Washington, DC: NASA.
Pizzo, V. J., T. Holzer, and D. G. Sime (eds.). 1987. *Solar Wind Six.* National Center for Atmospheric Research technical note NCAR/TN-3064 + proceedings, 2 vols. Boulder: NCAR.

PROBLEMS

4.1. Using the equation for the critical radius, find the coronal temperature for which a supersonic solution would be impossible.

4.2. The mean distances from the sun to Mercury, Venus, Earth, Mars, and Jupiter are 0.39, 0.72, 1.0, 1.5, and 5.2 AU. The average properties of the solar wind at 1 AU are as follows: density, 7 proton · cm^{-3}; velocity, 440 km · s^{-1}; proton temperature, 0.9×10^5K; electron temperature, 1.3×10^5K; magnetic-field strength, 7 nT, lying along a 45° Archimedean spiral angle to the flow. Calculate the following quantities at each of the planets: magnetic-field strength and spiral angle, proton-number density and temperature, electron temperature, and plasma beta, or ratio of thermal pressure to magnetic pressure. Assume that the proton temperature varies with heliocentric radius as r^{-1}, and electron temperature as $r^{-\frac{1}{2}}$.

4.3. Using the information calculated in Problem 4.1, calculate the Debye length and number of particles in a Debye cube at Mercury and Jupiter. Calculate the electron plasma frequency and proton gyrofrequency at Venus and Mars. Calculate the gyroradius of a 1-keV proton moving perpendicular to the magnetic field at the earth.

5 COLLISIONLESS SHOCKS

D. Burgess

5.1 INTRODUCTION

THE UNIVERSE IS INTERWOVEN by plasmas in motion. Between the planets, between the stars, between the galaxies there are flows of plasmas and field energy. Wherever these flows are, there will also be shock waves. In the solar system there are shocks in front of all the planets, shocks in their magnetotails, and shocks formed in the solar corona and solar wind. Shocks are the most extensively studied nonlinear waves of plasmas. They are places where the plasma and field go through dramatic changes: changes in density, temperature, field strength, and flow speed. These changes, combined with the collisionless nature of space plasmas and the wide variety of wave modes, produce a rich collection of different shock types. Shocks in collisionless plasmas are interesting both in their own right and because they are involved in a massive range of plasma phenomena, and their study can stretch and expand our knowledge of basic plasma processes.

Most of our everyday notions of the nature of a shock wave come from our experience of supersonic aircraft and explosive blasts. The study of shocks began with ordinary gas dynamics in the late nineteenth century and reached its maturity during the 1940s, at the time of the development of high-performance aircraft. The study of plasma shocks surfaced during the 1950s, with interest in fusion plasmas and shocks caused by explosions in the upper atmosphere. Also, it was realized that shocks would exist in collisionless plasmas, such as found in interplanetary space (the solar wind) and in other, more exotic astrophysical objects.

Collisions in an ordinary gas serve to transfer momentum and energy among the molecules, and they provide the coupling that allows the basic wave, the sound wave, to exist. In a *collisionless plasma,* such collisional coupling is absent. In other words, the mean free path between collisions is greater than the size of the system. For example, in the solar wind the collisional mean free path, as calculated from gas kinetic theory, is about one astronomical unit (1 AU is the distance from the earth to the sun, 1.5×10^8 km). In contrast, from observations, the thickness of the earth's bow shock, which forms in front of the magnetosphere, is only 100–1,000 km. Thus, whatever is happening at the shock, collisions cannot be important. Instead, there are processes in operation that are unique to collisionless plasmas. Since the beginning of spaceflight, we have collected many observations of shocks, to such an extent that laboratory studies of collisionless shocks have been eclipsed. The ability to make detailed observations of the particle distributions as well as the fields means that we have the opportunity to study, close up, a phenomenon that we know exists in similar forms throughout the universe.

In their behavior, collisionless plasmas are far removed from ordinary gases. Nevertheless, our concepts about what a shock is, and why there should be one, are based mostly on ideas from gas dynamics. Thus, we shall first examine some fundamental concepts without worrying about plasmas. Then we shall go on to see the implications of energy and momentum conservation for what can happen at a shock. Next we shall pick several of the wide selection of ongoing research topics to illuminate the special nature of collisionless shocks. By necessity, we are going to discuss only very few of the topics that constitute shock physics.

5.1.1 Information, Nonlinearity, and Dissipation

Our understanding of shocks at a fundamental level is based on a cluster of concepts involving the speed at which information can travel, nonlinearity (i.e., responses not proportional to input), and dissipation (i.e., entropy increases). We transmit information by making a disturbance that propagates. We clap our hands, and a pressure pulse in the air travels outward. At increasingly later times afterward, people increasingly farther away will be able to hear the hand clap. This defines an "information horizon," which is determined by the speed of the pressure pulse, that is, the speed of sound. Inside the information horizon, the clap has been heard, and outside it hasn't.

A sound wave in air is a compressive wave; that is, the density increases as the pressure increases. Further, at audio frequencies, the compression is adiabatic, which thermodynamics tells us is the way to compress the gas in such a way that when it again expands it will return to its original state, because no heat is taken away. (This is the case because the pressure changes are too fast for thermal conduction to be important.) From the fact that the compression is adiabatic, the speed of sound can be calculated as $c_s = (\gamma p/\rho)^{\frac{1}{2}}$, where γ is the ratio of specific heats, p is the gas pressure, and ρ is the mass density. This is the expression for the sound speed that everybody uses, but in fact it rests on several assumptions: that the wave is "small" in the sense that viscosity, friction, and heat conduction are not important; that the compression is adiabatic (actually, *isentropic,* the entropy does not change); and that the gas follows the ideal-gas law $p = nkT$.

The sound speed is the *typical speed* at which information is transmitted through the gas, and it is the same for all frequencies. Its actual value depends on the gas parameters (i.e., its temperature). The central problem of shocks arises when we ask a simple question: Can we transmit information faster than the speed of sound? Or we can ask, What will happen if when I clap, I move my hands together faster than the speed of sound? Let us look at this problem. When I move my hand *subsonically,* the gas molecules are pushed out of the way; the moving hand imparts momentum and energy to molecules striking it, which are conveyed to other molecules that propagate away from my hand as a pressure pulse. My hand can move through the air because the pressure changes cause flows that allow air in front of my hand to move aside to be replaced by my hand, and similarly for air that comes in behind my hand.

What happens when my hand moves faster than the speed of sound? Once again, my hand must be giving energy and momentum to any molecules hitting it; also, there must be flows of air around my hand to allow it to move forward. The latter point means that air in front of my hand knows that a hand is approaching, or, in other words, information is propagating ahead of my hand. How can this be when we have said that the sound speed defines an information horizon? Quite simply, there must be information (a "wave") moving ahead of my hand *faster* than the speed of sound. How can this be achieved? The clue is that an ordinary sound wave involves a reversible process – the compression is adiabatic (isentropic), without any dissipation (i.e., viscosity or friction). What happens when we look for a wave that is *not* reversible? If the wave involves a compression that is not reversible, then it will involve an increase of entropy at the wave, which implies dissipation. Also, an irreversible process will mean that whatever the wave does to the gas, it will leave it in a different state (i.e., it will change its density, temperature, etc.). If the wave changes the gas temperature, it will change the sound speed as well.

At this point we have defined a wave that travels faster than the sound speed (or typical information speed) and that changes the state of the medium through which it travels. This is a shock wave. We can now ask how much it changes the gas. For this, let us move into the frame of the shock where the shock is stationary. (We shall discuss more about different frames later.) In this frame there is gas flowing into the shock, from the direction that has not yet had any information about my hand moving. This is the upstream side, or *low-entropy side,* and the flow speed on this side is supersonic. We call the other side of the wave the downstream side, which is the high-entropy side. At the shock, irreversible processes make the gas compress, and the sound speed changes. A change in the density means that there will be a change in the flow speed, because we have conservation of mass (so the mass flux across any surface is constant). If the density is increased, the flow speed is reduced.

It is clear that if the flow downstream of the shock is still supersonic, then we haven't really solved our problem. In that case I would simply be trying to move my hand supersonically through the air, and information could not move fast enough upstream of my hand. If the shock changes the flow so that it is subsonic downstream, it is possible to move my hand through the air with the necessary flows caused by pressure changes that propagate ahead of my hand. In other words, we have reduced the problem to the subsonic case by having a transition from supersonic to subsonic flow. This is our basic definition of a shock: It is an irreversible (entropy-increasing) wave that causes a transition from supersonic to subsonic flow. Because I can, hypothetically, move my hand as fast as I like, it is a wave whose propagation speed can increase without limit. The ratio of the flow speed to the sound speed is called the Mach number, M. Upstream, $M > 1$; downstream, $M < 1$. The Mach number of a shock refers to the Mach number of the upstream flow in the shock frame.

There is an important consequence of this transition from supersonic to subsonic flow. Downstream, the sound speed is greater than the flow of gas away from the shock (this is in the shock frame). Thus, every pressure disturbance in the downstream region can propagate against the flow to the shock. If we take the impulse of my hand moving through the air and consider its Fourier components, then because the sound speed is independent of frequency (for small waves), the shock position represents the information horizon for *all* frequencies, all wavelengths. Therefore the shock will have at its position the shortest supportable wavelengths, which is like saying that it will have the thinnest possible width. Indeed, for an ordinary gas, the equations show the shock as a discontinuity, which is interpreted as meaning that the shock width is limited by viscosity and friction and will be only a few collisional mean free paths thick. In a rarefied plasma, such as the solar wind, the distance of even a few collisional mean free paths is very large, and other processes intervene to control the thickness of such shocks.

We have discussed the background ideas for the formation of shocks and have related their existence to the transition from supersonic to subsonic flow, but in a plasma the situation is complicated by the fact that there can be several "typical" information speeds. In magnetohydrodynamics (MHD) there are three fundamental small-amplitude waves: the fast, intermediate, and slow modes. Because of the different characteristic speeds, there are corresponding Mach numbers: the Alfvén Mach number M_A, the sonic Mach number M_{cs}, the fast and slow Mach numbers M_f and M_s, which are the ratios of the flow speed to the Alfvén, sound, fast, and slow magnetosonic speeds, respectively. It becomes a more difficult task to discuss shock formation when there are several different wave modes (i.e., different ways to convey information) and corresponding different information speeds. Generally speaking, there will be different types of shocks that will affect a transition from super–wave speed to sub–wave speed, where the actual wave speed will depend on which of the fundamental waves of the plasma is important. Which one is important will depend on the boundary conditions, because these conditions determine what information has to propagate.

5.1.2 An Unusual Shock

Our discussion has been in terms of the relation between shock waves and the corresponding (small-amplitude) fundamental wave. As an illustration of shock processes, let us imagine a situation where there is a shock wave, but no corresponding small-amplitude wave. Consider a stream of equally spaced vehicles on a straight freeway at night, in pitch-black fog, driven by people wearing blindfolds, all traveling at the same speed. Nobody knows what is ahead. Then, suddenly, one stops in the middle of the lane. The one behind it soon crashes into its rear and comes to a stop as well. The road is made of quick-setting cement, so that each vehicle comes to an abrupt stop. The vehicles in the stream successively hit the increasing pile of stationary vehicles. There is then an interface between moving vehicles and stationary vehicles that propagates opposite to the flow of vehicles. This interface satisfies one of our basic criteria for a shock: There is a dissipative, irreversible transition in the state of the medium. The shock is a so-called traveling shock. We have made sure it is dissipative by not allowing any elasticity in the collisions. Entropy clearly increases!

In this example, we have arranged it so that there is *no* small-amplitude wave corresponding to the shock wave: the vehicles are either moving or stationary. The shock wave is, in fact, the only way that information can propagate in this artificial system. Drivers know that there is a pileup only when they are in the pileup! The speed of the shock's propagation depends on the speed the vehicles are traveling, their separation while moving, and the extent by which their length reduces when they are in the pileup. In other words, the shock propagation speed depends on the "compressibility" of the stream of vehicles, but it also depends directly on the speed of the moving vehicles. That is, the shock speed can increase without limit, which was another criterion for a shock.

5.1.3 Same Shock but Different Frame

Shocks arise when there are supersonic flows. (We are being loose in our use of "sonic" here; we probably should say "basic information speed.") There are many different situations that can lead to shocks: a supersonic object moving in a stationary gas flow; a fast gas flow overtaking a slow one; a stationary object being hit by a supersonic gas flow. There are different words to describe these ways of making shocks: A

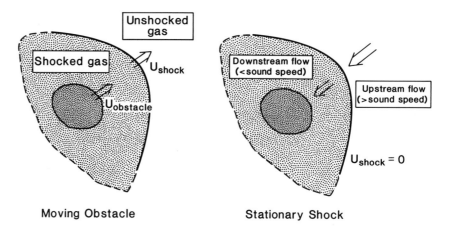

FIG. 5.1. Two ways of looking at the same shock.

standing shock is a shock formed by a stationary object in a supersonic flow, so that the shock stays in the same position. A propagating shock is made by a moving obstacle. A driven shock occurs when there is always energy being fed into the gas flow, or obstacle motion. A blast shock is when there is only a certain amount of energy available, deposited in a fixed time, such as in an explosion. Irrespective of whether the gas is flowing or stationary, whether the obstacle is fixed or moving, if the shock parameters are the same, the shocks formed are the same. It is just a question of which frame is chosen, so long as only the addition of a fixed velocity is needed to transform between them (Figure 5.1). That is, all frames are Galilean equivalent. (We are going to ignore relativity in our discussion.) For example, an aircraft traveling supersonically produces a shock wave that propagates over an observer on the ground. To the aviatrix there is a shock wave in front of her that keeps the same distance from the nose of the aircraft. To the pilot the shock is stationary; to the observer on the ground it is propagating. Both would agree that there was a flow of gas through the shock, because they both see a compression at the shock.

Of course, some frames are more useful than others, and, in particular, frames in which the shock is at rest are often used; such a frame is called a shock frame. If the shock surface is a plane (i.e., the shock is one-dimensional), then there is an infinity of shock frames, but there are two that are especially useful: the normal incidence frame, and the de Hoffman–Teller frame. These are discussed in Appendix 5A.

5.2 SHOCKS WITHOUT COLLISIONS

The plasmas that are found in the magnetosphere, in interplanetary space, and elsewhere in the universe are very different from an ordinary gas. A plasma can support several different types of waves, involving fields and particles, unlike the single sound wave of a gas. The greatest difference is that most space plasmas are *collisionless*. This means that they are either so rarefied (i.e., under-dense) or hot that Coulomb collisions between the constituent particles happen so infrequently that they do not play an important role. In a normal gas, collisions between molecules ensure that they all have the same temperature, irrespective of type; collisions provide the mechanism to propagate pressure and temperature changes; dissipation in the form of viscosity is also an outcome of collisions. Collisions also ensure that in equilibrium the distribution of molecular speeds is Maxwellian.

Knowing what collisions do in an ordinary gas, we can easily list what an absence of collisions will produce in a plasma. Different types of particles (e.g., protons and electrons) can have different temperatures. The particle distribution functions can be very different from Maxwellian, so that the concept of temperature has to be broadened to that of "kinetic temperature" (i.e., a velocity moment of the particles' distribution function). The important roles of the magnetic and electric fields in a plasma also mean that the distribution functions may no longer be isotropic in velocity space. All these effects, and more, lead to a superabundance of phenomena involving particles and fields.

Examples of collisionless shocks are spread throughout the universe. The most widely studied is the earth's bow shock, which is a standing shock in the solar wind ahead of the earth's magnetosphere. The magnetic field of the earth forms an obstacle to the supersonically flowing solar wind. The bow shock slows the solar wind to subsonic speeds, so that the solar wind can flow around the magnetosphere. The bow shock has a curved shape, symmetrical about the sun–earth line, close to a paraboloid of revolution. The position of the nose of the bow shock (the most sunward part) is at about 14 earth radii (R_E) from the center of the earth ($1R_E = 6{,}371$ km). A simple formula for the average location of the earth's bow shock is

$$R = K/(1 + \epsilon \cos \theta)$$

where K is about 25 R_E, and ϵ is about 0.8, with r measured from the center of the earth, and θ being the angle between this radius vector and the solar direction. The shock is a few R_E ahead of the magnetosphere, and this distance is called the standoff distance. The exact position of the bow shock relative to the earth depends on the solar-wind dynamic pressure (i.e., momentum flux), because the magnetosphere is slightly "spongy." The standoff distance is a function of the shape of the obstacle; blunt obstacles have larger standoff distances. The region of subsonic solar wind behind the bow shock is called the magnetosheath. Typically the interplanetary magnetic field (i.e., the solar magnetic field as carried by the solar wind) is at an angle of 45° to the sun–earth line. Thus, the magnetic field intersects the shock at different angles around its curved surface, with corresponding changes in the character of the shock (see Figure 1.14 in Chapter 1). All of the planets with magnetospheres or ionospheres have bow shocks in front of them, and most have been crossed at least once or twice by interplanetary probes, such as the Voyager and Pioneer probes.

Shocks occur in the solar atmosphere (the corona) during solar flares and other active solar events. Flares and coronal mass ejections inject energy and material into the solar wind, causing interplanetary shocks, which are traveling shocks propagating out through the solar system. The solar wind has high-speed and low-speed streams, coming from different source regions on the sun. Shocks can form at the interface between a slow stream being overtaken by a fast stream. In more exotic astrophysical bodies, one finds jets of material from active galactic nuclei, and there probably are shocks formed at the interface between the jet material and the interstellar medium. In supernovae, massive amounts of energy are deposited in a very short time, and shocks are formed as the supernova remnant piles up material as it expands away from the newly formed pulsar.

It would seem that an absence of collisions could make our problem impossibly complex. However, to an extent we are saved by the fact that in certain limits we can find a description of magnetized plasma that corresponds fairly closely to the more familiar gaslike behavior. The action of the magnetic field replaces that of collisions to "bind" the particles of the plasma together, and the resulting description is called magnetohydrodynamics (MHD). MHD describes the plasma at the level of the macroscopic fields (electric and magnetic) and quantities such as the density and bulk-flow velocity (i.e., averages over velocity moments of the particle distribution functions). Because MHD does not include, except by trickery, effects due to individual particles ("kinetic" effects), it cannot tell us anything about how a shock provides dissipation, or what the structure of the shock will be. Nevertheless, MHD will be suitable to describe the plasma far upstream and downstream of the shock itself.

5.2.1 The Shock-conservation Relations

However bizarre a shock may be, we know that mass, energy, and momentum will be conserved. We can use MHD to relate the plasma states upstream and downstream of the shock using these conservation relationships. In this description the shock is very much like a "black box" that changes the state of the plasma (field, density, etc.), although we cannot tell anything about what is actually happening at the shock. Most importantly, we do not know anything about how thick the shock is, because MHD has no fundamental length scale, and the scales that should be important (e.g., particle gyroradius) have been left out of

MHD. In the case of an ordinary gas or fluid, the relations between the upstream and downstream states were first derived by Rankine and Hugoniot toward the end of the nineteenth century. In the case of a collisional gas, these Rankine-Hugoniot relations determine *uniquely* the downstream state in terms of the upstream state. This has the important result that in an ordinary gas there is a unique transition between a given supersonic flow and a subsonic state. Thus the shock structure is determined by the dissipation mechanism, which is usually just viscosity. This result produces a shock that has a thickness of just a few collisional mean free paths. Actually, even an ordinary gas can have a more complicated shock structure, especially when the shock conditions produce chemical reactions (e.g., dissociation of molecules) or partial ionization.

In the case of a collisionless plasma, the conservation relations (also known as the shock-jump conditions or Rankine-Hugoniot relations) do not provide a unique prescription for the downstream state in terms of the upstream parameters, mainly because energy conservation gives information only about the total pressure (and hence temperature), not about how it is divided between the different types of particles in the plasma. In other words, we need to know about the shock structure, about how the shock works, in order to know how much the ions and electrons heat in passing through the shock.

In deriving the MHD Rankine-Hugoniot relations we make certain assumptions, such as that the shock is, on average, stationary in the shock frame, that the energy in waves is not important, and that the particle distributions can be described by Maxwellians. One of the basic observational tests for a shock is a comparison with the Rankine-Hugoniot relations. This is often a basis for a discussion about whether or not the observed discontinuity is actually a shock. Often, what are really being tested are the assumptions that go into our derivation of the jump conditions. In this case it is best to return to our basic definitions of a shock, which we discussed earlier. The Rankine-Hugoniot relations provide only one possible expression of the conservation of energy and momentum.

Let us consider the simple case of a one-dimensional, steady shock (Figure 5.2). We shall work in a frame where the shock is stationary. The n-axis will be aligned with the shock normal, so that the plane of the shock is parallel to the l–m plane. We shall also assume a uniform upstream magnetic field. We can think of the shock as a discontinuity, but in reality it will have some thickness, given by the kinetic processes at the shock.

The shock separates two regions of steady flow. Plasma flows into the shock on one side (upstream) and out the other side (downstream). We label these two regions u and d, respectively. The shock causes a change in the plasma description from mass density ρ_u, velocity $\mathbf{u}_u$, magnetic

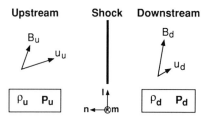

FIG. 5.2. Configuration for shock-conservation relations.

field $\mathbf{B}_u$, pressure p_u, and so forth, to the downstream values ρ_d, $\mathbf{u}_d$, $\mathbf{B}_d$, and p_d. The jump across the shock in any quantity X can be expressed using the following notation:

$$[X] = X_u - X_d$$

The MHD description gives us a set of conservation equations for the mass, energy, and momentum. For any quantity, a conservation equation has the form

$$\frac{\partial Q}{\partial t} + \nabla \cdot \mathbf{F} = 0$$

where Q and $\mathbf{F}$ are the density and flux, respectively, of any conserved quantity. If the shock is steady ($\partial/\partial t \equiv 0$) and one-dimensional (i.e., there are variations only along the n-axis, so that $\partial/\partial l \equiv \partial/\partial m \equiv 0$), then this implies that

$$\frac{\partial}{\partial n} F_n = 1 = 10$$

which in turn implies that $(\mathbf{F}_u - \mathbf{F}_d) \cdot \hat{\mathbf{n}} = 0$, where we denote the unit vector normal to the shock surface by $\hat{\mathbf{n}}$. Therefore, the component of the conserved flux normal to the shock remains constant, and this can be written

$$[F_n] = 0$$

The subscript n indicates the normal component.

For MHD, we have the equation for the conservation of mass or continuity equation (2.17), which, written in one dimension, is

$$\frac{\partial}{\partial n} (\rho u_n) = 0$$

which leads to the jump condition for the shock:

$$[\rho u_n] = 0 \tag{5.1}$$

This tells us, as we would expect, that if the shock slows the plasma, then the plasma density increases.

Next, there is conservation of momentum normal to the shock surface. From the momentum equation (2.19), ignoring the gravitational force, we obtain

$$\rho u_n \frac{\partial u_n}{\partial n} + \frac{\partial p}{\partial n} + \frac{\partial}{\partial n}\left(\frac{B^2}{2\mu_0}\right) = 0$$

The first term is the rate of change of momentum, and the second and third terms are the gradients of the gas and magnetic pressures, respectively. The corresponding jump condition for the normal momentum is then

$$\left[\rho u_n^2 + p + \frac{B^2}{2\mu_0}\right] = 0 \tag{5.2}$$

where the form makes use of the fact that $[B_n^2] = 0$. The transverse momentum also has to balance, and this gives

$$\left[\rho u_n \mathbf{u}_t - \frac{B_n}{\mu_0}\mathbf{B}_t\right] = 0 \tag{5.3}$$

The t subscript indicates the components transverse to the shock (i.e., parallel to the shock surface). We have assumed an isotropic pressure, to simplify things. Also, we have neglected the electric-stress term (from the $\mathbf{E} = -\mathbf{u}\times\mathbf{B}$ electric field), which is less than the magnetic stress by a factor u^2/c^2.

The final conservation equation from MHD is for the energy. We are going to assume that the plasma has an adiabatic equation of state, so that $p\rho^{-\gamma} = $ constant. For a normal monatomic gas, $\gamma = \frac{5}{3}$. Actually, the real equation of state probably will not be adiabatic, but the results we obtain will still be qualitatively correct. The shock-jump condition from the energy is

$$\left[\rho u_n\left(\frac{1}{2}u^2 + \frac{\gamma}{\gamma-1}\frac{p}{\rho}\right) + u_n\frac{B^2}{\mu_0} - \mathbf{u}\cdot\mathbf{B}\frac{B_n}{\mu_0}\right] = 0 \tag{5.4}$$

The first two terms are the flux of kinetic energy (flow energy and internal energy). The last two terms come from the electromagnetic energy flux $\mathbf{E}\times\mathbf{B}/\mu_0$, where we have used the ideal MHD result that $\mathbf{E} = -\mathbf{u}\times\mathbf{B}$.

Equations (5.1)–(5.4) are the jump conditions for the gas, but there are also purely electromagnetic boundary conditions. From Maxwell's equation $\nabla\cdot\mathbf{B} = 0$, the normal component of the magnetic field is continuous ($B_n = $ constant):

$$[B_n] = 0 \tag{5.5}$$

From $\nabla\times\mathbf{E} = -\partial\mathbf{B}/\partial t$, with the assumption that $\partial\mathbf{B}/\partial t = 0$, the tangential component of the electric field must be continuous. Using $\mathbf{E} = -\mathbf{u}\times\mathbf{B}$, this becomes

$$[u_n\mathbf{B}_t - B_n\mathbf{u}_t] = 0 \tag{5.6}$$

The conservation relations are referred to as the Rankine-Hugoniot relations. In a slightly modified form, which uses U, the internal energy

TABLE 5.1. Rankine-Hugoniot Relations

$$\frac{U_u}{\rho_u} - \frac{U_d}{\rho_d} + \frac{1}{2}\left(\frac{1}{\rho_u} - \frac{1}{\rho_d}\right)\left((p_u + p_d) + (B_{tu} - B_{td})^2/2\mu_0\right) = 0$$

$$(\rho u_n)^2[B_t/\rho] = (B_n^2/\mu_0)[B_t]$$

$$\left(\frac{1}{\rho_u} - \frac{1}{\rho_d}\right)(\rho u_n)^2 = [p] + [B_t^2/2\mu_0]$$

$$[v_t] = (B_n/\mu_0 \rho u_n)[B_t]$$

per unit mass of equation (2.42), they are collected in Table 5.1. These equations have been found with the intention of using them to calculate shock-jump conditions, but in fact we have not explicitly forced the solutions of equations (5.1)–(5.6) to be shocklike. The solutions of these equations describe a number of different types of MHD discontinuities, including shocks. For a discontinuity to be a shock, there must be a flow of plasma through the shock surface ($u_n \neq 0$), and there must be some dissipation and compression across the shock. A further distinction can be made between discontinuities that are threaded by a magnetic field (i.e., $B_n \neq 0$) and those that are not. Shocks with a normal magnetic-field component are called oblique, which refers to the angle between the shock normal and the upstream magnetic field (see Section 5.3.1). Table 5.2 summarizes the usual classification of MHD discontinuities. We shall give some specific examples of the use of the conservation relations later, but there are a few general points to be made first.

The conservation relations are a set of six equations. If we wish to find the downstream state in terms of the upstream state, then there are six unknowns: ρ, u_n, u_t, p, B_n, and B_t. This means that the downstream state is specified uniquely by the conservation equations, as in ordinary fluid theory. However, we have only to introduce either an anisotropic pressure (e.g., pressures parallel and perpendicular to the magnetic field are often used) or another fluid (e.g., electrons or heavy ions), and then there will be more unknowns than equations. In such cases we are forced to use additional relations, from theory or observation, to provide the missing information.

As seen in Table 5.2, the oblique shocks are divided into three categories: the fast, slow, and intermediate (sometimes known as Alfvén), which correspond to the three modes of small-amplitude waves in MHD (see Chapter 11). The intermediate shock is really a special case, because it is shocklike only under special circumstances. In an isotropic plasma (as we have been dealing with) it is not a shock, but is rightfully called a rotational discontinuity; there is flow through the boundary, but there is no compression of the plasma or dissipation. We shall not discuss this discontinuity further.

The fast and slow shocks have the same behavior in terms of plasma pressure and magnetic-field strength as do the corresponding MHD lin-

TABLE 5.2. Possible Types of Discontinuities in Ideal MHD

Contact discontinuity	$u_n = 0$, $B_n \neq 0$	Density jump arbitrary, but pressure and all other quantities are continuous
Tangential discontinuity	$u_n = 0$, $B_n = 0$	Plasma pressure and field change, maintaining static pressure balance
Rotational discontinuity	$u_n = B_n/(\mu_0 \rho)^{\frac{1}{2}}$	Large-amplitude intermediate wave; in isotropic plasma, field and flow change direction, but not magnitude
Shock waves	$u_n \neq 0$	Flow crosses surface of discontinuity accompanied by compression and dissipation
Parallel shock	$B_t = 0$	Magnetic field unchanged by shock
Perpendicular shock	$B_n = 0$	Plasma pressure and field strength increase at shock
Oblique shocks	$B_t \neq 0$, $B_n \neq 0$	
Fast shock		Plasma pressure and field strength increase at shock; magnetic field bends away from normal
Slow shock		Plasma pressure increases; magnetic-field strength decreases; magnetic field bends toward normal
Intermediate shock		Magnetic-field rotation of 180° in plane of shock; density jump only in anisotropic plasma

ear waves, but the shock-jump conditions are fully nonlinear. This can be explained by observing that the shock-jump conditions are valid for shocks of any strength. In particular, they must be true for very weak shocks. For consistency, then, in the weak-shock limit the shock-jump relations must correspond to the modes of small-amplitude MHD waves. Thus, even in the fully nonlinear (large-amplitude) limit the shock relations have the heritage of the linear modes. Another explanation is that the formation of a shock can be produced by the steepening of a large-amplitude wave. Such a wave could steepen because the speed of waves with shorter wavelengths could be changed by the wave itself. This would eventually lead to sharp gradients and hence shock formation. Such shocks would again be expected to retain the characteristics of the original mode of the wave.

Across a fast-mode shock, the field strength increases, but the normal

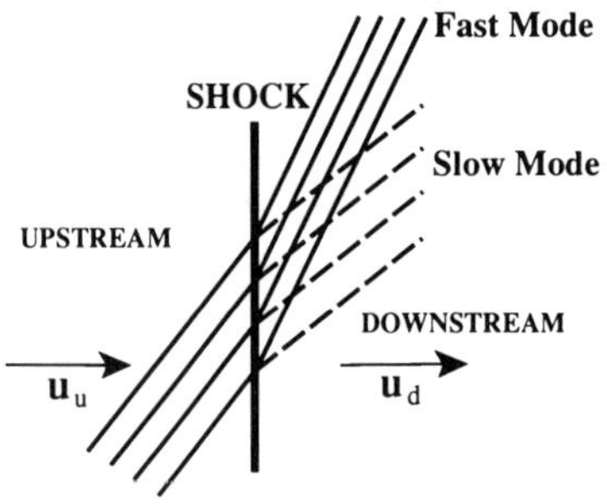

FIG. 5.3. Configuration of magnetic-field lines for fast and slow shocks. The field lines are closer together for a fast shock, indicating that the field strength increases when the field is bent away from the shock normal.

component is constant, so that the increase is all in the transverse component. Therefore, at a fast shock, the downstream field turns away from the shock normal (Figure 5.3). Conversely, at a slow shock, the downstream field bends toward the shock normal. It seems, from observation, that fast shocks are by far the most frequent types of shocks observed in solar-system plasmas. Planetary bow shocks are fast-mode shocks, as are most interplanetary shocks in the solar wind. Most of our discussion later will concentrate on fast shocks. Slow shocks are rarer, although they have been observed, and they play an important part in some theories of magnetic reconnection (see Chapter 9).

Because the slow wave does not propagate across the magnetic field, the limit of a slow wave for perpendicular propagation is a nonpropagating tangential discontinuity whose magnetic and thermal pressures are anticorrelated. Tangential discontinuities are ubiquitous in the solar system. Many solar-wind discontinuities appear to be tangential discontinuities. This is to be expected, because the perpendicular pressure in the solar wind has many days to come to an equilibrium. When there is no reconnection occurring, both the magnetopause and cross-tail current are thought to form tangential discontinuities. The most critical test of a tangential discontinuity is to check if the perpendicular pressure ($B^2/2\mu_0 + nkT$) is conserved.

5.2.2 The Coplanarity Theorem

The compressive fast-mode and slow-mode oblique shocks have the interesting and useful property that the upstream and downstream magnetic-field directions and the shock normal ($\hat{n}$) all lie in the same plane. This is called the coplanarity theorem and can be expressed in vector notation as

$$\hat{n} \cdot (\mathbf{B}_u \times \mathbf{B}_d) = 0 \tag{5.7}$$

That is, the cross product of $\mathbf{B}_u$ and $\mathbf{B}_d$ is perpendicular to the shock normal. This property is used in the observational calculation of shock normals (see Appendix 5C).

We can prove the coplanarity theorem by noticing that the jump condition for parallel momentum (5.3) and the condition on the transverse electric field (5.6) imply that both $[\mathbf{B}_t]$ and $[u_n\mathbf{B}_t]$ are parallel to $[\mathbf{u}_t]$ and thus parallel to each other. Consequently,

$$[\mathbf{B}_t] \times [u_n\mathbf{B}_t] = 0$$

[The special cases $u_n = 0$ and $u_n = B_n/(\mu_0\rho)^{\frac{1}{2}}$ are excluded.] In terms of upstream and downstream values (subscripts u and d), this means

$$u_{un}\mathbf{B}_{ut} \times \mathbf{B}_{ut} + u_{dn}\mathbf{B}_{dt} \times \mathbf{B}_{dt} - u_{dn}\mathbf{B}_{ut} \times \mathbf{B}_{dt} - u_{un}\mathbf{B}_{dt} \times \mathbf{B}_{ut} = 0$$
$$(u_{un} - u_{dn})(\mathbf{B}_{ut} \times \mathbf{B}_{dt}) = 0$$

Thus, if $u_{un} \neq u_{dn}$, $\mathbf{B}_{ut}$ and $\mathbf{B}_{dt}$ are parallel. The plane containing one of these vectors and the shock normal $\hat{\mathbf{n}}$ contains both the upstream and downstream fields. This means that the shock-normal direction can be determined in terms of observed fields on either side of the shock as

$$\hat{\mathbf{n}} = (\mathbf{B}_u - \mathbf{B}_d) \times (\mathbf{B}_u \times \mathbf{B}_d)/|(\mathbf{B}_u - \mathbf{B}_d) \times (\mathbf{B}_u \times \mathbf{B}_d)|$$

5.2.3 The Exactly Parallel Shock

Provided that $B_n \neq 0$, the elimination of $\mathbf{u}_t$ from equations (5.3) and (5.6) yields

$$\left[\left(1 - \frac{B_n^2}{\mu_0\rho u_n^2}\right)u_n\mathbf{B}_t\right] = 0 \qquad (5.8)$$

This equation has interesting consequences for the exactly parallel shock. The parallel shock has the upstream magnetic field parallel to the shock normal; that is, $\mathbf{B}_u = B_n\hat{\mathbf{n}}$, $\mathbf{B}_{ut} = 0$. We shall use the conservation relations in the frame where the upstream flow is also parallel to the shock normal, so that $\mathbf{u}_u = u_n\hat{\mathbf{n}}$, $\mathbf{u}_{ut} = 0$. In the case of a parallel shock, the quantity in parentheses in (5.8) is nonzero. If $\mathbf{B}_{ut} = 0$, then, to satisfy (5.8), $\mathbf{B}_{dt}$ is also zero. Thus the direction of the field is unchanged by the shock. Because B_n is the only nonzero component of the field, and because it does not change, it follows that the total magnetic field is also left unchanged by the shock. There is a compression in the plasma, but not in the field. Feeding this result back into the conservation relations removes all mention of the magnetic field. From the MHD perspective, this means that the shock is like an ordinary fluid shock, and the magnetic field does not play a role. However, in the context of a collisionless plasma, the only way for dissipation to occur is via field-particle processes, and it is certain that here the fields will play a crucial role (e.g., Quest, 1988).

5.2.4 The Exactly Perpendicular Shock

At the exactly perpendicular shock, the upstream field is perpendicular to the shock normal. In this case, $B_n = 0$, and $\mathbf{B}_u = \mathbf{B}_{ut}$. Again we examine the case where the upstream flow is parallel to the shock normal: $\mathbf{u}_u = u_{un}\hat{\mathbf{n}}$. To ensure shocklike solutions, there will be a nonzero mass flux through the shock:

$$\rho_u u_{un} = \rho_d u_{dn} \neq 0$$

We define a density *compression ratio*, $r = \rho_d/\rho_u$. Using the foregoing equation, we can write $u_{dn} = (1/r)u_{un}$. We can now begin to apply our jump conditions. Equation (5.3) becomes

$$\rho_u u_{un}\mathbf{u}_{ut} - \rho_d u_{dn}\mathbf{u}_{dt} = 0 \tag{5.9}$$

which implies that $\mathbf{u}_{dt} = 0$, because $\mathbf{u}_{ut} = 0$ and $\rho u_n \neq 0$.

From (5.6), using $B_n = 0$, the jump condition becomes

$$u_{un}\mathbf{B}_{ut} = u_{dn}\mathbf{B}_{dt} \tag{5.10}$$

This tells us that the upstream and downstream fields are parallel. Because there is no normal magnetic field or transverse flow velocity throughout the system, from now on we shall simply use B_u, B_d, u_u, and u_d. Using the compression ratio r, we see that $B_d = rB_u$. In other words, the field compresses as much as the flow.

In the perpendicular case, (5.2) reduces to

$$\rho_u u_u^2 + p_u + \frac{B_u^2}{2\mu_0} = \rho_d u_d^2 + p_d + \frac{B_d^2}{2\mu_0}$$

which can be rewritten as

$$\rho_u u_u^2\left(1 - \frac{1}{r}\right) + (p_u - p_d) + \frac{B_u^2}{2\mu_0}(1 - r^2) = 0 \tag{5.11}$$

After substituting for u_d and B_d, the energy-jump condition, equation (5.4), in the perpendicular case, becomes

$$\frac{1}{2}\rho_u u_u^2\left(1 - \frac{1}{r^2}\right) + \frac{\gamma}{\gamma - 1}\left(p_u - \frac{1}{r}p_d\right) + \frac{B_u^2}{\mu_0}(1 - r) = 0 \tag{5.12}$$

Equations (5.11) and (5.12) can be used to eliminate p_d, and we are left with an equation for r, the compression ratio, in terms of the upstream parameters:

$$(r-1)\left\{r^2\frac{2-\gamma}{M_A^2} + r\left(\frac{\gamma}{M_A^2} + \frac{2}{M_{c_s}^2} + \gamma - 1\right) - (\gamma + 1)\right\} = 0 \tag{5.13}$$

In order to reach this equation, we have introduced the Alfvénic Mach number M_A, which is the ratio of the upstream flow speed (along the shock normal) to the upstream Alfvén speed; similarly, we define the sonic Mach number M_{c_s} as the ratio of the upstream flow speed to the upstream sound speed:

$$M_A = \frac{u_u(\mu_0\rho_u)^{\frac{1}{2}}}{B_u}, \qquad M_{c_s} = u_u\left(\frac{\rho_u}{\gamma p_u}\right)^{\frac{1}{2}}$$

One of the solutions of (5.13) is clearly $r=1$, which represents downstream field, velocity, and density unchanged from the upstream values. Obviously, this doesn't correspond to a compressive shock. The quadratic in (5.13) leaves two other solutions, one of which is negative if $\gamma < 2$. This negative-r solution is unphysical, and so we are left with one solution for the compression ratio. An interesting limit is the high-Mach-number limit, when $M_A \gg 1$ and $M_{cs} \gg 1$. In this case, equation (5.13) becomes $r(\gamma - 1) - (\gamma + 1) = 0$, or

$$r(\gamma - 1) - (\gamma + 1) = 0 \Rightarrow r = \frac{(\gamma + 1)}{(\gamma - 1)}$$

Therefore, there is a finite limit to the compression, which depends only on γ. Remember that γ is just an indication of how the plasma heats and is not dependent on the upstream parameters. In particular, for a collisional monatomic gas, $\gamma = \frac{5}{3}$, and this gives a limiting compression of 4. At a high-Mach shock, the maximum jump that can be expected in the field, density, and velocity is by a factor of 4. We should remember, though, that this factor, which is much quoted, depends on the details of how the plasma heats at the shock.

Another consequence of the high-Mach-number limit (i.e., large upstream flow energy) can be obtained from (5.12), where the terms dependent on B_u and p_u can be neglected, and remembering that r is independent of the upstream parameters, which implies a direct proportionality between the ram energy of the flow $\frac{1}{2}\rho u^2_u$ and the downstream thermal pressure p_d. This is just an illustration of the operation of the high-Mach-number shock: It takes the flow energy upstream and converts it to thermal energy downstream. Finally, at the perpendicular shock, both the field strength and the density increase. From the foregoing discussion it follows that the shock is a fast-mode shock and that there is no slow-mode perpendicular shock. This is unsurprising, given the result from Chapter 11 (Figure 11.3) that slow-mode waves do not propagate perpendicular to the magnetic field.

5.3 SHOCK STRUCTURE: HOW SHOCKING?

Thus far we have described some fundamental concepts about shocks and mentioned some examples of where shocks will be formed. But in the case of shocks in space we are in an almost unique position, because we have direct observations of naturally occurring shocks. It is possible to generate collisionless shocks in the laboratory, and shock heating was one of the hopes for controlled nuclear fusion. But laboratory experiments cannot approach the scale or global nature of the naturally occurring space shocks. Space observations are also unique in that the smallest plasma scale (the Debye length) is usually larger than the spacecraft. This means that we can truly make a point measurement, because

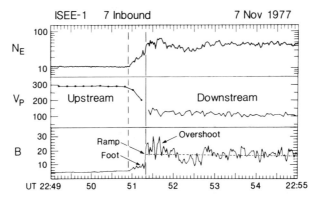

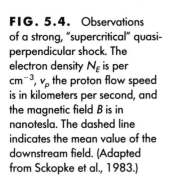

FIG. 5.4. Observations of a strong, "supercritical" quasi-perpendicular shock. The electron density N_E is per cm^{-3}, v_p the proton flow speed is in kilometers per second, and the magnetic field B is in nanotesla. The dashed line indicates the mean value of the downstream field. (Adapted from Sckopke et al., 1983.)

our measurement devices do not affect the plasma (at least if we are careful). Because of continual improvements in space technology, we are making measurements of the space shocks at increasing high resolutions. Not only are we measuring the upstream and downstream states, but also we can measure how the plasma changes as it passes through the shock. In other words, we can study the collisionless dissipation mechanisms in action.

Figure 5.4 shows measurements made as a spacecraft passed through the earth's bow shock; instruments on board measured the magnetic field. From counts of the particles arriving at the satellite, the electron and proton distribution functions can be measured, and from these the density and the flow speed of the solar wind as it passes through the shock can be calculated. The satellite has a very low speed, and usually the observations of the bow shock are taken when the shock moves relative to the satellite. This can happen because of slight changes in the solar-wind speed or density. The observations consist of time series for the different quantities, and if the speed of the shock relative to the satellite is constant, then the profiles we observe in *time* will be the same as the shock's profile in *space*. In Figure 5.4, the satellite is initially in the solar wind, and then the shock moves outward, with the observations being taken as the shock moves over the satellite.

In passing from upstream to downstream, it is obvious that the velocity decreases and the density increases; that is, there is compression at the shock. Also, the magnetic field is shown, and this increases like the density; so it is a fast-mode shock. The next thing to notice is that although the shock is thin, it is not just a smooth transition. Instead, there is structure within the transition. We see a "foot," a "ramp," and an "overshoot," and we shall see later that these are controlled by the way in which the solar-wind protons heat at the shock. For example, the thickness of the foot equals the distance to which protons are reflected by the shock before drifting through it.

5.3.1 Different Parameters Make Different Shocks

The most important result from space observations of shocks is that there are many different types of shocks, depending on the shock parameters. We have already seen that the Rankine-Hugoniot relations lead to different modes of shocks, but even restricting ourselves to fast-mode shocks, all planetary bow shocks, we find different types of shock structures. Early on there were quite serious debates whether or not such shocks were actually stable; perhaps all the different profiles seen in observations were just fleeting glimpses of an ever-changing entity. A major contribution was the demonstration that there was a definite pattern to the observations of shocks and that the type of shock was determined by the complete set of shock parameters (i.e., the strength of the shock and upstream conditions). The most important factors in controlling the type of shock are the direction of the upstream magnetic field (relative to the shock surface) and the strength of the shock.

We shall first look at the influence of the direction of the upstream field, sometimes called the shock geometry. Figure 5.5 shows a one-dimensional shock, and marked on it is the unit vector **n̂**, normal to the shock surface. A convenient way to describe the direction of the upstream field is the angle θ_{Bn} between the magnetic field and the shock normal. Depending on θ_{Bn}, the shock can have dramatically different behaviors. When $\theta_{Bn} = 0$, the shock is called parallel, and when $\theta_{Bn} = 90°$, it is called perpendicular. The terms quasi-parallel and quasi-perpendicular are used to divide the range of possible θ_{Bn} values, with the actual dividing line usually chosen as 45°. Another term, *oblique*, refers to a shock that is neither exactly perpendicular nor parallel, but it is also sometimes used to refer to the no man's land between quasi-parallel and quasi-perpendicular.

The importance of the parallel/perpendicular distinction is clear when we consider the motion of a particle that is initially headed away from the shock. In the de Hoffman–Teller frame (Appendix 5A) the motion has two parts: an unimpeded motion along the direction of the magnetic field, and a gyration around (i.e., transverse to) it. In the case of a parallel shock, the field lines pass through the shock, and a particle's motion along the field will carry the particle through (and away from)

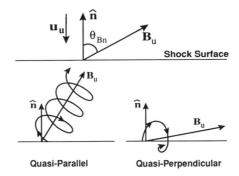

FIG. 5.5. The angle θ_{Bn} and examples of reflected-particle trajectories in the quasi-parallel and quasi-perpendicular cases.

the shock relatively easily. On the other hand, at a perpendicular shock, the field lines are parallel to the shock surface, and so particle motion along the magnetic field does not let the particle pass away from the shock. Indeed, the particle gyration at a perpendicular shock brings the particle back to the shock. There is a further simple conclusion to be drawn from the parallel/perpendicular distinction. Because particle motion in the normal direction is "easier" at the parallel shock, compared with at the perpendicular shock, then we can expect that the scale of the parallel shock will be larger than that for the perpendicular shock. Of course, this can be said with certainty only if we know the exact dissipation mechanisms. Also, our arguments about particle motion are really true only for the ions, which have much larger gyroradii than the electrons, but that is not so bad, because ions determine most of the structure of collisionless shocks. (Because of their much greater mass, the ions carry most of the mass and energy flows in a plasma.)

The shock angle θ_{Bn}, is the most important factor in controlling the shock type, but almost every plasma parameter can have an effect: temperature, composition (i.e., what types of ions present), and shock Mach number M. The Mach number indicates the strength of the shock, and is a measure of the amount of energy being processed by the shock. As might be expected, the higher the Mach number, the more dramatic the behavior of the shock. In the solar system, shocks can be found with Mach numbers between (almost) 1 and perhaps 20, but in more violent astrophysical objects, such as the shocks produced by supernova remnants, the Mach number could be of the order of 1,000. The processes operating at such shocks remain unclear, but for the solar-system shocks we believe that we have a fair understanding. However, even for solar-system shocks this statement is true only for *quasi-perpendicular shocks*. At the present time, quasi-parallel shocks remain a topic of debate, and unfortunately we shall have to neglect them.

In the category of quasi-perpendicular shocks there is a clear distinction, at least theoretically, between a type of low-Mach-number shock, the subcritical shock, and a higher-Mach-number shock, the supercritical shock. For most of the time, the earth's bow shock is supercritical, and to find subcritical shocks we usually have to wait for a suitable interplanetary shock. As the name suggests, there is a "critical" Mach number that separates the two types of shocks. This critical Mach number is at $M_A \sim 2.7$ for $\theta_{Bn} = 90°$, and it decreases as θ_{Bn} decreases. The earth's bow shock has values for M_A in the range of approximately 1.5–10.

Observationally, the difference between these two types of shocks is clear. Our original example (Figure 5.4) was of a supercritical shock. We have already mentioned the *structure* visible within the shock. The field has a single sharp jump (called the *ramp*), but it is preceded by a gradual rise called the *foot*. Also, the field right at and behind the ramp is higher than its eventual downstream value, and this is called the

5.3 SHOCK STRUCTURE: HOW SHOCKING?

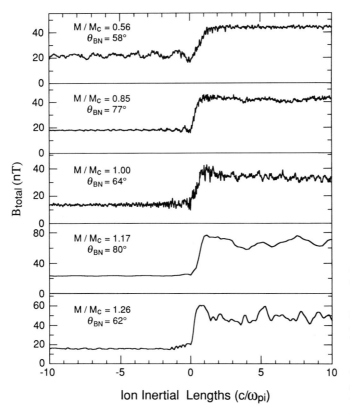

FIG. 5.6. Observations of magnetic-field strength for five low-Mach-number shocks ranging from subcritical to slightly supercritical. All shocks are quasi-perpendicular shocks. The measured temporal profile has been transformed to a distance scale using simultaneous measurements from a companion spacecraft that observed these shocks moments later. The low-frequency upstream waves in the two top panels are standing precursors that move with the shock. The higher-frequency waves are traveling waves in the shock frame. The appearance of these waves is sensitive to the value of θ_{Bn}.

overshoot. Like good biologists, we have arrived at this separation into component parts only after seeing many examples. We have ignored what we think is not important, which in this case includes all the small wiggles in the field. We label these wiggles as small-amplitude turbulence, which may play a role, but doesn't control or overwhelm the basic shock structure.

In contrast with the structure of the supercritical shock, the subcritical shock resembles much more the profile of a shock in an ordinary gas. Figure 5.6 shows examples of low-Mach-number shocks ranging from subcritical to slightly supercritical. The top two subcritical shocks have simple ramps from upstream to down, with little or no foot or overshoot. However, both these shocks have standing precursor waves that are stationary in the shock frame. At shorter wavelengths, both shocks have small-amplitude propagating waves. The appearances of all these waves are controlled by the θ_{Bn} angle. At higher Mach numbers, an overshoot and shock foot appear.

5.3.2 Shock Thermalization Mechanisms

In order to understand the basis for the distinction between subcritical and supercritical shocks, we shall discuss the basic mechanisms for heating the plasma at a shock. We have already emphasized how at a shock there is irreversible dissipation, which can be described as a transformation of ordered ram energy (i.e., kinetic energy of the flow into the shock) into random thermal energy in the particles. In an ordinary fluid, the dissipation is provided by viscosity due to collisions between the molecules. The collisions redistribute the energy among all the particles, so that they all have the same temperature, irrespective of their type. It is easy to explain the source of shock heating in a collisional fluid, but for a plasma it is the major problem of shock physics. The scale lengths of collisionless shocks are less than the mean free path between collisions. Therefore, the shock heating must use mechanisms unique to plasmas, and we shall now describe the major mechanisms. Because of the absence of collisions, the electrons and ions in the plasma, which have very different characteristics, may not be heated by the same process. Indeed, from the Rankine-Hugoniot relations, we know only that the *total* heating is specified from conservation laws, so that the heating at one type of shock could be dominated by either electrons or ions.

One way to describe the dissipation at a shock is to say that despite the lack of collisions, there is in fact an *effective* resistivity, producing heating that depends on the currents in the plasma, and an effective viscosity, producing heating dependent on velocity gradients. In this case the effective, or anomalous, resistivity and viscosity must be provided by changes in particle velocities caused by perturbations in the fields. In other words, waves in the fields replace collisions between particles. The waves are generated by some instability, which will be driven by some departure from equilibrium of the particle distribution function. This is a well-established idea in the physics of collisionless plasmas. For example, a common explanation for anomalous resistivity within the shock is as follows. A current must be the result of a distortion of the particle distribution function. This distortion can drive instabilities (we shall not describe which ones), which produce turbulence within the shock. The particles feel the turbulence and suffer small changes in their velocities, which has the same effect as if they were colliding with other particles.

This type of behavior is known as a collective dissipation mechanism, because the fields and the particles act together. This description seems attractive because we can use ordinary collisional equations, but replace the resistivity and viscosity by their anomalous counterparts. There are problems with this approach: The waves produced by instabilities tend to affect most strongly the particles responsible for the instability. In the case of current-driven instabilities in the shock layer, it is the electrons that mostly carry the current (they are of lower mass and therefore much more mobile than the ions), and so it is the electrons that should get heated. The problem is that observations show that it is the *ions* that are heated much more than the electrons at supercritical shocks. Of course, we could refine our model of collective dissipation to include instabilities that would depend on the ions, but then we would be confronted by another problem, namely, that the growth rate for ion instabilities tends to be fairly slow (on ion time scales), at least when compared with the time it takes for the plasma to pass through the shock ramp. Ion instabilities do not have enough time to heat up the ions by very much. These considerations lead us to look for other types of dissipation mechanisms.

We have just discussed the case in which the particles' velocities were changed by waves in the fields. In the case of a shock, there is a major change in the *macroscopic,* average fields. Because the plasma is collisionless, it may be possible to describe the behavior of the particles using the trajectories that they would follow in the macroscopic fields. In other words, we take the fields of the shock as fixed and ask what would happen to particles in those fields. Of course, the fields themselves are determined self-consistently with the particles, and our idea will work only if the particle trajectories are not disturbed too much by the microscopic fields (i.e., by instabilities). The dissipation mechanism relies on the particle dynamics in the shock fields and can be applied to both ions and electrons. In the case of ion heating at supercritical shocks, the ion dynamics is the major factor in determining the shock structure. This does not mean that collective processes do not play a role. In fact, they enter in new ways, because the particle dynamics provides new sources of instability.

What is the connection between particle dynamics (trajectories in the average shock fields) and heating? In a collisional plasma or gas, the temperature is easily related to the velocity spread of the particle distribution function. Because of the collisions, that distribution function is Maxwellian (or nearly so). In a collisionless plasma, the particle distribution function can take almost any form; it could be a superposition of Maxwellians of different temperatures, even drifting relative to each other. In this case our definition of temperature is expanded to be a measure of the "spread" of the particle distribution in velocity. This temperature is expressed by taking the second velocity moment of the distribution function, and it is usually called the kinetic temperature. If at a shock the magnetic and electric fields can spread out the particles' distribution functions, then that will increase the kinetic temperature of the distribution.

We are tempted to call this shock heating, but there is a second crucial stage. Just taking trajectories of particles in average fields means solving the equations of motion in steady fields. But these equations are time-reversible. If we replace t by $-t$, they will have the same form. For dissipation at a shock, we must have an irreversible process; entropy must increase. Thus, although particle dynamics can increase the kinetic temperature, there must be additional processes occurring to scatter the resultant distributions to make the whole process really dissipative. This additional scattering is caused by waves and turbulence, and it is the

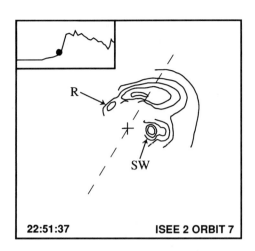

FIG. 5.7. Observation of the ion distribution function within the shock layer of a quasi-perpendicular, supercritical shock. In the top left-hand corner, the magnetic field through the shock is shown. The solid circle indicates the position in the shock where the measurement was made. The peak marked SW is the solar wind flowing into the shock. The component R consists of ions reflected at the shock and headed away from the shock. The orientation of the shock is shown by the dashed line, so that the reflection is about specular. The third component of the distribution consists of reflected ions that are headed back to the shock after having been turned around by the magnetic field. (Adapted from Sckopke et al., 1983.)

form of the distribution function that provides the free energy to generate this turbulence. So, in a way, particle dynamics does not work as a dissipation mechanism without collective phenomena! The shock fields can spread out a distribution function, which will increase the kinetic temperature and also provide a source of free energy to complete the dissipation process. Although ion instabilities do not have enough time to heat the ions at the shock ramp, they do have enough time *downstream* to ensure that the ion heating at the shock is irreversible.

Observations have revealed that the dissipation at a supercritical shock is dominated by the ion dynamics, so that the structure of the shock itself is controlled mainly by the ions. Figure 5.7 shows one such observation. The profile at the top is of the total magnetic field, and marked on it is the position in the shock where the velocity-space distribution function was measured. (We observe the shock because it passes over the spacecraft, and if we know the speed of the shock, we can convert the time of observation to spatial position.) The distribution is shown as a contour plot in a plane, in which the sun–earth line is aligned with the x-axis. The distribution is not just in one region of velocity space, but has several parts (i.e., several maxima). This fact immediately tells us that we are dealing with a distribution far from equilibrium. The part labeled SW is the solar wind, which is the upstream flow into the shock, with negative x-axis velocity. The part labeled R has positive x-axis velocity, so that it is headed away from the shock. This remarkable behavior can be explained if some of the particles in the solar wind are *reflected* by the shock ramp. This reflection is due to the average (macroscopic) fields at the shock. Some of the solar-wind ions are turned around at the shock, and this reflection happens in a small distance; they are very nearly turned completely around, that is, they are "specularly" reflected. What happens to these ions is determined by their dynamics in the shock fields. The reflection process gives them a velocity perpendicular to the magnetic field. Their reflected

5.3 SHOCK STRUCTURE: HOW SHOCKING? 153

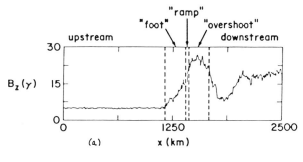

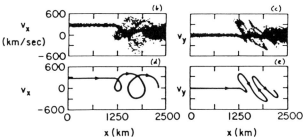

FIG. 5.8. Computer simulations of a supercritical perpendicular shock. The top panel shows the magnetic field, and the middle panels show the phase space (i.e., particle velocity versus particle position) for the protons in the simulation (each dot is a simulation particle). The bottom panels show the trajectory of a typical reflected particle. The magnetic field is in the z-direction. Note how the reflected particles have a negative v_x at the ramp, which results in motion in front of the ramp with large v_y, and then motion downstream with large v_x and v_y (i.e., with large gyrational motion). (From Wu et al., 1984.)

velocity is normal to the shock, but the magnetic field at a perpendicular shock is parallel to the plane of the shock. This perpendicular velocity means that they gyrate around the magnetic field, and at a perpendicular shock, as we noted earlier, the gyration takes them back to the shock, where, this time, they pass into the downstream region. The ions that are not reflected do get slowed down and pass into the downstream region.

In the downstream region there are then two different types of ions: a transmitted component and a reflected component. The reflected component still has its gyrational motion, and in velocity space it is on a ring of radius approximately $2u_u$, where u_u is the upstream flow speed. The reflection of some of the incoming ions has spread out the ion distribution function, thereby increasing the ion kinetic temperature. This is how the ion dynamics provides the shock dissipation, once some turbulence has made the process irreversible.

We have extrapolated from observations to describe the ion heating, but our understanding is based on many observations taken at various points through the shock structure, and these confirm what we have described. Further, we can use a computer to simulate, self-consistently, what happens at a supercritical shock. The simulation has the advantage that we follow the trajectories of individual ions, and these show us that some are reflected, just as we have described. The simulations can let us go even further, because they show that the *structure* of the shock itself is determined by the reflected ions. In particular, the foot of the shock mentioned earlier is caused by reflected ions that travel out in front of the shock as they are turned around. These points are illustrated in Figure 5.8.

Ion reflection at different types of shocks is quite common. Among various mechanisms, it has a great advantage, namely, that the heating it produces is proportional to the upstream flow speed. This can be seen from the fact that the gyrating ring of reflected ions has a spread proportional to u_u. Reflection takes the upstream flow speed and changes it to perpendicular (gyrational) velocity, which determines the spread of the downstream distribution. Thus, as the strength (i.e., the Mach number) of the shock increases, ion reflection can provide more and more dissipation. This is just what we need for a shock, because, as we discussed earlier, a shock is a wave that can have arbitrary speed (i.e., Mach number).

For a low-Mach-number shock, not much heating is required; there is not much dynamic pressure to convert. In this case, the anomalous resistivity η might provide sufficient dissipation. This dissipation is simply Joule heating, ηJ^2 per unit volume, where J is the current density. Suppose that the shock had a higher Mach number, so that there would be an increase in the dissipation required. One might think that one way to satisfy this would be to increase the resistivity. Increasing the resistivity would have the effect of spreading out the shock ramp. (In MHD, the diffusion of the field is proportional to the resistivity. If η is increased, any gradients in the field will be smoothed out.) That increase would have the effect of reducing the Joule heating. Actually, increasing the Mach number *can* increase the resistive heating, because the increase in the magnetic field increases with Mach number. Eventually this does not work, because, as we have seen, the jump in the field is limited to a factor of 4. What happens is that above a certain "critical" Mach number, anomalous resistivity cannot provide the required shock dissipation, however high the resistivity becomes. It is at this stage that ion reflection becomes important and provides the shock dissipation as the Mach number increases. This is a good example of the unique nature of collisionless shocks and how the shock parameters control the behavior of ions and electrons.

It is obvious from comparing Figures 5.4 and 5.6 that the supercritical and subcritical shocks have different structures. The difference can be traced to the important role of ion reflection at supercritical shocks. At sufficiently high Mach numbers, the dissipation required at the shock (i.e., the transformation of ordered flow energy to random thermal energy) is accomplished by the reflection of ions and their subsequent gyration downstream. The reflected ions travel out in front of the main ramp as they are turned around by the magnetic field. These ions are responsible for the foot seen in the magnetic field, and the size of the foot is determined by their turnaround distance. The overshoot also plays an important part in the reflection of some of the ions. In order to satisfy the requirements of dissipation, only some ions must be reflected, and this is determined self-consistently by the fields adjusting to the required ion heating. Because the ions are the dominant species in the thermalization at the shock, they determine the characteristic length scales of the shock. In summary, the difference between the supercritical and subcritical shock structures is the presence of reflected ions. However, there is evidence that reflected ions can be found even at low-Mach-number shocks, but only at very low fractional densities, so that they do not play a significant role in the shock structure. Also, the dissipation processes at subcritical shocks are expected to continue to operate at supercritical shocks, but with reduced relative importance.

We have concentrated on the importance of the ion dynamics at the high-Mach-number quasi-perpendicular shock. This is because it is the ions that mostly determine the length scale of the shock. The size of the foot is determined by the gyroradius of the reflected ions in the upstream magnetic field. Also, ion heating is much more important at high-Mach-number shocks, with a downstream-to-upstream temperature ratio of perhaps 50–100. However, even for the electrons, their dynamics in the average fields can be important for their heating. In this case it is not reflection (although that can happen), but rather that there is an electrostatic potential at the shock. This is in a sense to oppose the flow of protons through the shock and plays a role in their partial reflection. However, the electrons see it as a force that accelerates them into the downstream region. This produces, just behind the shock, a distribution of electrons passing from upstream to downstream that has been shifted in velocity. This means that the potential, by displacing part of the electron distribution function, spreads out the distribution, which increases the kinetic temperature and creates a source of free energy for instabilities. The situation is similar to that for the ions, but the nature of the spreading is different, and the instabilities are different. The overall shock dissipation is still essentially controlled by the ions.

5.4 THINGS THAT HAVEN'T BEEN MENTIONED

There are many interesting topics that haven't been discussed here. The microphysics of slow shocks has been completely omitted. Our discussion of quasi-perpendicular fast shocks has been limited, and we have had to leave out any description of fast-mode quasi-parallel and parallel shocks. Comets also form shocks in the solar wind, but they are sources of new mass, as cometary material ionizes and joins the already ionized solar wind. That mass addition makes cometary shocks very different from the bow shocks in front of the planets. Because the solar wind is collisionless, and because particle motion is controlled by the magnetic field, energetic particles can travel upstream of the bow shock. This creates a region called the *foreshock,* full of interesting waves and particles. Shocks are copious and efficient sources of energetic particles and are used to explain satellite observations and observations of cosmic rays. There are different theories of particle acceleration at shocks, and solar-system observations help to support the theories used to explain astrophysical observations. Some of the issues of the foreshock are briefly described in Appendix 5B.

It is for the interested reader to pursue these and other aspects of collisionless-shock physics.

Appendix 5A THE DE HOFFMAN–TELLER FRAME

A FRAME OF REFERENCE in which the shock is at rest is called a shock frame. In order to transfer from one shock frame to another that is in uniform relative motion, one should apply the appropriate relativistic transformations. However, for almost all work we can use a nonrelativistic transformation in which the velocity transformation is Galilean, the magnetic field is unchanged, and the electric field is obtained from the ideal MHD equation $\mathbf{E} = -\mathbf{u} \times \mathbf{B}$ (see Chapter 2) in the appropriate frame.

In Figure 5A.1 we have drawn the plane containing the shock normal $\hat{\mathbf{n}}$ and the upstream field $\mathbf{B}_u$, and we have the upstream flow velocity $\mathbf{u}_u$ in the same plane. We shall be considering only a one-dimensional shock. In the normal incidence frame (NIF, Figure 5A.1a) the upstream flow is parallel to $\hat{\mathbf{n}}$, and the origin has been marked O. In this frame there is a motional electric field $\mathbf{E}_u = -\mathbf{u}_u \times \mathbf{B}_u$ that is perpendicular to both $\mathbf{u}_u$ and $\mathbf{B}_u$ and thus is parallel to the plane of the shock. This motional electric field can be made zero by transforming to the frame

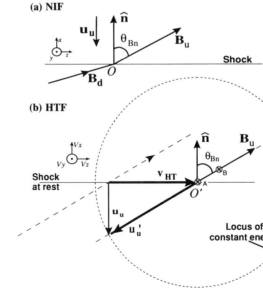

FIG. 5A.1. Configurations for (a) the normal incidence frame and (b) the de Hoffman–Teller frame. A typical coordinate system is marked. The HTF is shown in velocity coordinates. The dashed circle represents the locus of all particles with energy equal to the energy of the incident flow. Marker A represents a particle with zero velocity in the HTF; marker B represents a particle with enough field-aligned velocity to escape upstream from the shock.

where there is no flow velocity (the upstream flow frame), but in this frame the shock is moving. The alternative is to find a frame in which the upstream flow and magnetic field are parallel and the shock is stationary. This is obtained from any shock frame by adding a transformation velocity parallel to the shock plane. This is shown in Figure 5A.1b, where the transformation velocity from the NIF is marked $\mathbf{v}_{HT}$, and O' is the origin in the new frame, which is called the de Hoffman–Teller frame (HTF).

In the HTF, the upstream flow velocity is $\mathbf{u}'_u$, parallel to $\mathbf{B}_u$. ($\mathbf{B}_u$ is the same in all frames, because we are not using the full relativistic transformation.) From Figure 5A.1b we see that

$$\mathbf{v}_{HT} = u_u \tan \theta_{Bn}$$

so that the required transformation velocity increases rapidly as θ_{Bn} approaches 90°. Also, the assumption about a nonrelativistic transformation will break down if θ_{Bn} is sufficiently close to 90°. And when θ_{Bn} is exactly equal to 90°, it is not possible to find a de Hoffman–Teller frame at all.

We can get a more general expression for $\mathbf{v}_{HT}$ starting from the transformation of a velocity to a new frame in which all quantities are indicated by a prime (except $\mathbf{B}_u$):

$$\mathbf{u}'_u = \mathbf{u}_u - \mathbf{v}_{HT}$$

We specify that in the new frame the motional electric field is zero, $\mathbf{E}' = -\mathbf{u}'_u \times \mathbf{B}_u = 0$, and we relate this to the flow velocity in the original frame:

$$\mathbf{v}_{HT} \times \mathbf{B}_u = \mathbf{u}_u \times \mathbf{B}_u$$

We can solve for $\mathbf{v}_{HT}$ by crossing with $\hat{\mathbf{n}}$ [$\hat{\mathbf{n}} \times (\mathbf{v}_{HT} \times \mathbf{B}_u) = \hat{\mathbf{n}} \times (\mathbf{u}_u \times \mathbf{B}_u)$] and then expanding the left-hand side using the vector identity $\mathbf{a} \times (\mathbf{b} \times \mathbf{c}) = \mathbf{b}(\mathbf{a} \cdot \mathbf{c}) - \mathbf{c}(\mathbf{a} \cdot \mathbf{b})$, remembering that $\mathbf{v}_{HT} \cdot \hat{\mathbf{n}} = 0$, to keep the shock stationary in the new frame. The final expression is then

$$\mathbf{v}_{HT} = \frac{\hat{\mathbf{n}} \times (\mathbf{u}_u \times \mathbf{B}_u)}{\hat{\mathbf{n}} \cdot \mathbf{B}_u} \tag{5.14}$$

What are the properties of the HTF? Because there is no electric field, the particles upstream have a very simple motion, with two parts: motion parallel to the magnetic-field direction, and gyrational motion around it. Another consequence of $\mathbf{E} = 0$ is that the energy of a particle is constant, and surfaces of constant particle energy are spheres centered on the HTF origin. Although we have not marked in Figure 5A.1b the downstream field and flow velocity, there is another useful property of the HTF. From (5.14), we see that $\mathbf{v}_{HT}$ depends on the normal component of $\mathbf{B}$ and the transverse component of $\mathbf{u} \times \mathbf{B}$, but from the shock-jump conditions (5.5) and (5.6), both these components are continuous across the shock. Consequently, the de Hoffman–Teller transformation velocity is the same downstream as it is upstream.

Appendix 5B ENERGETIC PARTICLES AND FORESHOCKS

THE EXISTENCE OF A "FORESHOCK" is a unique property of collisionless plasma shocks. As the name implies, the foreshock is a region upstream of the shock that contains particles and waves associated with the shock. This appears to be a paradox; we earlier defined a shock as an "information horizon," so that information about the shock's existence could not travel upstream. The resolution is that the shock wave is associated with a characteristic wave speed of the medium; however, in a collisionless plasma a particle can have an arbitrarily high speed and thus can travel upstream faster than the shock. Such a particle will eventually scatter off some irregularities in the fields and thus eventually couple into the rest of the plasma. A fast particle that escapes into the foreshock carries energy, and therefore it generally produces waves. Observationally, the earth's foreshock is a rich "zoo" of different types of particles and the different waves they produce. Our restricted knowledge of other planetary foreshocks indicates similar phenomena.

We can maintain a distinction between the foreshock and the shock itself provided that the upstream flow is not too disturbed by the foreshock and that the characteristic wave speeds at the shock remain unchanged. In other words, the distinction holds if most of the changes in energy and momentum happen at the shock, not in the foreshock. Another view sometimes expressed is that our original definition of a shock as an information horizon is strictly correct and that any foreshock should be regarded as part of the shock system, and thus the "shock" is actually a subshock. Generally it is possible, and more useful, to consider the shock and foreshock as separate things, but there are important cases where the two are inextricably connected.

The foreshock exists because particles with sufficiently high energies can outrun the shock. There are two important issues. How do the particles gain these high energies? What is the structure within a foreshock? The subject of particle acceleration at shocks has a vast literature and would require at least another chapter. Thus, we shall have to be content with the statement that, observationally, shocks in the heliosphere seem to be good accelerators of both electrons and ions. At the earth's bow shock, ions are measured with energies up to several hundred kiloelectron volts (the kinetic energy of a proton in the solar wind is about 1 keV), and electrons with energies up to several tens of kiloelectron volts. Ultimately, the energy in these particles comes from the kinetic energy of the solar-wind flow, but for explanations of the various theories and their observational support, the reader should consult the monographs *Collisionless Shocks in the Heliosphere* (Tsurutani and Stone, 1985; Stone and Tsurutani, 1985).

In the case of the earth's foreshock, which has been studied in detail, there is a definite pattern of types of particles and waves. We shall now turn to this geometric configuration. The earth's bow shock is a curved surface, close to a paraboloid of rotation. It is finite in space, and the shock-normal angle θ_{Bn} varies everywhere across its surface. Planetary bow shocks thus have a foreshock structure very different from that of interplanetary shocks, which are essentially flat and extend over a very much larger area. In what follows, we concentrate on planetary foreshocks.

What particles we see, and where in the foreshock we see them, will depend on, in the first case, two conditions. The particle has to be fast enough to outrun the shock (we shall see that this depends on θ_{Bn}). Once in the foreshock, its path is given by particle dynamics, until it interacts with the background plasma in some way.

Escape from the shock is best treated in the HTF. In the HTF, the particle motion has only two components: motion of the guiding center in the magnetic-field direction, and gyrational motion around the magnetic-field direction. Thus, whether a particle is headed away from or toward the shock will depend simply on the direction of the component of its motion along the shock normal. For simplicity, consider a field-aligned particle with no gyration. The particle is at rest relative to the shock if it has zero velocity in the HTF. In Figure 5A.1b, particle A is at rest, but particle B will escape from the shock.

When calculating particle energies, it is most useful to use the upstream flow frame, in which, as in the HTF, the motional electric field is zero. In this frame, the speed of a marginally escaping particle u_{esc} is just u'_u. From Figure 5A.1b we see that $u_{esc} = u_u \sec \theta_{Bn}$. Another way of obtaining this escape velocity (with more gymnastics between reference frames) is to calculate how fast the shock is moving along a field line in the upstream flow frame. If the particle has some gyrational motion, it is more difficult to calculate its escape velocity, and indeed it becomes dependent on the phase of the particle motion as it leaves the shock.

The result $u_{esc} = u_u \sec \theta_{Bn}$ shows that at shocks increasingly close to perpendicular (θ_{Bn} closer to 90°) it becomes increasingly difficult for a particle to escape from the shock, because the escape velocity increases. On the other hand, this also means that the fastest, most energetic particles can be produced at shocks with θ_{Bn} close to 90°. For an exactly perpendicular shock, there is no particle escape, because the magnetic field keeps the particle motion parallel to the shock front.

What happens to upstream particles once they leave the shock? If the upstream field is reasonably uniform, their guiding center motion has two parts (in the observer's frame with respect to the rest frame of the whole bow shock). There is a parallel motion along the magnetic field $u_\parallel$ and a cross-field drift motion $\mathbf{u}_d = (\mathbf{E} \times \mathbf{B})/B^2$ (see Chapter 2). Here, **B** refers to the interplanetary magnetic field, and **E** is the motional electric field. The drift velocity is the same for all particles, regardless of their energies. Therefore, if particles with a range of energies escape from a point on the shock, the fastest particles will travel farther away from the shock before they have cross-field drifted the same distance as slower particles. In the case of a bow shock, this implies that, coming from upstream, one observes particles with higher energies before lower-energy particles. (The cross-field drift is always in the downstream sense, because $\mathbf{E} = -\mathbf{u} \times \mathbf{B}$). This effect is called velocity dispersion, or a time-of-flight effect.

At the earth, the interplanetary field has an angle to the sun–earth line of, on average, 45°. This is shown in Figure 5B.1. The first field line that touches the bow shock is the tangent field line, and, of course, the shock at this point has $\theta_{Bn} = 90°$. Field lines farther downstream are connected to the shock where θ_{Bn} is increasingly less than 90°. (One should be careful to think about the three-dimensional geometry, but we are going to skip over the details.) In principle, everywhere downstream of the tangent field line could contain foreshock particles. The first particles observed behind the tangent field line are electrons. As expected from velocity dispersion, the highest energies are observed closest to the tangent field line.

The region filled with upstream electrons is called the electron foreshock. There is a similar region of accelerated ions, the ion foreshock. However, it appears that for the earth, the ions farthest upstream are coming from the shock where θ_{Bn} is less than about 70°. This means that there is not an efficient enough acceleration mechanism for ions above $\theta_{Bn} = 70°$, but this might be a function of the size and shape of the shock. Although definite boundaries are shown in Figure 5B.1, it should be

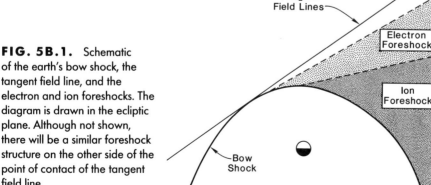

FIG. 5B.1. Schematic of the earth's bow shock, the tangent field line, and the electron and ion foreshocks. The diagram is drawn in the ecliptic plane. Although not shown, there will be a similar foreshock structure on the other side of the point of contact of the tangent field line.

noted that such boundaries usually are observational and are susceptible to factors such as instrument response and sensitivity.

Unfortunately, there is not sufficient space to further discuss the foreshock in all its glory, all the different particle distributions, the waves they produce (from 50 mHz to 100 kHz), and the theories to explain them. Finally, we have implied that the foreshock is filled with particles accelerated at the shock. However, the proximity of the magnetosphere, which is also filled with energetic particles, means that particles in the foreshock could make their way there from the magnetosphere. There is evidence that magnetospheric particles can travel all the way to the foreshock, but their relative importance, compared with shock-accelerated particles, is not yet generally agreed.

Appendix 5C DETERMINING THE SHOCK-NORMAL DIRECTION

As discussed in section 5.3.1, the angle between the upstream magnetic field and the normal to the shock is one of the most important parameters in determining the nature of the physical processes occurring at the shock. Whereas determination of the orientation of the upstream magnetic field is usually straightforward, determining the orientation of the shock surface may not be so simple. If the shock is standing in front of a relatively hard obstacle, the shock orientation may be obtained from the overall shape of the shock surface. If the location of the shock is oscillating, or if it is a propagating shock, this surface will not be geometrically constrained. In such a case we must use other constraints to determine the shock surface.

An example of such a constraint is provided by the relative timing and separation of multiple observations of the same planar shock surface. If $\delta \mathbf{x}_i$ ($i = 1, 2, 3, \ldots$) are the spatial separations between observations of a shock relative to a spacecraft at $\mathbf{x}_0$, and if δt_i are the time delays from t_0, the time of the observation at $\mathbf{x}_0$, then

$$\delta \mathbf{x}_i \cdot \hat{\mathbf{n}} = \mathbf{v}\, \delta t_i$$

where $\hat{\mathbf{n}}$ is a unit vector in the direction of the normal to the surface, and $\mathbf{v}$ is its velocity along the normal. If there are more than four observations (i.e., there are more than three baselines), then the equations are overdetermined and can be solved by multiplying each side of the equations by the transpose of the δx matrix and inverting to solve for $\hat{\mathbf{n}}$ and $\mathbf{v}$. However, we usually count ourselves lucky to obtain measurements of a shock with one spacecraft. In the past, there seldom were four or more such detections, but measurements made with the European Space Agency's Cluster mission will provide such data routinely.

A single-spacecraft method is the so-called coplanarity method, which uses the coplanarity theorem (Section 5.2.2) stating that at an oblique, compressive shock, the upstream magnetic field, the downstream magnetic field, and the shock normal are all coplanar. This is expected to be true for a time series of measurements in the absence of electric fields along the normal and is observationally determined to be true except close to the shock ramp. Thus, the cross product of the upstream and downstream magnetic fields should be perpendicular to the shock normal:

$$(\mathbf{B}_u \times \mathbf{B}_d) \cdot \hat{\mathbf{n}} = 0$$

Because the magnetic field is divergenceless,

$$(\mathbf{B}_u - \mathbf{B}_d) \cdot \hat{\mathbf{n}} = 0$$

With two different vectors perpendicular to the shock normal, we can calculate a vector along the normal:

$$\mathbf{N} = (\mathbf{B}_u \times \mathbf{B}_d) \times (\mathbf{B}_u - \mathbf{B}_d)$$

where $\mathbf{N}$ would be normalized to obtain $\hat{\mathbf{n}}$. This direction, of course, is ill-defined when $\mathbf{B}_u \parallel \mathbf{B}_d$, which occurs for both parallel and perpendicular shocks. The normal derived in this way is called the coplanarity normal.

Another constraint that can be used when three-dimensional velocity measurements are available is that

$$(\mathbf{B}_u \times \Delta \mathbf{v}) \cdot \hat{\mathbf{n}} = 0$$

and

$$(\mathbf{B}_d \times \Delta \mathbf{v}) \cdot \hat{\mathbf{n}} = 0$$

where $\Delta \mathbf{v}$ is the change in the flow velocity from upstream to downstream. When a velocity constraint such as this is combined with a magnetic constraint such as $\Delta \mathbf{B} \cdot \mathbf{n} = 0$, the resulting normal is called a mixed-mode shock normal. In fact, all these constraints can be combined into one overdetermined solution and solved for a best-fit shock normal (Russell et al., 1983).

Finally, we note that with knowledge of the shock normal $\hat{\mathbf{n}}$ and use of the continuity equation (5.1) we can determine the velocity of the shock relative to the measurement frame. It is possible to show that the shock velocity v_{sh} is

$$v_{sh} = \frac{\rho_d \mathbf{u}_d - \rho_u \mathbf{u}_u}{\rho_d - \rho_u} \cdot \hat{\mathbf{n}}$$

ADDITIONAL READING

Courant, R., and K. O. Friedrichs. 1948. *Supersonic Flow and Shock Waves*. New York: Interscience.

Stone, R. G., and B. T. Tsurutani (eds.). 1985. *Collisionless Shocks in the Heliosphere: A Tutorial Review*. Washington, DC: American Geophysical Union.

Tidman, D. A., and N. A. Krall. 1971. *Shock Waves in Collisionless Plasmas*. New York: Wiley-Interscience.

Tsurutani, B. T., and R. G. Stone (eds.). 1985. *Collisionless Shocks in the Heliosphere: Reviews of Current Research*. Washington, DC: American Geophysical Union.

PROBLEMS

5.1. Sudden changes are detected in the solar wind and interplanetary magnetic field. The radial velocity of the solar wind remains constant, but the density jumps from 5 to 10 cm^{-3}. The proton temperature jumps from 5 eV before the discontinuity to 13.8 eV afterward, but the electron temperature remains constant at 15 eV. The magnetic field of (0, −8, 6) nT before the discontinuity rotates and drops in strength to (0, 3, 4) nT. What type of discontinuity might this be, and why?

5.2. An interplanetary shock crosses the spacecraft *Space Physics Explorer*, and its magnetometer detects an upstream magnetic field of (6.36, −4.72, 0.83) nT and a downstream magnetic field of (10.25, −9.38, 1.74) nT. Using the magnetic coplanarity assumption, determine the orientation of the normal. The plasma analyzer detects an upstream velocity of (−378, 33.1, 19.9) km · s^{-1} and a downstream velocity of (−416.8, 7.3, 51.2) km · s^{-1}. Calculate the mixed-mode normal. If the upstream density was 7.5 cm^{-3} and the downstream density was 11 cm^{-3}, calculate the shock velocity.

5.3. A cold solar-wind proton encounters a strong collisionless shock and is reflected back into the solar-wind flow. If the magnetic field is perpendicular to the flow, and the flow and shock normal are aligned, how far backward from the shock does the proton move before its motion is reversed by the solar-wind electric field. This is the extent of the shock foot. Express this distance in terms of the gyroradius of a proton moving at the solar-wind velocity.

5.4. The dispersion relation for fast magnetosonic waves is $v_{ms}^4 - v_{ms}^2 (v_A^2 + c_s^2) + v_A^2 c_s^2 \cos^2\theta = 0$. Show that when the IMF is aligned with the solar-wind flow, the asymptotic Mach cone angle equals $\sin^{-1}(1/M_c)$, where $M_c = M_A M_s / (M_A^2 + M_s^2 - 1)^{\frac{1}{2}}$.

6 SOLAR-WIND INTERACTIONS WITH MAGNETIZED PLANETS

R. J. Walker and C. T. Russell

6.1 INTRODUCTION

As correctly foreseen by Chapman and Ferraro (1930) (see Chapter 1), a planetary magnetic field provides an effective obstacle to the solar-wind plasma. The solar-wind dynamic pressure, or momentum flux, presses on the outer reaches of the magnetic field, confining it to a magnetospheric cavity that has a long tail consisting of two antiparallel bundles of magnetic flux that stretch in the antisolar direction, as sketched for the earth in Figure 1.17. The pressure of the magnetic field and the plasma it contains establishes an equilibrium with the solar wind. When the solar wind "blows harder," the magnetosphere shrinks. When the solar wind abates, the magnetosphere expands. As discussed in the earlier chapters, the solar wind usually is highly supersonic before it reaches the planets. For supersonic flow, the wind velocity exceeds the velocity of any pressure wave that could act to divert the flow around the magnetosphere. In the case of a magnetized plasma, this wave is the fast magnetosonic wave introduced in Chapter 2. Because this wave is too slow to move upstream, it becomes nonlinear and forms a fast magnetosonic shock front, standing in the flow, well in front of the magnetopause. The physics of such collisionless shocks was discussed in Chapter 5 and will not concern us here. Rather, we shall examine the macroscopic configuration of the bow shock and the magnetopause and the region between, called the magnetosheath. To begin, we examine first the nature of the magnetic obstacle.

6.2 PLANETARY MAGNETIC FIELDS

Over 150 yr ago Gauss showed that the magnetic field of the earth could be described as the gradient of a scalar potential:

$$\mathbf{B} = -\nabla\Phi = -\nabla(\Phi^i + \Phi^e) \tag{6.1}$$

where Φ^i is the magnetic scalar potential due to sources inside the earth, and Φ^e is the scalar potential due to external sources. Gauss and his colleague Weber founded a chain of magnetic observatories around the world. With the data they collected they demonstrated that the field at the surface of the earth was almost entirely internal and principally dipolar. This latter fact is very useful in treating the terrestrial magnetosphere, because it allows simplification of otherwise complex problems. An example of this approach was given in Chapter 1: the mirror dipole approximation of the compression of the outer magnetospheric magnetic field by the solar wind.

The dipole moment of the earth is tilted about 11° to the rotation axis, with a present-day moment of about 8×10^{15} T·m³ or 30.4 μT·R_E^3. For more precise information on the earth's magnetic field, see Barker et al. (1986). The tilts of the magnetic dipoles of the other planets range from less than 1° to over 50°, and their moments span the wide range discussed in Chapter 15.

If we adopt a coordinate system that is fixed in this dipole moment, it will rotate as the planet rotates and in most cases will vary in orientation with respect to the solar-wind flow in the course of a day. However, near the planet, if the magnetic moment is strong, charged-particle motion will be dominated by the internal planetary field, not by the solar-wind interaction. Thus, this frequently is a useful approach.

In spherical coordinates,

$$B_r = 2Mr^{-3}\cos\theta \tag{6.2a}$$

$$B_\theta = Mr^{-3}\sin\theta \tag{6.2b}$$

$$B = Mr^{-3}(1 + 3\cos^2\theta)^{\frac{1}{2}} \tag{6.2c}$$

where θ is the magnetic colatitude, as defined in Figure 6.1, and M is the dipole magnetic moment.

An alternate means of expressing the magnetic field of a magnetic dipole is in cartesian coordinates. If we define a coordinate system with the z-axis along the magnetic-dipole axis, then

$$B_x = 3xzM_z r^{-5} \tag{6.3a}$$

$$B_y = 3yzM_z r^{-5} \tag{6.3b}$$

$$B_z = (3z^2 - r^2)M_z r^{-5} \tag{6.3c}$$

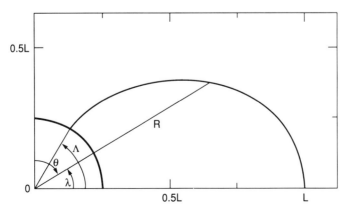

FIG. 6.1. Dipolar magnetic-field line. L designates the magnetic-drift shell and is equal to the distance in planetary radii from the center of the earth to the point where field line crosses the equator. The angle λ is the magnetic latitude of a point on the field line; θ is the colatitude, and R is the radial distance to that point. The angle Λ is the latitude of the point where the field line intersects the surface of the earth.

where M_z is the magnetic moment along the z-axis. This representation can be easily generalized to the case for a dipole moment with arbitrary orientation:

$$\mathbf{B} = \begin{pmatrix} (3x^2 - r^2) & 3xy & 3xz \\ 3xy & (3y^2 - r^2) & 3yz \\ 3xz & 3yz & (3z^2 - r^2) \end{pmatrix} \mathbf{M} \quad (6.4)$$

Given a series of observations $\mathbf{B}_i$, at locations $\mathbf{r}_i$ (x_i, y_i, z_i), this equation can be solved by standard matrix-inversion techniques.

6.2.1 Magnetic-Field Lines and the L Parameter

The spherical coordinate representation of a dipole magnetic field allows us to calculate readily the equation of a magnetic-field line. A field line is everywhere tangent to the magnetic-field direction. Thus,

$$r \, d\theta / B_\theta = dr / B_r \quad (6.5a)$$

and

$$d\varphi = 0 \quad (6.5b)$$

Integrating (6.5a), we obtain for the equation of a field line

$$r = r_0 \sin^2 \theta \quad (6.6)$$

where r_0 is the distance to the equatorial crossing of the field line, as illustrated in Figure 6.1. For historical reasons it became usual to write equation (6.6) in terms of L (with distance measured in planetary radii) and the magnetic latitude λ:

$$r = L \cos^2(\lambda)$$

A related parameter that is frequently used to organize low magnetospheric observations is the invariant latitude. It is the latitude where a field line reaches the surface of the earth and is given by

$$\Lambda = \cos^{-1}(1/L)^{\frac{1}{2}}$$

Thus, a dipole field line that extends to $4R_E$ in the equatorial plane of the magnetosphere maps to an invariant latitude of 60° at the earth's surface. A dipole field line that extends to $10R_E$ in the equatorial plane maps to a latitude of 71.6°.

The most intense portions of the earth's radiation belts are found in the near-earth part of the magnetosphere, where the magnetic field is primarily dipolar. In 1961, Carl McIlwain realized that the particle observations could easily be organized by using the properties of particle motion in a dipolar field. As we saw in Chapter 2, in addition to their gyromotion about a magnetic-field line, particles bounce between the northern and southern hemispheres. There is an invariant of motion associated with this bouncing along field lines. If the magnetic field is sufficiently slowly varying, a bounce adiabatic invariant will be valid. As discussed in more detail in Chapter 10, the bounce invariant is

$$J = \oint p_\parallel \, dl = 2 \int_{m_1}^{m_2} mv_\parallel \, dl \tag{6.7}$$

where $p_\parallel$ is the component of the momentum along the magnetic field, and the integral is along the field line. This is frequently called the second adiabatic invariant. A particle trapped on magnetic-field lines and conserving its adiabatic invariants will be confined to a surface specified by the distance L to the equatorial crossing of the field line over a half bounce from mirror point m_1 to m_2. McIlwain showed that a coordinate system with L and B (evaluated in the earth's real field or a model field) well organized the radiation-belt data.

6.2.2 Generalized Planetary Magnetic Fields

Although the dipole approximation is quite useful, often it is inadequate to express the significant complexities of magnetic fields in a planetary magnetosphere. Some planets, such as Jupiter, have very high contributions from nondipole fields, and in the case of the earth, our knowledge of the behavior of the radiation-belt particles may be too sophisticated for simple approximations to be useful. In such situations it is usual to express the scalar potential Φ in (6.1) as a sum of internal and external contributions to the field using associated Legendre polynomials:

$$\Phi^i(r, \theta, \varphi) = a \sum_{n=1}^{\infty} \sum_{m=0}^{n} [r/a]^{-n-1} P_n^m(\cos \theta)(g_n^m \cos(m\varphi) + h_n^m \sin(m\varphi)) \tag{6.8a}$$

and

$$\Phi^e(r, \theta, \varphi) = a \sum_{n=1}^{\infty} \sum_{m=0}^{n} [r/a]^n P_n^m(\cos \theta)(G_n^m \cos(m\varphi) + H_n^m \sin(m\varphi)) \tag{6.8b}$$

where a is the planet's radius, and θ and φ are the colatitude and east longitude, respectively, in planetographic coordinates. $P_n^m(\cos \theta)$ are associated Legendre functions with Schmidt normalization.

$$P_n^m(\cos \theta) = N_{nm}(1 - \cos^2 \theta)^{m/2} \, d^m P_n(\cos \theta)/d(\cos \theta)^m$$

where $P_n(\cos \theta)$ is the Legendre function, and $N_{nm} = 1$ when $m = 0$, and $[2(n-m)!/(n+m)!]^{\frac{1}{2}}$ otherwise. The coefficients g_n^m, h_n^m, G_n^m, and H_n^m are chosen to minimize the difference between the model field and observations. We can see the relation of this to our dipole approximation by examining the $n=1$, $m=0, 1$ terms in the series. The dipole moment becomes

$$M = a^3[(g_1^0)^2 + (g_1^1)^2 + (h_1^1)^2]^{\frac{1}{2}} \qquad (6.9)$$

and the tilt of the dipole moment to the rotation axis is

$$\alpha = \cos^{-1}(g_1^0/M) \qquad (6.10)$$

It is important to note that these coefficients are functions of time and have sizable secular or temporal variations. The dipole moment was about 9.54×10^{15} T·m^3 in 1550, but had decreased to only 7.84×10^{15} T·m^3 in 1990. Recently, the pace of this decrease has accelerated somewhat, and it is now about 0.1 percent per year. The geographic colatitude (tilt) of the dipole axis also varies. It was close to 3° in 1550 and rose to 11.5°, where it remained from 1850 to 1960. Since then, it has declined, reaching 10.8° in 1990. The other principal secular variation of the dipole field is a westward drift. The dipole axis lay at 334° east longitude in 1550, but in 1990 had drifted to 189° east longitude. This averages to a drift of 0.1° per year, but as with the other properties of the dipole field, this rate varies. We are not yet at a point where we can successfully predict the temporal variations of the internal magnetic field; it has to be measured.

Although it is common to draw the earth's magnetosphere with its dipole axis perpendicular to the solar-wind flow, it is seldom in this configuration. In addition to the 10.8° dipole tilt, the earth's rotation axis is inclined 23.5° to the ecliptic pole. Thus, in the course of its daily rotation and its annual journey around the sun, the angle between the direction of the dipole and the direction of the solar-wind flow varies between 90° and 56°. Because the interplanetary magnetic field is ordered in the ecliptic plane (or, more properly, the sun's equatorial plane), and because interplanetary magnetic fields that are opposite in direction to the earth's field more strongly interact with it, there are annual and semiannual variations in geomagnetic activity. These cycles are discussed in more detail in Chapter 13.

6.3 SIZE OF THE MAGNETOSPHERIC CAVITY

In the Chapman and Ferraro model of the solar-wind interaction with the earth's magnetic field, the boundary of the magnetic cavity was located halfway between the earth's dipole and its mirror image. The forces that determine where the boundary is to be drawn were not specified. The solar-wind plasma was treated solely as a superconductor. The real solar wind has mass and momentum. It exerts a force outward from the sun on every obstacle in its path. The earth's magnetic field is such an obstacle, because the magnetic fields of the earth and of the solar wind are "frozen in" to their respective plasmas by the high electrical conductivity of the plasmas. Hence, to first order, the major effect of the magnetized solar wind is to exert pressure, or normal stresses, on the magnetosphere.

In a steady-state situation, the force of the solar wind against the magnetosphere and the force of the magnetosphere against the solar wind are in balance, and an equilibrium is struck. The forces are exerted by pressure gradients. At the magnetopause, there is a pressure gradient in the magnetospheric magnetic field and plasma exerting an outward force, and a pressure gradient in the magnetosheath plasma and magnetic field exerting an inward force. The location of this equilibrium point is pressure-sensitive. If the magnetosheath plasma pushes harder, then the magnetopause moves inward to where the magnetic field is stronger and the magnetosphere can exert sufficient outward force to balance the new level of magnetosheath pressure. In order to determine the location of the magnetopause, we must determine how much force the solar wind exerts on the magnetosphere and how the counterbalancing magnetospheric force on the solar wind varies with the size of the magnetosphere.

6.3.1 The Pressure Exerted by the Solar Wind on the Magnetopause

In the solar wind, the pressure consists principally of the solar-wind dynamic pressure or momentum flux, ρu^2, where ρ is the mass density, which includes, on average, a 20 percent (by mass) contribution from double ionized helium, and u is the solar-wind velocity. The magnetic-field pressure and thermal plasma pressure add about 1 percent to the total. The balance among these different forms of pressure is altered in passing through the shock and the magnetosheath. At the magnetopause, the flow is tangential to the surface, so that the contribution of the dynamic pressure to the pressure balance across the boundary surface is zero. The pressure here must be totally due to magnetic and thermal contributions. It is proportional to the incident dynamic pressure, but even at the nose of the magnetosphere it is smaller than the incident pressure because of the divergence of the flow around the obstacle. To see this, we consider the momentum flux through unit area in the direction $\hat{n}$:

$$\rho \mathbf{u}(\mathbf{u} \cdot \hat{n}) + p\hat{n}$$

Integrating over the surface of a stream tube, we obtain the momentum-conservation equation

$$(\rho u^2 + p)S = \text{constant} \qquad (6.11)$$

Upstream (at infinity in the solar wind), p_∞ is small, and at the magnetopause, ρu^2 can be neglected. Thus,

$$K = \frac{p_s}{\rho_\infty u_\infty^2} = \frac{S_\infty}{S_s} \qquad (6.12)$$

Here the subscript s refers to measurements at the magnetopause, and ∞ refers to those in the solar wind. The parameter K tells how much the pressure has been diminished by the divergence of the flow. We can calculate K from Euler's equation for an ideal fluid with no viscosity or thermal conduction:

$$\frac{\partial \mathbf{u}}{\partial t} + (\mathbf{u} \cdot \nabla)\mathbf{u} = -\frac{1}{\rho}\nabla p \tag{6.13}$$

For an adiabatic fluid,

$$p\rho^{-\gamma} = \text{constant} \tag{6.14}$$

where γ is the ratio of specific heat or the polytropic index. Using the identity

$$\mathbf{u} \cdot \nabla \mathbf{u} = \tfrac{1}{2}\nabla u^2 - \mathbf{u} \times (\nabla \times \mathbf{u})$$

we obtain, in the steady state,

$$\frac{u^2}{2} + \frac{\gamma}{\gamma-1}\frac{p}{\rho} = \text{constant} \tag{6.15}$$

which is Bernoulli's equation for adiabatic flow. Substituting (6.14) in (6.15) and recalling that the sonic Mach number M_s is $u(\rho/\gamma p)^{\frac{1}{2}}$, we can relate the stagnation pressure to the pressure at any point upstream along the same streamline:

$$\frac{p_s}{p} = \left(1 + \frac{\gamma-1}{2}M^2\right)^{\gamma/(\gamma-1)} \tag{6.16}$$

From the Rankine-Hugoniot equation discussed in Chapter 5,

$$\frac{p}{p_\infty} = 1 + \frac{2\gamma}{\gamma+1}(M_\infty^2 - 1) \tag{6.17}$$

and

$$M^2 = \frac{1 + (\gamma-1)M_\infty^2}{2\gamma M_\infty^2 - \gamma - 1} \tag{6.18}$$

where M_∞ and p_∞ are measured upstream of the shock, and M and p downstream. Combining (6.16), (6.17), and (6.18), we obtain

$$K = \frac{p_s}{\rho_\infty u_\infty^2} = \left(\frac{\gamma+1}{2}\right)^{(\gamma+1)/(\gamma-1)} \frac{1}{\gamma[\gamma - (\gamma-1)/2M_\infty^2]^{1/(\gamma-1)}} \tag{6.19}$$

For $\gamma = \tfrac{5}{3}$ and $M_\infty = \infty$, we obtain $K = 0.881$; for $M_\infty = 4.5$, $K = 0.897$. For $\gamma = 2$, which corresponds to a gas with two degrees of freedom, and $M_\infty = \infty$, then $K = 0.844$. Because the effective polytropic index of the magnetosphere is found empirically to be about $\tfrac{5}{3}$, and the typical solar-wind Mach number is around 6 at 1 AU, the pressure exerted by the solar wind on the nose of the magnetopause is about 11 percent less than that of the momentum flux or dynamic pressure of the solar wind.

6.3.2 The Pressure Exerted by the Magnetosphere on the Magnetosheath Plasma

If the magnetosphere were a vacuum, then the magnetic field just inside the magnetopause would provide the total pressure to stand off the magnetosheath plasma. In practice there is a variable contribution from the magnetosphere plasma. At Jupiter, the magnetospheric plasma contributes so much pressure to the force balance that one can almost neglect the magnetic pressure, but we shall defer discussion of the implications of this to Chapter 15. At the earth, and all the other "intrinsic" magnetospheres, the plasma pressure is thought to be less than that of the magnetic field near the nose of the magnetosphere. Thus, a solution that ignores the plasma contribution to the pressure can be instructive.

To calculate the pressure at the boundary, we must determine how much the magnetic field at the nose of the magnetopause is compressed in the interaction. An image dipole model would cause a doubling of the field, but a more realistically shaped magnetosphere would produce a larger effect. For example, in the case of a dipole enclosed in a spherical superconductor, the equatorial field is tripled. Moreover, other current systems, such as the ring current, the tail current, and the field-aligned or Birkeland current, will also contribute to the magnetic field at the magnetopause. Thus, the magnetopause position depends on the "state of the magnetosphere" as well as the dynamic pressure of the solar wind. Despite this dependence on the state of the magnetosphere, it is instructive to proceed, leaving the compression factor as a free parameter, a, to be determined empirically. Balancing the solar-wind pressure against our assumed magnetospheric pressure, we obtain

$$K\rho_\infty u_\infty^2 = \frac{(aB_0)^2}{2\mu_0 L_{mp}^6} \tag{6.20}$$

where B_0 is the equatorial surface field of the planet, and L_{mp} is the distance to the magnetopause in planetary radii. For the earth, we obtain

$$L_{mp}(R_E) = 8.53 a^{\frac{1}{3}} (K\rho_\infty u_\infty^2)^{-\frac{1}{6}} \tag{6.21}$$

where the term in parentheses on the right is expressed in nanopascals. For a typical solar-wind momentum flux of 2.6 nPa (see Table 4.2), the magnetopause is observed to lie at about 10 R_E. Solving equation (6.21) for a, we obtain 2.44. Thus, the subsolar standoff pressure of the magnetosphere is approximately 2.44 times that expected for the vacuum dipole field at that distance. This value falls, as expected, between the value for the infinite planar magnetopause and that for the spherical magnetopause and is in accord with the compression factor developed for a vacuum magnetosphere with the observed boundary shape. In more practical units, we can rewrite (6.21) in terms of the measured solar-wind number density and velocity as

$$L_{mp}(R_E) = 107.4(n_{sw}u_{sw}^2)^{-\frac{1}{6}}$$

where n_{sw} is the proton-number density (adjusted for helium content, noting the mass factor of 4), and u_{sw} is the solar-wind bulk velocity in kilometers per second.

6.4 SHAPE OF THE MAGNETOSPHERIC CAVITY

Thus far we have concerned ourselves only with the nose of the magnetopause. The magnetospheric cavity is a three-dimensional object with a specific shape. It has a rather blunt nose region and an extended tail. One of the problems in calculating the shape is determining the pressure distribution in the magnetosheath. If there were no tangential stress, that is, if there were no viscosity or drag on the boundary, but only forces along the normal to the surface, one might expect a teardrop shape, because the magnetic field weakens with distance behind the earth, while the isotropic plasma pressure reaches a finite value downstream of the magnetosphere. The precise shape, even in this approximation, is difficult to determine because of the nonlinear nature of the postshock region. In 1960, David Beard developed a technique to determine the shape of the dayside magnetosphere. Figure 6.2 shows the results obtained using his approach for six slices through the magnetosphere. The outermost one is the equatorial plane. The innermost one is the noon–midnight meridian. Note that the magnetopause is closer to the earth over the poles than in the equatorial region, even though the magnetic-field strength over the polar regions at low altitudes is higher than that over the equator. Although that solution treated normal stresses correctly, it was unable to treat tangential stresses. As we shall see later, these stresses are nontrivial. Also, that approach ignored the contribution to the magnetic field of currents flowing within the magnetosphere.

The magnetosphere to which this treatment might best apply is that of the planet Mercury. Mercury fills up much of its magnetosphere, and hence it cannot have a significant ring current and radiation belt. The weakness of Mercury's atmosphere, and hence ionosphere, implies that many of the current systems that flow on earth, coupling the ionosphere to the magnetosphere, will not flow on Mercury. Beard and colleagues have applied their treatment to Mercury, solving for the magnetic-field line configuration and comparing with the *Mariner 10* observations. This model is shown in Figure 1.20 in Chapter 1.

6.4.1 The Effect of Tangential Stress

Tangential stress, or drag, transfers momentum to the magnetospheric plasma and causes it to flow tailward. This stress can be transferred by

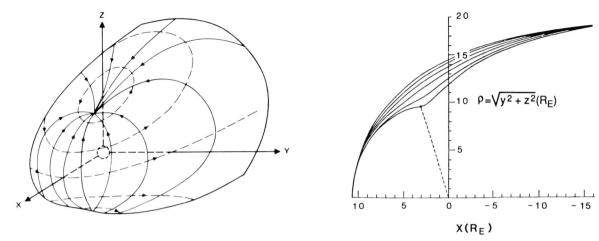

FIG. 6.2. Left: The magnetopause surface. Solid lines are magnetic-field lines. Dashed lines show the direction of currents on the boundary. (From Midgely and Davis, 1963.) Right: The shape of the magnetopause. The top curve shows the equatorial shape, and the other curves show the cross sections at higher latitudes. The bottom curve shows the shape in the noon–midnight meridian. (From Mead and Beard, 1964.)

diffusion of particles from the magnetosheath into the magnetosphere, by boundary-wave processes that cause motion in the magnetosphere, by the finite gyroradius of magnetosheath particles, and by the process known as "reconnection," which is discussed in detail in Chapter 9. In this latter process, interplanetary magnetic-field lines link with planetary ones when the interplanetary and planetary fields lie in opposite directions. This process is thought to provide the greatest tangential stress on the magnetopause, although it is believed to operate only half the time.

These processes act to transfer magnetic flux and plasma from the dayside to the nightside magnetosphere and hence have the potential to alter the shape of the magnetosphere. Because currently these effects are ill-defined, either observationally or theoretically, we shall mention only one such study, a two-dimensional treatment by Unti and Atkinson (1968) that used the amount of magnetic flux in the tail as a parameter in the model. Figure 6.3 shows how the shape of the magnetopause and the position of the inner edge of the tail neutral sheet (labeled 1–5 in the night magnetosphere) varied as the tail flux was altered while the solar wind was held constant.

6.4.2 The Limiting Width of the Tail

Eventually the tail expands to an asymptotic diameter, and there are no normal stresses associated with the solar-wind dynamic pressure, but only the normal stresses of the solar-wind plasma and magnetic field as expressed in the magnetosheath flow. If we assume that one tail lobe is a semicircle, then the magnetic flux in that tail lobe is

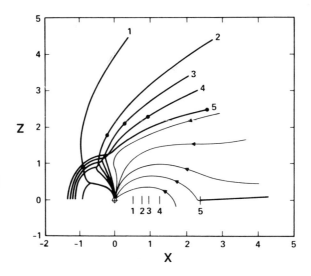

FIG. 6.3. Two-dimensional boundary shapes for a model magnetosphere in which the tail was represented by an equatorial current sheet, with return boundary currents. The five shapes correspond to different amounts of magnetic flux in the tail (1 contains the most flux, and 5 the least) for constant solar-wind conditions. (From Unti and Atkinson, 1968.)

$$F_T = \frac{\pi R_T^2}{2} B_T \tag{6.22}$$

where R_T is the lobe radius, and B_T the field strength. Because the tail magnetic pressure is in balance with the solar-wind thermal and magnetic pressure, the asymptotic radius of the tail is given by

$$R_T^2 = 2F_T/(\pi^2 \mu_0 p_{sw})^{\frac{1}{2}} \tag{6.23}$$

where p_{sw} includes both the thermal and magnetic pressures of the solar wind. The expansion of the tail to this asymptotic width is discussed in more detail in Chapter 9.

6.5 SELF-CONSISTENT MODELS

An alternate approach to modeling the magnetosphere is to attempt to solve the equations of magnetohydrodynamics (MHD) on a computer. Limitations exist in the accuracy with which the magnetosphere is represented in a computer model or in a laboratory-plasma model. A similar system can be established, but we are not certain how comparable are the computer solution and the magnetospheric process. Thus, computer models are continually being refined and continually compared with data. Computers have one very important advantage at the present time that is nontrivial. Computer power is increasing at a rapid rate, so that computer models can become more and more precise or sophisticated with time with little increase in cost.

In the MHD approach to modeling the magnetosphere, we solve the equations of a magnetized fluid and Maxwell's equations. The MHD equations were introduced in Chapter 2. The forms of the equations frequently solved in the magnetospheric models are as follows:

6.5 SELF-CONSISTENT MODELS

Continuity Equation

$$\partial \rho / \partial t = - \nabla \cdot (\mathbf{u}\rho) \tag{6.24}$$

Momentum Equation

$$\partial \mathbf{u}/\partial t = -(\mathbf{u} \cdot \nabla)\mathbf{u} - (\nabla p)/\rho + (\mathbf{J} \times \mathbf{B})/\rho \tag{6.25}$$

Pressure Equation

$$\partial p/\partial t = -(\mathbf{u} \cdot \nabla)p - \gamma p \nabla \cdot \mathbf{u} \tag{6.26}$$

Faraday's Law

$$\partial \mathbf{B}/\partial t = \nabla \times (\mathbf{u} \times \mathbf{B}) + \eta \nabla^2 \mathbf{B} \tag{6.27}$$

Ampère's law

$$\mathbf{J} = \nabla \times (\mathbf{B} - \mathbf{B}_d) \tag{6.28}$$

where ρ is the plasma density, $\mathbf{u}$ is the flow velocity, p is the plasma pressure, $\mathbf{B}$ is the magnetic field, and $\mathbf{B}_d$ is the earth's internal field. The polytropic index γ is taken to be $\frac{5}{3}$, and η is a magnetic diffusivity not found in the ideal MHD equations. Also note that in the model calculation the equations have been normalized so that distances are in earth radii, the magnetic field is in terms of the magnetic field of the earth at the equator, the densities are normalized to the value in the ionosphere, velocities are normalized to the Alfvén velocity at one earth radius, and times are given in terms of the time required to move one earth radius at this velocity. In a typical calculation, called a simulation, these equations are solved numerically in a three-dimensional box (Figure 6.4) as an initial-value problem. The boundary conditions are essentially those at infinity (values at the left-hand side are set to solar-wind values, and the other boundaries are assumed to be at infinity). The earth is assumed to be a conducting sphere. A major advantage of this approach is that the physical boundaries of the problem (i.e., the magnetopause and the bow shock) are calculated naturally without limiting assumptions. The

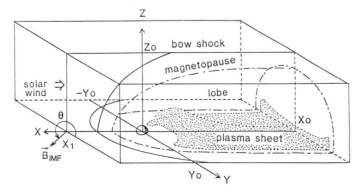

FIG. 6.4. Box used for global MHD simulation. The solar wind enters from the left and forms a magnetospheric cavity like that in the cartoon, within the box. (From Ogino et al., 1986.)

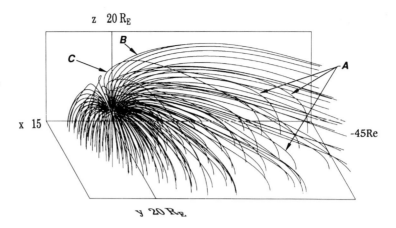

FIG. 6.5. Magnetic-field lines from a global MHD simulation for the case when the interplanetary magnetic field (IMF) was northward. Closed field lines (labeled A) have both ends attached to the earth. Open field lines (labeled B) have one end attached to the earth and the other exiting the simulation box before reaching the southern hemisphere. Reconnection between the IMF and open tail field lines is occurring just equatorward of point C. The field line just to the left of point C is newly closed by the reconnection process in which IMF field lines and tail-lobe field lines become newly closed field lines and IMF field lines. Lobe field line C is about to undergo reconnection. (From Ogino et al., 1992.)

limitation of the calculations is that they require numerical dissipation, which is not directly related to the physics of the solar-wind–magnetosphere interaction.

Figure 6.5 shows the magnetic-field lines from a global MHD solution to the problem of the solar wind with a northward interplanetary magnetic field interacting with the earth's magnetic field. In this plot the field lines labeled A close within the simulation box (both ends of the field go into the earth), while the field lines labeled B extend to the back of the simulation box. If the simulation box extended to infinity, some of the B field lines would also close. Interplanetary field lines can also reconnect with open field lines at position C (see also Chapter 9). The results on the dayside are very much like the results we saw earlier from the boundary calculations. The magnetosphere is bullet-shaped, and the magnetic null points occur at about 72° latitude.

MHD models such as that shown here are very useful for describing the magnetic configuration and the flow around it. However, they have significant limitations. Because the plasma is assumed to be a magnetized fluid, the models do not simulate kinetic effects and the plasma instabilities that could arise. Often parameters must be chosen for numerical stability rather than on the basis of physical constraints. Moreover, the spatial resolution often is too coarse to describe the phenomena of interest accurately. Hence, other techniques have been developed to address these problems. The hybrid technique treats ions as particles and electrons as a massless fluid in order to include some of the kinetic effects in a plasma. At the other extreme, gas-dynamic simulations have been used for situations in which the magnetic forces can be neglected, so that increased spatial resolution and faster computational speed can be obtained. Gas-dynamic simulations have been employed since the mid-1960s in studying the solar-wind interaction with the magnetosphere and have been very influential in guiding our understanding of this problem.

6.6 FLOW AROUND THE MAGNETOSPHERE

Perhaps the most productive application of the gas-dynamic model of the solar-wind interaction has been its use in determining the properties of the flow around the magnetosphere (i.e. the properties of the magnetosheath) (Spreiter et al., 1966). This model ignores all magnetic forces on the flow. It calculates magnetic-field lines, nevertheless, by convecting the field lines along with the fluid. Thus, this model is often called the convected-field gas-dynamic model. The results of the simulations depend on the shape of the obstacle, the Mach number of the flow, and the polytropic index γ. For computational-speed considerations, it is usual to solve for the flow parameters with a shape that is cylindrically symmetric about the sun–planet line and then convect the three-dimensional magnetic field through the flow field by using these flow parameters. The Mach number of the solar-wind flow that is appropriate for use in this gas-dynamic analogue is that of the fast magnetosonic wave, because the bow shock is a fast magnetosonic shock. The fact that the velocity of the magnetosonic wave is anisotropic is a complication, but one that is no more serious than the choice of an axisymmetric obstacle to the flow. For most situations found at 1 AU, the variation of the magnetosonic velocity around the shock front is a few percent. The polytropic index is usually chosen to be $\frac{5}{3}$, which is appropriate for a gas with three degrees of freedom.

Figure 6.6 shows the streamlines for a Mach-8 flow with $\gamma = \frac{5}{3}$ past a magnetosphere. The streamlines show the direction of the flow. Figures 6.7, 6.8, and 6.9 show lines of constant density, velocity, temperature, and mass flux normalized by the upstream solar-wind value. The density ratio immediately behind the shock is close to the maximum permitted by the Rankine-Hugoniot relations discussed in Chapter 5. This limiting value, $(\gamma+1)/(\gamma-1)$, equals 4 here. At the subsolar magnetopause, this ratio has increased to 4.23. In the actual magnetosheath, magnetic effects act to limit this density increase. As the gas expands around the obstacle, it decreases below the upstream value near the magnetosphere, but it is always compressed in the region just behind the shock.

The temperature contours are the same as the velocity contours in Figure 6.8, because the temperature ratio is related to the velocity ratio by the expression

$$\frac{T}{T_\infty} = 1 + \frac{(\gamma-1)M_\infty^2}{2}\left(1 - \frac{u^2}{u_\infty^2}\right) \tag{6.29}$$

which is obtained by integrating the energy equation (Spreiter et al., 1966). We note that the temperature rise in the magnetosheath is substantial. If the solar-wind temperature were 50,000K in the solar wind, it would be over 1,000,000K throughout much of the dayside mag-

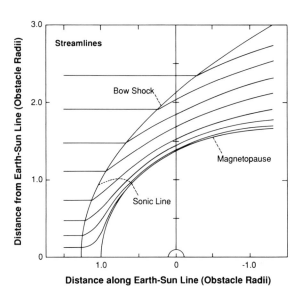
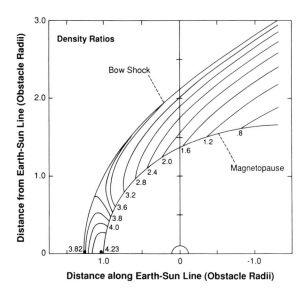

(*Left*) **FIG. 6.6.** Streamlines for supersonic flow past the magnetosphere for a Mach number of 8 and a polytropic index of $\frac{5}{3}$. The flow-line spacing has been chosen for convenience in illustration of the magnetosheath flow and is not an indication of mass flux. (Adapted from Spreiter et al., 1966.)

(*Right*) **FIG. 6.7.** Density contours for supersonic flow past the magnetosphere for a Mach number of 8 and a polytropic index of $\frac{5}{3}$. (Adapted from Spreiter et al., 1966.)

netosheath. We note that because the gas-dynamic temperature is the analogue of the sum of the electron and ion temperatures, and because the electron temperature in the solar wind often is more than twice that of the ions, but changes only slightly across the bow shock, the actual change in ion temperature in the magnetosheath can be many times that shown in Figure 6.9. In contrast to the case for the solar wind, the magnetosheath ion-to-electron temperature ratio is rather constant, averaging about 6 everywhere.

The mass-flux contours shown in Figure 6.9 were obtained by multiplying the velocity along the streamline by the mass density shown in Figures 6.7 and 6.8. The mass flux is important because in a sense it determines the location of the bow shock. All mass flux crossing the bow shock must flow around the obstacle in this model. The bow shock must assume a position that allows this flow. In particular, the subsolar shock location can be determined as follows. The ratio of (1) the distance from the magnetopause to the shock to (2) the distance from the center of the earth to the magnetopause has been found from experiments to be 1.1 times the density jump across the shock for a wide variety of conditions for the given shape of the magnetosphere. In this gas-dynamic solution, the density jump is solely a function of Mach number and γ and is equal to $[(\gamma-1)M^2+2]/(\gamma+1)M^2$ (Spreiter et al., 1966). For a Mach number of 8 and γ of $\frac{5}{3}$, the bow-shock nose should be 29 percent farther out than the magnetopause.

6.6 FLOW AROUND THE MAGNETOSPHERE

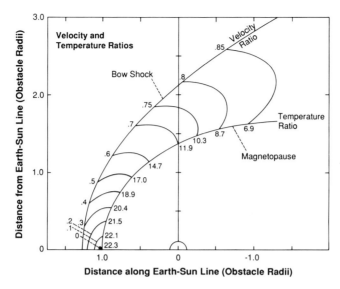

FIG. 6.8. Velocity and temperature contours for supersonic flow past the magnetosphere for a Mach number of 8 and a polytropic index of $\frac{5}{3}$. (Adapted from Spreiter et al., 1966.)

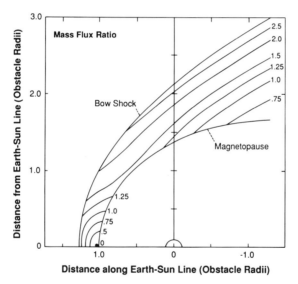

FIG. 6.9. Mass-flux contours for supersonic flows past the magnetosphere for a Mach number of 8 and a polytropic index of $\frac{5}{3}$. The mass flux is measured along the streamlines and is numerically equal to the product of the density ratio and the velocity ratios shown in Figures 6.7 and 6.8. (Adapted from Spreiter et al., 1966.)

Figure 6.10 shows the magnetic-field configuration obtained by carrying field lines in the gas-dynamic flow field, with no magnetic-pressure effects considered. This might be expected to be the case over most of the simulation volume for high-Mach-number conditions. Two situations are shown: one with the magnetic field perpendicular to the flow, and one with it at 45° to the flow. In both situations the magnetic field is seen to pile up in the subsolar region. We do not expect this to happen in the actual magnetosheath, because the associated magnetic-pressure gradients should alter the flow pattern there. At the stagnation point in the flow, the sums of the magnetic and thermal pressures of the magnetosheath and magnetosphere should be in balance.

6.7 CONCLUDING REMARKS

As discussed in this chapter, the solar-wind interaction with a magnetosphere is a complex process. The magnetosphere acts as a nearly impenetrable obstacle about which the solar wind must flow. The solar wind is supersonic, and so a shock must form and stand in front of the obstacle. We do not yet know how to solve for the properties of the flow throughout the magnetosheath, except under certain approximations. The simplest one is that the flow is a gas-dynamic one, in which magnetic-pressure forces can be ignored. This provides a simple and useful description of the magnetosheath that serves many purposes. However, it has many limitations. In particular, it predicts a density buildup and a magnetic-field buildup at the magnetopause that are not observed. We also expect the real magnetosheath to include at least weak Alfvén and slow-mode standing waves, in addition to the fast shock, because such waves are required, in general, to cause an arbitrary perturbation in the plasma. It is unlikely that the perturbation in the flow required to divert the solar wind around the magnetosphere could be accomplished with solely a compressional wave.

Another limitation of these techniques is their assumption of isotropic pressure. We expect anisotropies in the magnetosheath pressure if the pressure of the magnetic field is a significant fraction of the thermal pressure. These anisotropies will lead to interesting plasma instabilities, which in turn will attempt to restore the isotropy of the plasma. Finally, the MHD approach and, of course, the gas-dynamic approach do not simulate small-scale features of the gyroradius or smaller size. At the bow shock, these have been found to be of critical importance in provid-

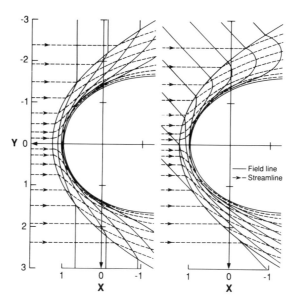

FIG. 6.10. Magnetic field (solid lines) in a plane containing upstream solar-wind velocity, magnetic field, and center of the planet for a Mach-number-8 gas-dynamic flow and $\gamma = \tfrac{5}{3}$. Two upstream magnetic fields are illustrated, a perpendicular field and a 45° field. Streamlines are shown by dashed lines. (Adapted from Spreiter et al., 1966.)

ing the dissipation required by the Rankine-Hugoniot equations. These processes may have equal import at the magnetopause.

ADDITIONAL READING

Lyons, L. R., and D. J. Williams. 1984. *Quantitative Aspects of Magnetospheric Physics*. Dordrecht: Reidel.
Siscoe, G. L., 1987. The magnetospheric boundary. In *Physics of Space Plasmas (1987)*, ed. T. Chang, G. B. Crew, and J. R. Jasperse (p. 3). Cambridge, MA: Scientific Publishers.
Spreiter, J. R., A. L. Summers, and A. Y. Alksne. 1966. Hydromagnetic flow around the magnetosphere. *Planet. Space Sci.* 14:223–53.

PROBLEMS

6.1. The magnetic field due to a dipole moment M along the z-axis and centered at $(0, 0, 0)$ is

$B_x = 3xzM/r^5$
$B_y = 3yzM/r^5$
$B_z = (3z^2 - r^2)M/r^5$

If the magnetic moment of the earth is $31{,}000$ nT $\cdot R_E^3$ and is aligned along the z-axis, calculate the magnetic field at the following points if an infinite flat plane of solar-wind plasma forms a magnetopause at $x = 10R_E$, as postulated by Chapman and Ferraro:

(a) at radial distances of 2, 4, 6, 8, and $10R_E$ along the earth–sun line
(b) at just inside the magnetopause at

$(10, 0, 0)R_E$
$(10, 0, 2)$
$(10, 0, 4)$
$(10, 0, 6)$
$(10, 0, 8)$
$(10, 2, 0)$
$(10, 4, 0)$
$(10, 6, 0)$

(c) If the neutral point is defined as the place on the magnetopause where the magnetic field is normal to the magnetopause, locate the neutral point.
(d) Graph these results in a manner that suitably illustrates the variation observed.

6.2. Use the image dipole model and a magnetopause distance of $10R_E$ to calculate the increase in surface magnetic-field strength on the equator as a function of local time.

6.3. A magnetic-field line crosses the magnetic equator of the earth at $4R_E$. Assuming that the earth's field is dipolar, where does this magnetic-field line intersect the surface of the earth?

6.4. Spacecraft Space Physics Explorer returns to its South Pole base from its repair mission at synchronous orbit and along the way makes the following measurements:

Position (Geomagnetic)	Field (Geomagnetic)
$(6.6, 0, 0)R_E$	$(0, 0, 111)$ nT
$(0, 2, 0)R_E$	$(0, 0, 4005)$ nT
$(0, 0, -1)R_E$	$(0, 0, -64233)$ nT

What is the magnetic moment of the earth (in $\text{nT} \cdot R_E^3$), according to these measurements? Do you consider them to be consistent? How could you obtain a best-fit solution that is based on all three observations instead of averaging the solutions for the individual observations?

6.5. If the magnetic moment of Mercury is 3×10^{12} T·m³, what is the subsolar distance to its magnetopause in planetary radii for a solar-wind velocity of 500 km·s⁻¹ and a density of 20 cm⁻³. The radius of Mercury is 2,440 km. What is the magnitude of the field at the subsolar point? Assume that the shape of the Mercury magnetopause is the same as that empirically found for the earth.

6.6. If the polar cap is defined by the open field lines that enter the tail lobe, and if it can be approximated by a circle of radius 15° centered on the magnetic-dipole axis, how much flux is there in a lobe of the geomagnetic tail? If none of this flux crosses the neutral or current sheet in the tail, if no solar-wind plasma enters the tail field lines, and if the strength of the interplanetary magnetic field (IMF) is 6 nT and $\beta = 1$, what is the radius of the distant tail lobe?

7 IONOSPHERES

J. G. Luhmann

7.1 INTRODUCTION

THE INGREDIENTS FOR a planetary ionosphere are simple: The only requirements are a neutral atmosphere and a source of ionization for the gases in that atmosphere. Sources of ionization include photons and energetic-particle "precipitation." The process involving the former is referred to as *photoionization,* and the latter is often labeled *impact ionization.* The photons come primarily from the sun. Ionizing particles can come from the galaxy (cosmic rays), the sun, the magnetosphere, or from the ionosphere itself if a process for local ion or electron acceleration is operative. Precipitating energetic electrons produce additional ionizing photons within the atmosphere by a process known as *bremsstrahlung,* or *braking radiation.* The only requirement on the ionizing photons and particles is that their energies ($h\nu$ in the case of photons, and kinetic energy in the case of particles) exceed the ionization potential or binding energy of a neutral-atmosphere atomic or molecular electron. In nature, atmospheric ionization usually is attributable to a mixture of these various sources, but one often dominates. Solar photons in the "extreme" ultraviolet (EUV) and ultraviolet (UV) wavelength range of approximately 10 nm to 100 nm typically produce at least the dayside ionospheres of most planets.

7.1.1 The Underlying Atmosphere

The density n_n of a constituent of the upper (neutral) atmosphere usually obeys a *hydrostatic equation:*

$$n_n m_n g = \frac{dp}{dh} = -\frac{d}{dh}(n_n k T_n) \tag{7.1}$$

which expresses a balance between the vertical gravitational force and the thermal-pressure-gradient force on the atmospheric gas. Here, m_n is the molecular or atomic mass, g is the acceleration due to gravity, h is an altitude variable, and p is the thermal pressure $n_n k T_n$ (k = Boltzmann's constant, T_n = temperature) of the neutral gas under consideration. If T_n is assumed independent of h, this equation has the exponential solution

$$n_n = n_0 \exp \frac{-(h-h_0)}{H_n} \tag{7.2}$$

where $H_n = kT_n/m_n g$ defines the *scale height* of the gas, and n_0 is the density at the reference altitude h_0. Note that the scale-height dependence on particle mass is such that the lightest molecules and atoms have the largest scale heights. (Most planetary atmospheres are dominated at high altitude by hydrogen and helium.) Of course, T_n may depend on h, so that this simple exponential distribution will not *always* provide an accurate description.

7.2 ION PRODUCTION

7.2.1 Photoionization

To "model" an ionosphere produced in a given neutral atmosphere, one must first calculate the altitude profile of the rate of ion production Q. For photoionization, this entails a consideration of the *radiative transfer* of photons through the neutral gas. When handled rigorously, this is a very complex problem, because it requires detailed knowledge of all of the photon absorption cross sections of the atmospheric constituents and some method of keeping track of absorption events that cause excitation of bound electrons as well as those that cause removal of *photoelectrons*. Fortunately, some simplifying assumptions can be made that together will allow an analytical approach to ionosphere modeling known as *Chapman theory*.

Before delving into Chapman theory, it is important to appreciate that the altitude profile of ion production will have a peak at some altitude, because the rate of ionization depends on both the neutral density (which decreases with height) and the incoming solar-radiation intensity (which increases with height). In Chapman theory, the goal is to describe ion production as a function of height for the simple case in which the details of photon absorption are hidden in a radiation-absorption cross section σ and in which ion production is assumed to depend only on the amount of radiative energy absorbed. We define the following variables:

n_n = density of neutrals (per cubic meter)
h = height
I = intensity of radiation (energy flux, electron volts per square meter per second)
σ = photon-absorption cross section (square meters)
q = rate of ion production (photoionization rate, electrons per cubic meter per second)
s = line-of-sight path length
χ = zenith angle

C = number of electrons produced in the absorber per unit energy absorbed (electrons per electron volt)

The path length s and zenith angle χ are illustrated in Figure 7.1. The atmosphere is presumed to be exponential, planar, and horizontally stratified (assumptions that are idealizations of the real world, where atmospheres are only approximately exponential and curved, and the scale height H_n depends on χ because of the effects of global circulation and chemistry). As radiation is absorbed, its intensity decreases as

$$-\frac{dI}{ds} = \sigma n_n I \tag{7.3}$$

Because the rate of ion production should be proportional to the rate at which radiation is absorbed, we can write

$$Q = -C\frac{dI}{ds} = C\sigma n_n I \tag{7.4}$$

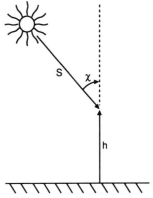

FIG. 7.1. Illustration showing the line-of-sight path length s, the solar zenith angle χ, and the altitude h.

where C is the constant of proportionality (≈ 1 ion pair per 35 eV in air). The production rate Q reaches a peak (along s) when

$$C\sigma\left(I\frac{dn_n}{ds} + n_n\frac{dI}{ds}\right) = 0$$

or

$$\frac{dQ}{ds} = 0$$

But s is related to h by $ds = -dh \sec \chi$ (Figure 7.1), and because

$$\frac{1}{n_n}\frac{dn_n}{ds} = -\frac{1}{n_n}\frac{dn_n}{dh}\cos\chi = \frac{\cos\chi}{H_n}$$

for the peak or maximum of production (subscript m), the foregoing equations give

$$\sigma H_n n_m \sec \chi = 1 \tag{7.5a}$$

or

$$\sigma N_{nm} = 1 \tag{7.5b}$$

where N_{nm} is the integrated density $N_{nm} = \int_\infty^{s_m} n_n \, ds$ along the line of sight up to the position of the peak s_m. A useful term is the *optical depth* τ, which describes the attenuation of the ionizing radiation. The optical depth arises naturally in the expression for intensity at position s along the line of sight relative to the intensity at infinity. From (7.3),

$$\frac{dI}{I} = d \ln I = -\sigma n_n \, ds$$

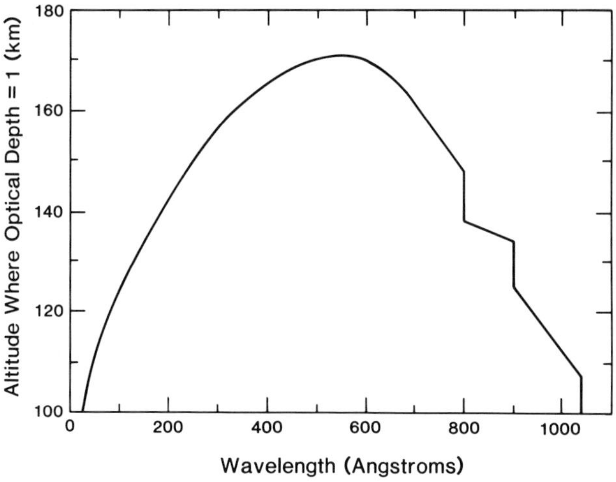

FIG. 7.2. Photon wavelength versus the altitude in the earth's atmosphere where the optical depth is equal to 1 (effectively, the depth of penetration of a photon). (Adapted from Rishbeth and Garriott, 1969.)

Integrating,

$$\ln\left(\frac{I(s)}{I(\infty)}\right) = -\sigma \int_0^s n_n \, ds = -\sigma N_{ns}$$

or

$$I(s) = I(\infty)\exp(-\sigma N_{ns}) = I(\infty)\exp(-\tau)$$

At the peak, $\sigma N_{ns} = \sigma N_{nm} = 1$; so the peak is the altitude where the optical depth is unity. Both the radiation intensity at the top of the atmosphere [which we shall call $I(\infty)$ instead of $I(0)$, because the former is standard nomenclature] and σ vary for different wavelengths, but it is σ that determines τ. The altitudes at which photons of various wavelengths reach unit optical depth in the earth's atmosphere are illustrated in Figure 7.2. Now we return to the analysis of the production rate Q.

The peak production rate is

$$Q_m = C\sigma n_m I_m = C\sigma(\sigma H_n \sec \chi)^{-1}(I(\infty)\exp(-1)) = \frac{CI(\infty)\cos \chi}{H_n \exp(1)} \tag{7.6}$$

Given the neutral-gas altitude profile $n_n = n_0 \exp(-h-h_0)/H_n)$, we can determine the height of the production peak h_m by writing

$$\sigma H_n n_m \sec \chi = 1 = \sigma H_n n_0 \exp\left(\frac{-(h_m - h_0)}{H_n}\right) \sec \chi \tag{7.7}$$

and solving for h_m. Similarly, we can determine the dependence of the radiation intensity I on h by using the earlier expression for $I(s)$ and noting that N_{ns} is the integrated density along the line of sight:

$$I(h) = I(\infty)\exp\left[-\sigma n_0 H_n \sec \chi \, \exp\left(\frac{-(h-h_0)}{H_n}\right)\right] \tag{7.8}$$

The dependence of the production rate Q on h is then given by

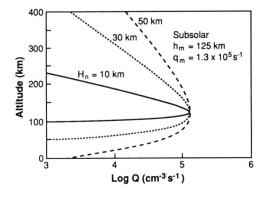

FIG. 7.3. Chapman production function for $h_m = 125$ km, $\chi = 0°$, and $q_m = 1.3 \times 10^5$ cm^{-3} s^{-1} for several values of the neutral scale height H_n.

$$Q = C\sigma n_n I = C\sigma n_0 I(\infty) \exp\left[\frac{-(h-h_0)}{H_n} - n_0 \sigma H_n \sec \chi \exp\left(\frac{-(h-h_0)}{H_n}\right)\right] \quad (7.9)$$

With $CI(\infty) = Q_m \exp(1) H / \cos \chi$, as before, we finally obtain

$$Q = Q_m \exp\left[1 + \frac{(h_m - h)}{H_n} - \exp\left(\frac{(h_m - h)}{H_n}\right)\right] \quad (7.10a)$$

This can be simplified by defining $y = (h - h_m)/H_n$, whence

$$Q = Q_m \exp[1 - y - \exp(-y)] \quad (7.10b)$$

which is the *Chapman production function*. Note that far above the peak ($y \geq 2$), a good approximation is

$$Q \propto \exp(-y) \quad (7.10c)$$

which says that because the radiation intensity is practically constant at high altitudes (not much absorption occurs), Q is proportional to neutral density. The rate Q is the rate of production of both ions and photoelectrons, because these usually are produced in pairs (most ions are singly ionized in this process). It is notable that in the expression for Q, the photon-absorption cross section does not explicitly appear. The properties of the absorber are contained in the constant C. The subsolar ($\chi = 0$) Chapman production function or ion production profile in atmospheres with several different scale heights is plotted in Figure 7.3.

Our derivation of the Chapman production function has assumed a flat earth in which the solar zenith angle has a fixed value. In practice, the solar zenith angle varies with position on the surface of the earth. We can reference the local production rate not just to the local production maximum but to the production maximum at the subsolar point Q_{mo} at a height h_{mo}. Doing this, we find that

$$Q = Q_{mo} \exp(1 - z - \sec \chi \exp(-z)) \quad (7.10d)$$
$$h_m = h_{mo} + \ln(\sec \chi)$$
$$Q_m = Q_{mo} \cos \chi$$

where

$$z = (h - h_{mo})/H$$

Thus, in an isothermal atmosphere, the altitude of the peak production rate increases with solar zenith angle, and the rate of ionization decreases.

7.2.2 Particle-Impact Ionization

In many situations of interest, solar photons can be assumed to compose the dominant source of atmosphere ionization. However, on occasion, energetic (energies $\geq$ 1 keV) precipitating charged particles are more important. This can occur, for example, during the night at high magnetic latitudes in the case of planets with dipolar magnetic fields, or on satellites with atmospheres submerged in planetary magnetospheres, such as Io and Titan. The altitude profiles of ion production from particle impacts are derived from considerations different from those for photoionization, except that the reasoning related to the production of a *peak* is the same. In this case it is the particle energy flux, rather than the photon flux, that is attenuated with decreasing altitude as the atmosphere density increases. The transport of charged particles and the loss of their energy to the atmosphere are unlike those processes for photons, because a particle gradually loses its energy in the excitation or removal of bound electrons from many particles by Coulomb collisions as it travels, whereas a photon is absorbed in a single event. To complicate matters further, a *primary* particle can produce *secondary electrons* that are energetic enough themselves to cause impact ionization, and both primary and secondary electrons can lose additional energy by radiating *bremsstrahlung* photons when their paths are deflected by the Coulomb collisions. In the case of ions, charge exchange may occur, in which case the energetic ion becomes an energetic neutral that distributes its energy by collisional processes other than Coulomb collisions. To deal with this problem, investigators may resort to such rigorous alternatives as Monte Carlo calculations. However, as for photoionization, there are some simpler ways to estimate the effects of energetic-particle absorption.

One simplified approach involves the use of an empirically determined function $R(\xi_0)$ called the *range–energy relation*. The range–energy relation gives the depth of penetration in a particular medium as a function of the incident-particle energy ξ_0. Because the important quantity is the amount of matter traversed by the particle, rather than the particle path length (which depends on the density distribution of the matter), the range typically is expressed in units of grams per square centimeter, instead of centimeters or meters. The relationship between this equivalent distance x and the matter density n_n along the path of the particle s is given by

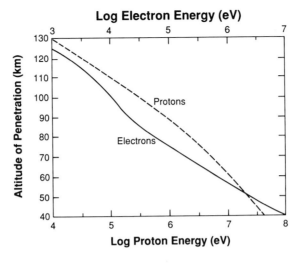

FIG. 7.4. Penetration altitudes for electrons and protons in the earth's atmosphere versus their incident energies.

$$x = \int_0^\eta n_n(s)\, ds \qquad (7.11a)$$

where η is the point of interest along s. Similarly, for vertical incidence, x is related to altitude h by

$$x = \int_\eta^\infty n_n(h)\, dh \qquad (7.11b)$$

where η is, in this case, the altitude of interest. For protons in air, for example, the range–energy relation given by Rees (1989),

$$R(\xi_0) = 5.05 \times 10^{-6} \xi_0^{0.75} \; g \cdot cm^{-2} \qquad (7.12)$$

is a good approximation for the incident-proton energy range between 1 keV and 100 keV. For electrons, another expression (Rees, 1989),

$$R(\xi_0) = 4.30 \times 10^{-7} + 5.36 \times 10^{-6} \xi_0^{1.67} \; g \cdot cm^{-2} \qquad (7.13)$$

applies for 200 eV $< \xi_0 <$ 50 keV. Once the range (in $g \cdot cm^{-2}$) is known, all that we need to do to calculate the corresponding stopping altitude is to perform the foregoing integration over the atmospheric-density distribution, that is, to find η in (7.11b) for which $x = R(\xi_0)$. For example, some stopping altitudes for protons and electrons incident on the earth's atmosphere are shown in Figure 7.4. The range–energy relation tells us about the depth to which the particle penetrates, but it does not give us the altitude distribution of its energy loss, which is what we require for the ionization-rate profile. At this point, we can make the following arguments.

The range–energy relation $R(\xi_0)$ is also defined by the integral

$$R(\xi_0) = -\int_0^{\xi_0} \frac{d\xi}{d\xi/dx} \qquad (7.14)$$

where $d\xi/dx$ is the energy lost per gram per square centimeter traversed. Suppose we postulate that the depth of matter x traversed at any point in the particle's transit can be approximated by

$$x = -\int_{\xi_{loc}}^{\xi_0} \frac{d\xi}{d\xi/dx} = R(\xi_0) - R(\xi_{loc}) \tag{7.15}$$

where ξ_{loc} is the energy at x. Because $R(\xi_0)$ has the general functional form

$$R(\xi_0) = A_1 + A\xi_0^\gamma$$

where A_1 and A are constants, this equation for x then gives

$$\xi_{loc} = \left[\frac{1}{A}(A\xi_0^\gamma - x)\right]^{\frac{1}{\gamma}} \tag{7.16}$$

It follows that the energy deposited at a given X is

$$\frac{d\xi_{loc}}{dx} = -\frac{\xi_{loc}^{1-\gamma}}{A\gamma} \tag{7.17}$$

The altitude profile of $d\xi_{loc}/dx$ times the local-atmosphere mass density $\rho(h)$ gives the *energy deposition* in electron volts per meter for the particle at x. The curve of $\rho(h)\, d\xi_{loc}/dx$ versus h is then the needed profile. Some examples obtained with the foregoing expressions for $R(\xi_0)$ in air, assuming the density profile for the earth's atmosphere, are shown in Figure 7.5. Because the incident particles have a distribution of energies described by a flux spectrum $J(E_0)$ (particles·cm^{-2}·s), a total energy-deposition profile can be built up by weighting the profiles for the single particles by $J(E_0)$, giving the energy deposition in electron-volt-seconds per cubic centimeter. Finally, for a particular gas mixture, one uses an empirically determined value for the energy required to produce an ion pair. As mentioned earlier, for air, approximately 35 eV will produce an ion pair. Division of the energy-deposition profile by this constant gives the ion production profile (in ions produced per cubic centimeter per second versus altitude) for the incident-particle flux described by $J(E_0)$. The first-generation secondary-electron production profile is the same (an electron avalanche can result if the secondary electrons are themselves capable of additional ionizing impacts).

One part of the energy deposition due to particle precipitation that has been neglected in the preceding discussion is peculiar to electrons, be they primary or secondary. Because of the small mass of electrons relative to that of the target particles, electrons traveling through a gas "scatter" much more than do ions. Their direction of motion can be significantly altered by a Coulomb collision. This causes radiation, because accelerating electric charges produce electromagnetic waves. In the case of the precipitating electron energies of interest here, this

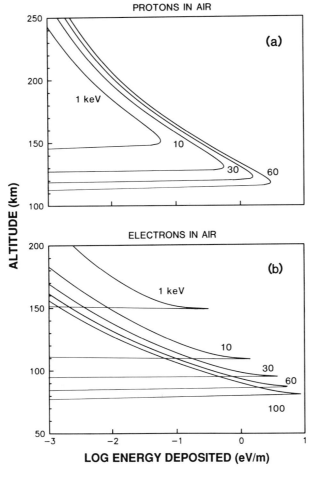

FIG. 7.5. Energy-deposition profiles for protons and electrons of various energies incident on the earth's atmosphere, as calculated from the method described in text.

bremsstrahlung or braking radiation tends to be in the x-ray range, at energies capable of additional photoionization of the gas. Thus, strictly speaking, one must deal with the complicated problem of keeping track of both the radiation losses of the precipitating electrons and the energy lost to the bound electrons, and then carry out a radiative transport calculation for the bremsstrahlung photons. The latter is not as straightforward as that for solar photons, because the bremsstrahlung photons are produced at different points within the same medium that is absorbing them. The bremsstrahlung transport calculation provides an additional ion production profile that must be added to that for the ionization loss. Fortunately, this nonsolar source of photoionization usually is unimportant, except at the lowest ionospheric altitudes, and as illustrated in Figure 7.6, the ion density it produces is well below the peak density produced by the precipitating particles. It is usually justifiable to neglect it except in special cases (such as in atmospheric-chemistry problems, which are sensitive to the local rates of ion production in the earth's middle atmosphere).

7.3 ION LOSS

Once the ion production rate is known, the next critical quantity that must be specified for an ionospheric model is the ion or electron loss rate L. Ionospheric electrons disappear by virtue of three types of *recombination*:

1. Radiative recombination $\qquad e + X^+ \rightarrow X + h\nu$
2. Dissociative recombination $\qquad e + XY^+ \rightarrow X + Y$
3. Attachment $\qquad e + Z \rightarrow Z^-$

The first two of these are most important throughout the bulk of the ionosphere. (Radiative recombination is responsible for many types of observed *airglows*. In contrast, most of the emission that one sees in an aurora occurs when atomic and molecular electrons are excited to higher energy levels by Coulomb collisions with the passing precipitating particles and then undergo radiative de-excitation.) Recombination occurs at rates that depend on the local concentrations of the ions and electrons:

$$L = \alpha n_e n_i$$

where n_e and n_i are the electron and ion densities, and α is a *recombination coefficient*. The recombination coefficient is determined by empirical and theoretical methods. For the more important atmospheric dissociative recombination reactions, such as

$$O_2^+ + e \rightarrow O + O$$

and

$$N_2^+ + e \rightarrow N + N$$

the recombination coefficients (in units of $m^{-3} \cdot s^{-1}$) are $1.6 \times 10^{-1}(300/T_e)^{0.55}$ and $1.8 \times 10^{-1}(300/T_e)^{0.39}$, respectively, where T_e is the electron

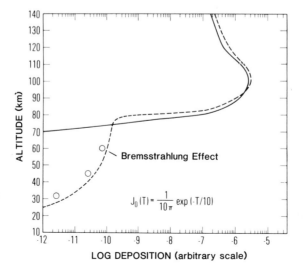

FIG. 7.6. Example of an energy-deposition profile for an incident electron spectrum of the form indicated, including the contribution from absorbed bremsstrahlung photons. (From Luhmann, 1977.)

temperature in the ionosphere. These quantities can generally be found in tables in the aeronomy literature (e.g., Banks and Kockarts, 1973; Schunk and Nagy, 1980). We see that the altitude profiles of the loss rates depend on altitude through the electron and ion densities and T_e. If there is one major ion constituent, such that $n_i \approx n_e$, the loss rate at a particular altitude is proportional to n_e^2. It is notable that α is often assumed constant, although T_e generally depends on altitude h.

7.4 DETERMINING IONOSPHERIC DENSITY FROM PRODUCTION AND LOSS RATES

Once the ion production rates and loss rates are established, we can consider the problem of finding the altitude distribution of the ionospheric electron density n_e. If the electrons and ions do not move very far from where they are produced (e.g., by virtue of a strong horizontal ambient magnetic field), we can say that n_e (and thus n_i) obeys the equilibrium-continuity or particle-conservation equation

$$\frac{\partial n_e}{\partial t} = Q - L = 0 \tag{7.18}$$

and if the loss rate is due to electron-ion collisions

$$Q = L = \alpha n_e^2$$

and hence

$$n_e = (Q/\alpha)^{\frac{1}{2}} \tag{7.19}$$

describes the spatial distribution of the electrons or ions. This particular distribution is called a *photochemical equilibrium distribution* because it involves only local photochemistry. *Chapman-layer models* of the ionosphere, for example, invoke the Chapman production function and the assumption of photochemical equilibrium. However, in many cases the electrons and ions move significantly from their points of creation before they recombine, and so we must consider a transport term in a more general continuity equation.

Because vertical transport is usually of most interest, given the relative horizontal and vertical scales of atmospheres (atmospheres often can be approximated by thin slabs), we can restrict our attention to the vertical transport attributable to the vertical velocity u_h. The equilibrium electron-density distribution must then satisfy the vertical-continuity equation

$$\frac{\partial n_e}{\partial t} = Q - L - \frac{\partial (n_e u_{eh})}{\partial h} \tag{7.20}$$

where the new influence on the electron density is the vertical-flux gradient, which describes the difference between the flux of electrons entering and leaving a given altitude. The subscript h is used to denote the vertical component of a vector. The flux gradient $\partial(n_e u_{eh})/\partial h$ can represent a source of electrons, if more arrive than depart, or a loss if the reverse is true. To determine the velocity u_{eh}, we need to invoke another equation, the momentum or force-balance equation.

Ionospheric electrons can be said to obey the following steady-state vertical-momentum equation:

$$-\frac{dp_e}{dh} - n_e m_e g - e n_e [E_h + (\mathbf{u}_e \times \mathbf{B})_h] = n_e m_e \nu_{en}(u_{eh} - u_{nh}) + n_e m_e \nu_{ei}(u_{eh} - u_{ih}) \tag{7.21}$$

where

- p_e = electron thermal pressure $n_e k T_e$
- g = gravitational acceleration
- $\mathbf{E}$ = electric field
- $\mathbf{B}$ = magnetic field
- $\mathbf{u}_e$ = electron velocity
- $\mathbf{u}_i$ = ion velocity
- ν_{en} = electron–neutral collision frequency
- ν_{ei} = electron–ion (Coulomb) collision frequency
- $\mathbf{u}_n$ = neutral velocity

From left to right, the terms represent the pressure-gradient force, the gravitational force, the force due to both externally applied and convection electric fields, and friction from collisions with other types of particles. This equation can be solved for u_{eh} in terms of all of the other variables, but it is particularly useful to obtain an expression that is independent of seldom-measured quantities like $\bar{\mathbf{E}}$. By adding the ion-momentum equation,

$$-\frac{dp_i}{dh} - n_e m_i g + q n_e [E_h + (\mathbf{u}_i \times \mathbf{B})_h] = n_e m_i \nu_{in}(u_{ih} - u_{nh}) + n_e m_i \nu_{ei}(u_{ih} - u_{eh}) \tag{7.22}$$

where p_i is ion pressure $n_e k T_i$, m_i is ion mass, q is ion charge, and ν_{in} is ion–neutral collision frequency, to the electron momentum equation, one can eliminate $\mathbf{E}$. It can further be assumed that the ions are singly charged ($q = e$) and that the ions and electrons vertically drift together at velocity u_{pl} (to maintain charge neutrality in the plasma), assumptions that eliminate the $\mathbf{u}_i \times \mathbf{B}$ and $\mathbf{u}_e \times \mathbf{B}$ terms. Finally, with the additional knowledge that $m_i \gg m_e$ and $m_i \nu_{in} \gg m_e \nu_{en}$, so that certain other terms can be neglected, we obtain

$$u_{pl} - u_{nh} \approx -\frac{1}{n_e m_i \nu_{in}} \left[\frac{d}{dh}(p_i + p_e) + n_e m_i g \right] \tag{7.23}$$

Furthermore, if the temperatures are all assumed to be independent of h, and the vertical neutral velocity u_{nh} is zero, the foregoing equation can be written in the form of a *diffusion equation* for n_e:

$$n_e u_{pl} = -D\left[\frac{dn_e}{dh} + \frac{n_e}{H_p}\right] \qquad (7.24)$$

where $D = k(T_i + T_e)/m_i \nu_{in}$ is called the *ambipolar diffusion coefficient*, and H_p is the "plasma scale height" $k(T_i + T_e)/m_i g$. The "ambipolar diffusion" nomenclature derives from the fact that in the absence of externally imposed **E** fields, the vertical drift given by u_{pl} is caused by the charge-separation (or polarization) electric field because of gravity acting on the different masses of the ions and electrons, which must maintain equal scale heights to conserve local charge neutrality.

In general, $\mathbf{u}_i$ and $\mathbf{u}_e$ have horizontal as well as vertical components. In this case, the $\mathbf{u}_e \times \mathbf{B}$ and $\mathbf{u}_i \times \mathbf{B}$ terms must be retained. For the special case where **B** is horizontal, Ampère's law for the current density **j**,

$$\mathbf{j} = n_e e(\mathbf{u}_i - \mathbf{u}_e) = \frac{\nabla \times \mathbf{B}}{\mu_0} \qquad (7.25)$$

can be used to derive the expression

$$u_{pl} = -\frac{1}{n_e m_i \nu_{in}}\left[\frac{dp_T}{dh} + n_e m_i g\right] \qquad (7.26)$$

where p_T = total pressure (thermal plus magnetic) = $n_e k(T_i + T_e) + B^2/2\mu_0$. When **B** is at an angle to the horizontal, other magnetic terms besides the vertical magnetic-pressure gradient must be considered. It should be noted here that in some planetary ionospheres, the magnetic and thermal pressures are comparable, whereas in others either thermal or magnetic pressure will dominate. At the earth, where the intrinsic magnetic field is strong, **B** can be assumed to be the planetary dipole field, but at weakly magnetized Venus (where, as we shall see in Chapter 8, the ionospheric field is of interplanetary origin), **B** must be calculated from Maxwell's equations. This greatly complicates matters. Temperatures can also be derived from another equation for the heat balance, but the most basic calculations generally assume constant temperatures or empirical values or other simple temperature models.

With the height-dependent vertical velocity $u_{pl}(h)$ put into the continuity equation, we can proceed to solve for $n_e(h)$. It should be appreciated that u_{pl} can be either upward or downward, depending on the sign of the total pressure gradient and the latter's size in comparison with the gravitational force. Large collision frequencies tend to keep u_{pl} small.

Ratcliffe (1972) considered the special case of zero vertical drift ($u_{pl} = 0$) to make the point that under such circumstances, for zero B, and for equal electron and ion temperatures, one can solve for the polarization electric field,

$$E = \tfrac{1}{2} gM/e \qquad (7.27)$$

This upward electric field makes both electrons and ions behave as if they had a mass of $\tfrac{1}{2}M$ in the gravitational field. Because $T_i = T_e$, the plasma scale height is twice that of a neutral gas made up of atoms of mass M. This is because the polarization electric field buoys the heavier ions up and "weights" the light electrons down. If different ion species are present, one can write individual ion-momentum equations with $u_{pl} = 0$ to see that for equal temperatures, the M in the plasma scale height becomes the mean ion mass $<M>$, and the vertical electric field is given by $E = \tfrac{1}{2}g<M>/e$. Under such circumstances, individual ion species can behave as if their mass were negative if it is less than $\tfrac{1}{2}<M>$. Thus, height profiles of individual species densities in a multicomponent ionosphere can have intervals in which the gradient is not what one would expect if that ion were the only constituent. Strictly speaking, one must simultaneously solve the continuity and momentum equations for all ion species in order to find the correct height profiles of their densities. In general, an ionosphere will not have the same composition as the neutral atmosphere, and the altitude of the peak electron density will not coincide with the peak in production. The ion composition and the peak altitude depend on production and loss rates and transport.

7.5 AN EXAMPLE: THE EARTH'S IONOSPHERE

In situ measurements with rocket- and satellite-borne instruments, combined with the remote-sensing data from topside and bottomside ionospheric sounders [wherein time delays of reflected radio signals, transmitted from a satellite or the ground, give the altitude profile of the plasma frequency (and thus n_e) on either side of the peak], have given the picture of the earth's dayside ionosphere shown in Figure 7.7. The density and composition of the neutral atmosphere are also shown, both

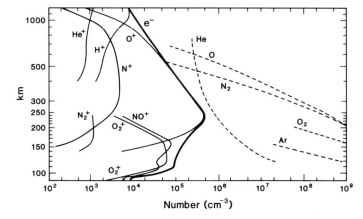

FIG. 7.7. International quiet solar year (IQSY) daytime ionospheric and atmospheric composition based on mass-spectrometer measurements. (From Johnson, 1969.)

to emphasize the weakness of atmospheric ionization at the earth and to illustrate the contrast in composition and vertical structure between the ions and the neutrals. It is seen that although there is one main peak in electron density at approximately 250 km altitude, there is considerable substructure. The discovery of this substructure led early observers to designate three major ionospheric layers or regions: the D region (below 90 km), the E region (between 90 and 130 km), and the F region (above 130 km). The F region is usually further divided into F_1 and F_2 layers, because a second ledge sometimes appears in its profile below the main (F_2) peak.

The concepts introduced earlier can be applied to an understanding of these layers as follows. One can think of the layers as being independently produced by the absorption of solar radiation by specific constituents of the neutral atmosphere, which respond differently to different parts of the incident solar photon spectrum. Two of these layers, the E and F_1 layers, are often thought to be fair approximations to Chapman layers. On the other hand, the F_2 layer, which is the highest, seems to require a more exotic explanation of its profile in terms of photochemistry or vertical movements driven by neutral drag or magnetospheric effects. The D layer, the deepest, is connected with the most energetic radiations (x-ray photons and cosmic-ray particles) and poorly understood loss processes.

The E layer is usually clearly noticeable as a change in slope in daytime electron density profiles near 110 km. The ions in this layer are mainly O_2^+ and NO^+ (Figure 7.7) that have been produced by ultraviolet radiation in the 100–150-nm range and solar x-rays in the 1–10-nm range. The peak density of this layer is close to the peak of the production rate Q for these ions. An effective recombination rate α can be deduced by dividing Q by the observed n_e^2 (see Section 7.5). Vertical transport of ions is considered to play a minor role in the formation of this layer.

The F_1 layer is composed primarily of O^+. The maximum electron density in this layer occurs near approximately 170 km, which is close to the level of the maximum ion production by photons in the spectral range from about 17 to 91 nm. This layer, which is more of a ledge, is not obvious in Figure 7.7, because it almost merges into the F_2 layer, which contains the main peak in the ionosphere density.

It is somewhat unfortunate (albeit not without interest) that the major peak in the terrestrial ionosphere cannot be described by simple Chapman-layer theory. This F_2 (or overall F-layer) peak density is also in a region dominated by O^+. However, at the altitudes where it occurs, other chemical processes besides simple direct recombination between the O^+ and surrounding electrons are important, and vertical drifts affect the ion distribution. Typically, recombination in the F_2 region may be preceded by the reaction of the ion with a nearby neutral molecule, with the net effect that the atomic (O^+) ion transfers its charge to a molecule that then dissociatively recombines. As long as the rates of such reactions exceed that of simple recombination, they dominate the loss term in the continuity equation. Further, collisional and ambipolar diffusions, described earlier, and vertical drifts driven by magnetospheric and atmospheric-dynamo electric fields (the latter driven by neutral drag forcing ions in the E region across the geomagnetic field), significantly affect the ion motion near the F-region peak. Thus, the reader is advised to consult more specialized references to fully appreciate all the features of the F region.

The lowest ionosphere, the D region, is of practical interest because of its role in commercial radio communication. The high ion–neutral collision frequencies make radio-wave absorption there important, and so the electron density is of primary concern. Only the most energetic ionization sources can penetrate to D-region altitudes. Between about 80 and 90 km, 0.1–1-nm x-rays from the sun are the primary sources; the very intense Lyman-α (121.6-nm) radiation from the sun has its peak ion production rate at about 70–80 km, and ionization of cosmic-ray particles dominates below. The predominant ions, NO^+ and O_2^+, can recombine with electrons, but at these low altitudes the electrons can also attach themselves to neutrals to form negative ions. Thus, a treatment of the D-layer "equilibrium" profile is not straightforward. Moreover, the aforementioned sources each undergo variations, depending on the prevailing solar activity and interplanetary conditions. These considerations leave the D region, like the F_2 region, a topic of ongoing research.

Finally, it is worth mentioning what happens at night when at least the solar photon sources are turned off (except for some scattered radiation that may make the disappearance of some wavelengths, such as Lyman-α, more gradual after sunset). At heights near 250 km, the effective time constant for recombination, which depends on the ion species, can be as short as 10 s for molecular nitrogen or as long as 300 h for atomic oxygen. However, the atomic oxygen ions can be removed much more quickly, as noted earlier, by virtue of charge transfer to a molecule with a faster recombination rate. In general, the diurnal variation of ionospheric density depends on altitude through both the local ion composition and the source diurnal variation. For instance, incident galactic cosmic rays that maintain the lower D region are present regardless of the local time, even though solar photons at all wavelengths undergo extreme diurnal fluctuations in intensity. There are also other sources that can appear at night on the earth, such as the draining of ionization stored at high altitudes in dipole flux tubes, and the highly spatially and temporally variable source of ionization from auroral-particle precipitation.

7.6 OTHER CONSIDERATIONS RELATING TO IONOSPHERES

7.6.1 High-Speed Outflow

A curiosity of particle conservation and transport reveals itself in the light elements at high altitudes in situations where production and loss can be ignored and the magnetic field is essentially vertical or absent. Schunk (1983) pointed out that the continuity equation for a minor ion of species α is then simply the law of flux conservation,

$$\partial/\partial h(n_\alpha u_{h_\alpha}) = 0 \tag{7.28}$$

and the momentum equation can be written

$$n_\alpha m_\alpha u_{h_\alpha}\frac{\partial u_{h_\alpha}}{\partial h} + n_\alpha q_\alpha \frac{k(T_i + T_e)}{n_e}\frac{\partial n_e}{\partial h} + n_\alpha m_\alpha g = n_\alpha m_\alpha \nu_{\alpha n} u_{h_\alpha} \tag{7.29}$$

where the *inertial term* on the left has been introduced in anticipation of its importance. It is usually negligible in lower ionospheres, where flows are generally subsonic. The polarization electric field set up by the major species is here expressed as

$$E = -\frac{1}{en_e}\frac{\partial p_e}{\partial h} = -\frac{kT_e}{en_e}\frac{\partial n_e}{\partial h} \tag{7.30}$$

under the assumption that all other forces but the pressure-gradient force are negligible for the electrons. The foregoing continuity equation gives

$$\frac{\partial n_\alpha}{\partial h} = -\frac{n_\alpha}{u_\alpha}\frac{\partial u_\alpha}{\partial h} \tag{7.31}$$

For singly charged ions ($q_\alpha = 1$), this can be used together with the thermal-velocity definition $v_{\text{th}} = (k(T_i + T_e)/m_\alpha)^{\frac{1}{2}}$ to cast the minor ion-momentum equation in the form

$$(u_\alpha^2 - v_{\text{th}}^2)\frac{1}{u_\alpha}\frac{\partial u_\alpha}{\partial h} + g = \nu_{\alpha n} u_\alpha$$

or

$$\frac{1}{M}\frac{\partial M}{\partial h} = -\left[\frac{g/v_{\text{th}}^2 - (\nu_{\alpha n} v_{\text{th}} M)}{v_{\text{th}}^2}\right]/(m^2 - 1) \tag{7.32}$$

where $M = |u_\alpha|/v_{\text{th}}$ is the species *Mach number*. This latter equation is reminiscent of the solar-wind equation derived in Chapter 4. In that chapter, the idea that a flow can theoretically undergo a subsonic-to-supersonic transition when the numerator and denominator in the flow equation become zero together was applied to the solar corona. The

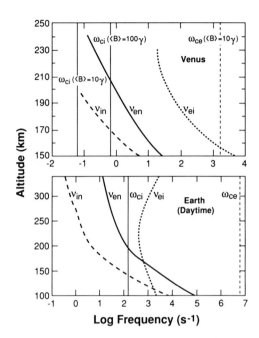

FIG. 7.8. Comparison of collision frequencies and gyrofrequencies in the ionospheres of Venus (top) and Earth (bottom). (From Luhmann and Elphic, 1985.)

density, obtainable from the preceding flux-conservation form of the continuity equation, decreases with altitude in accord with the velocity increase. Indeed, a supersonic polar wind of light ions is seen emanating from regions like the earth's polar cap. Of course, a detailed treatment of the polar wind is much more involved, requiring simultaneous solution of the equations for other species of ions and special treatments of collision frequencies for high-speed ions (Schunk and Nagy, 1980).

Also of interest here is the interplay between collisions with neutrals and the effects of the magnetic field. Between collisions, charged particles gyrate around the magnetic field at frequencies $\Omega_i = qB/m_i$ for ions and $\Omega_e = qB/m_e$ for electrons. If $\nu_{in} \gg \Omega_i$, the magnetic field exerts little direct influence on the motion of the ions (although it can indirectly exert influence through its effects on the electrons), and the ion motion in the absence of large pressure gradients and electric fields is practically the same as the neutral motion. On the other hand, if $\Omega_i \gg \nu_{in}$, gyromotion proceeds relatively uninhibited, but in doing so prevents transport across the magnetic field. Under such circumstances, collisions still play an important role in moving the ions parallel to the magnetic field. A comparison of some collision frequencies and gyrofrequencies in the ionospheres of Earth and Venus is shown in Figure 7.8. In calculating plasma velocities, the best policy is to evaluate all of the terms in the momentum equation before deciding which is negligible. Even a term that seems relatively small can be important if the other larger terms practically cancel. In addition, the relative velocities that enter into the collision term can determine whether or not collisions are important in the overall force balance at a particular altitude.

7.6.2 Conductivity

The competition between the magnetic field and collisions for control of the ion motion is particularly important in problems requiring an evaluation of the electrical conductivity of the ionosphere. If only the electric-field and collision terms participated in the force balance, the steady-state momentum equations for ions and electrons would be

$$q\mathbf{E} = m_i \nu_{in} \mathbf{u}_i \tag{7.33}$$

and

$$-e\mathbf{E} = m_e \nu_{en} \mathbf{u}_e \tag{7.34}$$

Thus, for this simple medium, $\mathbf{j}$ is related to the electric field by

$$\mathbf{j} = \sigma_0 \mathbf{E} \tag{7.35a}$$

where the conductivity σ_0 is a scalar that depends only on the collision frequencies.

If a magnetic field is present, magnetic terms are present in the momentum equations, and the velocities $\mathbf{u}_i$ and $\mathbf{u}_e$ are no longer expressed so simply in terms of $\mathbf{E}$. However, if we solve for $\mathbf{u}_i$ and $\mathbf{u}_e$ for the case where the magnetic field is in the z-direction (and $q = e$), we find that $\mathbf{j}$ can be written in the concise form

$$\mathbf{j} = \begin{pmatrix} \sigma_1 & \sigma_2 & 0 \\ -\sigma_2 & \sigma_1 & 0 \\ 0 & 0 & \sigma_0 \end{pmatrix} \begin{pmatrix} E_x \\ E_y \\ E_z \end{pmatrix} \tag{7.35b}$$

where

$$\sigma_1 = \left[\frac{1}{m_e \nu_{en}} \left(\frac{\nu_{en}^2}{\nu_{en}^2 + \Omega_e^2} \right) + \frac{1}{m_i \nu_{in}} \left(\frac{\nu_{in}^2}{\nu_{in}^2 + \Omega_i^2} \right) \right] n_e e^2 \tag{7.36}$$

$$\sigma_2 = \left[\frac{1}{m_e \nu_{en}} \left(\frac{\Omega_e \nu_{en}}{\nu_{en}^2 + \Omega_e^2} \right) - \frac{1}{m_i \nu_{in}} \left(\frac{\Omega_i \nu_{in}}{\nu_{in}^2 + \Omega_i^2} \right) \right] n_e e^2 \tag{7.37}$$

$$\sigma_0 = \left[\frac{1}{m_e \nu_{en}} + \frac{1}{m_i \nu_{in}} \right] n_e e^2 \tag{7.38}$$

The conductivity is thus a tensor quantity. It is a tensor because the magnetic field makes the medium anisotropic in its response to an applied electric field. It can be seen from the form of the tensor that if the electric field is applied perpendicular to the magnetic field ($\mathbf{E} = E_x \mathbf{i} + E_y \mathbf{j}$), σ_1 is the conductivity in the direction of the applied electric field. It is called the *Petersen conductivity*. The component σ_2 is the conductivity perpendicular to the direction of the applied field. It is called the *Hall conductivity*. If an electric field is applied parallel to the magnetic field, the conductivity is the same as that derived earlier for zero field. It is called the direct or longitudinal conductivity and depends only on the collision frequencies. Hence, we can deduce the values of the conductivities in an atmosphere once n_e is derived from the foregoing considerations, as long as the behavior of the magnetic field is known.

7.7 FINAL NOTES

In closing this chapter on the concepts central to the formation of a planetary ionosphere, it is worthwhile to caution the reader that although the simplified treatments described here are adequate for many investigations of interest, our omission of the details of atmospheric chemistry, time dependence, and horizontal structure limits their application. Careful consideration should always be given to the impact of these omissions on the problem at hand.

ADDITIONAL READING

Hargreaves, J. K. 1979. *The Upper Atmosphere and Solar Terrestrial Relations.* New York: Van Nostrand Reinhold.

Rishbeth, H., and O. K. Garriott. 1969. *Introduction to Ionospheric Physics.* New York: Academic Press.

PROBLEMS

7.1. Find the expression for h_m, the height of the production peak for the Chapman production function. [Hint: Use equation (7.7).] Then use that expression to derive equation (7.6).

7.2. Plot the Chapman production function versus altitude for solar zenith angles $\chi = 0°$, $30°$, and $60°$ and a neutral scale height H_n of 10 km. Assume $h_m = 125$ km and $q_m = 1.3 \times 10^5$ cm$^{-3} \cdot$s^{-1} at $\chi = 0°$. How does the height of the peak depend on χ?

7.3. Figure 7.4 shows some altitude profiles of energy deposition for protons in an exponential oxygen atmosphere. Using these curves, how would you obtain the deposition profiles for incident alpha articles (He^{2+}) in the same atmosphere? If you had a monoenergetic 100-keV proton flux of 10^5 cm$^{-2} \cdot$s^{-1} entering the atmosphere, what would the approximate peak production rate be? (Note: dE/dx is independent of the incident ion mass, but depends on the square of its charge.)

7.4. Compare the ranges of protons and electrons in air (see Figure 7.5). At a particular energy, which are most penetrating? How do the energies of photons that penetrate to the same altitudes compare (use Figure 7.2)?

7.5. Using the assumption that B is horizontal (along x) and h is along z, derive the equation for the vertical drift u_{pl} with the aid of Ampère's law.

7.6. Using the assumption that B is along z, derive the expression for the Petersen and Hall conductivities from the electron and ion-momentum equations.

8 PLASMA INTERACTIONS WITH UNMAGNETIZED BODIES

J. G. Luhmann

8.1 INTRODUCTION

WHEN A PLANET or satellite has a weak internal magnetic field or none at all, its interaction with the external plasma can be drastically different from that described in earlier chapters. Such interactions can vary widely, depending on whether or not the body has a significant atmosphere. The simplest case, which is an appropriate starting point for discussion, is that of a body similar to the earth's moon that has, at most, a negligibly thin atmosphere.

8.2 PLASMA INTERACTIONS WITH MOONLIKE BODIES

A body like our moon, composed of insulating material and submerged in a flowing plasma, simply absorbs the particles of the plasma that are incident on the body. Thus, the lunar soil contains a record of the composition and energy of the ancient solar wind. At the moon, a bow shock will not form upstream; there is virtually no obstacle as far as the oncoming flow is concerned. The magnetic field diffuses into the very weakly conducting outer layers at a rapid rate, so that it is barely perturbed from its upstream orientation. The most notable features are associated with the plasma-absorption wake left in the plasma behind the body.

Whereas the rest of the interaction is essentially the same as previously described for other bodies, regardless of the properties of the incident plasma flow, and regardless of the orientation of the ambient magnetic field, the wake structure will depend on the upstream variables. If the body's magnetic field is zero, and the flow speed is high compared with the thermal velocity, the wake will persist to large distances; but if the flow is slow relative to the thermal speed, thermal motions perpendicular to the flow direction can refill the empty space within a short distance downstream of the body. The introduction of a magnetic field can either inhibit the refilling of the wake, if the field is nearly parallel to the upstream flow, or have a minimal effect, if it is perpendicular. The cartoons in Figure 8.1 show two hypotheti-

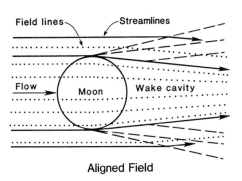

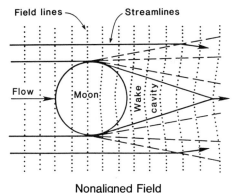

FIG. 8.1. Illustration of the interplanetary plasma flow and magnetic-field perturbation by the moon. (Adapted from Spreiter et al., 1970.) The wake created by solar-wind absorption closes more quickly when the magnetic field is not aligned with (parallel to) the undisturbed flow.

cal configurations for the lunar wake in the solar wind for these two magnetic-field orientations. Because the magnetic field in the solar wind is "frozen in," the field will be slightly perturbed as the flow closes behind the moon to fill the void created by the absorbing obstacle. A quantitative magnetohydrodynamic (MHD) treatment by Spreiter, Marsh, and Summers (1970) gives an idea of the extent and magnitude of the field and flow perturbations that form in the lunar wake when the moon is in the solar wind. The calculated field perturbations resemble the magnetic perturbations observed near the moon by *Explorer 33* and *35*. Asteroids and atmosphereless planetary satellites in the solar wind should have similar plasma interactions, although especially in the former case the irregular shape of the body will affect the details of the wake structure.

There will be additional effects if the body has a conducting core, but as illustrated by Figure 8.2, these effects will appear as magnetic perturbations of the field lines threading the body, rather than in the wake. The field perturbation by a small conducting core will occur primarily within the insulating layer of the body. Such perturbations will occur because currents are generated in the conducting core that will produce a magnetic field that will cancel any field internal to the conductor. For a spherical core, this field will take the form of an opposing dipole. These *shielding currents* will eventually decay, for a core is not infinitely conducting, but because they will decay much more slowly than the time between external-field changes, they will essentially al-

8.2 INTERACTIONS WITH MOONLIKE BODIES

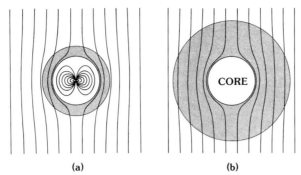

FIG. 8.2. Illustration of the field-perturbing effect of a conducting core when the insulating "mantle" is thin (a) and when it is thick (b). This perturbation will persist as long as the external field varies on a time scale that is short compared with the time required for penetration into the core.

ways be present in the lunar case. This scenario does not pertain to planetary satellites with cores that have been immersed in a steady magnetospheric field for millennia, but only to the time-varying fields. Cowling (1957) gives the diffusion-time constant as

$$\tau = R^2 \mu_0 \sigma \tag{8.1}$$

where R is the core radius, σ is its conductivity, μ_0 is the magnetic permeability. This is the decay time for the eddy currents that flow on the core to exclude the external field. This time constant for the lunar core is expected to be of the order of 1,000 yr. If the external magnetic field changes more rapidly than this time constant, the picture in Figure 8.2 will hold. Otherwise, the field will penetrate into the conductor. The moon, in its 60-earth-radius orbit, experiences constantly changing external conditions. The situation when the moon is inside the magnetosphere is somewhat similar to that when it is outside, except that the wake is modified by the different external flow and field conditions.

In the present context, it is important to appreciate that observations tell us that the moon and possibly some of the asteroids possess some atmosphere, although such atmospheres are thin and sometimes transient. The same is likely to be true of many planetary satellites. The existence of these weak atmospheres can have various causes: *sputtering* of the surface material by photons or impacting particles (from the primary external plasma flow itself, or from another, more energetic population that is a minor component of that plasma), outgassing from beneath the surface, or sublimation of ices. These atmospheres will have little effect on the plasma interaction as long as they remain rarefied, but they will produce observable signatures in the form of heavy ions in the vicinity of the body. The latter can be produced by photoionization, by particle-impact ionization, or by charge exchange (wherein an ion passes its charge to an atom) of the atmospheric atoms or molecules. As soon as they acquire their charge, the atmospheric ions start to gyrate in the ambient magnetic field, but because the ambient field is moving with the plasma, the particle is accelerated or picked up, as illustrated in Figure 8.3. This acceleration occurs because the motion of the magnetized plasma at velocity **u** creates an effective electric field $\mathbf{E} = \mathbf{u} \times \mathbf{B}$, where **B** is the ambient magnetic field. In the case shown, where **u** and **B** are

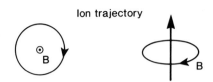

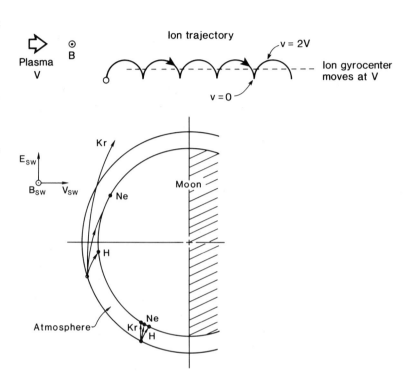

FIG. 8.3. Illustration of the manner in which ions can be picked up in a flowing, magnetized plasma. Here the gyrating ion is carried along, threaded on the magnetic field. Its maximum velocity is twice the velocity of the flowing plasma V.

FIG. 8.4. Fate of ions formed in the lunar atmosphere. (Adapted from Manka and Michel, 1973.)

perpendicular, the particle undergoes the cycloidal motion characteristic of drift in crossed electric and magnetic fields (as discussed in Chapter 2). The particle moves four gyroradii downwind in a gyroperiod, where the gyroradius is $mu_\perp/eB$. The ion's velocity oscillates between zero (at the cusps) and $2\mathbf{u}$ (at the peaks of the cycloids), while the center of gyration or *gyrocenter* moves at velocity $\mathbf{u}$. The orientations of these trajectories will change with the changing orientation of the ambient field. If the gyroradius of the *pickup ion* is small compared with the size of the body, it will generally be carried back into the body with the plasma flow. If the gyroradius is large compared with the body, the cycloids can carry the particles either away from the body or back into it, as illustrated in Figure 8.4 for sodium and potassium ions at the moon. Although these ions may not be important to the plasma interaction, their detection provides a means of studying weak atmospheres from orbiting spacecraft and also provides an example of the possible evolutionary effect of a flowing plasma on such bodies, because lost pickup ions represent loss of constituents.

An additional departure of the moon from an electrically passive obstacle is observed in the existence of some remanent magnetism near the surface. Russell and Lichtenstein (1975) described localized dipolar *arches* of magnetic field strong enough to extend into the solar wind near the flow terminator (the dividing line between the hemisphere where particles are incident and the wake hemisphere) and to deflect slightly the solar-wind plasma. However, these deflections represent only a small modification of the plasma interaction described earlier.

In general, quantitative treatments of flowing plasma interactions with even simple insulating bodies are not simple exercises. In the case of the moon, a fluid approach seems adequate, as mentioned earlier. Even then, the full MHD treatment has been carried out numerically by Spreiter and co-workers only for the case of the upstream magnetic field aligned with the flow. Intermediate and perpendicular field orientations have been treated only approximately using the convected-magnetic-field gas-dynamic code (see Chapter 6). Another class of problems arises when we deal with bodies comparable to or smaller than the Debye length for the flowing plasma (~10 m in the solar wind). Under such conditions, the details of the particle motions are important, as is the electrical potential of the body surface. Fluid approximations become inadequate, and we are forced to deal with the intricacies of matters such as the surface as an electron emitter. [To some extent, this was a consideration even for the moon (Manka and Michel, 1973).] Purely kinetic (particle) treatments of flowing plasmas interacting with bodies currently fall in the realm of spacecraft-charging studies. However, most of the analyses done in this field have been carried out with various assumptions (such as that of no magnetic field) that have limited their applicability. Future generations of researchers are left with the challenge of confronting these problems with new computational tools.

8.3 PLASMA INTERACTIONS WITH BODIES WITH ATMOSPHERES

8.3.1 Unmagnetized Planets and Satellites

An insulating body with an atmosphere residing in the darkness of space represents no more of an obstacle to the external plasma than does a bare insulating body, except for the fact that the gases of the atmosphere are somewhat more prone to sputtering and subsequent impact ionization than is a solid surface. What makes the difference is not only substantial ionization (by a source such as sunlight) but also the presence of a magnetic field in the flowing plasma. If the magnetic field is not steady, and moreover can be considered frozen into the external plasma, the scenario resembles that illustrated in Figure 8.5. Because the ionosphere is a conductor, the convecting magnetic field generates currents in it that at least initially keep the field from penetrating through the body by generating a canceling field. This situation persists as long as the magnetic field keeps changing its orientation (as it

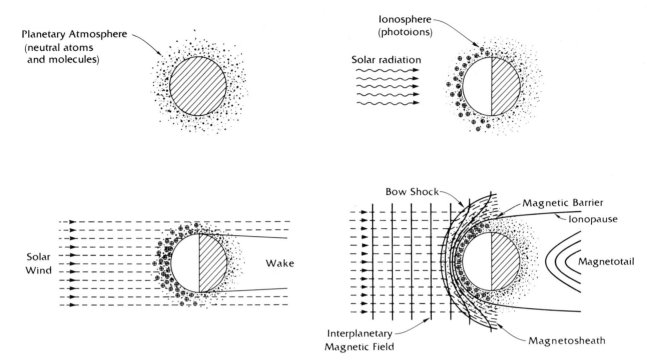

FIG. 8.5. Illustration of the steps that lead to the formation of an ionospheric planetary obstacle in a flowing plasma like the solar wind. Ionization by solar radiation, for example, is followed by diversion of the external plasma flow only if that flow is magnetized.

does in the solar wind). Otherwise, the field would eventually diffuse into the body on a time scale that would depend on the ionospheric conductivity. This basic picture is appropriate for the solar-wind interaction with the planets Venus and Mars, and perhaps for Saturn's satellite Titan when it is outside of Saturn's magnetosphere. It is not appropriate for a body inside of a magnetosphere where the magnetic field is relatively steady.

The details of the interaction shown in Figure 8.5 are based on a combination of observations and theoretical expectations. A bow shock is expected, since the solar-wind plasma flowing at supermagnetosonic velocity must be diverted around the conducting obstacle because the plasma is frozen to the draped magnetic field. (For a lunar core, this is not the case, because the diversion of field occurs inside of the absorbing body.) The position of this bow shock can be calculated approximately using numerical hydrodynamic or gas-dynamic models of flow past an impenetrable surface of specified shape. The results of one such calculation from a model of Spreiter and Stahara (1980) are shown in Figure 8.6. In this calculation, the position of the shock is sensitive only to the obstacle shape, the upstream Mach number, and an assumed ratio of specific heats for the solar-wind plasma. The fluid equations that are solved are those of continuity, momentum, and energy. The solutions give the gas or fluid parameters of density, velocity, and temperature

8.3 INTERACTIONS WITH BODIES WITH ATMOSPHERES

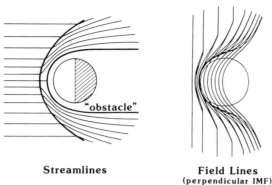

FIG. 8.6. Streamlines of plasma flow and projected magnetic-field lines calculated from the Spreiter and Stahara (1980) gas-dynamic magnetosheath model. In the right-hand panel, sets of field lines in the velocity–magnetic-field plane intersecting the planetary center and field lines slightly above this plane encounter the obstacle. The field lines above the central plane flow up and around the obstacle. The field lines in the central plane drape around the planet. (From Luhmann, 1991.)

throughout the region shown. The magnetic field in between the bow shock and obstacle, which is the same "magnetosheath" as that found in the solar-wind interaction with the earth's magnetosphere, is calculated separately from the velocity, under the assumption that it is frozen in [i.e., that it satisfies $\partial \mathbf{B}/\partial t = -\mathbf{u} \times \mathbf{E} = \nabla \times (\mathbf{u} \times \mathbf{B}) = 0$]. The field lines in the magnetosheath for a perpendicular upstream field seem to *hang up* on or *drape* around the obstacle, as illustrated in Figure 8.6.

The location of the obstacle boundary is another matter that deserves some elaboration. No results of a self-consistent (MHD) treatment of a plasma flow against a planetary ionosphere have been reported in the literature. However, analogous to the magnetospheric case, it is fair to assume that if the solar-wind flow is stopped (or stagnated) at the subsolar point of the ionospheric obstacle, there must exist a transformation there of upstream pressure (mostly dynamic pressure, ρu^2) to internal pressure. The three parts of Figure 8.7 schematically illustrate this pressure-balance situation for an internal pressure supplied by the thermal pressure of the ionospheric plasma $n_e k(T_i + T_e)$. At Venus during solar maximum, as illustrated in Figure 8.7c, the solar-wind pressure can only occasionally exceed the thermal pressure of the ionosphere. However, at Mars during solar minimum the solar-wind pressure generally exceeds the thermal pressure of the ionosphere. At Venus, the magnetosheath effectively stops at the altitude where the ionospheric plasma pressure is equal to that incident pressure. This obstacle boundary, which is called an *ionopause*, has the average solar-maximum location at Venus shown in Figure 8.8. The observed ionosphere of Mars, on the other hand, has insufficient pressure by itself to balance the typical incident solar-wind pressure. Although we lack in situ observations of the Venus ionosphere at solar minimum, we expect the same situation for Venus at that time based on observations at solar maximum. When the solar-wind pressure is extraordinarily high at solar maximum, the magnetosheath magnetic field is found to penetrate the ionosphere and contribute to the total obstacle pressure perceived by the solar wind.

Of course, the extent to which the field can contribute to the pressure must be limited by the rate of diffusion of field through the ionosphere.

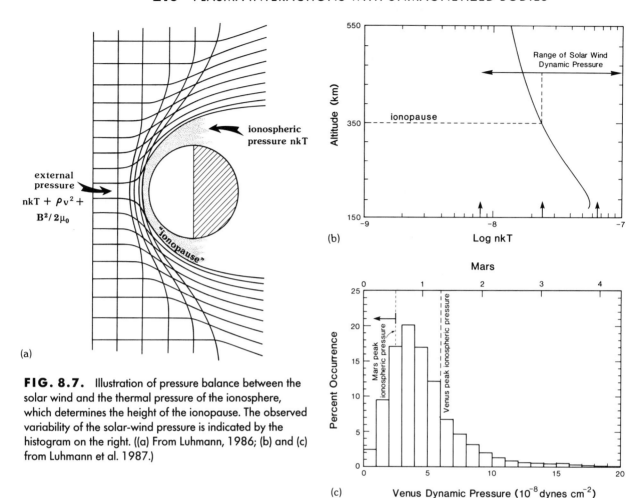

FIG. 8.7. Illustration of pressure balance between the solar wind and the thermal pressure of the ionosphere, which determines the height of the ionopause. The observed variability of the solar-wind pressure is indicated by the histogram on the right. ((a) From Luhmann, 1986; (b) and (c) from Luhmann et al. 1987.)

At the bottom of the ionosphere are the insulating neutral atmosphere and solid planet (Venus and Mars have no conducting oceans). Although it has not been verified experimentally, we can envision that when the ionosphere, by itself, can no longer form the obstacle, the interplanetary field will temporarily "hang up" in the ionosphere, then make its way through the solid mantle of the planet until it is diverted around a conducting core (if there is one) and finally pulled into the wake. The issue of ionospheric magnetization will be revisited later, but first it is necessary to consider an important consequence of the direct interaction of the solar wind with the atmosphere and ionosphere.

Figure 8.9 compares the size of the solar-wind interaction region of Venus with that of the earth's magnetosphere. A normalization of planet sizes would lead to a similar picture for Mars. The closeness of the effective obstacle boundary at Venus to the planet surface was also apparent in Figure 8.8. The Venus ionopause altitudes (subsolar distances of ~300 km) can be compared with the altitude profile of the

8.3 INTERACTIONS WITH BODIES WITH ATMOSPHERES

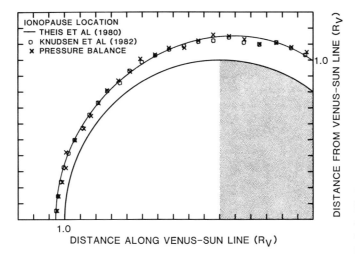

FIG. 8.8. Observed location of the "ionopause" surface at Venus as determined from the location of pressure balance between the ionosphere and the overlying magnetosheath magnetic field. (From Phillips et al., 1984.)

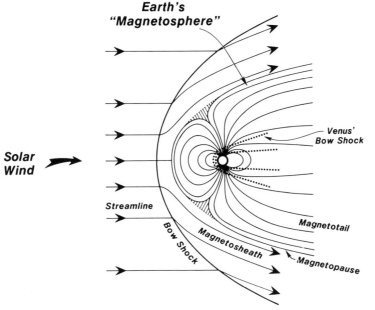

FIG. 8.9. Comparison of the size of the Venus–solar-wind interaction region with that of the magnetized earth. (Adapted from Luhmann and Brace, 1991.)

dayside Venus neutral ionosphere in Figure 8.10. It is apparent that whereas the earth's atmosphere lies protected deep inside its magnetospheric "bubble," much of the upper atmosphere of Venus is exposed to inflowing solar-wind plasma. The dayside ionopause ranges from about 350 km altitude at the subsolar point to 1,000 km at the terminator, and the bow shock is at an altitude as low as 2,000 km. As a result, photoions that might be produced above the ionopause can be picked up in the same manner as described for the lunar atmospheric ions. Impact ionization and charge exchange contribute to this population. The only difference in this case is that the magnetosheath **u** and **B** must be used to evaluate the accelerating electric field close to the planet. Figure 8.11 shows the gyroradius of picked-up oxygen at Venus relative to the scale

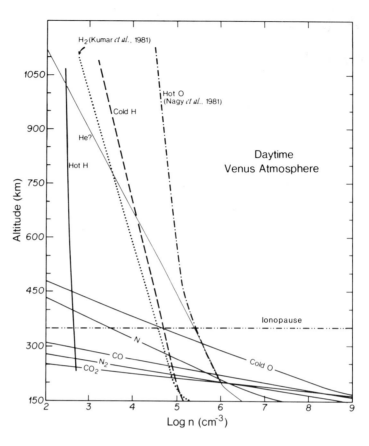

FIG. 8.10. Altitude profile of the Venus neutral upper atmosphere. (Adapted from Nagy et al., 1982.)

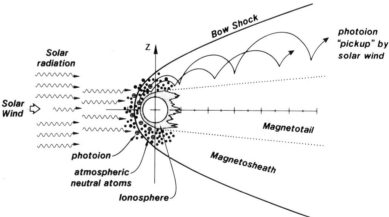

FIG. 8.11. Illustration of planetary pickup-ion trajectories at Venus. The cycloid sizes are approximately scaled for O^+ (oxygen is the main constituent of the Venus upper atmosphere, as shown in Figure 8.10).

of the obstacle. As shown, the pickup is asymmetric, because ions created below the planet spiral up into it. Oxygen ions have been observed above the ionospheres and downstream of both Venus and Mars, which have similar oxygen upper atmospheres. The result of this high-altitude ion-removal process can be seen in the measured ionospheric profiles in Figure 8.12, which are compared with profiles modeled from the neutral atmosphere under the assumption that no removal mechanism operates other than the usual recombination processes. The iono-

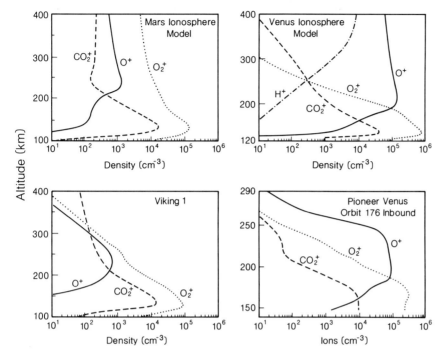

FIG. 8.12. Comparison of modeled (top) and observed (bottom) ionosphere altitude profiles at Venus and Mars. The difference between observation and theory is due to the loss of the upper ionospheres to the solar wind. The model profiles were obtained from Shinagawa and Cravens (1989) and Shinagawa et al. (1987). The observations were obtained by the spacecraft indicated on the figure.

spheric density is depleted above the ionopause, below which the flowing solar wind does not penetrate significantly. Integrated over the lifetime of the planet (~4.5 billion yr), this *scavenging* by the solar wind can have an impact on the atmosphere evolution.

Returning now to the ionospheric magnetic field, Figure 8.13 shows some altitude profiles of magnetic field and ionospheric electron density obtained at Venus during solar maximum by in situ measurements from the *Pioneer* Venus orbiter. The ionopause cutoff in the density is usually clear and sharp at solar maximum, as in the upper left corner. For such cases, the magnetosheath magnetic field stops abruptly where the ionospheric density rises. Evaluations of the pressures on both sides confirm that the external pressure of the magnetic field is very close to the internal thermal pressure just inside this boundary. This observation thus indicates that the piled-up, draped magnetic field in the magnetosheath ultimately accommodates most of the upstream dynamic pressure before the ionosphere takes over. For this reason, the lower magnetosheath is sometimes referred to as a *magnetic barrier* (Figure 8.5). Its existence, which is not taken into account in the gas-dynamic magnetosheath model, is inferred from observed shock positions that seem to require a larger obstacle than the ionopause presents. For the present, the small-scale structures in the magnetic field inside the ionosphere will be neglected, as they contribute negligibly to the total pressure. Because the ionopause nominally forms where the thermal pressure is equal to the incident solar-wind dynamic pressure (Figure 8.7), it gets lower when the solar-wind pressure gets higher. As this happens, as illustrated

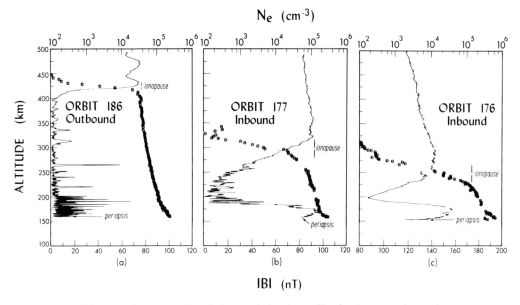

FIG. 8.13. Examples of observed altitude profiles for the ionospheric electron densities (points) and magnetic fields (solid line) at Venus. The ionopause is located where the magnetosheath field decreases and the plasma density increases. (From Elphic et al., 1980.)

in Figure 8.13, the boundary layer thickens, and a large-scale magnetic field appears inside of the ionosphere. This magnetic field is generally horizontal and has roughly the orientation of the overlying magnetosheath field. We can think of these fields as interplanetary fields that are incompletely canceled by the shielding currents in the upper ionosphere, but usually are canceled by the time they reach the altitude of the ionosphere peak near 140 km. Thus, ionospheric magnetic fields at Venus are related to higher solar-wind pressures and the associated low ionopauses. It should be noted that the ionopause at Venus is observed to have a minimum altitude of about 225 km. After this point, the increases in solar-wind pressure appear to manifest themselves only in the compensating increases of the ionospheric magnetic-field pressure.

The mechanism for ionospheric field generation at an unmagnetized planet can be thought of simply in terms of diffusion and convection of the magnetic field. Because the magnetic field is horizontal, the ionospheric plasma drift can be calculated from observations and the vertical steady-state momentum equation described in Chapter 7:

$$u_h = \frac{1}{n_e m_i \nu_{in}} \left[\frac{\partial p_T}{\partial h} + n_e m_i g \right] \quad (8.2)$$

The result for Venus, from a semiempirical model of the ionosphere, assuming a negligible magnetic pressure, as obtained by Cravens, Shinagawa, and Nagy (1984), is shown in Figure 8.14. Although ionosphere is

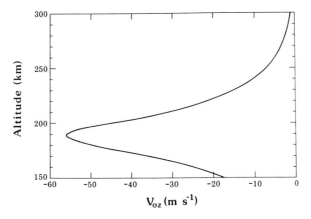

FIG. 8.14. Ionospheric-plasma drift-velocity profile at Venus derived from a semiempirical model by Cravens et al. (1984).

produced at all altitudes, it recombines most readily at lowest altitudes, where the collision frequencies are greatest. Thus the Venus ionosphere drifts downward. The steady-state ionospheric magnetic field, which has a boundary condition at the top controlled by the solar-wind magnetic field and dynamic pressure, to a first approximation obeys the one-dimensional *dynamo,* diffusion/convection, or induction equation:

$$\frac{\partial B}{\partial t} = 0 = \frac{\partial}{\partial h} D \frac{\partial B}{\partial h} - \frac{\partial}{\partial h}(Bu_h) \tag{8.3}$$

where the diffusion coefficient D is given by

$$D = \frac{m_e(\nu_{en} + \nu_{ei})}{n_e e^2 \mu_0} \tag{8.4}$$

The collision frequencies ν_{en} and ν_{ei} for Venus, shown in Chapter 7, are such that the diffusion coefficient is very small at high altitudes and large at low altitudes. This diffusion/convection equation is readily derivable from Maxwell's induction equation, $\partial \mathbf{B}/\partial t = -\nabla \times \mathbf{E}$, the ion- and electron-momentum equations, and Ampère's law (see Chapter 2). We can think of it as describing downward convection of the initially frozen-in field with the ionospheric plasma, followed by collisional diffusion to minimize gradients. The collisions also cause significant current (and thus field) dissipation at the lowest altitudes. It is simplest to solve the equation for **B** numerically, as a time-dependent problem from some assumed initial state (such as zero ionospheric magnetic field), until it converges to a steady solution. Some examples of the evolving solution are illustrated in Figure 8.15 for different upper boundary conditions characteristic of varying altitudes of the dayside ionopause of Venus. The same figure shows some comparable observations. Because the ionopause typically is lowest in the subsolar region, the ionospheric magnetic field is more frequently significant there.

Two caveats must be given here, one of which relates to the neglect of horizontal motions in the ionosphere. The upper ionosphere of Venus is observed to convect antisunward, as illustrated in Figure 8.16, be-

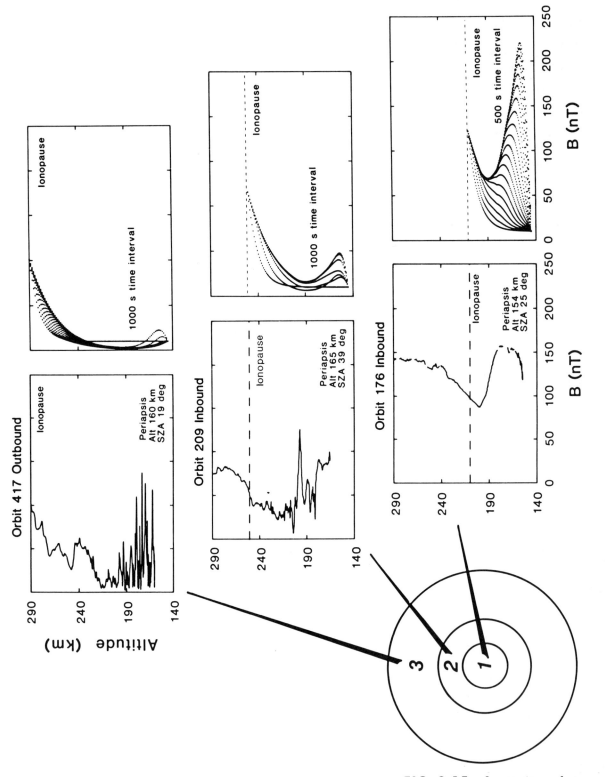

FIG. 8.15. See caption on facing page.

8.3 INTERACTIONS WITH BODIES WITH ATMOSPHERES

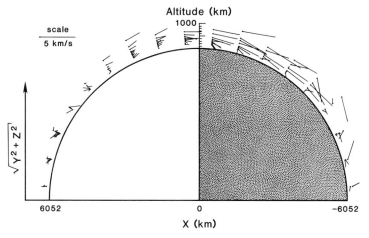

FIG. 8.16. Observed antisolar flow in the ionosphere of Venus. (From Knudsen et al., 1980.)

cause of the large day-to-night pressure gradients in the ionospheric plasma. This antisolar flow can also be inferred from the pressure gradients calculated from the global plasma density and temperature measurements. At low altitudes, where there is some collisional coupling of the plasma motion to the neutral atmosphere, antisolar flow can still prevail if the neutrals also flow antisunward, as models of the dynamics of the upper neutral atmosphere indicate (Bougher et al., 1986). In the one-dimensional model only dissipation removes field, but with horizontal convection there is another alternative. Thus far, the full three-dimensional problem of the ionospheric field structure is still incompletely solved. A three-dimensional solution could be different, because in three dimensions, dissipated ionospheric fields will cause a rerouting of some interplanetary field through the insulating lower atmosphere and solid planet. Another caveat is that the magnetic field will have some effect on the plasma velocity through the magnetic-pressure contribution to the plasma pressure in the equation for u_h. More fully self-consistent MHD models for the one-dimensional problem have been examined by Shinagawa and Cravens (1988, 1989), but these still suffer from the problems of properly defining the upper boundary condition (where solar-wind ion removal has an influence) and self-consistently describing the temperature. Nevertheless, from the good agreement of even the simple model with observations at Venus, it is considered that the major

FIG. 8.15. Examples of the converging solutions of the diffusion/convection equation for the ionospheric magnetic field given in the text. These magnetic profiles are determined by the plasma velocity profile shown in Figure 8.14 and the altitude profile for the collisional diffusion coefficient, together with the upper boundary condition on the field at the ionopause. The "bull's eye" represents the dayside of the planet as seen from the sun. The ionospheric magnetic field usually is largest in the subsolar region, where the ionopause is lowest and the overlying magnetosheath field is strongest.

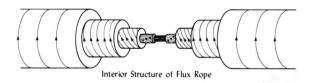
Interior Structure of Flux Rope

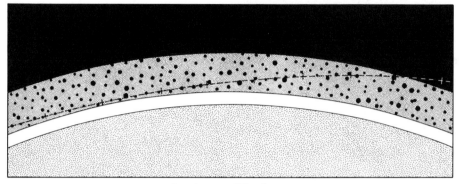
Distribution of Flux Ropes

FIG. 8.17. Illustration of the configuration of magnetic-field lines in a Venus ionospheric flux rope, as deduced by Russell and Elphic (1979), and their distribution in the ionosphere.

physical process of ionospheric magnetization is understood. This model is also expected to be quite relevant at Mars, where the ionosphere is almost always in a state where the ionopause should be near its minimum altitude, and thus it should be magnetized. The relevant in situ observations at Mars have yet to be made.

A fascinating phenomenon of ionospheric magnetization that is not so well understood is illustrated by the data in the first panel of Figure 8.13. The small-scale magnetic-field structures observed inside of the dayside ionosphere of Venus appear when the large-scale field is absent. They seem to be describable in terms of field lines that are twisted around an axis like ropes made of smaller strands. Hence they have been called *flux ropes,* which is standard nomenclature in solar physics and some other fields. One model for their structure, and their hypothetical ionospheric distribution, is illustrated in Figure 8.17. The problem with Figure 8.17 is that the individual rope axes do not appear to be parallel or even horizontal. It has been suggested that even the ropes themselves are twisted. In any event, there are numerous interpretations of these structures, but no widespread prevailing view at this time. Presently, ionospheric flux ropes are not a focus of much active research, both because they are so difficult to understand and because they appear to be a passive element of the overall solar-wind interaction. Nevertheless, by unraveling the physics behind their curious nature, we may learn some very basic information about space plasmas.

When the scavenging of ionospheric ions by the solar wind was discussed earlier, the fact that the draped magnetic fields of the magnetosheath can themselves exert a body force on the ionospheric plasma was ignored, as was the fact that the pickup ions can contribute a current that will perturb the surrounding magnetic field. In a fully self-consistent

8.3 INTERACTIONS WITH BODIES WITH ATMOSPHERES

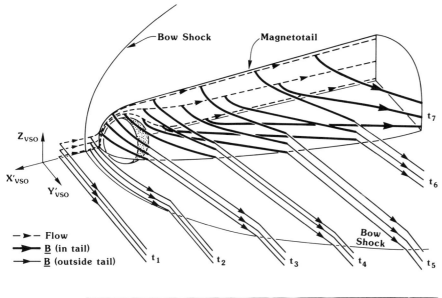

FIG. 8.18. Illustration of the manner in which draped magnetosheath field lines may sink into the wake to form the induced magnetotail of Venus. (From Saunders and Russell, 1986.)

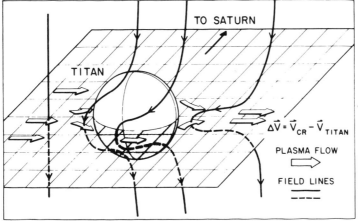

FIG. 8.19. Illustration of the magnetospheric magnetic-field perturbation produced by Saturn's satellite Titan, inferred from *Voyager 1* measurements. (From Ness et al., 1982.)

treatment of the problem, both of these effects would automatically be taken into account. Observations suggest that the pickup ions contribute to the formation of an "induced" magnetotail in the wake behind both Venus and Mars, as illustrated in Figures 8.5 and 8.18. There are $\mathbf{j} \times \mathbf{B}$ forces on the bulk plasma acting in the antisolar direction in the regions of sharp field draping, and there is a tail current sheet across which magnetic merging could conceivably occur. It is not at present certain how much additional planetary ion removal occurs in conjunction with the formation and evolution of this structure. In any case, an analogy can be made between the induced magnetotails of these weakly magnetized planets and cometary tails, which will be discussed a little later in this chapter. It is also worth noting that the similar picture in Figure 8.19 has been invoked to describe the magnetic-field draping observed near Saturn's moon Titan. However, because Titan was in Saturn's magnetosphere at the time of the observations, it is possible that both

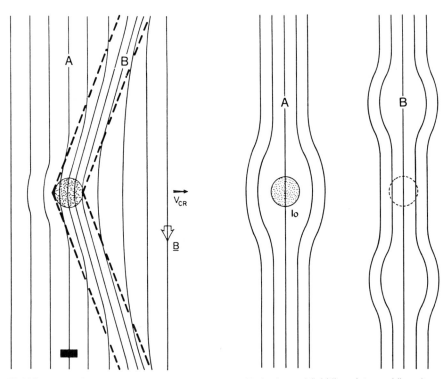

FIG. 8.20. Illustration of the magnetospheric-field perturbation expected to be produced by the Jovian satellite Io. (From Southwood et al., 1980.)

Field lines perturbed by conducting Io and location of Alfvén wings

Projections of field lines into meridian planes through A and B of previous diagram

Titan's ionosphere and Titan itself were penetrated by part of the magnetospheric magnetic field, rather than shielded by ionospheric currents like Venus and Mars. The relative velocity of the magnetospheric plasma is also too low to produce a bow shock, but if ionospheric shielding is weak or absent, the oncoming plasma flow will be absorbed in any event. Nevertheless, the observation of a severely draped tail field and evidence of flow deflection at Titan argue in favor of the importance of currents from picked-up ions and of at least some transient shielding currents. The Titan interaction in the magnetosphere might be expected typically to appear like the lunar interaction, except with a tail like that of Venus or Mars, or like that of Venus or Mars without the effective ionospheric shielding currents.

In the same vein as Titan is Jupiter's moon Io, which has a much thinner atmosphere, but nevertheless possesses an atmosphere that is ionized (by both sunlight and magnetospheric-particle impact) and subjected to a submagnetosonically flowing magnetospheric plasma. The Io–plasma interaction has the distinction of having been globally modeled, in three dimensions, using an MHD computer simulation that included an absorbing body and a fluidlike source term to represent the production of heavy ionospheric ions near the body (Linker, Kivelson, and Walker, 1988). The magnetic-field perturbation predicted by this model without the mass source resembles that illustrated in Figure 8.20, which is based on the assumption of partial shielding of the body. It takes the form of *Alfvén wings,* which occur as the field disturbance caused by Io travels up and down the magnetospheric-field lines at the Alfvén velocity v_A. The angle that this magnetic wake makes with the ambient flow is given by $\theta = \arctan(v_A/u)$. It is expected that any additional field perturbation from pickup ions at Io will be much less severe than that at Titan because of the less extensive atmosphere at Io. It is also worth mentioning that obstacles like Venus and Mars do not produce similar Alfvén wings in the interplanetary field because the external-plasma flow speeds greatly exceed the Alfvén velocity. Io thus represents yet another variation on the theme of flowing plasmas interacting with bodies that have atmospheres.

8.3.2 Comets

Perhaps the most extreme example of an atmosphere interacting with a flowing plasma occurs in the case of comets. When comets are near the sun, they have huge atmospheres and only small solid bodies. In discussing comets, it is convenient to introduce the term *mass loading* as a shorthand way of saying that the background flowing plasma becomes laden with heavy ions of atmospheric origin and slows down as a result (essentially, by conservation of momentum). In fluid treatments of the plasma interaction, mass loading typically is incorporated by putting a source term in the continuity equation. (It is assumed that the newly formed ions are created at nearly zero velocity, so that their initial momentum and energy can be neglected in the momentum and energy equations.) The production function for a comet is the product of two terms: an inverse-square dependence due to the spherical expansion of the outflowing gas in a vacuum, and an exponential decay due to the loss of gas to ionization processes such as photoionization. Thus the source term is

$$Q = \frac{Q_0}{r^2} \exp\left(-\frac{r}{u\tau}\right)$$

where u is in this case neutral gas outflow velocity, r is the distance from the nucleus, and τ is the ionization time.

If the gyration speed of the ionospheric ions at Venus or Mars is neglected, the mass-loaded solar-wind plasma is confined to the low-altitude magnetosheath because the atmosphere is gravitationally confined near the planet. In contrast, the gravitationally unbound, sublimated neutral atmosphere of a comet flows outward from the very small (a few kilometers in diameter) icy nucleus at speeds of about $1 \text{ km} \cdot \text{s}^{-1}$. The huge ionosphere that is produced (a model of which is illustrated, together with a model neutral atmosphere, in Figure 8.21) adopts the velocity of the expanding neutrals and creates a planet-sized cavity or obstacle. The cavity or obstacle in this case is created by the dynamic

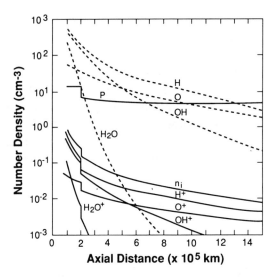

FIG. 8.21. Altitude profile of a model comet atmosphere and ionosphere. (Adapted from Ip and Axford, 1982.)

pressure of the outflowing plasma, instead of by the ionospheric thermal (and magnetic) pressures. The boundary of this cavity is called a *contact surface*. However, there is so much neutral atmosphere outside of the pressure-balance boundary (Figure 8.21) that ionization there produces a region of mass-loaded solar wind that can extend hundreds of thousands of kilometers from the nucleus. Whereas in the planetary-ionosphere case, the solar-wind plasma was slowed and deflected primarily by the presence of the effectively impenetrable obstacle of the shielded ionosphere, in this case the mass loading by itself substantially slows down the plasma flow long before the obstacle of the contact surface is encountered. In fact, because of ion pickup and charge exchange along the way, the composition of the flowing plasma is primarily cometary in origin by that time. The boundary above the contact surface where this composition transition takes place has been called the *cometopause*, following its first observation in the data from the comet Halley. The extended region where the flow is slowed down by the mass loading is permeated by draped interplanetary magnetic field, which in this case is draped because the plasma flow velocity has decreased, although there has been little deflection.

Figure 8.22 illustrates the full complement of features that compose the interplanetary disturbance produced by a comet. The approximate visual limit of the Lyman-α halo, which indicates the extent of the hydrogen component of the neutral atmosphere of the comet, is included for comparison. A bow shock is also indicated, although it is generally weaker than that found near the planets because it occurs in the mass-loaded plasma where the flow has already slowed down.

Like Io, comets have been modeled using self-consistent MHD simulations, with a source term in the continuity equation (Fedder, Lyon, and Giuliani, 1986). These simulations are different from those for Io because the incident flow is in this case supermagnetosonic, and the

8.3 INTERACTIONS WITH BODIES WITH ATMOSPHERES

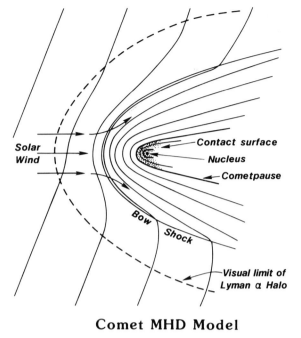

FIG. 8.22. Illustration of the features that make up the comet–solar-wind interaction.

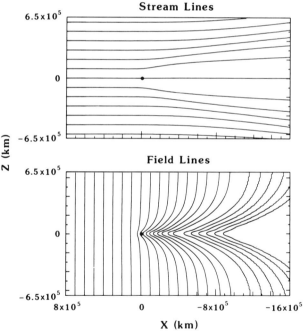

FIG. 8.23. Streamlines and magnetic-field lines derived from the MHD comet-model results for Giacobini-Zinner obtained by Fedder et al. (1986).

solid body of the nucleus itself is considered so small as to be negligible, but the basic equations are the same. The draped magnetic field and nearly undeflected plasma streamlines from the model of the solar-wind interaction with the comet Giacobini-Zinner are shown in Figure 8.23. The weak bow shock is practically invisible in these results. This illustration contrasts with that for the gas-dynamic planetary-magnetosheath model with no mass loading shown in Figure 8.6. In that case, the shock and deflection are the key features of the interaction and the features that produce the draped magnetic field. If we were to add mass loading by an atmospheric source, its effects would largely be confined to the low-altitude magnetosheath (and wake, if a more self-consistent treatment was carried out). For the comet, the extended mass-loading region plays the major role in defining the features of the solar-wind interaction. On the other hand, the magnetic field of an induced planetary magnetotail (Figure 8.18) inferred from observations very much resembles the draped field of a comet. Future modeling of the planetary case should allow us to compare the physical processes responsible for both structures. Of course, one can argue that these fluid models cannot accurately model these systems, because the behavior of individual particles and the details of ion chemistry are neglected, but in the case of both comets and planetary bodies, many aspects of the observed gross properties of the plasma and field are consistent with the fluid-model results.

8.4 CONCLUDING REMARKS

In this chapter, a largely qualitative description has been given of flowing plasma interactions with various types of weakly magnetized or unmagnetized bodies. We have considered the case of moonlike bodies without substantial atmospheres, the cases of Venus, Mars, Titan, and Io, which have enough of an atmosphere to make a major difference in their plasma interactions, and the case of comets, which are practically atmospheres without bodies. Each plasma interaction has distinctive features. The moon absorbs the incident plasma and leaves an empty wake, but produces relatively little distortion in the magnetic field. The weakly magnetized planets with substantial ionospheres deflect the incident plasma, thereby forming a bow shock and magnetosheath, but also produce some near-planet mass loading that probably leads to the formation of the induced magnetotail in the wake. The planetary satellites with atmospheres may represent a hybrid of these two types of interactions in that they absorb the flow, but are still mass-loading it. Comets show how an interaction involving mass loading, by itself, appears. Of these interactions, only those of the moon (under conditions of field-aligned flow), the satellite Io, and comets have been subjected to self-consistent global MHD modeling, but much about Venus and Mars can be understood from the gas-dynamic models with frozen-in magnetic fields. Future investigations of these and other plasma interactions (like those of Neptune's satellite Triton and Pluto) will no doubt involve further global MHD modeling, or rely on models that incorporate some of the single-particle aspects of the interactions. However, as we have learned from past experience, only comparisons between models and observations can cement our knowledge. For example, it has been suggested that both Mars and Io have weak intrinsic dipolar fields that are sufficiently strong to affect their plasma interactions (Kivelson, Slavin, and Southwood, 1979; Southwood et al., 1980; Slavin and Holzer, 1982), in which case our pictures described earlier will take on additional complications.

ADDITIONAL READING

Elphic, R. C., and C. T. Russell. 1983. Global characteristics of magnetic flux ropes in the Venus ionosphere. *J. Geophys. Res.* 88:2993.

Intriligator, D. S. 1989. Results of the first statistical study of Pioneer Venus Orbiter plasma observations in the distant Venus tail: evidence for a hemispheric asymmetry in the pickup of ionospheric ions. *Geophys. Res. Lett.* 16:167.

Kivelson, M. G., and C. T. Russell. 1983. The interaction of flowing plasmas with planetary ionospheres: a Titan–Venus comparison. *J. Geophys. Res.* 88:49.

Luhmann, J. G. 1986. The solar wind interaction with Venus. *Space Sci. Rev.* 44:241.

Luhmann, J. G., and T. E. Cravens. 1991. Magnetic fields in the ionosphere of Venus. *Space Sci. Rev.* 55:201.

Luhmann, J. G., C. T. Russell, L. H. Brace, and O. L. Vaisberg. 1990. Mars intrinsic field and solar wind interaction. In *Mars,* ed. H. Kieffer, C. Snyder, and B. Jakosky. Tucson: University of Arizona Press.

McComas, D. J., J. T. Gosling, C. T. Russell, and J. A. Slavin. 1987. Magnetotails at unmagnetized bodies: comparison of comet Giacobini-Zinner and Venus. *J. Geophys. Res.* 92:10111.

Mihalov, J. D., and A. Barnes. 1981. Evidence for the acceleration of ionospheric O^+ in the magnetosheath of Venus. *Geophys. Res. Lett.* 8:1277.

Ogino, T., R. Walker, and M. Ashour-Abdalla. 1988. A three-dimensional MHD simulation of the interaction of the solar wind with Comet Halley. *J. Geophys. Res.* 93:9568.

Schmidt, H. U., and R. Wegmann. 1982. Plasma flow and magnetic fields in comets. In *Comets,* ed. L. L. Wilkening (p. 538). Tucson: University of Arizona Press.

Wolf-Gladrow, D. A., F. M. Neubauer, and M. Lussem. 1987. Io's interaction with the plasma torus: a self-consistent model. *J. Geophys. Res.* 92:9949.

PROBLEMS

8.1. Calculate the maximum energy (in keV) of potassium and sodium ions picked up near the moon if the solar-wind velocity is 400 km·s^{-1} (assume the ions miss the moon). What is the size of their gyroradius compared with the radius of the moon if the external magnetic-field strength is 5×10^{-5} G?

8.2. There is no bow shock in front of the moon. Explain why. When the interplanetary magnetic field is aligned with the solar-wind flow, what is the radius of the lunar wake if the upstream beta of the plasma is 3 ($R_M = 1,738$ km).

8.3. The moon moves from the magnetosheath into the relatively steady field, near vacuum conditions of the geomagnetic tail lobe. Assuming that the average magnetosheath field is effectively zero, in the tail-lobe field how large a disturbance would be expected in the magnetic measurements of a satellite in a 100-km circular orbit if there were an infinitely conducting core with a radius of 200 km? 400 km? 800 km? Assume that the external field lies in the orbit plane of the spacecraft.

8.4. Assume that the conductivity of a spherical body of radius 1,000 km is 10^{-3} mho·m^{-1} (that of an insulator), and its magnetic permeability is $\mu_0 = 1.26 \times 10^{-6}$ H·m^{-1}. How long would it take for an externally imposed magnetic field to diffuse into the body? How long if the conductivity were 10^5 mho·m^{-1} (that of a good conductor)?

8.5. If the ionosphere of unmagnetized planet x had constant electron and ion temperatures, both equal to 10^5K, and with an exponential electron-density altitude profile given by

$$n_e(h) = 10^5 \exp\left(\frac{-(h-h_0)}{H_p}\right)$$

where the reference altitude h_0 is 130 km, and the plasma scale height H_p is 50 km, near what altitude would the ionopause be found for an upstream solar-wind pressure of 3×10^{-8} dyn·cm^{-2}? What if the pressure were reduced to 1×10^{-8} dyn·cm^{-2}? (If you care to work in SI units, 1 dyn·cm$^{-2} = 0.1$ N·m^{-2}.)

8.6. Using the steady-state momentum equations for electrons and ions in an ionosphere (see Chapter 7), Maxwell's equation $\partial \mathbf{B}/\partial t = -\nabla \times \mathbf{E}$, and Ampère's law $\mu_0 \mathbf{j} = \nabla \times \mathbf{B}$, derive the equation describing the evolution of a horizontal ionospheric magnetic field

$$\frac{\partial B}{\partial t} = \frac{\partial}{\partial h} D \frac{\partial B}{\partial h} - \frac{\partial}{\partial h}(B u_{\text{pl}})$$

where u_{pl} is the vertical velocity, and h is the vertical spatial coordinate. Assume that $\mathbf{B} = B \hat{\mathbf{e}}_x$ depends on h only. The equation for the diffusion coefficient D will naturally come out of this derivation. (Hint: Terms multiplied by the electron mass m can be dropped in the latter stage of the derivation.)

8.7. If a water-ion cometary plasma has a density of 10^5 cm^{-3} at a distance of 10 km from the nucleus and expands outward from the nucleus at 1 km·s^{-1}, at what distance from the nucleus does the cometary-plasma dynamic pressure ρu^2 balance an incident solar-wind pressure of 3×10^{-8} dyn·cm^{-2}? (In other words, what is the subsolar distance of the contact surface?)

9 THE MAGNETOPAUSE, MAGNETOTAIL, AND MAGNETIC RECONNECTION

W. J. Hughes

9.1 INTRODUCTION

THIS CHAPTER DEALS with the magnetopause and the magnetotail, the two regions within the earth's magnetosphere where relatively thin sheets of electric current separate regions of distinctly different magnetic fields. Such current sheets are common features in the solar system. Examples are found from the solar corona to the heliopause; so by extension we expect that they are common throughout the space-plasma universe. Yet these two current sheets, the magnetopause and the tail current sheet or neutral sheet, have been, by far, the most extensively studied, as they are by far the most accessible to direct observation.

A current sheet can be defined as a thin surface across which the magnetic-field strength and/or direction can change substantially. From Ampère's law it follows that the surface must carry a substantial electric current. "Thin" is a relative term, but here we mean that the sheet thickness is very much smaller than the other dimensions of the sheet, or than the sheet radius of curvature. This means that the sheet can be described locally as a plane. The thickness of the two magnetospheric current sheets typically is several hundred kilometers, while they extend for tens of earth radii.

As we shall show later, current sheets arise naturally out of the frozen-in-flux concept introduced in Chapter 2. Collisionless plasmas do not mix easily; instead, they tend to form cells of relatively uniform plasma permeated by a magnetic field. These cells are separated by thin current sheets through which little or no magnetic flux crosses. The interaction between different plasma regimes occurs at these thin boundaries – hence their importance and the importance of the physical processes that occur there. Magnetic reconnection is one of these processes, arguably the most important, certainly in a magnetospheric context. This chapter will focus on magnetic reconnection and how it occurs at the magnetopause and in the geomagnetic tail.

Before we begin our description of magnetic reconnection and the theories that have been developed to describe it, we introduce the magnetopause and magnetotail in rather more detail than was possible in Chapter 6. Throughout these sections, basic physics is stressed and is used to derive the properties of the regions described. We begin with the magnetopause.

9.2 THE MAGNETOPAUSE

The magnetopause is the name given to the upper boundary of the magnetosphere (analogous to older terminology developed for the lower atmosphere). It separates the geomagnetic field and plasma of primarily terrestrial origin from solar-wind plasma, which originates in the sun. Such a boundary was first proposed by Chapman and Ferraro (1931), though they thought in terms of an intermittent "corpuscular stream" from the sun that was present only during periods of solar activity and hence theirs was an intermittent boundary. They argued that the compression of the geomagnetic field by this outflowing plasma caused geomagnetic disturbances seen on the earth's surface that were clearly correlated with solar activity. Biermann (1951), through the analysis of comet tails, showed that the solar wind is present at all times. Thus, some years before its discovery, the magnetopause was predicted to be a permanent feature (e.g., Dungey, 1954a,b).

In the simplest approximation, the magnetopause can be considered as a boundary separating a vacuum magnetic field from a plasma. We saw in Chapter 6 that the location of this boundary can be reliably calculated by requiring the total pressure on the two sides of the boundary to be equal. As a good approximation, the pressure in the magnetosphere, which is mainly magnetic pressure, must match the pressure in the magnetosheath, which is a combination of thermal plus magnetic pressures. The magnetosheath pressure is in turn determined by the solar-wind momentum flux or dynamic pressure. The dominant pressure terms in the solar wind and at the nose of the magnetosphere are in approximate equilibrium:

$$\rho_{SW} u_{SW}^2 = B_{MS}^2 / 2\mu_0 \tag{9.1}$$

where the subscripts SW and MS refer to solar wind and magnetosphere.

Figure 9.1 is a simple sketch of the magnetosphere in the noon–midnight meridian. The magnetopause is closed; that is, no magnetic-field lines cross the boundary, or, put another way, the component of **B** normal to the boundary is zero. We also assume for the moment that the interplanetary magnetic field (IMF) entrained in the solar-wind flow is zero. Ampère's law applied across the boundary shows that the boundary must carry the currents shown flowing into and out of the plane of this figure. In common with most plasma boundaries in space, the magnetopause is a current sheet. In the simplest case, where there is no coupling of energy or momentum across the magnetopause, these currents close on themselves and also through the current sheet that flows

9.2 THE MAGNETOPAUSE

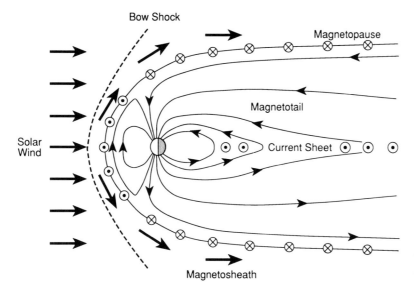

FIG. 9.1. Cross section of the simplest model of the magnetosphere in the noon–midnight meridian. In this so-called closed model, the geomagnetic field is perfectly confined by the sheet currents flowing on the magnetopause. A second current sheet flows across the midplane of the magnetotail and connects with the magnetopause currents at the flanks of the tail. The solar-wind flow (thick arrows) is deflected at the bow shock and flows around the magnetosphere, forming the magnetosheath.

in the center of the magnetotail. In addition to satisfying Ampère's law, those currents must satisfy a momentum equation, in that the $\mathbf{j} \times \mathbf{B}$ force must be that required to deflect the solar-wind plasma, as can be seen qualitatively from Figure 9.1.

The size and shape of the magnetopause have been treated in Chapter 6, and that discussion need not be repeated here. The field strength immediately inside the near-equatorial dayside magnetopause is slightly greater than twice that of the dipole field at the same location. This is because the magnetopause current, to first order, cancels the dipole field outside the magnetopause, and so it must create an equal but oppositely directed field just inside the magnetopause, which adds to the dipole field, doubling it. The curvature of the boundary enhances this effect. The effect of the magnetopause current is felt at the earth's surface. When a sudden increase in solar-wind dynamic pressure, as often follows the passage of an interplanetary shock, reaches the earth, the magnetosphere is compressed; the magnetopause moves nearer the earth, and at the same time the magnetopause current intensifies. The movement and intensification of the current are sensed at the earth's surface as a sudden increase in the geomagnetic-field intensity of a few tens of nanotesla. This feature is known as a sudden impulse (SI), or a storm sudden commencement (SSC) if a geomagnetic storm follows.

To gain some insight into the interface between a plasma and a magnetic field, let us consider the simplest form of such a boundary, shown in Figure 9.2. This is often called the Chapman-Ferraro current layer. We consider only a cold beam of electrons and ions on the left and a uniform magnetic field on the right. In reality, a boundary between an unmagnetized plasma and a vacuum magnetic field could occur only with a thermal distribution of particles instead of a cold beam on the left side of the boundary. Magnetohydrodynamic (MHD) pressure balance

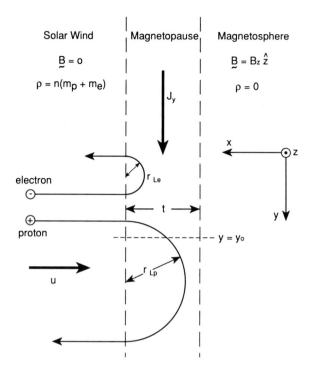

FIG. 9.2. A simple, planar magnetopause boundary separating an unmagnetized solar wind (left) from a magnetosphere containing no plasma (right). The magnetopause current, which flows down the page, is carried by the collective action of the solar-wind particles, each of which performs a half gyration in the geomagnetic field before returning to the magnetosheath.

[equation (9.1)] holds across the boundary, but we shall rederive the condition from a particle viewpoint.

When an unmagnetized proton and electron begin to penetrate the boundary, they sense a $\mathbf{u} \times \mathbf{B}$ Lorentz force, which causes them to gyrate. After half an orbit they exit the boundary, moving antiparallel to the solar-wind flow, as shown in Figure 9.2. In performing the half orbit, the proton moves $2\rho_{cp}$ down the page, and the electron moves twice the electron gyroradius, $2\rho_{ce}$, up the page, giving rise to a current (charges moving in opposite directions). As the inertia of the protons carries them much farther than the electrons, their motion constitutes most of the current. However, in a more realistic situation, the greater thermal velocity of the electrons would partially offset their smaller mass. We can compute the strength of the current by considering the number of protons that cross some particular x–z plane, say $y = y_0$, per unit time. From Figure 9.2 we see that any proton entering the boundary over a region $2\rho_{cp}$ wide (i.e., with a y-component between $y_0 - 2\rho_{cp}$ and y_0) will cross the $y = y_0$ plane. The flux of protons that encounters that section of the boundary is $2\rho_{cp} nu$ per unit length in the z-direction, where we have dropped the SW subscripts for convenience. As each proton carries a charge e, the current crossing $y = y_0$ per unit length in z is

$$I = 2\rho_{cp} nue = \frac{2nm_p}{B_z} u^2 \tag{9.2}$$

where we have used $\rho_{cp} = um_p/eB_z$. In a self-consistent treatment, the magnetic field would be modified by the currents carried by the electrons and protons, so that the gyroradius would not be given in terms of the unperturbed field. Now, applying Ampère's law across the boundary and noting that $I = \int j\, dx$, we get

$$B_z = \mu_0 I \tag{9.3}$$

Combining (9.2) and (9.3), we get

$$\frac{B_z^2}{2\mu_0} = nm_p u^2 = \rho_{SW} u_{SW}^2 \tag{9.4}$$

which brings us back to the pressure-balance criterion.

Reverting now to the fluid (MHD) picture, this current must provide the $\mathbf{j} \times \mathbf{B}$ force integrated across the boundary needed to balance the rate of change of solar-wind momentum or to divert the solar-wind flow. The momentum flux into the boundary is $2\rho_{SW} u_{SW}^2$, where the factor 2 comes from the fact that in our picture the plasma is perfectly reflected, so that its velocity change is $2u$. Equating these gives

$$2\rho_{SW} u_{SW}^2 = |\mathbf{I} \times \mathbf{B}| = B^2/\mu_0 \tag{9.5}$$

and we arrive back at the same formulation.

The key point here is that the particle and fluid pictures give us equivalent answers. They are two alternative ways of looking at plasma-physics problems. For the description of large-scale processes, neither is intrinsically better than the other. Each description has its merits, and in some situations it is easier to work with one, and in other situations the other, much as we sometimes choose to think of photons as particles and at other times as a wave. In this chapter we shall first treat magnetic reconnection using an MHD formulation. Later we shall treat the same problem from a particle viewpoint. Each approach will teach us something different about the reconnection process, but also by comparing the two descriptions we shall learn more about the duality of fluid and particle descriptions of a plasma.

Before we leave the magnetopause, a few words about the shortcomings of our simplified descriptions are in order. The picture of the boundary used in Figure 9.2 is not self-consistent. We took the field strength to increase discontinuously at the outer edge of the boundary, rather than matching the gradient in field strength with the local current density. Also, we neglected the electric field that arises when the protons penetrate farther than the electrons into the magnetic-field region, causing a charge separation. Such an electric field tends to drag the electrons in farther (and hold the protons back). When the calculation is done self-consistently, a solution is obtained that preserves charge neutrality, and the characteristic length scale turns out to be the electron inertial length c/ω_{pe}, not a gyroradius. Even this solution is of limited value, as it is unstable to the two stream instability because of the counterstreaming plasmas entering and leaving the boundary. More importantly, our initial model is grossly oversimplified, as we ignored the magnetosheath mag-

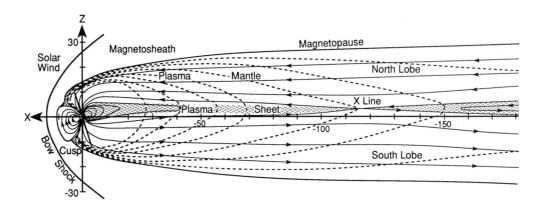

FIG. 9.3. Noon–midnight cross section of the magnetosphere and geomagnetic tail drawn to scale. The moon's orbit is near the center of the diagram. The plasma sheet, carrying the cross-tail current sheet, separates into two tail lobes. A magnetic x-line or neutral point is shown $115R_E$ behind the earth. The dashed lines show the trajectories followed by particles in the plasma mantle (cf. Section 9.6.2). (Adapted from Pilipp and Morfill, 1978.)

netic field and the magnetospheric plasma. In practice, the magnetopause turns out to be much thicker than either of these distances and has been measured to be several hundred or a thousand kilometers thick, which is several ion gyroradii.

9.3 THE GEOMAGNETIC TAIL

The geomagnetic tail is the name given to the region of the earth's magnetosphere that stretches away from the sun behind the earth. It is a region of great importance to the magnetosphere, for, as we shall see, it acts as a reservoir of plasma and energy. The energy and the plasma are released into the inner magnetosphere aperiodically during magnetically disturbed episodes called magnetospheric substorms (see Chapter 13). A current sheet lies in the center of the tail, embedded within a region of hot plasma, the plasma sheet, that separates two regions called the tail lobes. As Figure 9.3 shows, the two tail lobes connect magnetically to the two polar regions of the earth and are identified as the north and south lobes. The magnetic field in the north (south) lobe is directed toward (away from) the earth; hence the need for a current sheet to separate these two regions of oppositely directed magnetic fields.

In the early 1960s, spacecraft observations established the existence of the tail. The history of those early observations has been reviewed by Ness (1987). Those early measurements showed that the geomagnetic-field strength in the near-earth tail lobes is about 20 nT. Here we shall use this one observational parameter, together with parameters derived from observations of the earth's polar ionosphere, the polar cap, to derive the basic properties of the tail.

Figure 9.3 illustrates that the magnetic flux from each tail lobe connects to one of the polar caps. By requiring that the magnetic flux be conserved between the polar cap and the tail lobe, we can obtain an estimate of the radius of the tail. The polar cap is the area around the geomagnetic pole bounded by the auroral oval. The flux leaving the polar cap, given by the integral of the vertical component of the field strength over the area of the polar cap, is

$$\Phi_{PC} = 2\pi(R_E \sin\theta_{PC})^2 B_0 \tag{9.6}$$

where θ_{PC} is the colatitude of the equatorward edge of the polar cap (assumed circular), and B_0 is the equatorial field strength (half the polar field strength). This must equal the flux in a tail lobe (assumed to be a semicircle in cross section):

$$\Phi_T = \frac{1}{2}\pi R_T^2 B_T \tag{9.7}$$

where R_T is the tail radius, and B_T is the magnetic-field strength in the tail lobe. Equating the fluxes in the tail lobe and the polar cap gives

$$\frac{R_T}{R_E} = \left(\frac{4B_0}{B_T}\right)^{\frac{1}{2}} \sin\theta_{PC} \tag{9.8}$$

Taking $\theta_{PC} = 15°$, $B_0 = 31{,}000$ nT, and $B_T = 20$ nT, we find $R_T = 20 R_E$; if $B_T = 10$ nT (which is more typical of the distant tail), $R_T = 29 R_E$.

In a static tail there must be pressure equilibrium between the tail lobe and both the plasma sheet and the solar wind. We can use this to estimate the plasma-sheet properties and the geometry of the distant tail. Equating the magnetic pressure (which is much larger than the particle pressure) in the tail lobes with the particle pressure (which is much larger than the magnetic pressure) in the plasma sheet, we find

$$B_T^2/2\mu_0 = nk(T_i + T_e) \tag{9.9}$$

where n is the particle-number density in the plasma sheet, and T_i and T_e are the ion and electron temperatures. Again, using $B_T = 20$ nT, we get a plasma-sheet pressure of 0.24 nPa or 1,500 eV·cm^{-3}. This argument tells us nothing about n or T separately, but agrees well with typical plasma-sheet parameters of $n \sim 0.3$ cm^{-3}, $T_i \sim 4.2$ keV, and $T_e \sim 0.6$ keV.

We can calculate the current that must be carried by this plasma by applying Ampère's law across the current sheet. The total change in magnetic field across the plasma sheet is twice the lobe field strength, because the fields on either side are equal in size but oppositely directed. So

$$\Delta B = 2B_T = \mu_0 I \tag{9.10}$$

where I is the sheet current density. Again, using $B_T = 20$ nT gives $I = 30$ mA·m^{-1}. This may appear to be a rather small current, until we consider the length of the current sheet. This same current density can also be given as 30 A·km^{-1} or 2×10^5 A·R_E^{-1}. Thus, 10^6 A is carried in each $5R_E$ of the length of the tail, which means that diversion of only a small

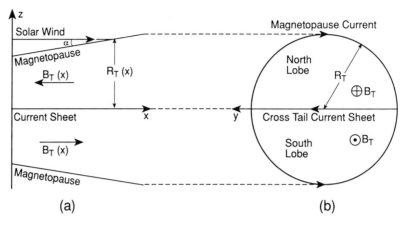

FIG. 9.4. Sketch of magnetotail geometry: (a) A cross section in the midnight meridian. When the geotail radius R_T increases with distance downtail, x, a component of the solar-wind dynamic pressure, acts normal to the magnetopause boundary. (b) A cross section normal to the tail axis. The tail current sheet closes via the magnetopause current in a θ configuration, forming two semicircular solenoids generating the oppositely directed magnetic fields in the two tail lobes.

part of the tail current is sufficient to explain the ionospheric auroral electrojet currents observed during substorms (Chapter 13).

Pressure balance across the tail magnetopause was used by Coroniti and Kennel (1972) to deduce how the tail radius increases with distance down the tail, or how the tail flares. If the tail flares, Figure 9.4 shows that there will be a component of the solar-wind dynamic pressure normal to the tail magnetopause. So the normal pressure used in the balance must include a component of the dynamic pressure plus an isotropic pressure p_0, which includes both thermal and magnetic pressures. Balancing this with the tail-lobe magnetic pressure gives

$$\rho u_{SW}^2 \sin^2\alpha + p_0 = B_T^2/2\mu_0 \tag{9.11}$$

The tail flaring angle α is related to the increase in tail radius with distance,

$$\frac{dR_T}{dx} = \tan\alpha$$

and the magnetic-field strength in the lobe, B_T, is given by

$$B_T(x) = \frac{2\Phi_T}{\pi R_T^2(x)}$$

where Φ_T is the total magnetic flux in one tail lobe. We shall assume Φ_T is not a function of downtail distance, (i.e., that no flux crosses either the tail magnetopause or the tail neutral sheet). Neither of these assumptions is strictly true, but as we shall see, over a limited range of distances along the tail axis the amount of flux crossing both these boundaries is small.

If α is small so that $\sin\alpha \sim \tan\alpha = dR_T/dx$, where x is positive in the antisolar direction, then

$$\rho u_{SW}^2 \left(\frac{dR_T}{dx}\right)^2 + p_0 = B_T^2/2\mu_0 = \frac{1}{2\mu_0}\left(\frac{2\Phi_T}{\pi R_T^2}\right)^2 \tag{9.12}$$

Rearranging, we get

$$M^2\left(\frac{dR_T^2}{dx}\right)^2 + 1 = \left(\frac{R_*}{R_T}\right)^4 \tag{9.13}$$

where $M = (\rho u_{SW}^2/p_0)^{\frac{1}{2}}$ is the solar-wind Mach number (sonic or Alfvén, depending on the dominant part of p_0), and

$$R_* = \left(\frac{2\Phi_T^2}{\mu_0 \pi^2 p_0}\right)^{\frac{1}{4}} \tag{9.14}$$

is the asymptotic radius of the tail. Rewriting (9.13) we get

$$\frac{dR_T}{dx} = \frac{1}{M}\left[\left(\frac{R_*}{R_T}\right)^4 - 1\right]^{\frac{1}{2}} \tag{9.15}$$

Integrating (9.15) and applying an earthward boundary condition that $R_T = R_0$ at $x = x_0$ gives

$$\frac{x - x_0}{MR_*} = \int_{R_0/R_*}^{R_T/R_*} \frac{dr}{(r^{-4} - 1)^{\frac{1}{2}}} \tag{9.16}$$

where $r = R_T/R_*$.

The distance downtail at which tail flaring ceases, x_*, is finite and can be estimated by evaluating (9.16) for the case $R_T = R_*$. Coroniti and Kennel (1972) show that this gives

$$x_* - x_0 \simeq MR_*\left[0.6 - \frac{1}{3}(R_0/R_*)^3\right] \tag{9.17}$$

Taking as the solar-wind parameters $M = 9$ and $p_0 = 3.2 \times 10^{-11}$ N·m^{-2} (Table 4.3), the estimate of Φ_T from (9.7) as 4.3×10^8 Wb, and the initial tail radius and distance as $R_0 = 18R_E$ and $x_0 = 10R_E$ yields $R_* \sim 27R_E$ and $x_* \sim 140R_E$ and an asymptotic lobe field strength of 9 nT. Thus, the tail reaches an asymptotic radius when the magnetic pressure of the lobes balances the thermal pressure of the solar wind. This occurs at around $150R_E$, but clearly depends on solar-wind conditions. Having obtained numerical estimates for geotail properties starting from rather simple assumptions of static equilibrium, we next show that these estimates agree remarkably well with observations.

In the early days of magnetospheric physics, the tail was observed out to lunar orbit ($\sim 60R_E$), but it was not until the 1980s that the more distant tail was surveyed. In 1983, *ISEE 3* reached an apogee of approximately $220R_E$ in the downstream direction, having been put into a deep-tail orbit using close lunar flybys. The projection of the orbit on the ecliptic plane is shown in Figure 9.5. Tick marks along the orbit path mark magnetopause crossings, which occur across the entire width of the tail, meaning that the tail flaps, moving up and down and sideways by at least its own radius. However, the statistics of the magnetopause crossings in Figure 9.6 show that the most probable magnetopause location is about $30R_E$ from the nominal axis of the tail, calculated assuming

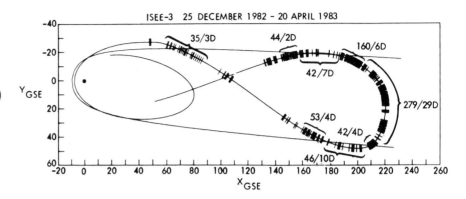

FIG. 9.5. Orbit of *ISEE 3* (Dec. 25, 1982–April 20, 1983) during part of its deep-geotail mission, projected onto the GSE x–y plane. Dashes indicate magnetopause crossings that occurred at all values of y, showing that the tail frequently moves at least its own radius in the y-direction. The model magnetopause is drawn for comparison with an aberration of 4° off the earth–sun line to account for the earth's orbital motion of 30 km·s^{-1}. (Adapted from Slavin et al., 1985.)

that the tail aligns with the average solar-wind flow direction. This is in excellent agreement with our estimate of the asymptotic radius of the tail.

Figure 9.7 shows the value of the magnetic-field strength measured by *ISEE 3* in the lobes as a function of distance down the tail. The lobe field strength falls off with distance out to somewhere in the vicinity of $140 R_E$, at which point it reaches an asymptotic value of 9.2 nT. Again, these two values agree with the Coroniti and Kennel calculations.

The plasma parameters of the near magnetotail have been established using data from *IMP 6* at distances around $30 R_E$. The three panels of Figure 9.8 show, from left to right, plasma particle-number density, average proton energy, and average electron energy. Within each panel, the data are sorted by plasma β. The lowest values of β correspond to the tail lobes, where the mean particle-number density is 0.03 cm^{-3}, and 66 percent of observations are below the detection threshold at 0.01 cm^{-3}. Thus, in spite of its high temperature, the lobe is an extremely good vacuum, with pressure about 10^{-15} Torr (10^{-10} Pa), much better than can be achieved in the laboratory. The two higher ranges of β correspond to the plasma sheet; $\beta > 2$ corresponds to the innermost parts of the plasma sheet near the current sheet, where the magnetic field is weakest. The intermediate range of $0.25 < \beta < 2$ corresponds to the so-called plasma-sheet boundary layer and to the outer parts of the central plasma sheet. We shall distinguish between these regions later. Here it suffices to note that average numbers are $n \sim 0.3$ cm^{-3}, $T_i \sim 4.2$ keV, and $T_e \sim 0.6$ keV, close to what we needed to balance the lobe magnetic pressure. Table 9.1 summarizes typical values of some parameters in the magnetotail.

9.4 MAGNETIC RECONNECTION

9.4.1 Concept of Magnetic Reconnection

The frozen-in-flux criterion was discussed in Chapter 2. Restated briefly, it is that if the electrical conductivity of a fluid is large enough,

9.4 MAGNETIC RECONNECTION

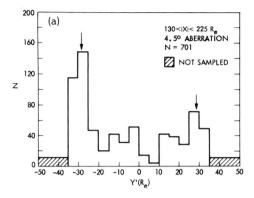

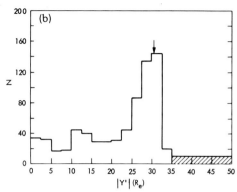

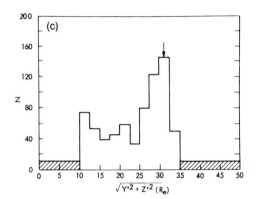

FIG. 9.6. Histograms showing the number of times *ISEE 3* encountered the magnetopause in the distant tail ($-130 > x > -225 R_E$) at different distances from the central axis of an aberrated tail (cf. Figure 9.5). The arrows mark the most probable location. Panel A shows distance along the aberrated *y*-axis, showing that the magnetopause can be encountered at any value of *y*. This point is emphasized in panel B, in which location is a function of $|y|$. Panel C adds the effect of the north–south motion of the spacecraft by plotting radial distance from the aberrated *x*-axis. (Adapted from Slavin et al., 1985.)

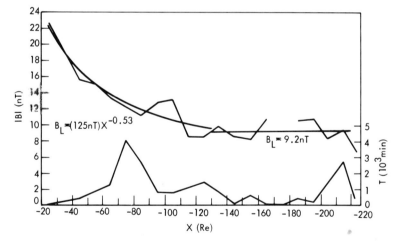

FIG. 9.7. Average magnetic-field strength measured by *ISEE 3* in the tail lobes as a function of distance down the tail. Fits to the data show it falling off as the square root of distance until it reaches a constant value around 140 R_E. The bottom trace shows the length of time *ISEE 3* spent at each distance. (Adapted from Slavin et al., 1985.)

the magnetic flux linking a particular fluid element remains fixed, which can be pictured as magnetic-field lines being convected with the fluid. Magnetic reconnection is a process that can take place only if this criterion breaks down, which at first sight seems very unlikely in the collisionless plasma of space. From Chapter 3, equations (3.1)–(3.8) allow us to examine the conditions under which the frozen-flux condition breaks down in the simple geometry of the magnetotail. As before, the flux is frozen into the field when the magnetic Reynolds number

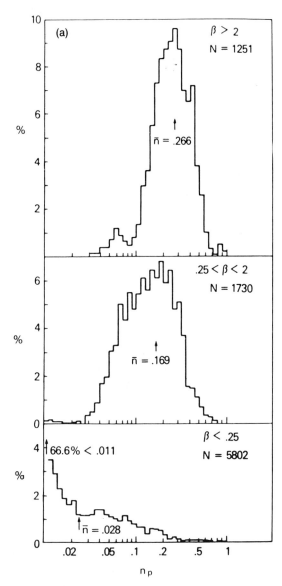
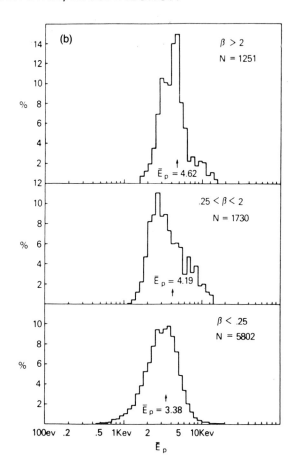
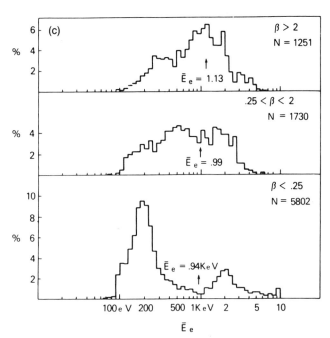

FIG. 9.8. Tail plasma parameters in the form of histograms showing *IMP 6* observations of the plasma-number density (a), average proton energy (b), and average electron energy (c). The top two plots in each panel are for moderate β (0.25–2) and high β (>2) and correspond to the plasma sheet. The bottom plots are for low β (<0.25) and correspond to the tail lobes. (Adapted from Fairfield, 1987.)

9.4 MAGNETIC RECONNECTION

TABLE 9.1. Typical Near-Tail Plasma and Field Parameters

	Magnetosheath	Tail Lobe	Plasma-Sheet Boundary Layer	Central Plasma Sheet
n (cm^{-3})	8	0.01	0.1	0.3
T_i (eV)	150	300	1,000	4,200
T_e (eV)	25	50	150	600
B (nT)	15	20	20	10
β	2.5	3×10^{-3}	10^{-1}	6

$R_m \simeq \mu_0 \sigma u L$ is large compared with unity. This is generally true in the magnetosphere.

For example, if L is a few R_E (the scale size of the magnetospheric cavity) and if u is 100 km·s^{-1} (a typical magnetospheric flow speed), then R_m is 10^{11}, implying that the frozen-in-flux condition holds to a very high accuracy. R_m must become very small, of order unity, for the frozen-in-flux condition to break down. But a direct consequence of the frozen-in-flux concept is that thin boundaries naturally form between different plasma regimes. Within these thin boundaries, scale lengths are small, and so MHD can and does break down locally.

If the frozen-in-flux condition holds, plasma can mix only along flux tubes, never across them. Hence, plasma properties should be relatively uniform along flux tubes, but abrupt changes in plasma conditions can occur from one flux tube to the next. If the plasma is all from the same source, large changes are unlikely. However, if two distinct plasma regimes interact, since they cannot diffuse into each other, a thin boundary will separate the plasmas from the two sources.

The magnetic field on either side of this boundary will be tangential to the boundary, but the two magnetic fields may otherwise have quite different directions and/or strengths. This is a description of a current sheet. Thus the frozen-in-flux concept leads to a picture of large, relatively uniform regions of plasma separated from neighboring regions (which usually originate from some different source) by relatively thin current sheets, such as the magnetopause. As we shall see, it is within these current sheets that the frozen-in-flux concept can break down.

As in Chapter 3, we approach the concept of reconnection by considering what happens over time at a one-dimensional boundary separating two regions of oppositely directed magnetic fields of equal strengths. Assume that there is no plasma flow; then the dynamo equation (3.3) becomes a pure diffusion equation:

$$\frac{\partial B_x}{\partial t} = \frac{1}{\mu_0 \sigma} \frac{\partial^2 B_x}{\partial z^2} \tag{9.18}$$

where we have taken the magnetic field to be in the x-direction, and z to be in the direction normal to the boundary, the coordinate system usually used by those studying magnetotails. If initially ($t=0$) the boundary is infinitely thin, then the solution of (9.18) takes the form of an error function:

$$B_x(z) = B_0 \text{erf}\left\{\left(\frac{\mu_0 \sigma}{2t}\right)^{\frac{1}{2}} z\right\} \tag{9.19}$$

where

$$\text{erf}(u) = \frac{2}{\pi^{\frac{1}{2}}} \int_0^u e^{-v^2} dv \tag{9.20}$$

The solution is illustrated in Figure 9.9a. The current density is a Gaussian whose width increases with time, but only as $t^{\frac{1}{2}}$. The magnetic flux diffuses down the field gradient toward the central plane, where it annihilates with the oppositely directed flux diffusing from the other side of the sheet. This reduces the field gradient, thus reducing the diffusion rate and slowing the whole process. Thus the process is self-limiting. However, magnetic-field energy has been converted into heat via Joule heating, and the resulting pressure increase is what is needed to balance the decrease of magnetic-field pressure.

To maintain this process in a steady state, flow must transport magnetic flux toward the boundary at the rate at which it is being annihilated, as is shown in Figure 9.9b. We impose this inflow from the outside as a boundary condition by imposing an electric field in the y-direction. We obtain equilibrium by matching the inflow and annihilation rates. We can see immediately that the faster the inflow, the steeper the equilibrium gradient becomes, in order to drive faster diffusion, which in turn requires a larger rate of annihilation or energy release.

In a steady state ($\partial/\partial t \equiv 0$), the imposed E_y must be spatially uniform. This follows from Faraday's law, which becomes

$$\nabla \times \mathbf{E} = \frac{\partial \mathbf{B}}{\partial t} = 0 \tag{9.21}$$

the x-component of which reduces to

$$\frac{\partial E_y}{\partial z} = 0 \tag{9.22}$$

implying that E_y is everywhere the same. We now apply Ohm's law in its two forms. Well away from the current, $\mathbf{E} = \mathbf{u} \times \mathbf{B}$, which gives

$$E_y = uB_0 \tag{9.23}$$

where u is the inflow rate, and B_0 is the magnetic-field strength at large z. In the center of the current sheet, B vanishes, and Ohm's law becomes

$$\mathbf{E} = \mathbf{j}/\sigma \quad \text{or} \quad E_y = j_y/\sigma \tag{9.24}$$

If the current-sheet thickness is taken as $2l$ (cf. Figure 9.9), Ampère's law applied across the current sheet gives

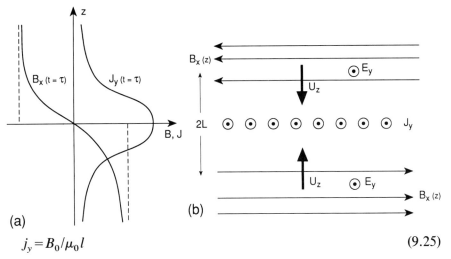

FIG. 9.9. (a) Solution to (9.21) showing the magnetic-field strength and current density across an idealized current sheet. As flux annihilates at the midplane ($z=0$), the Gaussian-shaped current distribution widens. (b) A sketch of the current sheet in the x–z plane showing inflow of magnetic flux toward the $z=0$ plane and the current flowing in the y-direction separating the two oppositely directed magnetic fields.

$$j_y = B_0/\mu_0 l \qquad (9.25)$$

Combining (9.23), (9.24), and (9.25) to get at expression for the current-sheet thickness gives

$$l = 1/\mu_0 \sigma u \qquad (9.26)$$

The magnetic Reynolds number derived using l as the length scale and the inflow speed as the characteristic velocity identically equals unity. This means that the current-sheet thickness adjusts to produce a balance between diffusion and convection at the edges of the current sheet. In other words, the current-sheet thickness is being determined by this balance. This results in current sheets that are very thin compared with global scale lengths, for which we obtained very large values of R_m.

In introducing plasma inflow, we have created an unphysical picture. There is plasma inflow, but we have not provided any means for plasma to escape from the system. In order to do this, we must introduce a second dimension by allowing annihilation to occur over only a limited extent of the field lines. This results in what we shall call the basic x-line reconnection picture, which is illustrated in Figure 9.10.

In Figure 9.10, rather than the magnetic-field strength being zero over an entire plane, as in Figure 9.9, it is zero only along a single line that intersects the plane of the figure in the center of the figure. This magnetic configuration is known as an x-type neutral line. A uniform E_y out of the page is still required for a steady state, as (9.22) still applies. This field drives flow inward from the top and bottom of the figure and outflow toward both sides. The small shaded region surrounding the x-line is the region within which the magnetic Reynolds number is less than unity; this is known as the diffusion region.

Here we have a fundamentally different picture. Magnetic-field lines enter the diffusion region from the top and bottom, and instead of being annihilated, they leave from both sides. In the process, they are "cut" and "reconnected" to different partners. Plasma originally on different

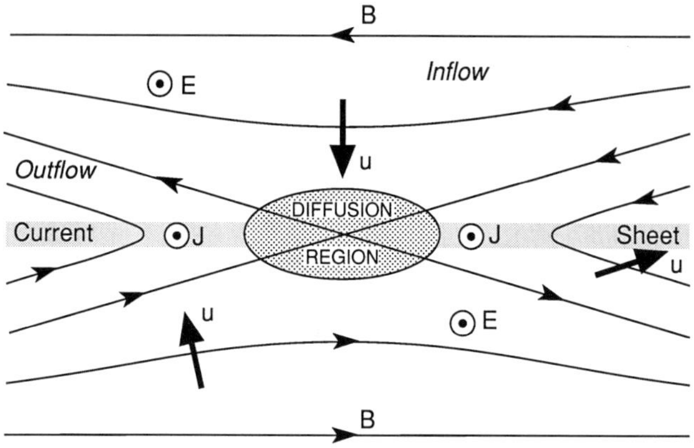

FIG. 9.10. Illustration of magnetic reconnection occurring at an *x*-type magnetic neutral line. Plasma and magnetic field flow in from the top and bottom of the figure and flow out toward both sides. Only in the diffusion region, where $R_m < 1$, is plasma not tied to magnetic-field lines.

flux tubes, coming from different regions, now finds itself on a single flux tube in total violation of the frozen-in-flux theorem. Previously the current sheet separated two magnetic-field regions; now magnetic flux crosses the current sheet. In other words, we now have an "open" boundary with a finite normal magnetic-field component, as opposed to the previous "closed" boundary. Plasma can cross the boundary by following flux tubes.

Another important consequence is that even though only a finite length of field enters the diffusion region, the process affects the entire flux tube. Plasma is now free to move from one region to another, fundamentally changing the nature of the boundary.

Reconnection was first proposed as a process to explain the rapid heating of plasma at the expense of magnetic-field energy that occurs in solar flares. And indeed, magnetic-field energy built up in the inflow regions is converted into heating and acceleration of the plasma in the outflow regions in the basic *x*-line picture (but we have yet to discuss how fast this occurs). However, it is perhaps the change in magnetic connection or topology that is the more profound effect, for this allows previously unconnected regions to exchange plasma readily, and hence mass, energy, and momentum.

9.4.2 Magnetic Reconnection and Magnetospheric Dynamics

By the late 1950s it was realized that the plasma flows seen in the polar and auroral ionospheres must map out to, and be driven by, magnetospheric flows. Magnetometer and other ground-based measurements showed that plasma flowed over the polar regions from noon toward midnight and then back toward the dayside at somewhat lower latitudes, corresponding very roughly to the auroral zone, on both the dawn and dusk sides. This flow pattern is roughly stationary in a sun-fixed coordinate system, so that any observatory will detect a diurnal variation as it rotates under it. This double-vortex flow pattern looks

9.4 MAGNETIC RECONNECTION

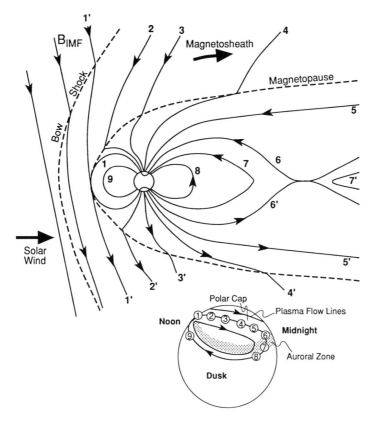

FIG. 9.11. Flow of plasma within the magnetosphere (convection) driven by magnetic reconnection. The numbered field lines show the succession of configurations a geomagnetic field line assumes after reconnection with an IMF field line (1′) at the front of the magnetosphere. Field lines 6 and 6′ reconnect at a second x-line in the tail, after which the field line returns to the dayside at lower latitudes. The inset shows the positions of the feet of the numbered field lines in the northern high-latitude ionosphere and the corresponding high-latitude plasma flows, an antisunward flow in the polar cap, and a return flow at lower latitudes.

similar to thermally driven flow cells, and so the term "convection" became attached to it, although it is in no sense thermally driven. It was natural to map this flow along flux tubes using the frozen-in-flux concept and consider the corresponding magnetospheric flow. If all flux tubes remain inside the magnetosphere, that is, if the magnetopause is a magnetically closed boundary in the sense that no magnetic-field lines cross the magnetopause, then this flow pattern is very similar to the familiar flow driven in a falling raindrop by the viscous drag at the drop/air interface. This analogy provided what seemed to be a natural explanation of the circulation, though the exact nature of the viscous interaction between the solar wind and magnetospheric plasma was a mystery (the classic viscosity is zero as long as particles are tied to individual flux tubes). However, an alternative explanation was proposed by Dungey (1961). He showed that if the geomagnetic and interplanetary magnetic fields reconnect near the front of the magnetosphere, the observed flow pattern will result. For simplicity, assume that the interplanetary field is directed predominantly southward, as illustrated in Figure 9.11. Then the magnetic field driven by the solar-wind flow against the front of the magnetosphere will be approximately antiparallel to the geomagnetic field on the other side of the magnetopause. Suppose that a magnetic x-line forms there and that reconnection occurs between the field lines labeled 1 and 1′. Then, instead of a purely geomagnetic-field line with both ends attached to the earth and an interplanetary-field line with both ends on the sun or at least stretching far from the earth, we obtain two field lines of a new type. These new field lines each have one end attached to the earth, one near the North Pole and one near the South Pole, and the other end stretching out into interplanetary space. The solar-wind flow will pull the solar-wind portion of the field line antisunwards, or, to put it another way, the plasma on the flux tube will sense an electric field $\mathbf{E} = \mathbf{u}_{SW} \times \mathbf{B}_{SW}$. In a steady state, the electric field must be sensed all along these now open flux tubes, as field lines are equipotentials. At the ionospheric end of the field line, this electric field, which is directed from dawn toward dusk, drives flow from noon toward midnight, as observed. Thus the field line moves successively through the numbered locations in Figure 9.11, and this process naturally draws out flux tubes to form the geotail.

If this process went on indefinitely without some method of returning magnetic flux to a "closed" state, the entire geomagnetic field would soon be connected with the interplanetary field. The return of flux is achieved by reconnection at another x-line in the geotail. Here, two open field lines, one from the northern polar region and northern tail lobe and the other from the south, reconnect to form a newly closed geomagnetic field line and a new, purely interplanetary field line. The new interplanetary flux tube contains some plasma of terrestrial origin and is distorted and stressed; it continues flowing to the right and ultimately rejoins the solar-wind flow. The new geomagnetic field line is also stressed and attempts both to flow and to relax earthward, though this may be hindered by the pressure of the plasma contained on the flux tube. The flow circuit is finally closed when the newly closed field lines flow around either the dawn or dusk side of the earth back to the dayside. The inset in Figure 9.11 shows the flow of the end of the field line in the northern ionosphere and shows how the return flow occurs at lower latitudes.

This description is grossly oversimplified. In practice, the entire pattern is inherently nonsteady (which can give rise to inductive electric fields that do not map along field lines), and although in a time-averaged sense the reconnection rates at the magnetopause and in the tail must be equal, on an instantaneous basis they probably rarely are. It would be fair to say that magnetospheric physics is largely about understanding the dynamics and transport associated with this flow. And although magnetic reconnection is the dominant means of momentum coupling to the solar wind, some sort of viscous interaction probably does account for some small fraction of momentum transfer, perhaps 10–20 percent on average. Nevertheless, the Dungey picture of an open magnetosphere provides a framework to which details may be added; it demonstrates the fundamental role that magnetic reconnection plays in magnetospheric physics.

We can use the open model of magnetospheric convection (Figure 9.11) to obtain an estimate of the length of the tail (Dungey, 1965). As field lines are swept tailward in the solar wind, their ionospheric footprints are swept across the polar cap from near noon to near midnight. The plasma flow across the polar cap can be measured from the ground; it flows at speeds of a few hundred meters per second. We can then estimate the time it takes plasma, and hence the foot of a field line, to cross the polar cap. Again assuming that the polar cap has a radius of 15° of latitude or 1,500 km, and a speed of 330 m·s^{-1}, we obtain a time of 10^4 s or about 3 h. The other ends of these polar-cap field lines are embedded in the solar-wind flow, and so are moving away from the sun at speeds typically of 400 km·s^{-1}. In 10^4 s they move 4×10^6 km or about $600 R_E$. So the solar-wind portion of field line 6 in Figure 9.11, which is on the point of being disconnected from the earth, is about $600 R_E$ downstream of the earth. This then provides an estimate for the length of that part of the tail that is still connected to the earth magnetically and shows that this part of the tail is over 10 times longer than its diameter. Cowley (1991) showed that tail field lines that have disconnected from the earth will retain a tail geometry several thousand R_E beyond the earth.

The same observations can be used to obtain an estimate of the total potential drop ϕ across the open fields, either across the polar cap or equivalently across the tail diameter. Using $\mathbf{E} = -\mathbf{u} \times \mathbf{B}$, we have

$$E = \frac{\phi}{2R_{PC}} = u_{PC} B_{PC} \qquad (9.27)$$

where R_{PC}, u_{PC}, and B_{PC} are the radius, plasma flow speed, and magnetic-field strength in the polar cap. This gives $\phi \sim 53$ kV, a value typical of the direct measurements that can now be made by low-altitude polar-orbiting spacecraft. The potential drop over a distance equal to the diameter of the tail ($50 R_E$) in the undisturbed solar wind, using $u_{SW} = 400$ km·s^{-1} and $B_{SW} = 5$ nT, is 640 kV. This implies that about 10 percent of the IMF magnetic flux that would impact the geometric cross section of the magnetosphere reconnects with the geomagnetic field. The remainder flows around the magnetosphere, carried by the magnetosheath flow.

The Dungey model was not immediately accepted, but two distinct types of evidence have shown it to be true. Fairfield and Cahill (1966) first showed that geomagnetic activity is modulated by the north–south component of the interplanetary magnetic field. Numerous subsequent studies have confirmed and quantified this relationship. The other evidence is that electrons that flow outward from the sun with a distinctive spectrum are also found on polar-cap field lines, but only in the polar cap that connects magnetically to the sun, not the one connected to the outer heliosphere. That is, they are seen only in the north when the magnetic field is directed away from the sun, and only in the south when it is toward the sun (see Chapter 3). This observation proves that polar-cap field lines are directly connected to interplanetary field lines.

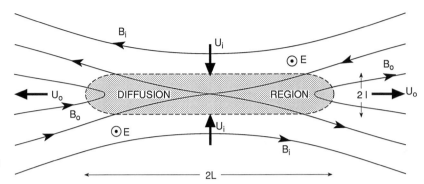

FIG. 9.12. Sweet-Parker reconnection geometry, in which all the reconnecting plasma flows through the shaded diffusion region. Very slow reconnection results.

Part of the reason that the Dungey model took so long to be accepted was that magnetic reconnection was poorly understood as a physical process. In the next two sections we shall examine reconnection from a fluid or MHD perspective and then from a particle-motion viewpoint and show that the two descriptions are equivalent. We shall then return to the magnetopause and tail and examine their more detailed morphology in the light of reconnection theory.

9.4.3 Fluid Descriptions of Reconnection

In this section we follow the development of MHD fluid descriptions of magnetic reconnection. The models will be time-stationary and two-dimensional. The main concern will be to establish that reconnection can proceed fast enough for it to have an effect in space-plasma systems. We find that it can, but only when the plasma far from the diffusion region adjusts to the new flow pattern.

We start with the Sweet-Parker solution shown in Figure 9.12. This is a basic x-line configuration introduced in Figure 9.10. The diffusion region, shown shaded, is $2L$ long and $2l$ wide, where $L \gg l$. For simplicity, we assume that the inflow and outflow regions (parameters that are identified by subscripts i and o) are symmetrical. This is appropriate for reconnection in the tail, where the inflowing plasma comes from the northern and southern lobes. At the magnetopause, the two inflow regions are quite different, as we discuss in Section 9.5. As before, the electric field E is spatially uniform and points out of the page, so that

$$E = u_i B_i = u_o B_o \tag{9.28}$$

We further assume that the flow is incompressible, that is, $\rho_i = \rho_o = \rho$; then conservation of mass gives

$$u_i L = u_o l \tag{9.29}$$

Next we equate the kinetic energy gained by the outflowing plasma with the electromagnetic energy flowing into the diffusion regions. The electromagnetic-energy inflow rate per unit area is given by the Poynting flux:

9.4 MAGNETIC RECONNECTION

$$|\mathbf{S}| = |\mathbf{E} \times \mathbf{H}| = \frac{EB_i}{\mu_0} = \frac{u_i B_i^2}{\mu_0} \tag{9.30}$$

The mechanical energy out is given by the gain in kinetic energy of the outflowing plasma. The mass flowing in per unit area per unit time, ρu_i, is accelerated to speed ρu_o, so that the rate of energy gain per unit area in the incident flow is

$$\Delta W = \frac{1}{2} \rho u_i (u_o^2 - u_i^2) \tag{9.31}$$

Equating the energy rates in (9.30) and (9.31), and using $u_o \gg u_i$, which follows from (9.29), gives

$$\frac{u_i B_i^2}{\mu_0} = \frac{1}{2} \rho u_i u_o^2 \tag{9.32}$$

so

$$u_o^2 = \frac{2 B_i^2}{\mu_0 \rho} = 2 v_{Ai}^2 \tag{9.33}$$

where v_{Ai} is the Alfvén velocity in the inflow region. From the magnetic-annihilation calculation we obtained the estimate of l, the thickness of the diffusion region, given in (9.26). Combining this with (9.29) and (9.33) gives us an expression for the inflow speed:

$$u_i^2 = 2^{\frac{1}{2}} v_{Ai} / \mu_0 \sigma L \tag{9.34}$$

so

$$u_i = v_{Ai} (2^{\frac{1}{2}} / R_{mA})^{\frac{1}{2}} \tag{9.35}$$

where

$$R_{mA} = \mu_0 \sigma v_{Ai} L \tag{9.36}$$

What this means is that in all solar-system plasmas for which R_{mA} is very large, the inflow into the reconnection site, which corresponds to the reconnection rate, is very, very slow. Using typical solar-corona parameters, a solar flare would take tens of days to grow, rather than a few minutes as observed. This result, derived independently by Sweet and Parker in the late 1950s, made people tend to rule out reconnection as a viable process.

However, a few years later, Petschek (1964) solved the rate problem by realizing that most of the plasma involved in the reconnection process does not need to flow through the diffusion region in order to be accelerated. Instead, it can be accelerated in the region where MHD is still valid, the so-called convection region. The acceleration occurs as the plasma passes through shock waves that are connected to the diffusion region and that remain fixed in space; that is, they stand in the flow. This process removes the bottleneck caused by requiring that all the

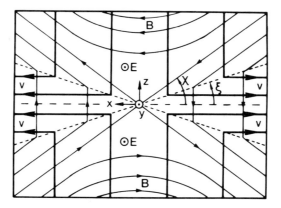

FIG. 9.13. Petschek reconnection geometry. The diffusion region has been shrunk to a dot in the center of the diagram, and the plasma is accelerated at four slow-mode shock waves inclined at angle ξ to the x-axis. (Adapted from Hill, 1975.)

plasma come within a length l of the midplane, and so a much larger inflow rate can be accommodated.

The diffusion region is still important in the sense that the actual process of reconnection must occur there, and so its presence is required, but the diffusion region may now become vanishingly small. The size of the diffusion region will not even enter our calculation. Rather, the entire calculation will be based on jump conditions across an MHD shock. So although at the heart of the process the MHD approximation is being violated, the rate at which reconnection proceeds is based entirely on constraints away from the diffusion region where MHD breaks down, and within the convection region where $\mathbf{E} = -\mathbf{u} \times \mathbf{B}$ holds.

The geometry of this new solution is shown in Figure 9.13. Now the diffusion region has been reduced to a dot at the center of the diagram, and only an insignificant fraction of the flow passes through it. Emanating from this dot are four shock waves shown as dashed lines, at an angle ξ to the x-axis. Both immediately upstream and downstream of these shocks, the magnetic-field and flow vectors are taken as uniform. (Geometric constraints make it impossible for the entire inflow region to be uniform; note that the field lines are curved where they first enter the diagram from top and bottom.) At the shocks, the magnetic field and flow abruptly change in both direction and strength. The magnetic-field strength decreases (field lines are farther apart in the outflow region), and although the flow speed increases (flow lines are closer together in the outflow region), the normal component of the flow velocity drops. These are slow-mode shocks (see Chapter 5), and they are also current sheets. The current is needed both to change the magnetic field and to accelerate the flow via the $\mathbf{j} \times \mathbf{B}$ force. Note also that a parcel of plasma reaches the shock some time after the flux tube it is on has reconnected at the diffusion region; this time delay depends on how far from the diffusion region the plasma crosses the shock.

In the frame of the figure, the shocks are stationary, but in the frame of the plasma the shocks travel along the magnetic field at the inflow Alfvén speed, $v_{Ai} = B_i/(\mu_0 \rho_i)^{\frac{1}{2}}$. We use this to relate the inflow speed u_i to the angle of the shock, ξ, and to the angle between the magnetic field immediately upstream of the shock and the x-axis, χ. The plasma inflow velocity normal to the shock must equal the shock speed normal to the shock front in the plasma frame for the shock to remain fixed. This gives

$$u_i \cos \xi = v_{Ai} \sin(\chi - \xi) \tag{9.37}$$

Again we can appeal to the steady state and hence require a uniform electric field in the y-direction, equation (9.22). Hence,

$$E_i = u_i B_i \cos \chi = u_o B_o = E_o \tag{9.38}$$

Next, the component of **B** normal to the shock must be conserved. This gives

$$B_i \sin(\chi - \xi) = B_o \cos(\xi) \tag{9.39}$$

Eliminating the magnetic field between (9.38) and (9.39) and using (9.37) gives

$$u_o = \frac{u_i \cos \xi \cos \chi}{\sin(\chi - \xi)} = v_{Ai} \cos \chi \tag{9.40}$$

So, as in the Sweet-Parker solution, the outflow speed is comparable to the inflow Alfvén speed. By conserving mass flow across the shock, we get

$$\rho_i u_i \cos \xi = \rho_o u_o \sin \xi \tag{9.41}$$

where we have now allowed for the plasma to be compressed at the shock (i.e., $\rho_o \geq \rho_i$). Eliminating u_o and u_i between (9.40) and (9.41), after some rearrangement, gives

$$\tan \chi = \frac{\tan \xi}{1 + \rho_o/\rho_i} \tag{9.42}$$

Further rearrangement of (9.40), (9.41), and (9.42) yields an expression for the inflow speed

$$u_i = v_{Ai} \sin \chi /(1 + \rho_i/\rho_o) \tag{9.43}$$

Now, because a slow-mode shock can only compress the plasma, $0 < \rho_i/\rho_o \leq 1$, so

$$\frac{1}{2} \leq u_i/(v_{Ai} \sin \chi) < 1 \tag{9.44}$$

So, in complete contrast to the Sweet-Parker solution, the inflow speed is some reasonable fraction of the inflow Alfvén speed. However, further analysis of the constraints of this solution shows that $u_i \leq 0.1 v_{Ai}$.

Physically the plasma acceleration occurs at the shock fronts and can be pictured as being due to the $\mathbf{j} \times \mathbf{B}$ force. The shock orientation adjusts to accommodate the inflow speed; the outflow speed also varies but remains close to v_{Ai}. This solution so speeds up the allowable reconnection rate that magnetic reconnection can be regarded theoretically as a viable process.

A further refinement was made by Sonnerup (1970) who, by introducing a further set of MHD shocks, overcame the limitation on the inflow speed in Petschek's solution. The shocks in this second set are fast-mode shocks that compress both the magnetic field and the plasma in the inflow region and begin to divert the flow direction, a process that is completed at the slow-mode shock, where the largest increase in flow speed occurs. In Sonnerup's solution, the location of the fast shocks is effectively determined by external boundary conditions. In some circumstances the expansion does not appear as discrete shocks, but occurs gradually throughout the inflow region.

The role of the fast-mode shocks imposed by external boundary conditions brings us to the final and perhaps most important point of this section. Once we have established that reconnection can happen, how it happens is governed more by external conditions than by conditions at the actual site of reconnection. Having started with reconnection as a microscopic problem, we find that it becomes a macroscopic or global problem in which the entire system must be considered. This was elegantly shown by Priest and Forbes (1986; Forbes and Priest, 1987), who found a general solution to steady-state MHD reconnection that includes the Petschek and Sonnerup solutions as special cases. They showed that both Petschek and Sonnerup implicitly applied particular boundary conditions in the inflow region. When the boundary conditions are generalized, the family of solutions illustrated in Figure 3.19 in Chapter 3 results. As their parameter b (which describes the form of the current at the upper boundary) changes, the flow within the inflow region changes from converging to diverging. For small b (≤ 0.5), they found that a maximum reconnection rate exists, but for larger b, reconnection can proceed arbitrarily quickly (within the validity of their model). Forbes and Priest also found that differences in boundary conditions explained previously puzzling differences between analytical and numerical reconnection calculations.

9.4.4 Particle Descriptions of Reconnection

In parallel to the development of MHD theories of reconnection, a particle picture of reconnection arose. It was not always clear that these were two equivalent descriptions of the same process, and that has given rise to some misconceptions. In this section we develop the particle description and then show how the two approaches give the same or very similar results.

We begin by describing particle orbits in a current sheet. The simplest analytical description of a one-dimensional current sheet is due to Harris (1962), who showed that this description is self-consistent using either

MHD or kinetic theory. In a Harris neutral sheet, the magnetic-field and plasma pressures are given by

$$B(z) = B_0 \tanh(z/h) \hat{x} \tag{9.45}$$

$$p(z) = p_0 \text{sech}^2(z/h) \tag{9.46}$$

It is simple to verify that total pressure is constant in this structure:

$$B^2/2\mu_0 + p = \frac{B_0^2}{2\mu_0} \tanh^2\left(\frac{z}{h}\right) + p_0 \text{sech}^2\left(\frac{z}{h}\right) \tag{9.47}$$
$$= 2p_0 \quad \text{if } p_0 = B_0^2/2\mu_0$$

So the plasma pressure at the center of the neutral sheet balances the pressure of the asymptotic value of the magnetic field B_0 far from the sheet, where the plasma pressure becomes negligible. This is very similar to the current sheet illustrated in Figure 9.9. To complete our description of this current sheet, we can obtain the current from Ampère's law,

$$(\nabla \times B)_y = B_0 \frac{d}{dz}\left(\tanh\frac{z}{h}\right) = \frac{B_0}{h} \text{sech}^2\left(\frac{z}{h}\right) = \mu_0 j_y(z) \tag{9.48}$$

and show that there is force balance between the gradient of pressure and the $\mathbf{j} \times \mathbf{B}$ force.

$$\mathbf{j} \times \mathbf{B} = \frac{B_0^2}{\mu_0 h} \text{sech}^2\left(\frac{z}{h}\right) \tanh\left(\frac{z}{h}\right) \hat{z} \tag{9.49}$$
$$= \frac{d}{dz} p_0 \text{sech}^2\left(\frac{z}{h}\right) \hat{z} = \nabla p$$

where we used (9.47).

The simplicity of this current sheet has resulted in its often being used in theoretical models. In particular, the orbits that charged particles make in this type of field are well known (and are easy to derive numerically). Some sample ion orbits are shown in Figure 9.14. Far from the current sheet, ions simply gyrate in the uniform magnetic field. Where the gradient in the magnetic field becomes larger, the ions move to the right at the gradient-drift velocity (see Chapter 2). At first sight this would seem to constitute a current to the right, but in fact the net current is to the left, as is required by the magnetic-field geometry. At any given point in z, ions with gyrocenters at larger z are moving to the right, and those with gyrocenters at smaller z are moving to the left. The gradient in pressure means that there are always more particles with gyrocenters nearer the z-axis, so that these dominate the local current flow, carrying current to the left. This is known as the diamagnetic drift current.

The most interesting orbits occur where the ions cross the neutral sheet across which the field changes direction. In Figure 9.14 a number

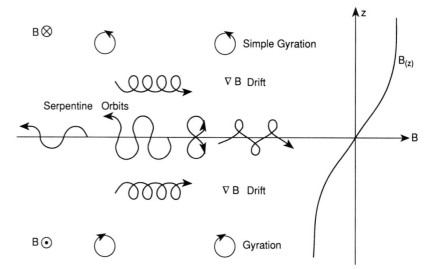

FIG. 9.14. Particle orbits in a Harris neutral sheet. The graph on the right shows how the magnetic-field strength varies across the neutral sheet. In regions of uniform field, the particles simply gyrate. Where the field has a gradient, particles drift to the right. Particles that cross the plane of the neutral sheet have more complex serpentine orbits. (Adapted from Cowley, 1986.)

of these orbits are shown; ions can drift to the right or to the left in serpentine orbits. The parameter that determines whether a particle will drift right or left is the angle at which it crosses the neutral sheet. There is one angle for which a particle orbits in a figure-eight pattern and has no net motion. This particle is moving right as it crosses the neutral sheet. So any particle that crosses the neutral sheet moving left or moving right, but with a velocity vector closer to the normal than the particle that stays still, will drift to the left. In an isotropic velocity distribution this must be a majority of the particles, resulting in a net leftward current, as required. Harris (1962) showed that exact equilibria exist for Maxwellian particle distributions. Particle pressure just balances magnetic-field pressure everywhere, and particle motions provide just the right current to satisfy Ampère's law.

An important point about these orbits is that particles have no net motion in the z-direction. Particles in the neutral sheet stay there, and no new particles enter the neutral sheet. This is fine as long as the system is infinite and has no boundaries, especially no boundaries in the y-direction. If there are boundaries in the y-direction, as, for example, the edges of the tail current sheet at the magnetopause, then either these boundaries must act as sinks and sources of particles, at just the rate required to maintain equilibrium, or a new source of particles must be found. One such source is particles drifting in from above and below. An electric field in the y-direction will cause particles to drift toward the current sheet from both sides. In the fluid picture, such an electric field implies plasma inflow that carries magnetic field with it, and so we are back to considering field annihilation. The conceptual difference between the particle and fluid descriptions is that in a particle description we do not use the device of moving magnetic-field lines.

The foregoing scenario, with flow toward a Harris neutral sheet,

9.4 MAGNETIC RECONNECTION

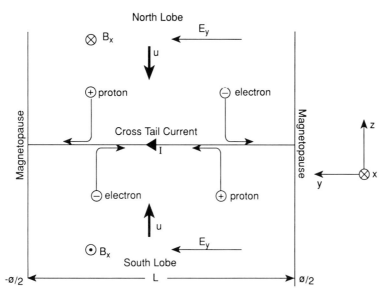

FIG. 9.15. View looking toward the sun showing the drift of particles into the tail current sheet from either lobe. After entering the current sheet, protons and electrons travel to the dusk and dawn magnetopauses, respectively, carrying the cross-tail current. The width of the tail is L, across which there is a potential drop of ϕ.

was considered by Alfvén (1968) and is illustrated in Figure 9.15. The coordinate system is the same as in Figures 9.9 and 9.13, but here we are looking in the y–z plane, with the magnetic field perpendicular to the page. The current flows in a narrow sheet across the diagram in the $z=0$ plane over a finite length L between two perfectly absorbing boundaries that Alfvén was picturing as the dawn and dusk edges of the plasma sheet in the geotail. There is a potential drop of ϕ between the two boundaries that results in an electric field.

$$\mathbf{E} = \frac{\phi}{L}\hat{\mathbf{y}} \tag{9.50}$$

This electric field induces a particle drift $u = E/B$ toward the central plane or neutral sheet from both above and below. On reaching the central plane, the particles become unmagnetized (the magnetic field is zero there) and are accelerated in the electric field, ions to the left, electrons to the right. This latter motion provides the current. The particles are lost when they reach the walls. Current continuity occurs because as many ions as electrons reach the current sheet and then move in opposite directions. So the total sheet current density can be calculated by calculating the total number of ions that reach the left boundary per second (or, alternatively, the number of electrons reaching the right boundary). This, in turn, must equal the total number of ions entering the current sheet per second. From this we get

$$I = 2nLue \tag{9.51}$$

where I is the sheet current (i.e., the total current per unit length in x), and the factor 2 arises because ions enter from both above and below. Applying Ampère's law across the current sheet gives

$$I = 2B/\mu_0 = 2nLue \qquad (9.52)$$

so

$$u = \frac{B}{\mu_0 neL} \qquad (9.53)$$

This inflow rate is equivalent to a merging rate, and so this expression should be compared to (9.34), the Sweet-Parker merging rate. We see immediately that both depend on a macroscopic length of the system (in both cases called L) but note that the lengths in the two cases are in different directions, in the x-direction for Sweet-Parker, and in the y-direction for Alfvén. Nonetheless, in both cases reconnection is limited by the system size.

We can rearrange (9.53) to make it appear more like (9.34). With $v_A^2 = B^2/\mu_0 nm$ and $\sigma = ne^2/m\nu$, the classic collisional conductivity, where ν is a collision frequency, we obtain

$$u^2 = \frac{v_A^2}{\mu_0 \sigma \nu L^2} \qquad (9.54)$$

Equation (9.54) is very similar to (9.34) if $v_A/\nu L \sim 1$, which it is if L is considered to be a mean free path and v_A a thermal speed. Again, this model results in a very slow merging rate that is kept small by a macroscopic length scale, just as in the Sweet-Parker solution.

Following Alfvén analysis further, Cowley showed that the difference in ion and electron masses results in a nonuniformly distributed potential drop, with the result that most of the plasma enters the current sheet very close to the left-hand boundary, and electrons carry the bulk of the current over most of the length of the current path, but this does not alter the basic result of the merging rate.

The resolution of the slow-merging-rate dilemma in the particle picture has much in common with the resolution for the MHD analysis. We consider what happens away from the strict neutral point by adding a small but finite B_z to the model. Field lines now close across the current sheet, as illustrated in Figure 9.16, but the model remains one-dimensional, as all parameters depend only on z. Following Cowley (1986), we work in two frames, first in the frame in which $E_y = 0$, which implies no particle inflow, and then in a frame with finite electric field and particle inflow. If the field lines in Figure 9.16 are pictured as those to the left of the x-point in Figure 9.10, then the latter frame can be chosen as the rest frame of the x-line. The orbits that particles follow in this field geometry, illustrated in Figure 9.16, were first studied by Speiser (1965) and are referred to as Speiser orbits. We now examine these orbits.

In the frame in which $E_y = 0$, the only force acting on particles is the Lorentz force, which cannot change a particle's energy or speed (as the force always acts perpendicular to the velocity). Consider the motion of a particle that starts out gyrating around a field line, with a parallel

9.4 MAGNETIC RECONNECTION

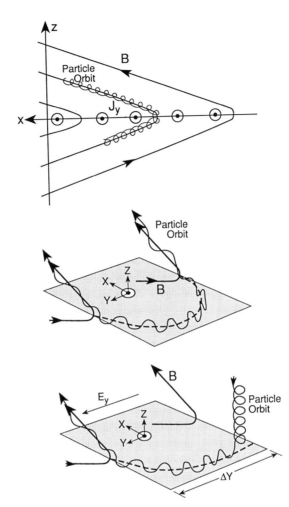

FIG. 9.16. Orbits of particles in the tail current sheet when a finite B_z is included. The top two panels show a particle trajectory in the frame in which $\mathbf{E} = 0$. In the top panel, the trajectory is projected onto the x–y plane, and in this plane the particle follows a field line. However, the middle panel shows that the particle moves in y as it crosses the field minimum plane. When an electric field is added (bottom panel), the particle is also accelerated. (Adapted from Cowley, 1986.)

velocity toward the current sheet. When the particle reaches the neutral sheet, it moves in the y-direction on a gyration orbit in the weak B_z field (Figure 9.16, middle panel), while bouncing between mirror points resulting from the stronger magnetic field above and below the current sheet. After half a gyration orbit, it exits the current sheet, gyrating about another field line, at a distance

$$\Delta y = 2v_x/\Omega_z \quad \text{where } \Omega_z = eB_z/m \tag{9.55}$$

from its original gyrocenter. During this motion, the particle's speed has never changed, but its velocity in the x-direction has been reversed. Whether its velocity in the z-direction is changed (i.e., whether it exits the current sheet moving up or down) depends critically on the initial gyrophase of the particle.

We can confirm equation (9.55) by considering the particle's equation of motion in the x-direction:

$$\frac{\partial v_x}{\partial t} = \frac{e}{m} v_y B_z = \Omega_z \frac{\partial y}{\partial t} \tag{9.56}$$

Integrating this, we find

$$v_x - \Omega_z y = \text{constant}$$

so

$$v_{xi} - v_{xo} = \Omega_z(y_i - y_o) \tag{9.57}$$

Now, if the inflow and outflow speeds are the same, but in opposite directions (i.e., $v_{xo} = -v_{xi} = -v_x$),

$$\Delta y = y_i - y_o = 2v_x/\Omega_z \tag{9.58}$$

as we surmised before (9.55).

Now let us consider a frame moving relative to our original frame at a speed v_f in the x-direction. This motion adds an electric field in the y-direction such that $v_f = E_y/B_z$. The speed at which the particle leaves the current sheet is now larger than that at which it entered the current sheet by $2v_f$. In particular, if the inflow speed is $v_x - v_f$, the outflow speed is $v_x + v_f$. The particle has gained kinetic energy, and the energy gain is given by

$$\Delta W = \frac{1}{2}m\{(v_x + v_f)^2 - (v_x - v_f)^2\} \tag{9.59}$$

This is just the energy gained by dropping through the electric field a distance Δy, which is

$$\begin{aligned} eE_y\Delta y &= 2eE_y v_x/\Omega_z \\ &= 2mE_y v_x/B_z \\ &= 2mv_x v_f \end{aligned} \tag{9.60}$$

So in this frame, the particle is accelerated by falling down a fraction of the potential of the electric field that provides the inflow. The energy gained depends on the particle mass, and so ions gain more energy than the electrons, and the heavy ions gain more energy than protons. Energy gain also depends on the initial particle velocity or energy. This is because fast-moving heavy particles have much larger gyroradii, and so they fall through a much larger potential drop. Thus, current sheets act preferentially as ion accelerators, and the ion temperature in the plasma sheet is several times higher than the electron temperature.

Although it is significant that the particles are accelerated in the process described, the important result here is that they do not travel indefinitely in the y-direction; instead, they travel only a finite, microscopic distance before being ejected from the current sheet. This greatly modifies the reconnection rate we derived for the Alfvén model. We can go back to (9.53) and replace the macroscopic distance L with the new Speiser distance Δy and get

$$\begin{aligned} v_d &= B_x/\mu_0 ne\Delta y \\ &= B_x B_z/2v_x\mu_0 nm \end{aligned} \tag{9.61}$$

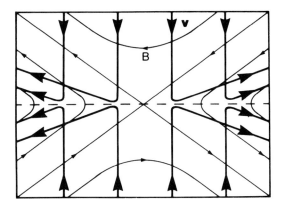

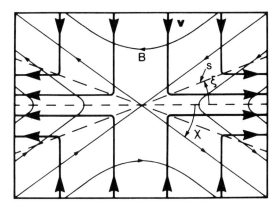

FIG. 9.17. Comparisons of particle trajectories (top panel) and equivalent fluid flows (bottom panel), both drawn as heavy arrows, in the vicinity of the magnetic x-line (magnetic-field lines are marked with small arrows). The top panel is a projection of the bottom panel of Figure 9.16 onto the x–y plane. Note that particles accelerated near the x-line move past particles entering the current sheet farther from the x-line. The bottom panel is obtained by integrating over the particle trajectories. Where the incoming and outgoing particles coexist, the two populations combine to give rapid outflow. Thus, even though the individual particles are all accelerated at the central plane, the fluid velocity changes abruptly along the planes marked s. (Adapted from Hill, 1975.)

Now, v_d in this model is v_z, and because particles are moving along field lines, v_z and v_x are related by

$$v_x = v_z B_z / B_x \qquad (9.62)$$

Substituting (9.62) into (9.61) gives

$$v_z^2 = B_x^2/\mu_0 n m = v_{Ai}^2 \qquad (9.63)$$

Thus we again find that the merging rate or inflow speed can approach the inflow Alfvén speed, just as in the MHD models, when macroscopic lengths are replaced by microscopic lengths that arise from the actual particle motions.

9.4.5 Comparison and Limitations of the Models

It should be apparent by now that although the fluid and the particle descriptions of reconnection begin with different concepts and approach the problem quite differently, they provide very similar answers. Each teaches us something about reconnection and about plasma physics in general. In this section we first compare the two solutions in more detail, to show that they are equivalent, and then discuss some of their shortcomings, notably that they assume a time-stationary process and relatively uniform boundary conditions.

The upper panel of Figure 9.17 shows single-particle trajectories resulting from Speiser-type particle orbits in an x-line magnetic geometry mapped into the x–z plane. The particles enter from the top and bottom and are accelerated at the field-reversal current sheet, marked by the horizontal dashed line. The accelerated particles then stream out of the sides of the figure, passing through the trajectories of incoming particles that reach the current sheet farther from the x-line. The lower panel shows the bulk-plasma streamlines obtained by summing over the incoming and outgoing flows in the upper panel. Although the individual particles are accelerated at the field reversal, the abrupt changes in bulk flow occur along the inclined planes marked s. These planes are geometrically similar to the slow shocks in Figure 9.13, but because the particles have been taken as cold (that is to say, their effect on the field has been ignored), no jump in magnetic field occurs here.

Within the region bounded by the planes s, the plasma must have a finite temperature because the velocity distribution has a finite width, as it includes contributions from both the inflowing and outflowing populations. This means the plasma has been heated. It is also denser than the inflowing plasma. In practice, the inflowing plasma will itself have a finite temperature, so that the particle population will be heated from a finite temperature as well as being accelerated at the field reversal. The higher-temperature plasma within the outflow region will induce a diamagnetic effect that will reduce the magnetic-field strength across the planes s, causing them to become current sheets also. The planes then become more akin to the slow-mode shocks in Figure 9.13.

Before we go on to discuss reconnection in the magnetosphere, let us consider the limitations in all that we have done. Some limitations are obvious. First, we assumed that the two inflow regions were identical, except for the magnetic-field directions. This is a good assumption in the tail. At the magnetopause, it is a poor assumption. The magnetopause separates the plasma and field of the magnetosphere from those of the magnetosheath, where the plasma is much denser, but cooler, and the field strength generally is weaker. Second, we assumed that the fields in the two inflow regions were strictly antiparallel. In practice, this does occur at a few points over a complex surface like the magnetopause, but it is far from clear that reconnection will preferentially occur where the fields are antiparallel. In the magnetotail, the fields are antiparallel to a much better approximation, but a small, cross-tail magnetic field has been observed that seems to be correlated to the direction of the interplanetary magnetic field (IMF). This means that the lobe fields are not exactly antiparallel. A more important limitation is that we have not allowed for any temporal variations in the reconnection models we have covered, nor have we discussed the stability of current sheets from either a fluid (MHD) or kinetic viewpoint. These are fundamental restrictions, but space does not allow for a full discussion here.

Implicit in what we did is that reconnection is driven by conditions external to the current sheet, perhaps by flow converging at the current sheet. This is generally referred to as *driven reconnection*. Driven reconnection will be time-variable if the external conditions (direction of IMF, velocity, density, etc.) vary. If external conditions change, then clearly it will take some finite time, at least on the order of Alfvén transit times, to set up a new equilibrium. *Spontaneous reconnection* is the name given to reconnection that begins once some threshold of instability is reached in the current sheet. This may be, but is not necessarily, what happens at the onset of a magnetospheric substorm in the tail, as discussed in Chapter 13. The most extensively studied instability applicable to reconnection is the tearing mode (e.g., Melrose, 1986). Strictly speaking, this is a resistive MHD instability that arises because of the introduction of a finite Ohmic resistivity. However, a "collisionless tearing mode" exists in which the conductivity is so large that the left side of equation (2.45a) vanishes, but other terms in the generalized Ohm's law introduce an effective resistivity; in some cases, the scattering of particles by waves or turbulence introduces an anomalous resistivity. The effect of the tearing mode is to break (tear) a uniform current sheet into a series of current filaments, resulting in a magnetic structure of an alternating series of x- and o-type neutral lines. The tearing mode was originally studied by the fusion-plasma community, as it is an important process limiting the stability of tokamaks. However, recently it has been invoked in studies of both magnetopause (e.g., Lee and Fu, 1985) and tail (e.g., Forbes and Priest, 1983) reconnection. Computer simulations play a dominant role in this work in treating the nonlinear saturation and subsequent behavior of the fields, but that is beyond the scope of this text.

Models of the sort that we have described cannot address when and where reconnection will occur. At both the magnetopause and in the tail, these remain open questions from both the observational and theoretical standpoints. Perhaps theoretically we shall get answers to these questions only from three-dimensional simulations, given the complex geometries and nonsteady inputs into the magnetospheric system.

9.5 RECONNECTION AT THE MAGNETOPAUSE

9.5.1 The Magnetopause Boundary Layers

Nature is rarely simple. When looked at more closely, the magnetopause is far from being the simple current layer described at the beginning of this chapter. Earthward of the magnetopause current layer, spacecraft have detected a boundary layer typically several thousand kilometers thick.

Three distinct types of magnetopause boundary layers are recognized. Figure 9.18 shows how these three types of boundary layers are distrib-

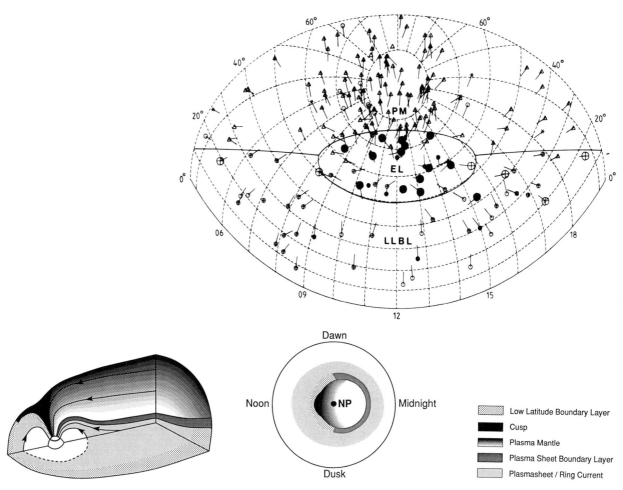

FIG. 9.18. (top) Polar projection of the magnetopause showing the types of magnetopause crossings observed by *HEOS 2*. Note how the observations of low-latitude boundary-layer plasma (open circles), entry-layer plasma (solid circles), and plasma mantle (triangles) divide into three distinct spatial regions on the magnetopause. (Adapted from Haerendel et al., 1978). (bottom) Vasyliunas's (1979) mapping of the plasma boundary layers down to the high-latitude ionosphere.

uted over the magnetopause. The top part of Figure 9.18 shows the *HEOS 2* observations of the northern magnetopause. This is the only spacecraft that has surveyed the magnetopause at all latitudes. During the magnetopause crossings, marked by open circles, the spacecraft encountered the low-latitude boundary layer (LLBL), a region that contains a mix of magnetosheath and magnetosphere plasma, and within which plasma flows can be found in almost any direction, but are generally intermediate between the magnetosheath flow and magnetospheric flows. The solid circles denote observations of a type of boundary layer known as the entry layer (or, alternatively, the high-altitude cusp). This type of boundary layer is found in the region where the magnetic null or cusp occurs in a closed magnetopause model (Figure 9.1) and extends about 3 h either side of noon. The plasma is characteristic of the magnetosheath, but the flows are low-speed, disordered, and probably turbulent. The final type of boundary layer, the plasma mantle or high-latitude boundary layer (HLBL), marked by triangles, is found at higher latitudes tailward of the cusp and entry layer. Within the plasma mantle, flows are always tailward, but flow speed, density, and temperature all decrease away from the magnetopause. The plasma mantle is, in general, spatially uniform, with a gradual transition from magnetosheath to lobe properties. It often has no distinct inner edge, in sharp contrast to the LLBL, which usually has a very distinct inner edge. The LLBL, as will be illustrated shortly, can be made up of distinctly different plasma regimes separated by sharp boundaries. Magnetopause motion moves these boundaries back and forth across a spacecraft, giving a pulsating effect to the observations. The different plasma on neighboring flux tubes suggests that the flux tubes have had very different recent histories, suggesting in turn that the processes responsible for the formation of the LLBL occur on a spatially and/or temporally short scale.

The lower part of Figure 9.18 shows Vasyliunas's (1979) interpretation of the boundary-layer topology. The entry layer is formed by direct entry of magnetosheath plasma along open field lines that form the magnetospheric cusp. The plasma mantle is found near that part of the magnetopause that separates the tail lobes from the magnetosheath, and, like the tail lobes, it presumably consists of open field lines. The LLBL is found everywhere else, over most of the dayside magnetopause, and at the boundary between the plasma sheet and magnetosheath. Vasyliunas also suggested how the magnetic-field lines in these regions map down to the ionosphere. Most of the auroral oval consists of field lines that map to yet another boundary layer, the plasma-sheet boundary layer (PSBL), at the boundary between the tail lobes and the plasma sheet. (The PSBL is discussed in Section 9.6.) The magnetopause boundary layers map, in this interpretation, to the dayside auroral zone, near local noon and to a small part of the adjacent polar cap. Other workers draw minor variations of this picture, but most would agree with the general topology. Notice that there is a region near the tail magnetopause where the plasma mantle, PSBL, and LLBL all come together.

The entry layer and plasma mantle are generally agreed to be on open magnetic-field lines. They are populated by a mixture of magnetosheath plasma that entered the magnetosphere along open field lines in the cusp and ionospheric plasma that flowed up from the cusp and polar cap in an upward flow known as the "polar wind." Figure 9.19 illustrates the process. Reconnection is assumed to occur at the nose of the magnetopause. Magnetosheath particles flow along the newly opened field lines, perhaps being accelerated in the reconnection process. The particles mirror as they encounter the stronger magnetic field nearer the earth. After mirroring, they move back up the field line, joined by lower-energy ionospheric particles. The field line has meanwhile convected tailward to become a lobe field line, still near the magnetopause, or a

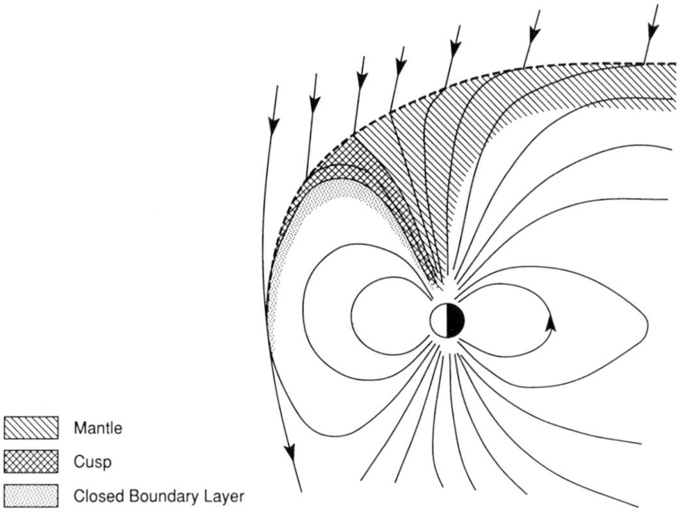

FIG. 9.19. Connection between the cusp or entry layer and the plasma mantle. Field lines forming the cusp convect tailward to later form the mantle, carrying into the mantle magnetosheath plasma that mirrored at low altitudes near the cusp. Ionospheric plasma can also be drawn up cusp field lines and into the mantle. (Adapted from Crooker, 1977.)

mantle field line. Lower-energy particles move slower and thus take longer to reach a given distance downtail. In this longer time, the field line will have convected farther from the magnetopause. This explains the decrease in mantle temperature and flow speed with distance from the magnetopause.

The origin of the LLBL plasma is far less clear. Some magnetosheath plasma may have crossed the magnetopause on open field lines, perhaps being accelerated in the reconnection process as it did so. Other magnetosheath plasma may have crossed the magnetopause via some diffusive process, perhaps being pitch-angle-scattered in the small-scale current structure of the magnetopause current sheet. Yet other LLBL plasma is of magnetospheric origin. It is not known how much of the LLBL is on open or on closed field lines, nor whether or not there are distinct regions of open and closed field lines within the LLBL. During periods of northward IMF, the LLBL plasma distribution function appears to be formed from a simple mixture of magnetosheath and magnetopause plasma. However, during southward IMF, some acceleration or heating of the magnetosheath plasma is required to explain the observed plasma distribution functions. On a small fraction of magnetopause crossings, the LLBL is entirely absent. The observations strongly suggest that reconnection plays some role in the formation of the LLBL, but that other processes may also be important. We emphasize that the role of reconnection need not be restricted to times of southward IMF. Reconnection between the magnetosheath magnetic field and the terrestrial magnetic field is thought to occur beyond the polar cusp when the IMF is northward. Computer simulation and some observations suggest that such reconnection plays a role in boundary-layer formation.

Figure 9.20 suggests how part of the LLBL might be predominantly on open field lines and part predominantly on closed field lines. This figure probably is altogether too simple to depict the true situation in the LLBL, which probably consists of mixtures of open and closed field

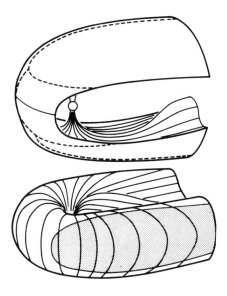

FIG. 9.20. Volumes occupied by open field lines (top) and closed field lines (bottom) in a model magnetosphere. The top volume contains the tail lobes, including the mantle, the cusp or entry layer, and the open portion of the low-latitude boundary layer. The lower volume contains the plasma sheet, the quasi-dipolar magnetosphere, and the closed portion of the low-latitude boundary layer. The shaded area shows that part of the magnetopause that abuts closed field lines. This idealized model is symmetric about the equatorial plane and the noon–midnight meridian, symmetries generally broken by dipole tilt and IMF direction. (Adapted from Crooker, 1977.)

lines. However, it is a useful diagram to help us focus on the overall topology of the open and closed portions of the magnetosphere. The upper part of the diagram portrays the volume occupied by open flux tubes. These tubes originate from the two polar regions and form the two tail lobes, as well as a thin layer over the dayside magnetopause, the dayside LLBL, and the entry layer. The lower part shows the closed-flux-tube volume. This is the quasi-dipolar region of the magnetosphere, which includes most of the dayside magnetosphere, together with the plasma sheet, which is thinner in the midtail than near the flanks. These two volumes fit together. When the volumes are combined, the shaded area in the lower part is the only region of closed field lines exposed to the magnetopause and magnetosheath. It is across this part of the magnetopause that some process other than magnetic reconnection could be responsible for mass and momentum transfer across the magnetopause.

Before going any further, we must emphasize that Figure 9.20 is greatly oversimplified. The magnetosphere is drawn symmetrically about the noon–midnight plane and the equatorial plane. In practice, the dawn–dusk component of the IMF creates significant dawn–dusk asymmetries, as we shall see later. North–south asymmetries are created by the varying tilt of the dipole, which can tilt by up to 35° from perpendicular to the earth–sun line. If reconnection occurs away from the vicinity of the equatorial plane, as it well may, further asymmetries arise. Nevertheless, this figure is useful in helping us picture the complex magnetic topology of the magnetosphere and magnetopause, and it probably is correct in a gross topological sense.

Reconnection at the magnetopause is not the only process that drives convection within the magnetosphere. Observations suggest that averaged over time some small fraction, perhaps 10–20 percent, of the potential drop across the polar cap could be the result of a viscous

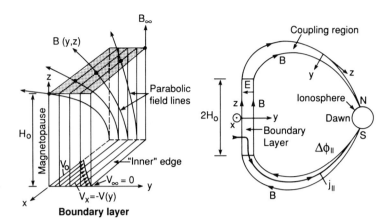

FIG. 9.21. Model of the closed portion of the low-latitude boundary layer. On the left, a detail where the magnetopause meets the equatorial plane shows the field lines dragged into parabolas by the velocity shear across the boundary layer. The velocity shear drives the field-aligned currents shown on the right that link the boundary layer and ionosphere. Field-aligned potential drops $\Delta\phi_\parallel$ along these field lines further complicate the coupling. (Courtesy of B. U. Ö. Sonnerup and W. Lotko.)

interaction. The term "viscous interaction" is used loosely to include any process other than magnetic reconnection that can transport momentum across the magnetopause. It can include kinetic processes, such as anomalous particle diffusion via wave–particle interactions, and large-scale fluid interactions such as the Kelvin-Helmholtz instability, or magnetopause distortions resulting from traveling pressure variations in the magnetosheath. A third concept, called impulsive plasma penetration (e.g., Lemaire, Rycroft, and Roth, 1979), has also been proposed to contribute to this viscosity.

Figure 9.21 depicts the results from a model of that part of the LLBL on closed field lines where the flow is driven by some viscous interaction with the solar wind. The figure illustrates, among other things, two reasons why it is difficult to map the LLBL to low altitudes. The left part of the figure shows the flow-velocity field and magnetic-field geometry close to the equatorial plane. The $y = 0$ plane represents the magnetopause, where the tailward flow is largest. Farther from the magnetopause, the flow speed decreases, until, in this diagram, it becomes zero. More realistically, it should reverse and merge with the sunward flow in the magnetosphere proper. The magnetic-field lines are carried tailward by this flow and stretched into parabolas in planes parallel to the magnetopause. Clearly, the increased stretching nearer the magnetopause results in field lines whose equatorial crossing points are separated radially, mapping to quite different local times at low altitudes. The shear in this flow drives field-aligned currents that flow into the ionosphere on the dawn side and out of the ionosphere on the dusk side, shown on the right (region-1 currents). These currents distort and twist the field lines, further complicating the mapping. These currents, and the cross-field potential drop associated with them, are the sources of "viscously driven" convection within the magnetosphere. The field-aligned currents create field-aligned potential drops, $\Delta\varphi_\parallel$ in the figure. The potential drops are different on the dawn and dusk sides, as they depend not only on the magnitude but also on the direction of the current. The existence of these potential drops complicates the mapping of electric fields, and

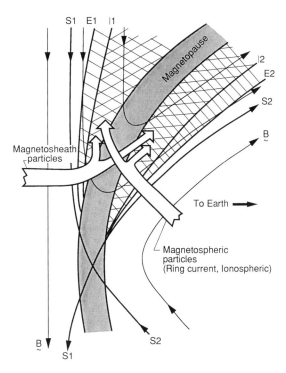

FIG. 9.22. Geometry of magnetopause reconnection at an x-line. Magnetosheath and magnetospheric particles enter from the left and right, respectively. The field lines marked S1 and S2 form the x-line at this instant. The shaded regions bounded by lines E1 and E2 and I1 and I2 are those regions accessible to accelerated electrons and ions, respectively. (Adapted from Gosling et al., 1990b.)

hence flows, between the ionosphere and magnetosphere. In general, the point at which flow reverses between sunward and antisunward in the ionosphere will not map to the flow-reversal boundary observed in the LLBL, even in a steady state. If the magnetic-field geometry is changing in time in response to changing solar-wind conditions or inherently unsteady coupling processes, inductive electric fields further complicate the picture.

9.5.2 Signatures of Magnetopause Reconnection

Taking the ideas we have developed about reconnection geometries and applying them to the dayside magnetopause results in the sketch shown in Figure 9.22. Here the basic x-line reconnection figure has been rotated through 90° to show the magnetopause in the orientation in which it is usually depicted. Plasma flows in from the sides, from the magnetosheath on the left, and from the magnetosphere on the right. The plasma, on reconnected flux tubes, flows out of the figure toward the top and bottom to form the boundary layer.

The situation at the magnetopause differs from the idealized cases we discussed earlier in that the plasma in the two inflow regions is quite different. The magnetosheath consists of shocked solar-wind plasma and is relatively dense ($n_e \sim 10$ cm^{-3}) and warm ($T \sim 100$ eV). Inside the magnetosphere there are two particle populations: the trapped energetic particles that constitute the ring current (which have energies of several kiloelectron volts), and cold plasma ($T \sim 1$ eV) of ionospheric origin.

When combined, their density typically is 1 cm^{-3}. The quite different properties of the plasma entering from the two sides of the boundary means that the origin of the plasma in the boundary layer can, to a large extent, be inferred from its properties. By far, the greatest number of particles come from the magnetosheath.

The two field lines reconnecting (which therefore instantaneously form the x-line) are labeled S1 and S2 (for separatrix). Also shown are the boundaries where electrons (E1 and E2) and ions (I1 and I2) that have been accelerated in the reconnection process would first be encountered by a spacecraft crossing the magnetopause and boundary layer some distance from the x-line. These boundaries are not field-aligned, because the particles move along the field lines with a finite speed as they also drift or convect with the magnetic field. Electrons move along the field line much faster than ions; yet both species drift across the field at the same speed (assuming pure $\mathbf{E} \times \mathbf{B}$ drift). This results in the electron boundaries being more closely field-aligned than the ion boundaries, and hence being outside of the ion boundaries, as shown. These particle boundaries are equivalent to the particle trajectories of the particles that passed closest to the x-line in the upper panel of Figure 9.17.

Particles on the Speiser-type orbits shown in Figure 9.16 have equal probabilities of exiting the neutral sheet on either side. Thus, we should observe particles of both magnetosheath and magnetospheric origin throughout the boundary layer. This is shown schematically in Figure 9.22 by the two wide arrows entering from either side of the diagram that bifurcate and merge in the magnetopause.

The direct evidence for quasi-time-stationary reconnection at the magnetopause comes from observations of high-speed plasma streams at the magnetopause (e.g., Sonnerup et al., 1981; Paschmann et al., 1986). Figure 9.23 shows data acquired on a typical outbound magnetopause crossing made by *ISEE 2* on 12 August 1978 that was studied by Gosling et al. (1990a,b). The two vertical lines bracket the actual magnetopause crossing when *ISEE 2* was at (8.3, 2.4, 3.8) R_E in geocentric solar ecliptic (GSE) coordinates, that is, about an hour postnoon at 25° latitude. From top to bottom of the figure are shown plasma-number density, ion temperature, the bulk-plasma flow speed and its dominant GSE y-component, and the GSE z-component of the magnetic field, that is, the component perpendicular to the ecliptic plane. Focusing on the bottom trace first, we see that this component of the magnetic field changes sign across the magnetopause, and in the magnetosheath the field has a southward component, which favors reconnection. However, the magnetic field was primarily in the GSE y-direction in the magnetosheath; the total magnetic-field rotation at the magnetopause was about 140°. Looking now at the ion data, on the left, in the magnetosphere, the ions have a small number density ($\sim$1 cm^{-3}) and high temperature ($\sim 5 \times 10^7$K $\sim$ 5 keV) and essentially zero flow velocity, charac-

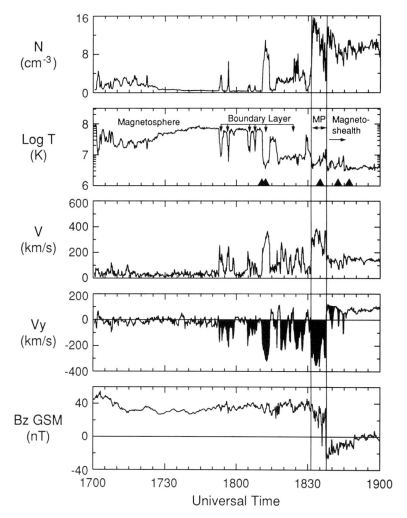

FIG. 9.23. Plasma and magnetic-field data from *ISEE 2* surrounding an outward magnetopause crossing on 12 August 1978. From top to bottom are shown ion-number density (cm^{-3}), ion temperature (K), the magnitude and GSE y-component of ion-flow velocity (km·s^{-1}), and the GSE z-component of the magnetic field. The vertical lines show the actual magnetopause crossing. (Adapted from Gosling et al., 1990a.)

teristic of the ring current. On the right, in the magnetosheath, the plasma has a density of about 10 cm^{-3} and a temperature of 3×10^6 K (300 eV) and flows at about 150 km · s^{-1} northward ($v_z > 0$), duskward ($v_y > 0$), and antisunward ($v_x < 0$), consistent with flow over the magnetopause away from the subsolar point. Beginning at 1743 (UT), *ISEE 2* began to make brief incursions into the LLBL. Those incursions gradually increased in length until *ISEE 2* encountered the magnetopause. The LLBL is characterized by ion-number densities and temperatures between those for the magnetosphere and magnetosheath and, on that occasion, strong plasma flows in the northward, dawnward, and antisunward directions. The dawnward flows are shaded for emphasis. These flows are observed both in the LLBL and in the magnetopause current sheet, and are at times more than twice as fast as the magnetosheath flow. Notice how patchy the LLBL appears, especially evident in the anticorrelation of n and T. The flow speed v is even more irregular than n and T, but high-speed flow is seen only when n is high and T low.

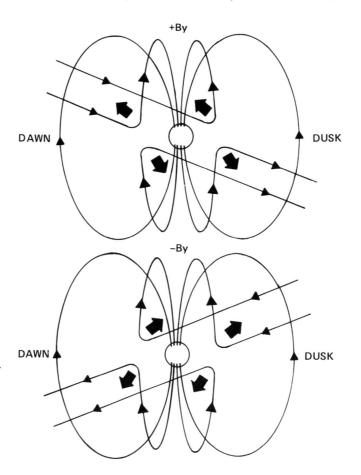

FIG. 9.24. Shapes of newly reconnected field lines for the cases when the IMF is southward and (upper panel) has a positive B_y-component (an away sector) and (lower panel) has a negative B_y-component (a toward sector), viewed from the sun. The short broad arrows show the direction that magnetic tension will tend to pull these field lines. Undisturbed magnetosheath flow is radially away from the earth everywhere in this projection. (Adapted from Gosling et al., 1990a.)

The direction of flow in the LLBL on this occasion is explained by the left part of Figure 9.24. This sketch shows two magnetospheric field lines that have recently reconnected with the magnetosheath field, which is directed duskward (an "away" sector in the solar wind) and a little southward, much as was the case on 12 August 1978. Although magnetosheath flow will be everywhere radially away from the earth in this projection, the curvature force will tend to move the reconnected flux tubes in the direction of the large arrows, that is, dawnward north of the reconnection line and duskward south of it. The opposite happens in the lower sketch, when the IMF is directed toward dawn, as is the case in a "toward" sector. Returning to the upper diagram, we expect to see magnetosheath and LLBL flow roughly aligned prenoon north of the reconnection line and postnoon south of the reconnection line. Postnoon north of the reconnection line, which is where *ISEE 2* was on 12 August, flow should be directed northward and dawnward, and also sunward if the flow remains parallel to the magnetopause, as we would expect. This is just what was observed.

Before returning to the observations, we follow the consequences of the picture presented in Figure 9.24. In the upper diagram, recently

9.5 RECONNECTION AT THE MAGNETOPAUSE

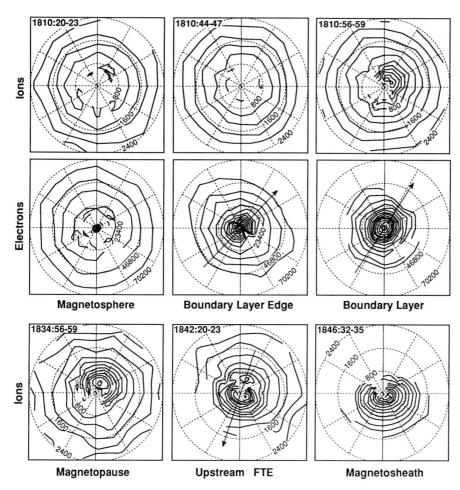

FIG. 9.25. Velocity distributions of ions (top and bottom rows) and electrons (middle row) measured by *ISEE 2* in its spin plane (approximately parallel to the ecliptic plane) at various times during the 12 August 1978 magnetopause crossing. Indicated are the times at which the distributions were measured and the regions where the measurements were made. The long arrows in several panels indicate the direction of **B** projected onto this plane. Sunward velocities are to the left, and duskward toward the bottom. (Adapted from Gosling et al., 1990a,b.)

reconnected flux tubes are pulled from postnoon across to the dawnside of the northern polar cap, and vice versa in the southern hemisphere. This IMF-dependent asymmetric flow in the polar caps is readily observed (e.g., Heppner and Maynard, 1987) and gives rise to the field-aligned currents that produce the Svalgaard-Mansurov effect, a signature seen in high-latitude magnetograms near noon that is the best indirect monitor of IMF sector polarity (e.g., Cowley, 1981).

Let us now return to the observations and compare detailed particle observations with the model presented in Figure 9.22. Figure 9.25 shows a series of 3-s snapshots of the two-dimensional velocity distributions of ions (top and bottom rows) and electrons (middle row) observed during the 12 August 1978 magnetopause crossing. The distributions were measured in the spacecraft spin plane, which was approximately the ecliptic plane, and all particles within about 30° of that plane were measured. The velocity scale is shown by the dotted circles. Sunward velocities are to the right, and duskward toward the bottom, and the large arrows in the three electron panels and one ion panel show the magnetic-field direction projected into this plane. Contours represent constant phase-space density, spaced logarithmically, with two contours per decade. These diagrams merit careful study, for not only do they show explicitly the mixing of magnetosheath and magnetospheric populations, but also they illustrate how misleading it can be to use fluid quantities obtained by taking moments of a distribution, such as the temperature and velocity shown in Figure 9.23 (see Chapter 2).

The top two rows of velocity distributions were observed a few seconds apart during the entry into the LLBL at about 1810 UT, marked by two black triangles in the second panel of Figure 9.23. The bottom row of distributions are taken in the magnetopause current layer itself, upstream of the magnetopause during a so-called flux-transfer event (FTE), and in the magnetosheath proper. These last three times are also marked with black triangles in Figure 9.23.

We begin with the earliest and latest ion distributions, those at the top left and lower right. The earliest ion distribution is of a pure ring-current or plasma-sheet ion population. The contours are circles centered on the origin, that is, a symmetric particle distribution with no detectable flow speed with respect to the spacecraft (which moved very slowly, ~2 km·s^{-1}). The characteristic width of this distribution is 1,000 km·s^{-1} or so, a thermal velocity corresponding to a temperature of several kiloelectron volts. In the center of the diagram, at low velocities (low energies) the flux of ions is too small to give a reliable measurement. The last ion distribution is representative of the magnetosheath proper. Again the contours are relatively circular, but now the circles are centered on a point below and to the right of the origin. This indicates a net ion flow with respect to the spacecraft (and hence to the earth). The displacements of the centers of the contours from the origin are between 100 and 200 km · s^{-1}, and a displacement to the lower left indicates flow antisunward and duskward, that is, a flow around the earth in the magnetosheath, as is shown in Figure 9.23. The elliptical shape of the contours indicates that the ions have a temperature anisotropy. Nevertheless, the characteristic velocity is smaller, corresponding to a temperature of a few hundred electron volts. These two distributions are relatively simple and are reasonably well described by the calculated moments for density, temperature, and velocity shown in Figure 9.23.

The earliest electron distribution (left, center row) taken simultaneously with the ion distribution above it, is also indicative of hot, isotropic electrons. The additional population, indicated by the clustered contours in the very center of the distribution, is a mixture of photoelectrons (produced by sunlight hitting the spacecraft) and cold electrons from the ionosphere. The hot electrons show no obvious asymmetry in the direction of the magnetic field, shown by the long arrow. The next pair of distributions, taken 24 s later, shows a dramatic change in the electron distribution, but not in the ions. The hot-electron distribution, which was isotropic moments earlier, is now peaked at 90° to the field direction (indicated by elliptical contours with major axis perpendicular to the field direction). In addition, a new, much colder population has appeared in the center of the distribution. This is peaked along the field direction, with deep minima perpendicular to the field. Turning back to Figure 9.23, we can see what has happened. *ISEE 2* is passing outward through the magnetopause north of the reconnection line. As it crosses the boundary between the magnetosphere proper (on closed field lines) and the LLBL, it moves onto field lines that have been open for successively longer. In turn it will cross boundaries S2, E2, and I2. There is no particle signature or indeed any signature at all associated with S2. Between the first two pairs of distributions, *ISEE 2* crossed the E2 electron boundary, but not the I2 ion boundary. The field line is open, and the first signatures appear in the electrons with field-aligned velocities. Some of the previously trapped magnetospheric electrons have crossed the magnetopause and been lost, leaving behind electrons with pitch angles near 90°, thus flattening the contours of the hot-electron population. Meanwhile, magnetosheath electrons with field-aligned velocities moving in the direction of the large arrow have reached *ISEE 2*. Some have even gone down to near the foot of the field line, have mirrored, and are on their way back to the magnetopause, moving antiparallel to the large arrow. However, magnetosheath electrons with pitch angles near 90°, which move more slowly along the field line, have yet to reach the spacecraft. Note that the magnetosheath electrons have parallel velocities of several thousand kilometers per second.

In the next pair of distributions, taken only 12 s later, the electrons look entirely different, and the ions have begun to change. *ISEE 2* has now crossed the I2 ion boundary. Cool magnetosheath ions have appeared in the upper right part of the distribution. These are flowing along the magnetic field from the direction of the magnetopause. No ions have yet had time to mirror and return to the spacecraft. Note that the ions move along the field at a few hundred kilometers per second and that the magnetosheath ion contours are not symmetrical about the magnetic-field direction, but rather they have a net flow perpendicular to **B**, which is antisunward and dawnward. It is this aspect of the distribution that results in the reversed ion flow shown in Figure 9.23. The electrons have undergone a dramatic change. Most of the hot magnetospheric electrons have escaped along the open field line. Meanwhile, the contours of the cool distribution have become much more nearly circular, either through pitch-angle scattering or because there has now been time for near-90°-pitch-angle magnetosheath electrons to reach the spacecraft. Some 30 s later (not illustrated), ions that have mirrored near the foot of the field line appear at the reconnection point. Some distributions show a parallel velocity cutoff in both the reflected particles and those arriving at the spacecraft directly. Using these two velocities and estimating the distance to the mirror point, the distance along the field line from the spacecraft to the reconnection point can be calculated (see Problem 9.3).

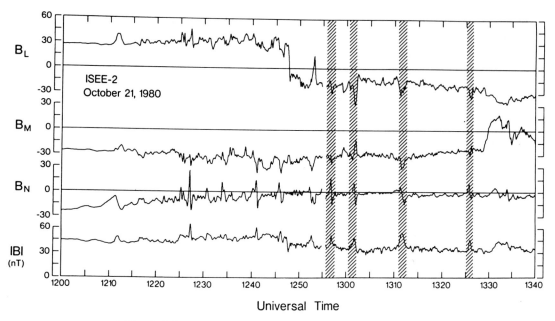

FIG. 9.26. Magnetometer data from *ISEE 2* during an outward magnetopause crossing on 21 October 1980, showing a series of FTEs. The data are in boundary-normal coordinates. *ISEE-2* crossed the magnetopause at 1247 UT, marked by the sign change in B_L. At least seven FTEs, each marked by a positive and then negative excursion in B_N, occurred. The four observed in the magnetosheath are marked by cross-hatching. (Adapted from Elphic, 1990.)

Using values obtained at the crossing of a boundary-layer edge some 10 min earlier, Gosling et al. (1990b) estimated that that particular field line reconnected 10 s before the observation at a point about $5R_E$ south of *ISEE 2*, which would place it close to the equatorial plane.

Within the current layer itself (Figure 9.25, bottom row, left), both magnetosheath and magnetospheric ions are again present. On this particular crossing, most of the field rotation occurs at the outer boundary of the current sheet. When this distribution was measured, the magnetic field was depressed, but primarily in the direction it had in the LLBL, and the magnetosheath ions were moving in the same direction along the field.

In Figure 9.23, just beyond the magnetopause, there are three brief intervals of negative v_y between 1840 and 1845 UT. They are accompanied by increases in temperature and reductions in flow speed and density. The magnetic field is not illustrated here (Gosling et al., 1990a, Fig. 3). However, bipolar signatures are seen in the component normal to the magnetopause, and there are slight depressions in the field magnitude. These are examples of FTEs, which were first identified as characteristic signatures seen close to either side of the magnetopause current layer by Russell and Elphic (1978). Three different ion populations can be identified within the FTEs (Figure 9.25, bottom row, center). First, there is the ordinary magnetosheath population flowing toward dusk

9.5 RECONNECTION AT THE MAGNETOPAUSE

in the $+y$-direction. Second, there is a somewhat warmer population counterstreaming along **B** in the $-y$-direction. Finally, there are hotter ions also streaming along **B** in the $-y$-direction. The presence of these three distinct populations shows that this flux tube crosses the magnetopause (i.e., *ISEE 2* is on the magnetosheath part of an open magnetospheric flux tube). The magnetosheath ions were present on the field line before it reconnected. As the IMF is directed away from the sun, and *ISEE 2* is north of the reconnection site (as evidenced by the flows in the boundary layer and the time-of-flight calculations), this flux tube is still connected to the sun (cf. Figure 9.24, upper panel); so the source of magnetosheath particles is not switched off. The somewhat warmer counterstreaming ions are magnetosheath ions that have entered the magnetopause current sheet, have been accelerated, and have left on the magnetosheath side (cf. Figures 9.17 and 9.23). These ions are thus flowing back along the field line, having been "reflected" at the magnetopause. The hotter ions are magnetospheric ions that have crossed the magnetopause and are escaping. Integrating over this combined distribution function results in the $-v_y$ velocity shown in Figure 9.23. However, moment calculations cannot adequately describe what is happening in situations such as this, where several distinct particle populations coexist on the same flux tube. The flows of all three populations are primarily field-aligned. Clearly, we would also like to know how the flux tube itself is moving (i.e., the $\mathbf{v}_\perp$ component of the ion flow), but because the field is directed almost exactly in the y-direction, $\mathbf{v}_\perp$ is directed primarily antisunward and northward; so we cannot tell if this flux tube will be pulled to the dawn or dusk side of the earth or straight back over the top of the polar cap.

The principal characteristic of FTEs identified by Russell and Elphic (1978) is a bipolar signature in the component of the magnetic field normal to the magnetopause. A series of FTEs observed on both sides of the magnetopause is apparent in the *ISEE 2* magnetometer data shown in Figure 9.26. The data are presented in boundary-normal coordinates. In these coordinates, B_N points along the outward normal to the magnetopause; B_L and B_M are tangential to the magnetopause, with B_L along the projection of the earth's dipole axis onto the magnetopause (positive northward); B_M is directed dawnward. The sharp change in the sign of B_L at 1247 UT marks the passage of *ISEE 2* through the magnetopause from the magnetosphere to the magnetosheath. Four FTEs observed outside the magnetopause are indicated by cross-hatching. At least three more are seen inside the magnetopause. All are marked by a positive and then negative (outward then inward) perturbation in B_N and a local field-strength maximum. Within the FTE, the magnetic field points in a direction different from both the field in the LLBL and that in the magnetosheath. Note also that the magnetic-field strength is larger within FTEs than anywhere else.

The fact that the polarity of the B_N perturbation, positive and then

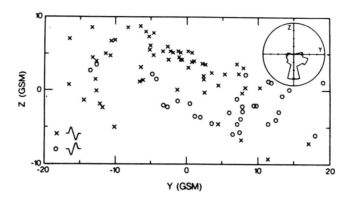

FIG. 9.27. Distribution of FTEs over the dayside magnetopause projected onto the geocentric solar magnetospheric (GSM) y–z plane. The FTE B_N polarity is indicated by crosses (+/−) and circles (−/+). During a single crossing, FTE polarity is the same on both sides of the magnetopause, but there is a clear tendency for +/− polarity north of the equator, and −/+ polarity south of the equator. The inset shows the probability of FTEs being observed as a function of IMF direction. FTEs are almost never observed when the IMF is northward. (Adapted from Elphic, 1990.)

negative, is the same on both sides of the boundary is significant. This type of signature can be produced only by some sort of bubble or blister on the magnetopause. The perturbation produced by a wave on the boundary would result in opposite polarities being observed inside and outside the boundary. The polarity shown here is that usually observed when the spacecraft is north of the magnetic equator. The opposite polarity, B_N first negative and then positive, is usually observed on both sides of the magnetopause south of the equator (Figure 9.27).

These observations, together with plasma observations of the type discussed in relation to Figures 9.23 and 9.25, are consistent with the picture in Figure 9.28, which is based on Russell and Elphic's original interpretation. A magnetospheric flux tube has reconnected somewhere near the nose of the magnetosphere in a spatially and temporally limited burst of reconnection. The result is two open flux tubes, which lie along the magnetopause, partially inside and partially outside the current sheet, distorting the current sheet either inward or outward. The purely magnetospheric and magnetosheath field lines that drape around the sides of the flux tubes are not shown in this diagram. These two open flux tubes are pulled upward and downward much in the same manner as in Figure 9.24. The passage of either the magnetosheath or magnetopause portion of the flux tube past the spacecraft produces the characteristic magnetic signature. Note that the passage of the upper flux tube will cause a positive/negative B_N perturbation, whereas the lower one causes the opposite polarity perturbation. If a spacecraft merely passes through the region of the magnetosheath or magnetospheric field lines draped around the open flux tube, the characteristic B_N and $|B|$ signature is still observed, but of course the mixed ion and electron populations characteristic of open flux tubes are then not seen.

So far, the evidence for magnetopause reconnection we have presented, though very strong, is qualitative. We have shown that the expected plasma populations exist in the correct magnetic topology. However, quantitative comparisons with theory are also possible. These involve matching the accelerated flows seen in the LLBL with the acceleration required for stress balance within the current sheet. The

9.5 RECONNECTION AT THE MAGNETOPAUSE

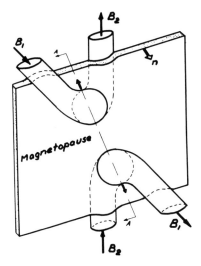

FIG. 9.28. Sketch of Russell and Elphic's (1978) original interpretation of an FTE. Both parts of an isolated, recently reconnected flux tube are shown crossing the magnetopause. B_1 is the magnetosheath field, and B_2 the geomagnetic field. Flux crosses the magnetopause only in the two circular areas; the rest of the magnetopause is closed. All parts of the reconnected flux tubes push against the magnetopause, bending it slightly. (Adapted from Sonnerup, 1984.)

particles are accelerated by a $\mathbf{j} \times \mathbf{B}$ force in the current sheet, which will always be present unless the boundary is closed ($B_N = 0$). The observational difficulty is in measuring the small B_N component, which requires that the orientation of the boundary must be well determined. As we discussed in Section 9.4, single particles do not get accelerated in the frame in which $\mathbf{E} = 0$, the so-called de Hoffman–Teller frame, which moves with the reconnected field lines. In transforming to the frame stationary with respect to the earth (or the spacecraft), particles gain twice the velocity of the frame transformation. Sonnerup et al. (1981), Paschmann et al. (1986), and others have checked this velocity gain quantitatively for a large number of magnetopause crossings and have found excellent agreement between both the direction and magnitude of the observed flows.

9.5.3 Patchy, Unsteady Reconnection

As we have just seen, the evidence that reconnection occurs at the magnetopause is overwhelming. But questions remain: Where does reconnection occur? On what time and space scales does it occur? Is reconnection able to explain all the observed properties of the magnetopause boundary layers and the flow, current, and precipitating-particle patterns observed at low altitudes? Definitive answers to these questions have not yet been found, though they remain topics of active research. Three factors are believed to influence the onset and rate of reconnection: the relative orientation of the magnetic fields, or the magnetic shear; the flow speed in the magnetosheath; and the plasma β in the magnetosheath (e.g., Quest and Coroniti, 1981). Reconnection should occur most easily where the magnetosheath and magnetosphere magnetic fields are antiparallel. Where on the magnetopause this occurs depends, of course, on the orientation of the IMF (and, to a lesser extent, on the earth's dipole tilt). Crooker (1979) mapped these regions and argued on the basis of the so-called antiparallel-merging theory that these locations are where reconnection is most likely to occur. However, reconnection is also more likely to occur where the magnetosheath flows slowly over the magnetopause, as this gives more time for the instability to grow. Magnetosheath flow is slowest near the stagnation point at the nose of the magnetosphere. The magnetosheath flow accelerates away from this point, becoming super-Alfvénic by the time it reaches the flanks. This would thus seem to favor reconnection near the nose of the magnetosphere. Finally, lower plasma β in the magnetopause probably favors reconnection. The lowest plasma pressures are found, again, near the stagnation point, as flux tubes tend to partially empty via field-aligned plasma flow while they are hung up in the slow-flow region. This creates a region known as the plasma-depletion layer, found just outside the front side of the magnetopause in which flux tubes pile up. Here plasma pressure is decreased, and magnetic pressure increased (thus maintaining a uniform total pressure), and so β is reduced. Thus, this effect will also tend to favor nose reconnection. So two of these factors favor reconnection near the nose of the magnetosphere, and the third favors two sites that move around the magnetopause in response to the IMF direction.

It seems most probable that magnetopause reconnection occurs, at least for southward IMF, most frequently near the nose of the magnetosphere close to the flow stagnation point. The reconnection rate probably varies on a time scale of minutes, but this is sufficiently long for quasi-steady reconnection stress balance to prevail. The time variability also gives rise to the FTE signatures observed near the magnetopause. Because a single x-line is sufficient to explain the observations, there seems to be no reason to invoke multiple reconnection sites, but this does not mean they do not exist. Several authors have invoked either multiple or turbulent reconnection to explain various observations. Whether the time variability is inherent to the reconnection process or is a result of ever-changing conditions in the magnetosheath remains an open and challenging question.

9.6 RECONNECTION AND THE PLASMA-SHEET BOUNDARY LAYER

The PSBL forms the boundaries between the plasma sheet and the two tail lobes. The northern PSBL is illustrated in Figure 9.18; a similar boundary layer separates the southern lobe and the central plasma sheet (CPS). Within this boundary layer, plasma-number density and temperature are intermediate between lobe and CPS values, but its most characteristic feature is sustained field-aligned ion and electron flows, directed both earthward and tailward.

9.6 RECONNECTION AND PLASMA-SHEET BOUNDARY LAYER

The PSBL is formed of plasma that has just been accelerated in a reconnection region. Such boundary layers exist elsewhere (e.g., LLBL), but the PSBL provides a clear example of the relevant physics. Reconnection in the magnetotail starts with inflow of plasma from the two lobes, which have very similar plasma properties. This makes tail reconnection closer to the classic models we examined in Section 9.4. One major difference from the models is that following reconnection in the magnetotail, the two outflow regions are different (Figure 9.11). Tailward of the x-line, the field lines are unattached to the earth and are pulled away to rejoin the solar wind by magnetic tension. But earthward of the x-line, flux tubes are restrained by the slower-moving dense plasma on the closed flux tubes that compose the CPS. This results in a bifurcation of the outflow region. The flux tubes themselves move slowly, but the particles that have been accelerated in the current sheet (Figure 9.16) move rapidly along the now-closed field lines toward the earth in both directions, that is, toward the southern and northern poles. These are the earthward field-aligned particle flows seen in the PSBL both north and south of the CPS. On reaching the near-earth region, these particles mirror and return along the boundary layer, thus creating the tailward field-aligned flows. After a very small number of bounces, the counterstreaming particles will thermalize, and the flux tube, now containing a more isotropic plasma, will merge into the CPS.

The observed ion beams have energies ranging from a few kilovolts to a few tens of kilovolts and consist of a mix of species, H^+, He^+, He^{2+}, O^+, indicating that the ions come originally from both the solar wind (He^{2+}) and the ionosphere (He^+, O^+). The electron energies typically are lower by a factor of 2 or 3. We now examine the motion of particles, particularly ions, in the tail as a whole to show how this can come about.

The primary source of lobe plasma is the mantle. In Section 9.5.1 we saw that the mantle is formed from a mixture of magnetosheath plasma that entered the magnetosphere through the cusp and ionospheric plasma. The magnetosheath plasma has a temperature of approximately 300 eV, and the suprathermal escaping ionospheric plasma has a temperature of a few electron volts. Imagine a magnetosheath H^+ that entered the magnetosphere and has just mirrored at low altitude on the field line labeled 2 in Figure 9.11. This ion moves up the field line; as it encounters weaker magnetic fields, its field-aligned velocity $v_\parallel$ increases. Soon it has a pitch angle close to zero, so that most of its velocity is directed along the field. If it is moving quickly, it will soon reach the magnetopause again and reenter the magnetosheath, perhaps by the time the field line has convected to the position of the field line labeled 3. A more slowly moving ion will exit the magnetosphere farther down the tail, where field line 4 or 5 crosses the magnetopause. However, if the ion moves sufficiently slowly, it will still be on the magnetospheric portion of the field line when the field line reaches position 6. Now what happens to the particle depends on whether it is tailward or earthward of the x-line when the field line reconnects. If the ion is tailward of the reconnection point, it will continue to move away from the earth. It will never cross the current sheet unless the most earthward point of the field line moves tailward more rapidly than the particle, in which case the ion will be accelerated. But in any case, this particle is now on a purely IMF field line, so that ultimately it will rejoin the solar wind, whether accelerated or not. If, on the other hand, the ion is earthward of the x-line when the field line reconnects, it will enter the current sheet on a closed field line. It will follow a Spieser-type orbit (Figure 9.16), be accelerated, and leave the current sheet moving earthward with a much increased $v_\parallel$. It becomes one of the earthward-streaming ions seen in the PSBL. The velocity dispersion of mantle particles we have just described is illustrated in Figure 9.3. The dashed lines show the paths followed by the mantle ions with different $v_\parallel$ as the field lines convect to the center of the tail.

Quantitatively we can derive a relation among the speed at which the particle moves (or its energy), the cross-tail potential drop, and the distance to the x-line to find what energy particles will enter the plasma sheet earthward of the neutral line. Let the x-line be distance L_x from the earth. Once the particle has left the near vicinity of the earth and is in the lobe where $B \sim 20$ nT, a factor of about 10^3 less than at its mirror point, $v_\parallel \sim v$, and the field line is roughly parallel to the tail axis. So, to a good approximation, the particle will take a time

$$t = L_x / v_\parallel \tag{9.64}$$

to reach the point on the field line where reconnection will occur.

In Section 9.4.2 we derived a value for the cross-tail potential drop ϕ of 53 kV. This provides the electric field that convects the flux tubes from near the lobe magnetopause to the center plane of the tail. The electric-field strength is ϕ divided by the diameter of the tail, $2R_T$, and the field line has to convect to distance R_T, which will take time

$$t = \frac{2R_T^2 B_T}{\phi} \tag{9.65}$$

where B_T is the lobe field strength. Note that the numerator is proportional to the total magnetic flux in either lobe.

Equating these two times, (9.64) and (9.65), gives us the parallel velocity of the particle that reaches the point on the field line where it reconnects just as it reconnects (i.e., this particle goes through the x-line). This critical velocity is

$$v_{\parallel c} = \frac{\phi L_x}{2 R_T^2 B_T} \tag{9.66}$$

Hence the critical particle energy is

9.6 RECONNECTION AND PLASMA-SHEET BOUNDARY LAYER

$$W_c = \frac{m}{2}\left(\frac{\phi L_x}{2R_T^2 B_T}\right)^2 \tag{9.67}$$

Particles of lower energy will enter the plasma sheet earthward of the x-line, and particles of greater energy tailward of the x-line. Taking the ion to be a proton, $L_x = 100 R_E$, $\phi = 53$ kV, and $R_T = 26 R_E$ and $B_T = 10$ nT, as derived in Section 9.3, we obtain $W_c = 21$ eV, somewhat low for a typical magnetosheath proton, but high for an ionospheric one. But both ϕ and L_x can be smaller or larger by at least a factor of 2, so that W_c could be larger or smaller by more than an order of magnitude, depending on the cross-polar-cap potential and the distance to the x-line. (The denominator, which, as we noted, is proportional to the flux in either lobe, varies by a much smaller factor.) Note that it is much easier to trap the heavier ions, such as ionospheric O^+, because these must be more energetic to escape, as this is in essence a velocity filter, not an energy filter.

What we learn from this calculation is that ionospheric ions escaping in the polar wind are mostly retained in the magnetosphere, for they will convect into the plasma sheet earthward of the x-line unless the x-line is very close to the earth. Magnetosheath ions that enter the cusp and populate the mantle will usually escape from the tail, except perhaps the coldest ones, but during periods of rapid convection they may become trapped onto closed field lines if the x-line is far down the tail.

If the ion enters the current sheet earthward of the x-line, it will be accelerated. From (9.60), using $W = \frac{1}{2}mv_x^2 = \frac{1}{2}mv_\parallel^2$, we find that the gain of energy is

$$\Delta W = \frac{4E_y}{v_\parallel B_z} W \tag{9.68}$$

where now the x, y, and z subscripts refer to directions in a tail coordinate system in which the plasma sheet is in the $z = 0$ plane and x points toward the earth. (This is roughly GSE coordinates.) So the energy of the ion after its acceleration, W', is

$$W' = \left(\frac{4E_y}{v_\parallel B_z} + 1\right) W \tag{9.69}$$

where E_y is the cross-tail electric field. Distributing a 60-kV potential uniformly across the width of the tail gives an electric field of about 0.2 mV·m^{-1}. In practice, the field can be several times larger locally if fast reconnection is confined to some fraction of the tail width, say the midsection of the tail. The component of the magnetic field normal to the current sheet, B_z, is a very poorly known parameter. It could be as high as 1 nT. More probably, near the x-line it is much smaller, perhaps as small as 0.1 nT. The quantity $v_\parallel$ is the speed of the incoming particle, for which we can use the critical velocity just found, 64 km·s^{-1}. Substituting these numbers into (9.69) yields an energy increase of anywhere from a factor of 8 to several hundred, sufficient to accelerate the 21-eV

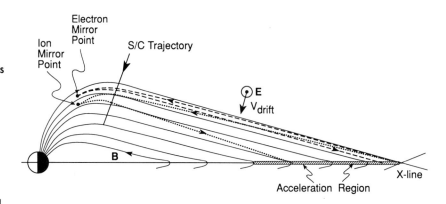

FIG. 9.29. Motions of ions (short dashes) and electrons (long dashes) after their acceleration in the portion of the tail current sheet earthward of the tail x-line. Both species travel along the field line and mirror near the earth, as they also **E** × **B** drift toward the current sheet. The faster-moving electrons drift far less in one bounce than do the ions, resulting in a spatial separation of earthward- and tailward-streaming electrons and ions that would be observed by a spacecraft following the trajectory shown.

proton into the kilovolt range. An O^+ with the same critical velocity has an initial energy of 340 eV and so could be accelerated to an energy of tens of kilovolts.

These particles are now on closed field lines. Figure 9.29 shows their subsequent motion, a combination of an electric-field drift toward the current sheet and a field-aligned motion. The finite particle speed results in a layering of the boundary, much as in the LLBL (Figures 9.22 and 9.25), with an electron boundary inside the separatrix and an ion boundary still farther inside. On reaching the stronger fields near the earth, the particles will mirror and return, forming return-flow boundaries inside the earthward-flowing boundaries. Thus a spacecraft entering the plasma sheet from the lobe earthward of the reconnection region, say 15–20R_E from the earth, will see first earthward-flowing energetic electrons, next returning electrons, then earthward-flowing ions, and finally tailward-flowing ions, which is what is observed (e.g., Onsager et al., 1991). If the acceleration region were small, this time-of-flight particle dispersion would result in particles of only one particular value of $v_\parallel$ being observed at any location in the boundary layer, with the earthward-moving particle having a smaller $v_\parallel$ than the reflected tailward-moving one. What is in fact observed is that ions (and electrons) are observed at all energies (i.e., velocities) above some lower cutoff. The energy at the cutoff for the earthward-moving ions is lower than that for the tailward-moving ions. This can be explained only by an extended source, such as is shown in Figure 9.29. Ions of different energies come from different places in the acceleration region, with the high-energy ions being accelerated nearer the spacecraft (or the earth).

Onsager et al. (1991) have modeled the observed ion and electron distributions in some detail using simple adiabatic particle trajectories. Figure 9.30 compares spacecraft observations with their model results. The four distributions on the left (*a–d*) were measured by *ISEE 2* during a passage from the lobe through the PSBL into the CPS on 1 March 1978. Several minutes elapsed between successive distributions. On the right (*e–h*) are four model distributions at four positions successively closer to the current sheet in the time-stationary model of Onsager et al.

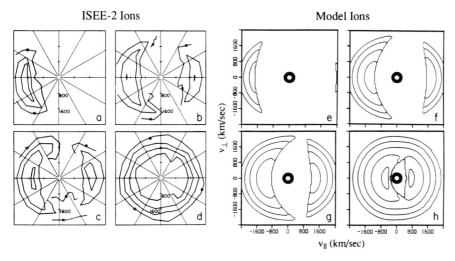

FIG. 9.30. Left: Ion-velocity distributions observed by *ISEE 2* as it crossed from the tail lobe into the CPS on 1 March 1978. The distributions are measured in the spacecraft spin plane (approximately parallel to the ecliptic plane). Earthward-moving (tailward-moving) ions are on the left (right). The magnetic field is along the horizontal axis. The distributions *a–d* were obtained at 0405, 0408, 0414, and 0434 UT as *ISEE 2* passed deeper into the plasma sheet. Right: Ion distributions successively closer to the center of the plasma sheet in a time-stationary model. (Adapted from Onsager et al., 1991.)

(1991). The distance between distributions *e* and *h* is about $1.5 R_E$. In distributions *a* and *e*, only earthward-streaming ions are seen. The lower energy cutoff is a result of the time-of-flight filter. The higher energy falloff in the model is the result of an assumed Maxwellian distribution, with a temperature of 30 keV at the acceleration point. In the next pair of distributions (*b* and *f*), returning ions are observed. Note that the low energy cutoff of the reflected ions is higher than that of the earthward-moving ions and that the cutoff of the earthward-moving ions is at a lower energy than in the first distributions. In the next pair of distributions (*c* and *g*), locally mirroring (pitch angle 90°) ions of the highest energies have arrived, and in the final pair the highest-energy ions appear almost isotropic. Throughout, the fluxes of ions with speeds below about 900 km·s^{-1} are too low to observe, which results in the observed distribution in panel *d* appearing isotropic, suggesting that *ISEE 2* is now in the CPS. However, model distribution *h* shows that at velocities below 800 km·s^{-1} (energies below 3 keV), the distribution is far from isotropic and still shows a clear boundary-layer shape. The dense set of contours near the origin of the model distribution represents unaccelerated lobe ions (i.e., those that have yet to cross the current sheet, cf. Figure 9.17) that are unobservable by the spacecraft.

The fastest-moving reflected particles will return to the current sheet at a point where it is still thin and will then be accelerated for a second time. Electrons may even return more than once and get accelerated to large energies. The limiting energy will be one for which the particle gyroradius in the current sheet becomes so large that the particles exit the tail at the flanks.

Green and Horwitz (1986) considered the fate of the earthward-streaming ions in a more realistic near-earth field model. On the left in Figure 9.31 are shown the trajectories of two ions, each with a pitch angle of 2.6°, and with speeds of 500 and 1,500 km·s^{-1}. The particles are started at a point $40R_E$ downtail and $7R_E$ above the current sheet,

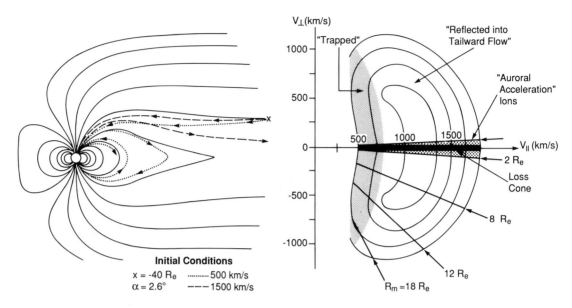

FIG. 9.31. Left: Trajectories of two plasma-sheet ions in a model geomagnetic field. Both ions start at a point $40R_E$ downtail and $7R_E$ above the current sheet, with an initial pitch angle of 2.6°. One ion has an initial velocity of 500 km·s^{-1}, and the other 1,500 km·s^{-1}. Right: A velocity distribution typical of those observed by spacecraft in the PSBL (cf. Figure 9.30). The subsequent fates of ions in different parts of the distribution are indicated by the shading. (Adapted from Green and Horwitz, 1986.)

and they mirror at an altitude of $2R_E$ in the magnetic-field model used. Green and Horwitz imposed an electric field equivalent to a cross-tail potential of 150 kV (typical of a disturbed time). In this field, the faster particle returns to the tail beyond $40R_E$, but the slower particle next crosses the current sheet at about $20R_E$ and remains within $10R_E$ of the earth after three bounces. The particle distribution on the right in Figure 9.31 is typical of an earthward-directed flow (cf. Figure 9.30). Green and Horwitz shaded this distribution to show what happens to the various parts of it. Particles with pitch angles less than 1° mirror at altitudes below 100 km and are lost. Particles with pitch angles less than 2.6° mirror within $2R_E$ of the center of the earth and are liable to enter the region of field-aligned potential drops that accelerate electrons downward to generate bright aurora. The sense of the potential drops is to accelerate ions upward, so that these ions will be reflected. The bulk of the distribution mirrors at higher altitudes. (The radial distances of mirror points, R_m, are shown around the bottom of the distribution.) Ions composing the darker-shaded region recross the current sheet earthward of $40R_E$; the remainder travel back beyond $40R_E$, but as they are on closed field lines, the entire distribution is trapped.

In this section we have seen how single-particle trajectories and the imposition of a simple cross-tail electric field are sufficient to explain the migration of the plasma mantle toward the current sheet, the acceleration of these particles in the current sheet, the subsequent beaming of these particles to form the PSBL, and the bouncing of these ions to form the CPS. Single-particle orbit theory cannot describe the collective behavior of the particles, either as a fluid or through kinetic effects. Fluid and kinetic instabilities act to remove sharp gradients in particle distributions tending to thermalize them. For example, the very sharp gradients at the lower-energy cutoffs in the model ion distributions in Figure 9.30 and in similar electron distributions will be smoothed via the two stream and other instabilities and will give rise to the plasma waves seen in the PSBL (e.g., Elphic and Gary, 1990). In the next section we shall comment on tail convection from a fluid standpoint.

9.7 IS STEADY-STATE CONVECTION POSSIBLE IN THE TAIL?

In the preceding section we considered tail convection from the standpoint of single-particle dynamics. Such a treatment is, of course, not self-consistent, for the collective behavior of the particles that give rise to currents and pressure gradients is not considered. The magnetic-field structure is imposed and is not consistent with the plasma conditions. The problem that emerges is that if the plasma convects inward from the deep tail, it compresses adiabatically. The convection of closed plasma-sheet flux tubes earthward may set up a gradient in plasma-sheet pressure with distance downtail that is inconsistent with a reasonable tail magnetic-field model. The dilemma was first pointed out by Erickson and Wolf (1980) and has subsequently generated much discussion.

Three solutions have been put forward to this dilemma: (1) Pressure balance can be maintained if flux tubes lose a substantial fraction of their plasma (Kivelson and Spence, 1988). (2) Some sort of global or average steady state might be maintained if the plasma sheet is very patchy (i.e., if it consists of small regions of very different temperature and density that convect differently) (Pontius and Wolf, 1990; Goertz and Baumjohann, 1991). (3) A steady state never occurs; instead, tail structure continually evolves (Erickson, 1984; Erickson, Spiro, and Wolf, 1991). Modeling studies (e.g. Erickson, 1984) suggest that a field-strength minimum develops in the near tail. This could provide a site for a new x-line to form, at which reconnection could begin, disconnecting much of the distended plasma sheet from the earth and allowing the formerly trapped plasma sheet to escape downtail. This picture is similar to the so-called near-earth x-line model of substorms, which is discussed at length in Chapter 13.

9.8 CONCLUSION

This chapter is about magnetic reconnection as a physical process, about the direct evidence for its occurrence at the two principal current sheets in the near-earth environment, the magnetopause and the tail current sheet, and about some of its consequences. The importance of this process cannot be overemphasized. In the collisionless-plasma regime, magnetic reconnection provides the primary means by which distinct plasma regions are able to interact and exchange mass, momentum, and energy.

Magnetic reconnection is the key to understanding magnetospheric dynamics, but in turn the magnetosphere is our primary laboratory for understanding how reconnection works. The essence of what we learn locally must apply to all astrophysical plasma systems in which collisions are negligible.

Throughout this chapter I have tried to stress the duality of the fluid (MHD) and particle approaches to the analysis of plasma processes. Each approach has its merits and drawbacks, and they provide slightly different understandings of what is happening, but in most situations they reach the same or very similar answers. For reconnection, in particular, the particle picture provides a means of understanding the process without the need to "cut" and "retie" field lines, which for some is a very difficult concept. It also explains the acceleration of a few particles to very high energies. The MHD models, on the other hand, allow us to take account of the finite pressure of the plasma and how this affects the flows. But a full understanding comes only from considering both pictures and from the realization that both approaches have something to teach us.

ACKNOWLEDGMENTS

This chapter is based on lecture notes for a space-plasma physics course I teach at Boston University. Those notes, in turn, derive from a multitude of sources, but the sections on reconnection draw heavily on articles by Cowley (1986) and Hill (1975), which have been a great help to me in understanding reconnection. I thank the editors for extreme patience in waiting for this chapter to be finished. I dedicate this chapter to Professor J. W. Dungey.

ADDITIONAL READING

The best collection of review and research papers on reconnection, including papers on reconnection at the magnetopause and in the tail, resulted from a topical conference on reconnection:

Hones, E. W., Jr. (ed.). 1984. *Magnetic Reconnection in Space and Laboratory plasmas*. AGU Geophysical Monograph 30.

Other good review articles on reconnection are those by Cowley (1986), Forbes and Priest (1987), Hill (1975), and Vasyliunas (1975).

The geotail was the subject of a topical conference in 1986 which resulted in the following:

Lui, A. T. Y. (ed.). 1987. *Magnetotail Physics*. Baltimore: Johns Hopkins University Press.

There is an older but still excellent conference proceedings that covers both magnetopause and tail physics:

Battrick, B. (ed.). 1979. *Proceedings of the Magnetospheric Boundary Layers Conference*, ESA SP-148. Noordwijk: European Space Agency.

There has been no equivalent conference on the magnetopause. However, the review articles by Cowley (1980, 1982, 1984), Siscoe (1987), Russell (1990), and Lee (1991) are recommended.

PROBLEMS

9.1. Show that
$$c^2 \omega_{ce} \omega_{ci} = \omega_{pe}^2 v_A^2$$
and
$$\beta = \frac{2}{\gamma} \frac{c_s^2}{v_A^2}$$

9.2. Add four more rows to Table 9.1 showing the Alfvén speed, the sound speed, and the gyroradii for a proton and an electron with energy equal to the average thermal energy (temperature) of that species in each of the four regions. Discuss the implications of these values, both relative to each other and in absolute terms. In what region is the MHD approximation most likely to break down?

9.3. The electron-velocity distribution measured by a spacecraft just as it enters the LLBL exhibits low-energy cutoffs in both earthward-streaming and returning electrons with 0° pitch angle. No earthward-streaming electrons are measured with velocities less than $v_e = 5{,}000$ km·s^{-1}, and no returning electrons with velocities less than $v_m = 22{,}000$ km·s^{-1}. Assuming that the returning electrons are from the same source and that they mirrored close to the earth, and that the distance from the spacecraft to the mirror point x_m can be obtained from a magnetic-field model, derive a formula to estimate the distance from the spacecraft to the acceleration point. If the distance to the mirror point is $12 R_E$, use your formula to estimate the distance to the acceleration region, and deduce how long prior to the observation the acceleration began on this field line.

9.4. A spacecraft is moving poleward in a low-altitude earth orbit at 7 km·s^{-1}. As it crosses the auroral zone, it detects precipitating protons with the temporal/spatial dispersion in energy shown in Figure 9.32. This proton dispersion occurs because the more energetic protons travel faster and

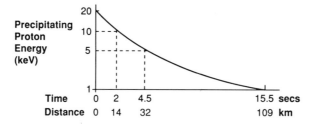

FIG. 9.32. Observed precipitating-proton energy versus time and distance on a poleward-moving pass of a low-altitude satellite.

hence arrive at the spacecraft altitude from above sooner than do less energetic protons. From the electric field measured on the spacecraft, it is known that the ambient plasma is convecting (flowing) poleward at 1 km · s^{-1}. This allows two possible explanations for the observation:

(a) The dispersion observed by the spacecraft is temporal. That is, a short-lived acceleration event occurred over a broad region of space.

(b) The dispersion observed by the spacecraft is spatial. That is, a spatially confined acceleration region exists for an extended period of time. As the plasma flows poleward, each field line spends only a short time in the acceleration region. This results in a spatial-dispersion pattern at the altitude through which the spacecraft flies.

For each of these scenarios, calculate the height of the acceleration region above the spacecraft. Use your knowledge of the magnetosphere to discuss which explanation is more probable.

9.5. In Section 9.4 we saw that reconnection can proceed at a fraction of the inflow Alfvén speed. Calculate the cross-tail potential difference required to convect tail-lobe flux toward the center of the tail at a speed equal to $0.1 v_A$ in the lobes (use parameter values given in Table 9.1 and derived in the text). How does this potential drop compare to observed cross-polar-cap potential drops? Discuss any differences.

9.6. During the growth phase of a geomagnetic substorm, the polar cap is observed to expand. If the polar cap were to expand from an initial mean radius of 15° to a mean radius of 20° latitude, how would the geomagnetic tail change? Assume that the tail radius $10R_E$ behind the earth remains fixed at $R_T = 18R_E$ ($x = 10R_E$), and find how the tail radius, the lobe field strength, and the plasma-sheet pressure change as functions of distance downtail.

9.7. The cross-tail current is carried primarily by plasma-sheet ions, which have a density of $n = 0.3$ cm^{-3}. If the current sheet is $1R_E$ thick, and the lobe magnetic-field strength is 20 nT, calculate the average ion velocity required for the ions to carry the current. Compare this velocity to the average ion thermal velocity if the average ion energy is 4 keV. Discuss whether or not a plasma detector on board a spacecraft could measure the current directly.

9.8. It has been suggested that a substorm onset might be triggered when the current-sheet thickness becomes less than the gyroradius of a thermal proton in the current sheet. During the growth phase, increasing lobe magnetic-field strength compresses the plasma sheet. Assume that the plasma sheet compresses adiabatically (i.e., follows an equation of state $pV^\gamma = $ constant with $\gamma=\frac{5}{3}$) and that the initial current-sheet thickness is $1R_E$, that the initial ion temperature is 4 keV, and that B_z in the current sheet is 2 nT and does not change. By what fraction must the lobe magnetic field increase to reach this triggering threshold? Is this answer reasonable?

9.9 Write a computer code to compute ion orbits in a Harris neutral sheet (see Section 9.4.4); that is, integrate the equation

$$m_i \frac{d\mathbf{v}_i}{dt} = e\mathbf{v}_i \times \mathbf{B}$$

where $\mathbf{B} = B_0 \tanh(z/h)\hat{\mathbf{x}}$. Reproduce the orbits sketched in Figure 9.15.

10 MAGNETOSPHERIC CONFIGURATION

R. A. Wolf

10.1 INTRODUCTION

THIS CHAPTER ATTEMPTS to describe the essential features of the interior of the earth's magnetosphere and its coupling to the ionosphere. The first half of the chapter is observational. It describes the configuration of the magnetic field, the plasma density and temperature, the electric field, and the plasma flow velocity. The second half is more theoretical. It describes the application of the theory of adiabatic charged-particle drifts to the earth's magnetosphere and to the explanation of basic features of the observed particle and electric-current distributions. It then discusses the theory of ionosphere–magnetosphere coupling and its relationship to observed electric-field and current distributions. Mechanisms for loss of particles from the magnetosphere are described last.

10.2 MAGNETIC-FIELD CONFIGURATION OF THE EARTH'S MAGNETOSPHERE

Figure 10.1 shows magnetic-field lines of the earth's magnetosphere, as represented by an observation-based computer model. The most important currents that give rise to the magnetospheric magnetic field are, of course, those that flow inside the earth. The earth's intrinsic field is effectively a perfect dipole at distances more than about two earth radii ($1R_E = 6{,}371.2$ km) from the center of the earth. As discussed in Chapter 6, the strength and orientation of the dipole vary slowly with time; as of 1985, the dipole moment was approximately 0.30438×10^{-4} T·R_E^3 and was tilted by about 11° relative to the earth's rotation axis. For more precise information on the earth's internal field, see Barker et al. (1986).

Figure 10.2 shows conventional names for the various types of large-scale currents that flow in the earth's magnetosphere. Currents flow on the magnetopause, the boundary that separates the shocked solar wind (magnetosheath) from the magnetosphere; these boundary currents are called *Chapman-Ferraro currents,* after the authors of the famous papers in which their existence was proposed (Chapman and Ferraro, 1930, 1932). The highly stretched magnetic-field geometry of the earth's

10.2 CONFIGURATION OF THE EARTH'S MAGNETOSPHERE

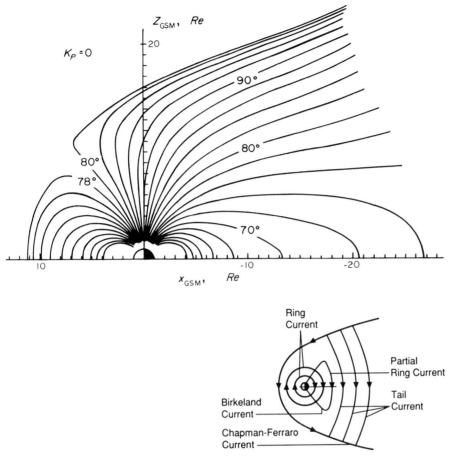

FIG. 10.1. Magnetic-field lines viewed in the noon–midnight meridian plane, for magnetically quiet conditions. Some field lines are labeled to indicate the magnetic latitude at which they cross the surface of the solid earth. The earth's dipole is parallel to the z_{GSM}-axis. The solar wind is assumed to flow in the $-x_{GSM}$ direction, perpendicular to the dipole. (From Tsyganenko and Usmanov, 1982.)

FIG. 10.2. Major types of currents in the earth's magnetosphere. The view is of the magnetic equatorial plane, as seen from high above the North Pole. The sun is to the left.

magnetotail (Figure 10.1) implies a westward *tail current* across the center of the tail, near the equatorial plane. The inner magnetosphere contains a *ring current,* which flows westward in approximate circles centered on the earth. Additional *partial-ring currents* also flow partway around the earth in the middle magnetosphere. *Birkeland currents* connect the ends of the partial rings to the ionosphere, where conduction currents can complete the circuit. Additional Birkeland currents (not shown) connect to the tail and magnetopause boundary layer.

The results of some of these currents can readily be discerned in the observational plot shown in Figure 10.3, which shows contours of constant differences in the observed field strength and that of a dipole in the noon–midnight meridian plane. The average has been taken over many spacecraft magnetic-field measurements, all for low levels of geomagnetic activity. The contours indicate strongly positive changes in field strength (ΔB) on the dayside of the earth, near the equatorial plane, which is a result of the eastward Chapman-Ferraro currents flowing along the magnetopause; these currents cause a northward perturbation, and thus a strengthening, of the basic northward field due to the earth; in other words, the pressure on the dayside magnetopause due to the

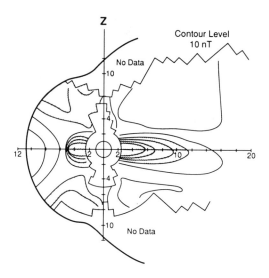

FIG. 10.3. Observed ΔB contours for $K_p = 0-1$ in the noon–midnight meridian plane. The sun is to the left. Shading indicates regions in which the magnetic field is depressed below the value of the dipole field. (Adapted from Sugiura and Poros, 1973.)

solar wind compresses the magnetic field in the dayside magnetosphere. There is a negative ΔB (i.e., a depression in field strength) at high latitudes on the dayside, where field lines approaching the magnetopause have difficulty deciding whether to turn equatorward and sunward along the magnetopause or poleward and tailward (Figure 10.1). The region near the equatorial plane and within about $5R_E$ of the earth's center shows negative ΔB, an effect of the westward ring current, which causes a southward magnetic perturbation inside itself. The lobes of the magnetotail are regions of positive ΔB, where the westward tail currents (directed out of the figure) cause a sunward magnetic field north of the central plane of the tail, and an antisunward field south of the central plane.

Because charged particles move relatively freely along magnetic-field lines in a low-density plasma like the magnetosphere, it is frequently useful to consider a whole field line as a unit and to try to map features from magnetosphere to ionosphere. In Figure 10.1, the compression of the dayside of the magnetosphere causes field lines from a given magnetic latitude in the ionosphere to map somewhat closer to the earth than would be estimated from the dipole formula [Chapter 6, equations (6.2)–(6.4)], whereas on the nightside the stretching of the tail causes field lines to extend farther out in the equatorial plane.

10.3 PLASMA IN THE EARTH'S MIDDLE AND INNER MAGNETOSPHERE

In this chapter we give a brief observation-based description of the major plasma populations of the inner and middle magnetosphere, which are shown in Figure 10.4.

10.3.1 Plasma in the Earth's Near Magnetotail: The Tail Lobe, the Plasma-Sheet Boundary Layer, and the Plasma Sheet

Aside from the magnetopause boundary layers, which were discussed in Chapter 9 and will not be discussed in this chapter, three major plasma regions exist in the near part of the earth's magnetic tail. Experts are not in complete agreement as to the precise definitions of these regions, but we generally follow those of Eastman et al. (1984).

Tail Lobes. Plasma densities are low, generally less than 0.1 cm^{-3} and sometimes below the level of detectability. Ion and electron spectra are very soft, with very few particles in the 5–50-keV range. Cool ions are often observed flowing away from the earth, and their composition often suggests an ionospheric origin. There is strong evidence that the tail lobe normally lies on open magnetic-field lines.

Plasma-Sheet Boundary Layer. Ions in this region typically exhibit flow velocities of hundreds of kilometers per second, principally parallel to or antiparallel to the local magnetic field. Frequently, counterstreaming ion beams are observed, with one beam traveling earthward, and the other traveling tailward along the field line. Densities typically are of the order of 0.1 cm^{-3}, and thermal energies tend to be smaller than the flow energies. For more details on the observational properties of the plasma-sheet boundary layer, see Eastman et al. (1984). The plasma-sheet boundary layer probably lies on closed field lines.

Plasma Sheet. This region, often referred to as the "central plasma sheet" to emphasize its distinctness from the plasma-sheet boundary layer, consists of hot (kilovolt) particles that have nearly symmetric velocity distributions. Number densities typically are 0.1–1 cm^{-3}, a little bit higher than those of the plasma-sheet boundary layer. Flow velocities

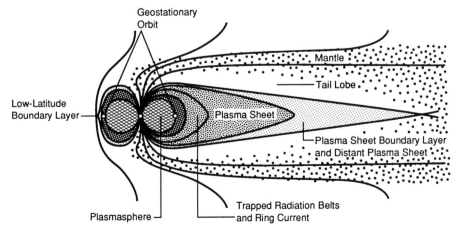

FIG. 10.4. Schematic diagram of plasma regions of the earth's magnetosphere as viewed in the noon–midnight meridian plane. The plasmasphere typically occupies much of the same region of space as the radiation belts. Frequently there is little or no gap between the inner edge of the plasma sheet and the outer boundary of the trapped radiation belts.

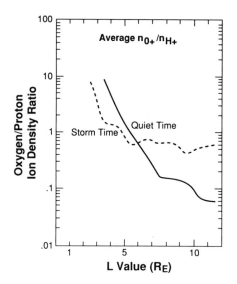

FIG. 10.5. Averaged radial profiles of O^+/H^+ density ratio, as measured from the ISEE 1 spacecraft. (Adapted from Lennartsson and Sharp, 1982.)

typically are very small compared with the ion thermal velocity. The term "plasma sheet" was originally coined to describe just the kilovolt electrons. In this chapter, however, we adopt the common modern practice of using the term "plasma sheet" to describe both the electrons and the ions, appending the word "electron" or "ion" when discussing only one species of the plasma. The ion temperature in the plasma sheet is almost invariably about seven times the electron temperature (Baumjohann, Paschmann, and Cattell, 1989). For the most part, the plasma sheet lies on closed field lines, although it may sometimes contain "plasmoids," which are closed loops of magnetic flux that do not connect to the Earth or solar wind. The plasma-sheet boundary layer is generally observed as a transition region between the almost empty tail lobes and the hot plasma sheet.

As discussed in Chapter 9, the plasmas of the tail are dynamic. Reconnection in the distant tail converts antisunward-streaming mantle plasma into beams that stream along the field lines, toward the earth; these beams mirror in the region of strong magnetic field near the earth, creating antisunward streams. The counterstreams tend to be unstable to various plasma waves, which eventually convert the streaming energy to thermal energy, creating the hot, slow-flowing plasma sheet.

Figure 10.5 summarizes the composition of the plasma sheet and the outer part of the radiation belts, including two components: H^+, which is abundant in both the solar wind and the earth's upper ionosphere; O^+, which is abundant in the ionosphere, but not in the solar wind. The ionospheric ion O^+, while present in modest concentrations during quiet times, is almost as abundant as H^+ during active times. These results clearly indicate that the plasma-sheet ion population is a mixture of solar-wind and ionospheric particles, being mostly of solar-wind origin in quiet times and mostly ionospheric in active times.

10.3 PLASMA IN THE EARTH'S MAGNETOSPHERE

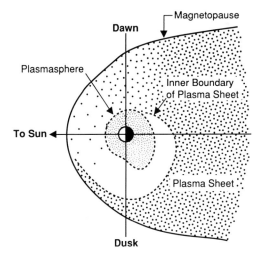

FIG. 10.6. Average distribution of low-energy electrons in the magnetosphere. The diagram shows the magnetic equatorial plane, viewed from high above the North Pole. The sun is to the left. (Adapted from Vasyliunas, 1968b.)

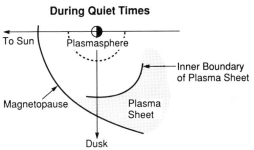

FIG. 10.7. Sketch of changes in the spatial distribution of low-energy electrons during magnetospheric substorms. View is of the equatorial plane, with the sun to the left. (Adapted from Vasyliunas, 1968b.)

By examining data from the *OGO 1* spacecraft, V. M. Vasyliunas found that the electron plasma sheet generally has a well-defined inner edge on the duskside; as indicated in the figure, the boundary is less distinct on the dawnside. The average distribution of kilovolt electrons in the inner magnetosphere is sketched in Figure 10.6. The tendency of the sheet to move earthward near local midnight during substorms is sketched in Figure 10.7.

10.3.2 Geostationary-Orbit Region

The geostationary-orbit (or geosynchronous-orbit) region, at approximately $6.6 R_E$ geocentric distance and in the geographic equatorial plane, is a particularly important one for practical purposes, because it is the home for many operational spacecraft. In this orbit, centrifugal force just balances gravity for a spacecraft that is in a circular orbit with a period of one orbit per day. It is a particularly complex region, from the viewpoint of magnetospheric physics.

Part of the complexity of the geosynchronous-orbit region results from the fact that the geographic equatorial plane (where the spacecraft reside) is tilted by about 11° relative to the equatorial plane (which is approximately the symmetry plane for the magnetic-field configuration). A geostationary spacecraft lies between −11° and 11° magnetic latitude. In times of high magnetic activity, when the inner edge of the plasma sheet is substantially inside geostationary orbit, the field lines may be extremely stretched and tail-like; a spacecraft at 11° magnetic latitude at $6.6R_E$ geocentric distance sometimes lies on a field line that stretches far out into the tail.

The geosynchronous orbit is also complex because it tends to skim the inner boundary of the plasma sheet. In quiet times, geosynchronous spacecraft are generally earthward of the plasma sheet. In a major magnetic storm, the same orbit is generally bathed in the hot plasma sheet. During a typical substorm, a spacecraft at geosynchronous orbit will skim the edge of the plasma sheet and observe the injection of fresh ring-current plasma. (These substorm phenomena will be discussed in Chapter 13.)

Extensive statistical studies of synchronous-orbit particles were carried out by Garrett, Schwank, and DeForest (1981a,b) using data from the *ATS 5* and *ATS 6* spacecraft. They found that neither the electrons nor the ions could be adequately described as simple Maxwellians. For the purpose of their statistical study, they fitted each observed distribution to the sum of two Maxwellians, one relatively cold (labeled component 1) and one hot (labeled component 2). No attempt was made to model pitch-angle distribution. They plotted average fitting parameters versus local time and also versus K_p. The index K_p is a measure of overall geomagnetic activity. As discussed in more detail in Chapter 13, the index runs from 0 (extreme quiet) to 9 (extremely active).

The following features can be noted in Figures 10.8 and 10.9:

1. For both electrons and ions, kT for the cooler component (T_1) ranges from about 100 eV to about 1 keV and clearly increases with increasing magnetic activity. It is clearly hottest near local midnight, where a geostationary spacecraft is most likely to encounter the plasma sheet. The number density of the cool component is typically about 1 cm^{-3}.

2. The temperature of the hot ions exceeds the temperature of the hot electrons.

3. The number densities for hot electrons and hot ions increase with increasing magnetic activity.

4. Number densities for hot electrons are highest in the postmidnight

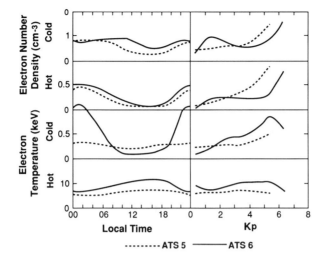

FIG. 10.8. Average local-time dependence and K_p dependence of electron parameters observed at synchronous orbit by ATS 5 and ATS 6. Differences may be due to the different magnetic latitudes of the two spacecraft. (Adapted from Garrett et al., 1981a.)

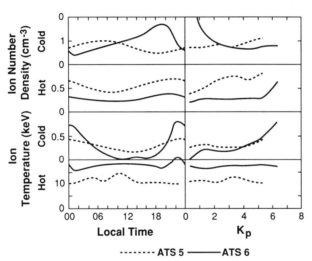

FIG. 10.9. Average local-time dependence and K_p dependence of ion parameters observed by ATS 5 and ATS 6. These values were obtained from instruments that could not measure ion composition, and all ions were assumed to be hydrogen. Consequently, the number densities shown probably are slightly lower than they should be. (Adapted from Garrett et al., 1981b.)

sector (i.e., toward the dawnside from local midnight), whereas hot-ion densities peak in the premidnight sector (i.e., toward dusk from local midnight); for both kinds of particles, densities are higher on the dayside than on the nightside.

5. There are remarkably large statistical differences between the observations by the two spacecraft, probably the result of their being at different longitudes and thus different magnetic latitudes.

10.3.3 The Trapped Radiation Belts and the Ring-Current Particles

The radiation belts consist of particles in orbits that circle the earth, from about 1,000 km above the surface to a geocentric distance of about $6R_E$. Because the particles move easily along magnetic-field lines, the radiation belts are basically magnetic-field-aligned structures. However,

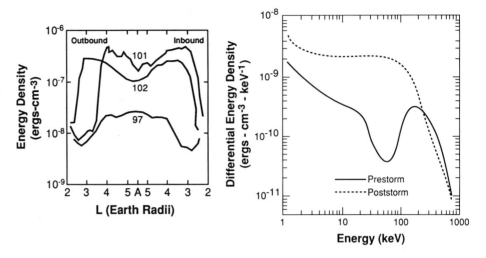

(*Left*) **FIG. 10.10.** Radial profiles of ion-energy density as measured by *Explorer 45* during the prestorm quiet (orbit 97), the main-phase injection (orbit 101), and the recovery phase (orbits 102 and 103) for the 17 December 1971 magnetic storm. These energy densities were computed using the assumption that all of the ions were H^+. (Adapted from Smith and Hoffman, 1973.)

(*Right*) **FIG. 10.11.** Ion-energy-density spectrum as measured by *Explorer 45* during prestorm quiet and early in the recovery phase, for the 17 December 1971 magnetic storm. Energy densities were estimated under the assumption that all ions were H^+. (Adapted from Smith and Hoffman, 1973.)

the number density and energy density of radiation-belt particles are not constant along field lines. The particles are most intense near the equatorial plane, and less intense at lower altitudes, where they are subject to loss by interaction with the neutral atmosphere.

Figures 10.10 and 10.11 show two views of the low-energy part of the outer portion of the trapped radiation belts for a specific event. Figure 10.10 shows radial profiles of ion energy density, as measured from the *Explorer 45* satellite (also called S^3, for "small scientific satellite"). Several orbits are shown: one for a quiet period before a geomagnetic storm, and several in the main phase of the storm. (Geomagnetic storms are discussed in detail in Chapter 13.) Within the energy range 1–138 keV, the storm obviously causes a dramatic increase in the energy density. (Data from this early spacecraft were processed under the assumption that all of the ions were protons, an assumption that probably caused some underestimation of energy densities.) Figure 10.11 shows an energy spectrum for a location near $L=4$, before the storm and during the main phase. In the prestorm quiet condition, there is a peak in the spectrum at around 100 keV, due to a population known as the "Davis-Williamson protons." The storm injects a broad spectrum of ions below a few hundred kiloelectron volts. Note that the population of particles above a few hundred kiloelectron volts is not immediately affected by the storm.

For some years there was considerable uncertainty about the composition of ring-current ions. In the late 1970s, observations from the *GEOS 1* and *GEOS 2* spacecraft established the importance of O^+ for the geosynchronous region, for energies less than about 20 keV. Shortly thereafter, ISEE missions provided similar information for a wider range of geocentric distances (as summarized in Figure 10.5). However, neither of those instruments provided information on composition in the 20–300-keV range, which contributes most of the ring current and most of the energy density in the trapped plasma near the earth. The Active

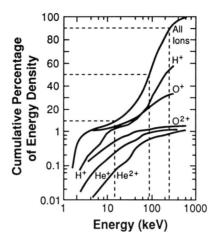

FIG. 10.12. Accumulated percentage of ring-current-energy density versus energy, for $L = 3.9$, during the 5 September 1984 magnetic storm. A logarithmic scale has been used below the 1% level to highlight the minority species. (Adapted from Williams, 1987.)

Magnetospheric Particle Tracer Explorer (AMPTE) mission of the mid-1980s was the first to measure the composition of the main part of the ring current. Results for one storm are presented in Figure 10.12, which shows the fraction of the total ring-current energy at a specific location ($L = 3.9$) contributed by ions of various types having energy less than E. Below 10 keV, O^+ is the dominant ion, but H^+ begins to dominate above a few tens of kiloelectron volts. In terms of total energy density due to ions below 50 keV, O^+ and H^+ contribute about equally; H^+ clearly dominates above about 50 keV, and the total contribution of H^+ exceeds that of O^+ by nearly a factor of 2. Of course, it should be remarked that the events studied so far in the AMPTE mission have been from only part of the solar cycle. It is, however, clear that both O^+ and H^+ contribute substantially to the ring current.

Note also that the energy midpoint in terms of contribution to the ring current is about 85 keV, according to Figure 10.12. That is, about half of the ring current comes from particles below 85 keV, and half comes from particles above 85 keV.

We now turn the discussion to the "trapped radiation belts" or "Van Allen belts." There is no clear distinction between these belts and the "ring-current particles." In fact, there is a strong overlap in the terminology. Most of the "ring current" is carried by trapped particles, and all of the trapped particles contribute to the ring current. However, the term "ring current" emphasizes those components of the particle distribution that contribute importantly to the total current density, which, for non-relativistic particles, is proportional to the energy density (see Section 10.5.5). The terms "trapped radiation belts" and "Van Allen belts" emphasize penetrating radiation, specifically particles that can penetrate deep into dense materials and thus cause radiation damage to spacecraft instrumentation or to humans. (Van Allen identified these particles in the earliest spacecraft measurements, because those were the particles that could be measured by his simple Geiger-counter detectors.)

Electrons, which contribute relatively little to the ring current and to

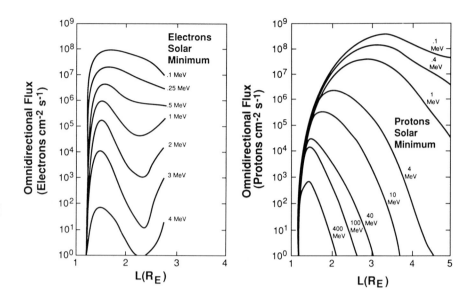

(*Left*) **FIG. 10.13.** Equatorial omnidirectional electron flux versus L shell for the AE5 solar-minimum radiation-belt model. The flux curves are labeled by threshold energy. Each curve gives the total electron flux above the specified threshold. (Adapted from Spjeldvik and Rothwell, 1983.)

(*Right*) **FIG. 10.14.** Radial distribution of proton omnidirectional fluxes in the equatorial plane, according to the AP8 solar-minimum radiation model. The curves give total fluxes above various threshold energies from 0.1 to 400 MeV. (Adapted from Spjeldvik and Rothwell, 1983.)

the total trapped-particle energy, contribute importantly to the penetrating radiation. Figure 10.13 shows fluxes of trapped electrons estimated from the NASA radiation-belt model (AE5-MIN). ("MIN" refers to solar minimum.) This model, and other very useful quantitative empirical models of the radiation belts, can be obtained from the National Space Science Data Center (NSSDC). One obvious feature in the figure is a "slot" centered at about $L = 2.2$, a minimum in fluxes for energies above about 1 MeV. Electrons on the low-L side of the slot are called the "inner belt," and those on the high-L side are called the "outer belt." Figure 10.14 shows corresponding plots for ions in the equatorial plane. Note that there is no slot in the ion spectra.

10.3.4 The Plasmasphere

We now turn to the dense cold-plasma population, the "plasmasphere," which coexists in approximately the same region of space as the radiation belts. Through analysis of "whistler" waves observed on the earth's surface, it has been known for many years that high densities ($\sim 10^3$ cm^{-3}) of cold (~ 1 eV) plasma exist up to altitudes of several earth radii. Early satellite work by K. I. Gringauz of the Soviet Union and approximately simultaneous research carried out by D. L. Carpenter of the United States using ground-based measurements of plasma waves called "whistlers" showed that these regions of dense cold plasma frequently terminated at a sharp boundary at an altitude of 3–5R_E, which is called the "plasmapause." Detailed whistler measurements and in situ observations from several spacecraft established the basic dynamics of the plasmasphere and plasmapause during the late 1960s and early 1970s. The region of low cold-plasma densities just outside the plasmapause was usually referred to as the "trough."

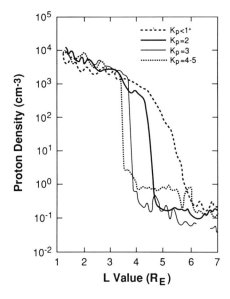

FIG. 10.15. Variation of the nightside plasmasphere to varying levels of geomagnetic activity as measured by *OGO 5*. (Adapted from Chappell, 1972.)

The theoretically proposed mechanism for populating the plasmasphere is the "polar wind," which was postulated by W. I. Axford in 1968 and developed theoretically in some detail over the succeeding few years (Banks and Holzer, 1969). The existence of an upper-ionospheric plasma with a temperature that is not very small compared with the gravitational binding energy led to the prediction of a steady flow of plasma upward from the planet. This long-predicted polar wind was finally directly observed with the *Dynamics Explorer* spacecraft (Nagai et al., 1984).

Figure 10.15 shows plots of proton density versus L value for several different days, to illustrate how the plasmasphere typically depends on magnetic activity. The innermost part of the plasmasphere does not depend strongly and systematically on magnetic activity. The density declines gradually with increasing L. However, the plasmapause is systematically closer to the earth during times of high activity (e.g., $K_p = 4-5$) than during times of low activity ($K_p < 1+$). In times of strong magnetospheric activity, the strong convection strips flux tubes away from the outer plasmasphere. After a magnetic storm or other sustained period of high activity, the outer plasmasphere is observed to refill gradually, over a period of a few days.

Figure 10.16 shows an equatorial view of the average shape of the plasmapause, as determined by whistler measurements. Note the bulge on the duskside, which is a persistent feature, though it can rotate somewhat in local time, depending on magnetospheric condition.

The plasmasphere is basically a magnetic-field-aligned structure that can be traced all the way from the equatorial plane to the ionosphere. In observations of the topside ionosphere (~1,000 km altitude) one finds characteristic density structures near the region that maps to near the

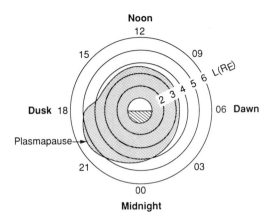

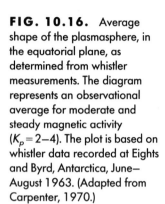

FIG. 10.16. Average shape of the plasmasphere, in the equatorial plane, as determined from whistler measurements. The diagram represents an observational average for moderate and steady magnetic activity ($K_p = 2-4$). The plot is based on whistler data recorded at Eights and Byrd, Antarctica, June–August 1963. (Adapted from Carpenter, 1970.)

plasmapause in the equatorial plane. Proceeding poleward from low latitudes on the nightside of the earth, one typically moves from a region of stable, high density to a region with markedly lower density, and then to a highly structured region of high density. The low-density region is called the "midlatitude trough." The high-latitude edge of this trough corresponds approximately with the low-latitude edge of the auroral zone (see Chapter 14). (Auroral precipitation ionizes neutrals and raises ionospheric plasma densities.) The low-latitude edge of the midlatitude trough seems to correspond roughly to the high-L boundary of the "permanent plasmasphere" (i.e., the region that is not constantly being emptied and refilled). The actual plasmapause, as would be identified from the plasma density gradients near the equatorial plane, lies at a slightly higher L value. The duskside bulge is usually not evident from observations of the topside ionosphere.

Beginning gradually in the late 1970s, and more quickly in the 1980s with the *Dynamics Explorer* mission, more detailed observations of the plasmaspheric plasma have provided information on composition, temperature, and pitch-angle distributions of the low-energy plasma near the plasmapause. As would be expected, H^+ is the dominant ion species, but He^+, O^+, O^{2+}, N^+, and N^{2+} are also observed. There seems to be a torus of warm, heavy ions in the vicinity of the equatorial plasmapause, perhaps a result of the collision of upstreaming plasma from the two hemispheres. In general, ion temperatures are much higher and more erratic in the vicinity of the plasmapause than in the inner plasmasphere, and the situation is quite complex. For more details, see, for example, Olsen et al. (1987).

10.4 ELECTRIC FIELDS AND MAGNETOSPHERIC CONVECTION

Electric fields in a low-density space plasma can be measured either directly by measuring the potential difference between the ends of a long

10.4 ELECTRIC FIELDS AND MAGNETOSPHERIC CONVECTION

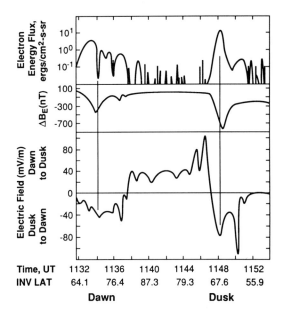

FIG. 10.17. Observations made by the $S3-2$ spacecraft from a dawn-to-dusk South Pole pass on 19 September 1976. The top panel shows the energy flux of downcoming electrons. The middle panel shows the transverse magnetic fluctuation. The bottom panel shows the component of electric field in the direction of the spacecraft's motion, which is essentially from dawn to dusk. The data, which are displayed with even spacing in universal time (UT), are also presented as a function of invariant latitude. (Adapted from Harel et al., 1981.)

boom or indirectly by measuring the $\mathbf{E} \times \mathbf{B}$ drift of cold plasma and using the formula $\mathbf{E} = -\mathbf{u} \times \mathbf{B}$. The measurement of plasma flow provides only the perpendicular components of $\mathbf{E}$. Further, the magnetic field $\mathbf{B}$ must be either measured or otherwise known. Both types of measurements are much easier at low altitudes, in the upper ionosphere, than they are in the magnetosphere proper. Consequently, our most comprehensive information concerning electric fields and $\mathbf{E} \times \mathbf{B}$ drift in the magnetosphere comes from measurements made at low altitudes.

It is similarly of interest to measure the electric currents that flow along magnetic-field lines connecting the ionosphere and magnetosphere. Again, there are two ways to make the measurement: by observing all components of the particle distribution and computing the current by summing charge × velocity × density for each particle type, or by using a magnetometer to measure $\nabla \times \mathbf{B}$. The first method is often difficult in practice, because a large fraction of the current is frequently carried by particles of very low energy, which are quite difficult to measure accurately, especially in the presence of high fluxes or more energetic particles. The most comprehensive and reliable information concerning the magnetic-field-aligned component of $\nabla \times \mathbf{B}$ comes from spacecraft in low polar orbits at relatively low altitudes above the ionosphere. Such spacecraft observe a qualitatively consistent pattern of deflections of $\mathbf{B}$ in the east–west sense, apparently the result of large-scale sheets of current that flow along field lines up from and down into the auroral ionosphere.

Figure 10.17 shows electric- and magnetic-field measurements made from a dawn-to-dusk pass over the region near the South Pole. The spacecraft first entered the region of strong auroral precipitation at about 64° invariant latitude and about 0650 magnetic local time (i.e., slightly past local dawn). The observed electric field was equatorward (i.e., opposite to the forward motion of the spacecraft) from 64° to about 80° latitude; then it reversed abruptly, near the poleward edge of the dawn-side auroral precipitation. Over the polar cap, the highest-latitude region where the precipitating fluxes are low, the electric field was in the direction of spacecraft motion (i.e., dawn-to-dusk). The electric field reversed again near the dusk edge of the polar cap and remained consistently dusk-to-dawn until it became extremely weak equatorward of the duskside auroral zone. Because the magnetic field is basically downward over the high-latitude polar cap, it is clear that the electric-field pattern shown in Figure 10.17 corresponds to sunward $\mathbf{E} \times \mathbf{B}$ drift in the morningside and afternoonside auroral zones, and antisunward $\mathbf{E} \times \mathbf{B}$ drift over the polar cap. This is classic magnetospheric convection. Figure 10.17 also shows the associated east–west deflections in the magnetic field, indicating sheets of magnetic-field-aligned current. A ramping-up of the ΔB curve indicates a downward current, and a ramping-down represents an upward current. One crossing of the auroral zone usually displays a pair of such currents.

Figure 10.18 shows sketches of the classically observed patterns of Birkeland current, electric field, horizontal ionospheric current, and $\mathbf{E} \times \mathbf{B}$ drift for the northern ionosphere. The Birkeland-current pattern was taken from Iijima and Potemra (1978). Near the poleward edge of the auroral zone we have the region-1 Birkeland currents, down into the ionosphere on the dawnside, and up from the ionosphere on the duskside. These currents, which connect far out in the magnetosphere to the boundary layers and the distant plasma sheet, are the currents that drive convection in the ionosphere. In the more equatorward part of the auroral zone is an oppositely directed set of currents, called the region-2 currents, which flow up from the ionosphere on the dawnside, and down into it on the duskside. The physical nature of the region-2 currents will be discussed later in this chapter. The regions where currents flow down into the ionosphere tend to charge up positively, whereas regions of upward current charge up negatively, giving rise to the electric-field pattern shown. Ionospheric currents flow in response to the electric field, consisting of a Pederson current that flows in the direction of $\mathbf{E}$ and a Hall current that flows in the direction of $-\mathbf{E} \times \mathbf{B}$. Only weak currents are shown for the nightside polar cap, because with little auroral bombardment and no sunlight, there is little ionospheric conductivity. The final diagram shows the equipotentials of the electric-field pattern, and it is also a flow diagram for $\mathbf{E} \times \mathbf{B}$ drift, because $\mathbf{E} \times \mathbf{B} = \mathbf{B} \times \nabla \phi$ is perpendicular to curves of constant electrostatic potential ϕ.

The total strength of convection is characterized by the polar-cap potential drop. To calculate this parameter from the electric fields measured by a polar-orbiting spacecraft on a pass over the high-latitude ionosphere, one first calculates the indefinite path integral

10.4 ELECTRIC FIELDS AND MAGNETOSPHERIC CONVECTION

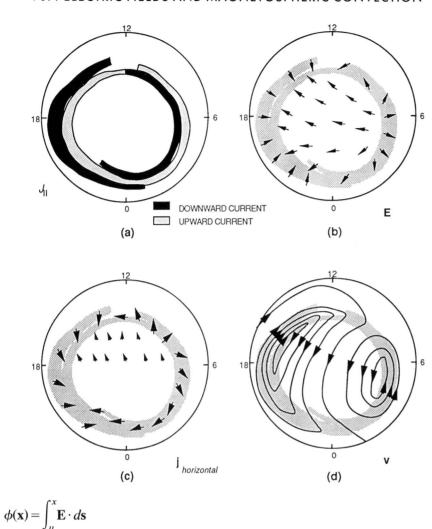

FIG. 10.18. Typical patterns of (a) Birkeland current, (b) electric field, (c) horizontal ionospheric current, and (d) **E** × **B**-drift velocity observed in the earth's ionosphere, as viewed from high above the North Pole. Local noon is toward the top of the page, local dusk to the left, and so forth. These patterns represent the primary ionospheric effects of magnetospheric convection. The Birkeland-current pattern was adapted from Iijima and Potemra (1978). Note how the ionospheric currents act to connect the upward and downward Birkeland currents. The flow diagrams shown in (d) are equipotentials.

$$\phi(\mathbf{x}) = \int_{ll}^{x} \mathbf{E} \cdot d\mathbf{s}$$

with the integral starting from a low-latitude point where the electric field is negligible and proceeding to a general point **x** along the path of the spacecraft. The difference between the maximum and minimum values of $\phi(\mathbf{x})$ along the path is defined to be ϕ_{pc}, the polar-cap potential drop. The measured values typically range from about 20 kV in quiet times to about 150 kV in very active times. The observed average of ϕ_{pc} is about 50 kV. See Reiff and Luhmann (1986) for detailed information on observations of the polar-cap potential drop.

If we visualize magnetic-field lines as moving at velocity $\mathbf{E} \times \mathbf{B}/B^2$, then

$$\int_{a}^{b} \mathbf{E} \cdot d\mathbf{s}$$

is the number of field lines per unit time that cross the curve segment that connects a and b. A potential drop can therefore be considered to represent the rate of transport of magnetic flux.

The approximation of frozen-in flux (or, equivalently, perfect conductivity) implies that $\mathbf{E} \cdot \mathbf{B} = 0$. Further, if the magnetic-field configuration is independent of time, then the electric field can be written as $\mathbf{E} = -\boldsymbol{\nabla}\phi$, where ϕ is constant along each magnetic-field line. In that case, the distribution of ionospheric potential can simply be mapped out to the magnetosphere, if we have a reliable model of the magnetospheric magnetic field. The basic equipotential pattern can then be mapped to the equatorial plane. There are, however, several difficulties involved in accurately mapping ionospheric electric fields to the equatorial magnetosphere:

1. It may be difficult to find a reliable model of the magnetospheric magnetic field for a given time.
2. If the magnetospheric magnetic-field configuration is changing in time ($\partial \mathbf{B}/\partial t \neq 0$), then $\boldsymbol{\nabla} \times \mathbf{E} \neq 0$, and there is an induction electric field present that cannot be represented in terms of the gradient of a potential; induction electric fields are frequently large in the outer and middle magnetosphere.
3. Frequently there is a substantial potential drop along magnetic-field lines that connect the auroral zone and the plasma sheet, and these potential drops apparently are responsible for most of the bright auroral forms. The potential drops involved typically are several kilovolts, which, though small compared with the total potential drops involved in magnetospheric convection, are large enough to complicate the relationship between ionospheric and magnetospheric electric fields.

Despite the subtleties involved in mapping electric fields between the ionosphere and the magnetosphere, there is a basic consistency between inner-magnetospheric electric fields and the ionospheric patterns shown in Figure 10.18c,d. The lower-latitude part of the northern auroral ionosphere generally flows sunward and clearly lies on closed field lines; the flow pattern shown in Figure 10.18d can be mapped to the equatorial plane, as sketched in Figure 10.19. The mapping becomes uncertain near the high-latitude edge of the sunward-flow region, but the electric-field-reversal region of the ionospheric pattern presumably maps near the magnetopause boundary layer.

10.5 ADIABATIC INVARIANTS AND PARTICLE DRIFTS

10.5.1 Introductory Comments

The past four sections have described the electric and magnetic fields observed in the magnetosphere, as well as the observed particle distributions. Sections 10.5 and 10.6 are theoretical and attempt to explain how

10.5 ADIABATIC INVARIANTS AND PARTICLE DRIFTS

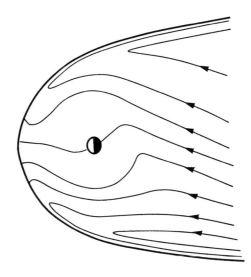

FIG. 10.19. Qualitative sketch of the result of mapping the observed ionospheric flow pattern shown in Figure 10.18d along magnetic-field lines to the equatorial plane. The view is from high above the North Pole, with the sun to the left. No correction has been made for the earth's rotation, a point that is discussed in Section 10.5.7. There is assumed to be little electric field near the earth on field lines that map to the middle- and low-latitude ionosphere.

the patterns of the observed **E** and **B** and particles are related physically. We cannot hope to explain all of the crucial connections here, because some of them still are not understood. Rather, we shall concentrate on aspects of the theory that are relatively elegant and simple and shall find that they shed light on some major features of the **E**, **B**, and particle patterns.

We begin by discussing the elegant theory of adiabatic particle drifts in the earth's magnetosphere.

10.5.2 Time Scales

An ion or electron moving through the earth's magnetosphere executes three motions: (1) cyclotron motion about the magnetic-field line, (2) bounce motion along the field line, and (3) drift perpendicular to **B**. For particles below 1 MeV in energy, which are the ones that we shall be considering, these motions take place on three different time scales. The gyromotion, which occurs at the cyclotron frequency Ω_c (also called the Larmor frequency or gyrofrequency), is the fastest. From the definition of Ω_c given in Chapter 2, we have the following expression for the period of the gyromotion:

$$T_c = \frac{2\pi}{|\Omega_c|} = \frac{2\pi m}{|q|B} \cong (0.66 \text{ s})\left(\frac{100 \text{ nT}}{B}\right)A \tag{10.1}$$

where A is the ratio of the particle mass to the mass of a proton ($= 1/1{,}836$ for an electron), and 1 nanotesla (nT) $= 10^{-9}$ tesla (T) $= 10^{-5}$ gauss (G). In writing (10.1), we assume that the particle is singly charged (i.e., $|q| = e$). Because B averages about 30,438 nT at the earth's surface at the equator and the field is very roughly dipolar, which implies that the field strength declines like the inverse cube of the geocentric distance r^{-3}, we expect $B \sim 240$ nT at $r = 5R_E$, 30 nT at $r = 10R_E$. For hydrogen ions, T_c

TABLE 10.1. Gyroradii (kilometers)

L (R_E)	Energy			
	10 eV	1 keV	100 keV	10 MeV
Electrons				
1.5	0.001	0.012	0.12	3.9
2	0.003	0.028	0.29	9.2
3	0.009	0.095	0.99	31
4	0.022	0.22	2.4	74
5	0.044	0.44	4.6	140
6	0.076	0.76	7.9	250
7	0.12	1.2	13	400
8	0.18	1.8	19	590
Protons				
1.5	0.051	0.51	5.1	51
2	0.12	1.2	12	120
3	0.40	4.0	40	410
4	0.96	9.6	96	960
5	1.9	19	190	1,900
6	3.2	32	320	3,200
7	5.1	51	510	5,200
8	7.7	77	770	7,700

ranges from about 1 ms at low altitudes to about 2 s in the equatorial plane at $r = 10 R_E$. The radius of the particle's cyclotron motion (the Larmor radius, see Chapter 2) is given by

$$\rho_c = \frac{m v_\perp}{|q| B} \cong (46 \text{ km}) A^{\frac{1}{2}} \left(\frac{E_\perp}{1 \text{ keV}} \right)^{\frac{1}{2}} \left(\frac{100 \text{ nT}}{B} \right) \quad (10.2)$$

where $v_\perp$ is the velocity of the particle's cyclotron motion, which is perpendicular to **B**, and $m v_\perp^2 / 2$ ($W_\perp$) is the energy involved in the cyclotron motion. The Larmor radius is typically several orders of magnitude smaller than magnetospheric dimensions, and it is smaller for electrons than for ions of the same energy. Table 10.1 lists Larmor radii for electrons and hydrogen ions in the earth's dipole field for various values of L and for various values of $W_\perp$ [the table was computed from the relativistic form of equation (10.2) and therefore differs from (10.2) for very high energies].

The particle's pitch angle α is defined to be the angle between **v** and **B**, specifically

$$\alpha = \tan^{-1}(v_\perp / v_\parallel) \quad (10.3)$$

The particle's motion parallel to **B** consists of bouncing between mirror points, as illustrated in Figure 10.20. The period of the bounce motion is given, in order of magnitude, by

10.5 ADIABATIC INVARIANTS AND PARTICLE DRIFTS

$$\tau_B \sim \frac{2l_b}{\langle v_\parallel \rangle} \sim (5 \text{ min})\left(\frac{l_b}{10 R_E}\right) A^{\frac{1}{2}} \left(\frac{1 \text{ keV}}{W_\parallel}\right)^{\frac{1}{2}} \quad (10.4)$$

where l_b is the distance along the field line between mirror points. Comparison of equations (10.1) and (10.4) indicates that $\tau_B \gg T_c$ (i.e., the bounce motion is generally much slower than the cyclotron motion).

The third type of motion that the particles undergo is drift perpendicular to **B**, as discussed in Chapter 2. The drift formula that we shall use is

$$\mathbf{v}_D = \frac{\mathbf{E} \times \mathbf{B}}{B^2} + \frac{\mathbf{F}_{\text{ext}} \times \mathbf{B}}{qB^2} + \frac{W_\perp \mathbf{B} \times \nabla B}{qB^3} + \frac{2W_\parallel \hat{\mathbf{r}}_c \times \mathbf{B}}{qR_c B^2} \quad (10.5)$$

where $\mathbf{F}_{\text{ext}}$ represents an external (nonelectromagnetic) force (left unspecified for the moment), R_c is the radius of curvature of the magnetic-field line, and $\hat{\mathbf{r}}_c$ is a unit vector outward from the center of curvature. The terms in (10.5) are called, respectively, $\mathbf{E} \times \mathbf{B}$ drift, external-force drift, gradient drift, and curvature drift. For a typical particle with pitch angle $\alpha \sim 45°$, the gradient and curvature drifts are of comparable magnitude. Also, in a dipole field, $\hat{\mathbf{r}}_c$ and ∇B are in opposite directions, so that the gradient and curvature drifts are in the same direction. Therefore, we can make an order-of-magnitude estimate of the sum of the gradient- and curvature-drift terms by just estimating one of them. The time scale for gradient/curvature drift for the case of the earth is therefore given, in order of magnitude, by

$$\tau_D = \frac{2\pi r}{v_{GC}} \sim \frac{2qBr^2}{W} \sim (56 \text{ h}) \left(\frac{r}{5R_E}\right)^2 \left(\frac{B}{100 \text{ nT}}\right) \left(\frac{1 \text{ keV}}{W}\right) \quad (10.6)$$

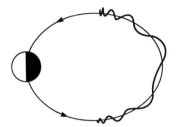

FIG. 10.20. Sketch of the bounce motion of a charged particle on an inner-magnetospheric field line. Note that the particle spends a large fraction of its time near the mirror points, where $v_\parallel$ drops to zero.

Note that for particle energy of about 1 keV, the three time scales given by equations (10.1), (10.4), and (10.6) are separated by approximately equal factors of about $600 A^{-\frac{1}{2}}$, which ranges from more than two orders of magnitude for O^+ to more than four order of magnitude for electrons. Electrons gyrate and bounce much faster than ions. However, the gradient/curvature drift rate of a particle is independent of mass for a given energy.

Figure 10.21 displays gyrofrequencies (T_c^{-1}), bounce frequencies (τ_B^{-1}), and drift frequencies (τ_D^{-1}) for equatorially mirroring protons and electrons. [One might think that the bounce period would go to zero as the length of the bounce path $l_b \to 0$, but it doesn't. The average velocity $\langle v_\parallel \rangle$ actually goes to zero linearly as $l_b \to 0$, so that in equation (10.4), the ratio τ_B goes to a finite constant, whose inverse is displayed in Figure 10.21.] The figure shows results for electrons and protons only. To find gyration, bounce, and drift frequencies for O^+, for example, multiply the proton values shown in Figure 10.21 by $\frac{1}{16}$, $\frac{1}{4}$, and 1, respectively.

10.5.3 Bounce Motion and the First and Second Adiabatic Invariants

The wide separation of these time scales allows theoretical separation of the three motions through the use of adiabatic invariants. The general theory of adiabatic invariants (Landau and Lifshitz, 1960) implies that $\oint p \, dq$ is conserved under slow changes in a system that exhibits periodic motion in the coordinate q, and p is the corresponding momentum. To apply this to the case of the cyclotron motion of a particle in an approximately uniform magnetic field, we let the magnetic field be in the z-direction, $q = x$, and $p = mv_x$. The corresponding adiabatic invariant is then given by

$$\oint p \, dq = \oint p_x v_x \, dt = m <v_x^2> T_c = \frac{1}{2} mv_\perp^2 \frac{2\pi m}{qB} = \frac{2\pi m \mu}{q} \tag{10.7}$$

where

$$\mu = \frac{mv_\perp^2}{2B} \tag{10.8}$$

is the magnetic moment of the particle's cyclotron motion. Equation (10.7) specifies the "first adiabatic invariant" of the motion of a charged particle in a magnetic field. For the normal situation where the ratio (m/q) for the particle is constant, equation (10.7) implies that the magnetic moment μ is an adiabatic invariant. Specifically, μ is conserved if the magnetic field experienced by the gyrating particle changes slowly (i.e., on a time scale that is long compared with the cyclotron period T_c).

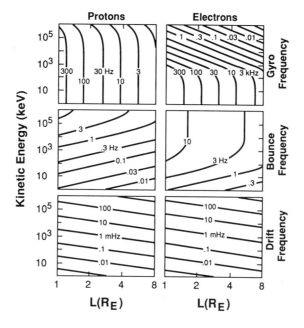

FIG. 10.21. The gyration, bounce, and drift frequencies for equatorially mirroring particles moving in the earth's dipole field, as a function of L and particle energy W. (Adapted from Shulz and Lanzerotti, 1974.)

Because the time scale for changes in the magnetic-field strength as the particle moves along a field line ($\sim \tau_B$) is in fact very long compared with the cyclotron period, we can use the fact that the magnetic moment is an adiabatic invariant to formulate a simple description of the bounce motion of the particle. We make the further simplifying assumption that there is no electric field parallel to **B**. In that case, because the magnetic field cannot change the kinetic energy of the particle, the kinetic energy W of the particle must remain constant as it moves along a field line in its bounce motion. We can then write

$$W = \frac{1}{2}m(v_\perp^2 + v_\parallel^2) = \frac{1}{2}mv_\parallel^2 + \mu B = \text{constant} \tag{10.9}$$

Therefore, $v_\parallel$ decreases as the particle moves along the field line into regions of stronger field B, and the parallel velocity goes to zero when B reaches a critical value B_m:

$$B_m = \frac{W}{\mu} \tag{10.10}$$

In terms of pitch angle α, the kinetic energy associated with the motion parallel to the field line is $W \cos^2\alpha$, and the motion perpendicular to the field line carries energy $W \sin^2\alpha = \mu B$.

Because the bounce motion of the particle on a given flux tube is periodic, we can define another adiabatic invariant $\oint p\, dq$ associated with that motion:

$$J = \oint p_\parallel\, ds = 2\sqrt{2m} \int_{m_1}^{m_2} \sqrt{(W - \mu B(s))}\, ds \tag{10.11}$$

where s is distance along the field line, and m_1 and m_2 are the locations of the particle's mirror points. The parameter J should be conserved if the particle experiences magnetic-field changes on a time scale that is long compared with the bounce period τ_B. The constant energy W can be pulled out of the integral as follows:

$$J = 2\sqrt{2m\mu} I \tag{10.12}$$

where

$$I = \int_{m_1}^{m_2} \sqrt{B_m - B(s)}\, ds \tag{10.13}$$

Note that I does not depend on the energy of the particle, but only on the particle's mirror point and on the field line that is being considered.

10.5.4 Bounce-averaged Gradient/Curvature Drift

Consider a particle that has a nonzero first adiabatic invariant μ, but a zero second adiabatic invariant J. This particle can gradient-drift,

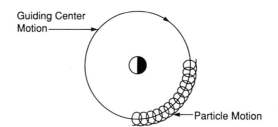

FIG. 10.22. Gradient-drifting charge circling the earth.

but cannot curvature-drift. It is trapped in a magnetic-field minimum. Assume, for simplicity, that the magnetic-field minimum on each field line occurs in an equatorial plane, as would be true, for example, for the configuration shown in Figure 10.1. Substituting equation (10.8) in the gradient-drift term of equation (10.5) gives

$$\mathbf{v}_{GD} = \mu \frac{\mathbf{B} \times \nabla B}{qB^2} = \frac{\mathbf{B} \times \nabla W}{qB^2} \tag{10.14}$$

where, in writing the second equality, we use the fact that $W(\mu, \mathbf{x}) = \mu B(\mathbf{x})$ for this equatorially mirroring particle. In taking the gradient of W, we hold μ constant. If the particle is also $\mathbf{E} \times \mathbf{B}$-drifting in a potential electric field $\mathbf{E} = -\nabla \varphi$, then we can write the combined $\mathbf{E} \times \mathbf{B}$-drift and gradient-drift velocity as follows:

$$\mathbf{v}_D = \frac{\mathbf{B} \times \nabla(q\varphi + W)}{qB^2} \tag{10.15}$$

The formula implies that the particle drifts perpendicular to the gradient of its total energy (potential + kinetic), so that equation (10.15) is an expression of conservation of energy for the particle. Because of the requirement of consistency with conservation of energy, we can, in fact, write the general expression for bounce-averaged gradient/curvature drift in the form

$$\mathbf{v}_{GCD} = \frac{\mathbf{B} \times \nabla(W(\mu, J, \mathbf{x}))}{qB^2} \tag{10.16}$$

where the gradient is now taken at a constant μ and J. [For a more rigorous and general discussion of bounce-averaged gradient/curvature drift, see Roederer (1970) or Wolf (1983).]

10.5.5 Particle Drifts and the Ring Current

Both gradient drift and curvature drift cause positive (radiation-belt) particles to drift westward in the earth's dipole field (Figure 10.22), and negative particles to drift eastward. The radiation belt thus forms a ring of westward current circling the earth, which tends to decrease the strength of the basic northward field observed at low latitudes on the earth's surface.

10.5 ADIABATIC INVARIANTS AND PARTICLE DRIFTS

There is, in fact, a simple theoretical relationship between the depression of the magnetic field at the earth's surface and the total energy in the trapped particles. In what follows, we derive that relationship for a special case.

The magnetic field caused at the earth's surface by the gradient drift of a single positive charge q that mirrors in the equatorial plane, has energy W, and drifts in a circle at geocentric distance LR_E is given by

$$\Delta \mathbf{B}_{\text{grad drift}} = -\frac{3}{4\pi} \frac{\mu_0 W}{R_E^3 B_0} \hat{\mathbf{e}}_z \qquad (10.17)$$

where B_0 is the magnetic-field strength of the dipole field at the earth's surface at the equator. [Equation (10.17) is derived in Problem 10.2.] However, formula (10.21) does not give the complete magnetic effect of the single charge q on the field at the earth. Because of its gyrational motion about the magnetic field, the particle has a magnetic moment

$$\boldsymbol{\mu} = -\hat{\mathbf{e}}_z \frac{W}{B_e(L)} \qquad (10.18)$$

where $B_e(L) = B_0 L^3$ is the equatorial field strength at LR_E geocentric distance. That dipole produces a northward field

$$\Delta \mathbf{B}^q_{\text{dipole}} = \frac{\mu_0}{4\pi} \frac{\boldsymbol{\mu}}{(LR_E)^3} \hat{\mathbf{e}}_z = \frac{\mu_0}{4\pi} \frac{W \hat{\mathbf{e}}_z}{R_E^3 B_0} \qquad (10.19)$$

at the center of the earth. Therefore, the total magnetic perturbation at the center of the earth due to the charge q is

$$\Delta \mathbf{B}^q = -\frac{\mu_0}{2\pi} \frac{W \hat{\mathbf{e}}_z}{B_0 R_E^3} \qquad (10.20)$$

Note that $\Delta \mathbf{B}$ does not depend on L, but only on particle energy. Consequently, we can write the total magnetic perturbation at the earth's center as

$$\Delta \mathbf{B}_{\text{particles}} = -\frac{\mu_0}{2\pi} \frac{W_{\text{particles}} \hat{\mathbf{e}}_z}{B_0 R_E^3} \qquad (10.21)$$

where $W_{\text{particles}}$ is the total energy in trapped particles. This result can be written in another form using the result of Problem 10.3, which shows that the total energy $\int B^2/(2\mu_0)\, d^3x$ in the earth's dipole magnetic field above the earth's surface is given by

$$W_{\text{mag}} = \frac{4\pi}{3\mu_0} B_0^2 R_E^3 \qquad (10.22)$$

so that

$$\frac{\Delta \mathbf{B}_{\text{particles}}}{B_0} = -\frac{2}{3} \frac{W_{\text{particles}}}{W_{\text{mag}}} \hat{\mathbf{e}}_z \qquad (10.23)$$

Equation (10.23) is called the Dessler-Parker-Sckopke relationship, after the people who first derived it (Dessler and Parker, 1959; Sckopke, 1966). (The derivation presented here was suggested by A. J. Dessler.) Equations (10.21) and (10.23) are valid in a dipole field for arbitrary equatorial pitch-angle distributions, not just for equatorially mirroring particles.

Because of the Dessler-Parker-Sckopke relation, the observed change in the magnetic field at the earth's surface is used as an indication of the amount of energy in ring-current particles. As an approximation to $\Delta \mathbf{B}$ at the center of the earth, we use the average of $\Delta \mathbf{B}$ values equally spaced in longitude around the earth. After certain corrections for neutral-wind-driven currents, this average is defined to be the D_{st} index, which is the standard measure of magnetic storms (for more details, see the discussion in Chapter 13). The calculation outlined earlier assumes a nonconducting earth. If we account for the exclusion of externally applied magnetic fields by a realistic model of the interior conductivity, we find that a 100-nT depression in the H component on the surface of the earth corresponds to a ring current whose total energy is 2.8×10^{15} joules (J).

10.5.6 South Atlantic Anomaly and Drift-Shell Splitting

In a pure dipole field, gradient- and curvature-drift velocities are in the same direction, namely, westward for positively charged particles and eastward for negatively charged particles. However, the actual magnetic field in the inner magnetosphere is not exactly dipolar, and the departures from a dipolar configuration have interesting effects on the radiation environment.

For the inner radiation belt, the most important departures from a dipole field result from the higher moments of the earth's internal field (i.e., the quadruple, octupole, . . . moments). The point on the earth's surface where the field strength is weakest is near the east coast of South America. This feature of the earth's field is called the "South Atlantic anomaly." Inner-radiation-belt particles that mirror near the equator drift on contours of constant magnetic field, according to equation (10.14). They have to come closer to the center of the earth when they pass the weak point in the field that is above the South Atlantic, which means that they move to lower altitudes there and that they have an increased chance of precipitation. A large fraction of the loss from the inner radiation belts is associated with the South Atlantic anomaly.

For outer radiation belts, the most important departures from a dipole magnetic field result from magnetospheric currents. The day–night asymmetry of the magnetic-field configuration has particularly interesting effects on the radiation belts. Chapman-Ferraro current flowing eastward along the magnetopause, which compresses the magnetic field in the inner magnetosphere, has its strongest effect on the dayside. The

10.5 ADIABATIC INVARIANTS AND PARTICLE DRIFTS

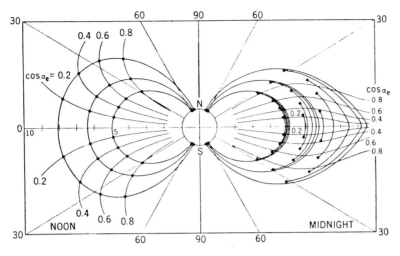

FIG. 10.23. Computed drift-shell splitting for particles starting on common field lines in the noon meridian. Dots represent particles' mirror points. Curves giving the positions of mirror points for constant equatorial pitch angle α_0 are shown. (Adapted from Roederer, 1967.)

westward tail current, which tends to expand and weaken the inner magnetospheric field, has its strongest effect on the nightside. The two currents together cause a systematic difference in the shapes of dayside and nightside field lines, as shown, for example, in Figure 10.1. This day–night asymmetry has different effects on particles with different pitch angles. For example, a particle that has $J = 0$ and thus gradient-drifts in the equatorial plane follows contours of constant equatorial magnetic field strength. Such a particle has to come closer to the earth on the nightside, where the field tends to be weaker, than on the dayside, where the field is stronger. On the other hand, consider a particle with very small μ and relatively large J. On its bounce path, the field strength B is very small compared with the mirror field B_m, except very close to the end of the path, where B is increasing rapidly. Its geometric invariant I is approximately equal to $(B_m)^{\frac{1}{2}}S$, where S is the length of the bounce path. Assuming that electric-field effects are negligible, the particle maintains constant kinetic energy μB_m as it drifts. Because μ also stays constant, it follows that B_m is also a constant along the drift path, and the only way for it to retain constant I is for S to remain constant. In other words, such a particle curvature-drifts on a path of constant field-line length S. It is clear from Figure 10.1 that if we compare dayside and nightside field lines that go equally far from the center of the earth, nightside field lines tend to have shorter overall lengths than dayside field lines. Therefore, particles with near-zero equatorial pitch angles tend to drift farther from the earth on the nightside than on the dayside, which is opposite from the behavior of particles with 90° equatorial pitch angle. Consequently, the outermost parts of the trapped radiation belts tend to have small equatorial pitch angles on the nightside of the earth, and to have large equatorial pitch angles on the dayside of the earth. This phenomenon is called "drift-shell splitting."

Figure 10.23 shows the results of a quantitative calculation of the drift-shell-splitting effect. As an example of a small-J particle that mirrors near the equatorial plane, consider a particle that passes local noon with $\cos \alpha_0 = 0.2$, on a field line that crosses the equatorial plane at $8R_E$ geocentric distance. Scaling off the figure, we find that it passes local midnight bouncing on a field line that extends only to about $7.1R_E$; its bounce path has also been shortened by about a factor of 2, a result of the fact that $B(s)$ increases much more rapidly with distance from the equatorial plane on the nightside than on the dayside. As an example of a large-J particle that mirrors near the earth, consider one that passes local noon with $\cos \alpha_0 = 0.8$, on the field line that extends to $8R_E$. The figure indicates that the particle passes local midnight on a field line that extends to about $9.6R_E$; its bounce length remains approximately constant, as expected.

10.5.7 The Plasmapause and Alfvén Layers

We now apply the theory of adiabatic drifts to a very simple model of the magnetosphere in order to interpret certain basic observed features of the magnetospheric plasma distribution, as discussed in Section 10.3.

For the purpose of the present discussion, assume that the magnetic field is a simple dipole, with equatorial field strength given by

$$B_e^{\text{dipole}} = \frac{B_0 R_E^3}{r_e^3} \tag{10.24}$$

where r_e represents distance from the center of the earth in the equatorial plane, and B_0 is the field strength at the earth's surface at the equator. We pretend that the particles being considered mirror in the equatorial plane (i.e., have $J=0$), so that

$$W = \mu B_e \tag{10.25}$$

As discussed in Section 10.4, particles on the dawnside and duskside of the main auroral zone generally $\mathbf{E} \times \mathbf{B}$-drift toward the sun, and this sunward drift maps to a sunward drift of plasma-sheet plasma in the equatorial plane of the magnetosphere. This maps to generally sunward $\mathbf{E} \times \mathbf{B}$ drift of plasma in the plasma sheet, which corresponds to a dawn-to-dusk electric field, as shown in Figure 10.24. The potential representing the uniform dawn-to-dusk electric field in the equatorial plane is

$$\phi_{\text{convection}} = -E_0 r \sin \varphi \tag{10.26}$$

The dawnside of the sunward-flow region is charged positively, while the duskside is charged negatively.

One more physical element must be added to the convection electric field to make a reasonable first approximation to the electric field in the equatorial magnetosphere, namely, the effect of the rotation of the earth. The basic ionospheric $\mathbf{E} \times \mathbf{B}$-drift pattern that is illustrated in Figure 10.23, with antisunward flow over the polar caps, sunward flow in the main auroral zone, and very weak flow at low latitudes, represents the

10.5 ADIABATIC INVARIANTS AND PARTICLE DRIFTS

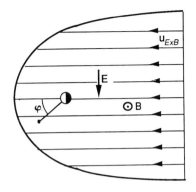

FIG. 10.24. E × B drift in a simple model magnetosphere. The view is of the equatorial plane, with the sun to the left. The electric field is assumed to be uniform and directed from dawn (top of diagram) to dusk (bottom). The lines with arrows on them are equipotentials and also E × B-drift paths. The coordinate angle φ is zero at local noon, $\pi/2$ at local dusk, and π at local midnight.

drift relative to the solid earth. The corresponding electric-field pattern (Figures 10.17 and 10.18b) represents the electric field that is measured in a reference frame that rotates with the earth. On the other hand, the basic nose–tail structure of the magnetosphere is organized by the solar wind, and the natural reference frame for representing drift paths of magnetospheric particles is one that does not rotate with the planet. To find the electrostatic potential in a frame that does not rotate with the earth, we have to add an additional term, called the corotation potential, to the potential given in equation (10.26). The corotation electric field causes particles to rotate eastward with the earth. The corresponding E × B-drift velocity is given by

$$\frac{(-\nabla \phi_{\text{corotation}}) \times \mathbf{B}}{B^2} = \omega_E r \, \hat{\mathbf{e}}_\varphi \tag{10.27}$$

where ω_E is the angular velocity of the earth ($2\pi/24$ h), and $\hat{\mathbf{e}}_\varphi$ is an eastward unit vector. Substituting equation (10.24) for B leads to the following expression for the corotation potential:

$$\frac{d\phi_{\text{corotation}}}{dr} = \frac{\omega_E B_0 R_E^3}{r^2} \tag{10.28}$$

or

$$\phi_{\text{corotation}} = \frac{-\omega_E B_0 R_E^3}{r} \tag{10.29}$$

We can combine the effects of gradient drift, convection, and corotation and write the total-drift velocity in the form

$$\mathbf{v}_D = \frac{\mathbf{B} \times \nabla \phi_{\text{eff}}}{B^2} \tag{10.30a}$$

where

$$\phi_{\text{eff}} = -E_0 r \sin \varphi + \frac{\mu B_0 R_E^3}{q r^3} - \frac{\omega_E B_0 R_E^3}{r} \tag{10.30b}$$

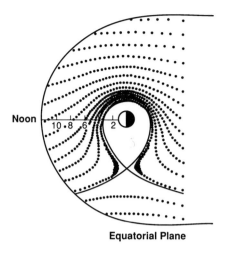

FIG. 10.25. Drift paths in the equatorial plane for particles with $\mu=0$ and a uniform electric field of 0.3 mV·m^{-1}. The positions of the magnetopause and the forbidden region are indicated by solid lines. The distance between successive dots is the distance a particle drifts in 10 min. (Adapted from Kavanagh et al., 1968.)

The particles drift along paths of constant ϕ_{eff}. Consideration of some simple examples of this drift sheds light on some basic characteristics of magnetospheric plasmas.

Consider first the motion of cold particles (i.e., $\mu=0$), for which case

$$\phi_{\text{eff}}^{\text{cold}} = -E_0 r \sin \varphi - \frac{\omega_E B_0 R_E^3}{r} \tag{10.31}$$

Contours of constant ϕ_{eff} computed for this case are shown in Figure 10.25.

Near the earth, the corotation term dominates the effective potential, which decreases toward the earth as the corotation term gets more and more negative. The electric field thus points toward the earth, and the $\mathbf{E} \times \mathbf{B}$ drift is eastward (counterclockwise). However, far out in the magnetosphere, the convection potential dominates, and the flow is toward the sun. At intermediate distances, convection and corotation compete for control of the plasma. On the dawnside of the earth (the top of the figure), both corotation and convection are trying to make the plasma drift eastward (sunward). On the duskside, however, corotation and convection fight each other. There is one point on the dusk meridian where the flow velocity is zero, representing an exact balance between the competing processes of convection and corotation. In our simple time-independent model, no cold plasma crosses the boundary (called a "separatrix") between the trajectories that circle the earth and the trajectories that lead from the tail to the dayside magnetopause. The region on the earthward side of the separation is called the "forbidden region" because it cannot be entered by particles drifting in from the tail.

We can find the point where the flow velocity is zero by setting $\partial \phi_{\text{eff}}/\partial r = 0$ and $\partial \phi_{\text{eff}}/\partial \varphi = 0$ in equation (10.31), which leads to the condition

$$r_{\text{zero-flow}}^2 = \frac{\omega_E B_0 R_E^3}{E_0} \tag{10.32}$$

Cold charged particles that lie inside the separatrix go continuously round and round the earth. They form the plasmasphere, which, as described in Section 10.3.5, consists of dense, cold plasma that evaporated up from the ionosphere. The shape of the plasmasphere implied by this simple model has a bulge on the duskside, as indicated in Figure 10.25. This theoretically predicted feature is in qualitative agreement with the average observed shape, as displayed in Figure 10.16. Plasmasphere particles drift more slowly as they pass dusk because of the competition between convection and the earth's rotation. Consequently, their flow lines have to spread out, which causes the bulge. Our simple theory also makes it clear why the plasmasphere shrinks in times of high magnetospheric activity, as indicated in Figure 10.15. In such times, the rate of magnetospheric convection, as indicated by the strength of the dawn–dusk field E_0, increases, which, according to equation (10.32), reduces the geocentric distance to the zero-flow point and the overall size of the plasmasphere.

It should, of course, be noted that when the strength of convection changes in time, the plasmasphere generally will not line up with the separatrix. Suppose that the strength of convection suddenly increases. The separatrix will instantaneously move earthward, but, of course, the physical plasmapause will not move instantaneously. However, the particles that used to lie on the outermost closed trajectories, just inside the separatrix, suddenly find themselves on open trajectories. As they rotate around and approach local dusk, they make a right turn and head for the dayside magnetopause. Consequently, increasing convection causes the bulge of the plasmasphere to rotate toward local noon and causes plasma to be stripped off the outer plasmasphere and to drift to the dayside magnetopause. Strands of cold plasma have in fact been observed outside the plasmapause in the local afternoon after convection has increased (e.g., Grebowsky, 1970).

Now consider the case of particles that are energetic enough that the gradient-drift term in equation (10.30b) is much larger than the corotation term. (This is true for most plasma-sheet particles, for example.) For these particles, the effective potential is given approximately by

$$\phi_{\text{eff}}^{\text{hot}} = -E_0 r \sin \varphi + \frac{\mu B_0 R_E^3}{qr^3} \tag{10.33}$$

Close to the earth, the motion of these "hot" particles is dominated by gradient drift [the second term in equation (10.33)]. For positive hot particles, $\phi_{\text{eff}}^{\text{hot}}$ becomes large and positive near the earth, and the particles drift in the direction of $\mathbf{B} \times \nabla \phi_{\text{eff}}$, which is westward. For negative hot particles, $\phi_{\text{eff}}^{\text{hot}}$ becomes large and negative near the earth, and the particles drift eastward. At large distances, both kinds of particles convect toward the sun. The configuration of hot-particle drift trajectories is shown qualitatively in Figure 10.26. In each case, the region earthward of the separatrix represents the trapped-radiation region for the

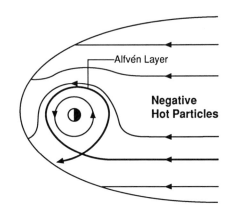

FIG. 10.26. Drift paths for positive and negative hot magnetospheric particles. The view is of the equatorial plane, with the sun to the left. The separatrix between the trajectories that lead from the magnetotail to the dayside magnetopause and those that circle the earth is called the "Alfvén layer."

species in question. Note also that plasma-sheet electrons penetrate closer to the earth on the dawnside than do positive ions from the plasma sheet; ions penetrate closer on the dusk side; both are observed general characteristics of the inner edge of the plasma sheet.

We can find the position of the zero-flow point by setting the radial derivative of $\phi_{\text{eff}}^{\text{hot}}$ equal to zero. The result is

$$r_{\text{zero-flow}} = \left[\frac{3\mu B_0 R_E^3}{|q|E_0}\right]^{\frac{1}{4}} \tag{10.34}$$

This formula implies that the size of the trapped-particle region is larger for more energetic (higher-μ) particles than for less energetic particles. Thus, more energetic plasma-sheet particles cannot penetrate as close to the earth as can the less energetic particles. Notice also that $r_{\text{zero-flow}}$ decreases with increasing E_0. Thus, plasma-sheet particles can penetrate closer to the earth during times of strong convection than during times of weak convection, which agrees with observations. Mapping this conclusion to the ionosphere, the corresponding prediction is that the auroral particles can penetrate to lower ionospheric latitudes during times of strong convection, which agrees with observations. During large magnetic storms, the auroral zone gets very large and extends to much lower latitudes than it normally does.

It is now easy to understand the essential features of the observation-based Figures 10.6 and 10.7. The plasmasphere lies earthward of the inner edge of the plasma sheet, because the geocentric distance of an electron Alfvén layer increases with increasing particle energy. Plasmaspheric particles have zero energy, but plasma-sheet electrons are much hotter. The inner edge of the electron plasma sheet is closer to the earth on the dawnside than on the duskside because, for electrons, the zero-flow point is on the duskside, as shown on the right side of Figure 10.26. The inward motion of the plasma sheet during a time of increased convection results from the decrease of the Alfvén-layer radius.

However, the Alfvén layer for particles of given μ and J does not actually correspond to a mathematical discontinuity in the population of those particles, for at least two reasons: (1) Consider the simple case where the electric field remains constant for many days (this never happens in reality, but it is a useful simple case to consider theoretically). Particles on closed orbits just earthward of the Alfvén layer presumably will decay away by loss, there being no sources available in this idealized steady-state situation. Population levels just outside the Alfvén layer must earlier have drifted very close to the stagnation point in the flow (shown as an x-point in Figure 10.26). Because the drift velocity goes to zero at the stagnation point, it must be very small near that point. An infinitely long time is required for a particle to reach a point that is an infinitesimal distance outside the Alfvén layer. Thus the population just outside the layer must be reduced by loss, and therefore we expect population levels to rise continuously with distance outside the Alfvén layer. (2) Time variations in the convection electric field make the boundary between the plasma sheet and the trapped particles indistinct. Suppose, for example, that the cross-tail electric field is very strong when a certain ion drifts in from the tail. Its Alfvén layer will, according to (10.34), be close to the earth, and the particle will start to drift west, passing by the earth on the duskside, as shown in the left diagram of Figure 10.26. But suppose that the cross-tail electric field suddenly decreases as the ion passes local dusk. The Alfvén layer will then suddenly move out, and the particle will now be trapped on a closed-loop trajectory. This type of process creates a symmetric ring of westward current in a magnetic storm. However, this type of conversion of untrapped particles to trapped ones, and vice versa, happens all the time. A result of these fluctuations is that flux-tube population levels, expressed in terms of the plasma distribution function, gradually decreases with decreasing radial distance in the inner edge of the plasma sheet. The gradient may be more or less steep, depending on the particle species and the loss rate.

The process in which fluctuating fields spread the L shells of particles that drift near the stagnation point is called *radial diffusion*. Even particles on drift orbits that remain far from the stagnation point can be affected by fluctuating fields that cause their drift paths to move toward and away from the earth, so that they do not follow the idealized steady-state drift patterns that we have illustrated in this chapter. Radial diffusion always has the effect of reducing the radial gradients of the distribution function at fixed μ and J. This gives us a very important tool for identifying the source of a population of particles. If we plot the distribution function at fixed μ and J versus L, we can identify the L value of the peak as the source of the particles found on other L shells.

Table 10.2 compares the energies and magnetic moments of several major ion populations in the magnetosphere. Each population dominates the total particle energy density in its region of space. Note that all of these populations correspond to roughly comparable values of the magnetic moment, which suggests the same source. In terms of absolute population level (particles per unit magnetic flux), the inner-magneto-

TABLE 10.2. Properties of Some Major Ion Populations in the Magnetosphere

Population	$<r>$ (R_E)	$<E>$ (eV)	B_{dipole} (nT)	μ (eV·nT^{-1})
Plasma-sheet ions	13	9,000	14	640
Hot synchronous-orbit ions	6.6	22,000	110	200
Davis-Williamson protons	4	100,000	480	210

spheric populations have lower population levels than the plasma sheet, which, of course, suggests that the plasma sheet is the source, but that the transport mechanism connecting the plasma sheet to the inner magnetosphere is not highly efficient.

Plasma-sheet particles are injected deeper into the magnetosphere during magnetic storms. Fluctuations in the electric and magnetic fields also cause variations and irregularities in the idealized drift paths shown in Figure 10.26, which causes radial diffusion of already existing radiation-belt particles. However, the most energetic particles in the radiation belts (those above $\sim$ MeV) may result from cosmic-ray processes, not from radial transport from the plasma sheet.

10.6 IONOSPHERE–MAGNETOSPHERE COUPLING

The key electrodynamic element in the physical coupling between the ionosphere and the magnetosphere is the current that flows along field lines between the two regions. Certain field-aligned currents are, in fact, implied by the plasma distributions that were discussed in the preceding section. Here we shall present only a very simple theoretical discussion of these currents and electric fields. A more complete discussion is given, for example, by Wolf (1983).

We continue to picture the plasma sheet as consisting of particles that mirror in the equatorial plane. Because the electrons are considerably cooler than the ions, they make a relatively minor contribution to the total gradient/curvature-drift current, and we shall neglect their contribution. The density of current flowing in our equatorial plane is given by Nqv_{GD}.

$$\mathbf{j}_D = \frac{N\mu \mathbf{B} \times \nabla B}{B^2} \tag{10.35}$$

where N is the number of these gradient-drifting particles per unit area in the equatorial plane. For the case of particles drifting in the equatorial plane, the other significant contribution to the current density is the magnetization current $\mathbf{j}_M = \nabla \times \mathbf{M} = \nabla \times (-N\mu \hat{\mathbf{e}}_B)$, where $\hat{\mathbf{e}}_B$ is a unit vector in the direction of the magnetic field (e.g., Jackson, 1975). However,

10.6 IONOSPHERE–MAGNETOSPHERE COUPLING

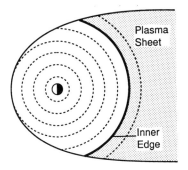
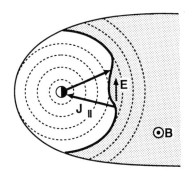

FIG. 10.27. Field-aligned currents and electric fields generated by a bump in the inner edge of the plasma sheet. The dotted curves are contours of constant equatorial magnetic field. Gradient-drift currents flow westward along these curves. The left diagram shows a configuration in which the inner edge is parallel to the gradient-drift current, which results in no diversion of current through the ionosphere. The right diagram shows the result of a bump in the inner edge of the plasma sheet. The extra westward current in the bump has to be completed through the ionosphere, which causes the westward (bottom) edge of the bump to charge up negative, and the east (top) side positive. This causes a tailward **E** × **B** drift in the bump, which tends to cause the bump to disappear.

the magnetization current, which results only from gradients in the density of magnetic moments in gyrating particles, not from the systematic drift of any particle, has no divergence and consequently is not directly involved in coupling to the ionosphere.

Figure 10.27 shows the drift-current flow lines implied by equation (10.35). Because ∇B is earthward, the current is westward (obviously, because we have already concluded that positive particles drift westward, and negative particles eastward). In the figure, we have assumed a sharp inner edge for the plasma sheet, as indicated by a heavy curve. In the left panel of the figure, the inner edge is exactly aligned with a current flow line, which is also a gradient-drift trajectory. Such a configuration implies no interruption in the gradient-drift current. The right panel shows an inner edge with a bump in it. Note that this implies extra gradient-drift current across the bump, a partial ring current that has to be completed in some manner. As the diagram suggests, current flows up from the ionosphere to the east side of the bump, across the bump, and then down from the west side of the bump to the ionosphere.

Electrodynamically, the ionosphere acts like a complicated conductor, as we shall discuss in more detail later. To transport current from point A to point B, the region around A has to charge up positively, and point B negatively. In the present case, the part of the ionosphere that maps to the west side of the bump has to charge up positively, and the east side negatively. The west side is then at a higher electrostatic potential than the east side, and there is an electric field from A to B. If we assume that the magnetic-field lines are good conductors, which is a reasonable first approximation, then the west side of the bump in the magnetosphere will be at a higher potential than the east side. This implies an eastward electric field across the bump, as indicated in the figure. Eastward electric field corresponds to $\mathbf{E} \times \mathbf{B}$ drift away from the earth. Therefore, if a bump forms in the inner edge of the plasma sheet, as shown in Figure 10.27, the electric field caused by the bump will tend to squash the bump back into line. In other words, the bump should decay with time, unless there is some powerful mechanism causing it. In still other words, the inner edge of the plasma sheet is *stable* against interchange of plasma-sheet and near-plasma-sheet flux tubes. The ob-

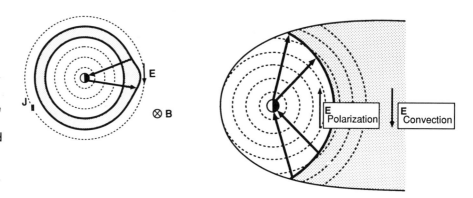

(Left) **FIG. 10.28.** Illustration of the interchange instability expected in the Io plasma torus. The view is of the equatorial plane. The magnetic field is directed into the page. Current due to centrifugal-force drift flows counterclockwise around the planet. The assumed bulge in the Io plasma torus carries more of this counterclockwise current than do other sectors of the torus, which requires diversion of current through the Jovian ionosphere. The resulting polarization electric field is then clockwise, which corresponds to an outward $\mathbf{E} \times \mathbf{B}$ drift and corresponds to growth of the bulge.

(Right) **FIG. 10.29.** Field-aligned currents and electric fields generated by large-scale magnetospheric convection. The format is the same as Figure 10.27.

served equatorward edge of the auroral zone, as displayed in auroral images, does, in fact, generally appear smooth.

Figure 10.27 exhibits the fact that the inner edge of the earth's plasma sheet is stable under interchange. Jupiter's magnetosphere, on the other hand, contains an example of interchange instability. The primary source of plasma for Jupiter's magnetosphere appears to be the volcanoes that occur frequently on the moon Io, which orbits the planet deep within Jupiter's huge magnetosphere; some of the material lifted up by the volcanoes escapes the moon's weak gravity and can become ionized magnetospheric plasma. The primary drift current in this plasma is an external-force drift, the second term on the right side of equation (10.5), the relevant external force being the centrifugal force due to the rapid rotation of the planet. (External-force drift due to centrifugal force is normally negligible for the case of the earth.) The drift current is in the $\mathbf{F}_{ext} \times \mathbf{B}$ direction, which is counterclockwise in the diagram. The required Birkeland current and the polarization electric field are shown in Figure 10.28. The resulting $\mathbf{E} \times \mathbf{B}$ drift is outward (i.e., in a direction corresponding to ripple growth). If the Io plasma torus has a heavier population of plasma than the region farther out, the outer boundary of the torus should thus be interchange-unstable.

Let us return to earth now, or at least to its immediate vicinity. Suppose now that the deformation of the inner edge of the earth's plasma sheet is due to sunward convection. Specifically, suppose that for $t<0$ the magnetosphere looks like the left diagram of Figure 10.27 and that there is no convection. Suppose further that at $t=0$, a dawn-dusk potential drop is imposed across the tail, which will cause plasma-sheet particles to start to drift sunward. The inner edge should then deform as indicated in Figure 10.29, with the nightside moving nearer to the earth, and the dayside moving out. This deformation results in field-aligned currents up out of the ionosphere on the dawnside, and down into the ionosphere on the duskside. A dusk-to-dawn polarization electric field forms across the inner magnetosphere, opposing the original dawn-to-dusk convection electric field. As the plasma-sheet inner edge continues to move sunward on the nightside, the polarization field keeps increasing and cancels an increasing fraction of the applied dawn-to-dusk convection electric field. The sunward motion of the inner edge consequently slows. If the applied convection potential is held constant, the inner edge finally comes to rest at an equilibrium position, where the polarization electric field very nearly balances the applied convection field near the inner edge. The electric field in the inner magnetosphere is then nearly zero. For typical steady conditions, the inner edge of the plasma sheet can effectively shield the inner magnetosphere, and the low-latitude and midlatitude ionosphere to which it is connected, from the convection electric field.

The Birkeland (field-aligned) currents shown in Figure 10.29 correspond to the more equatorward ring of currents shown in Figure 10.18a, the "region-2 Birkeland currents" defined by Iijima and Potemra (1978). They are directed down into the ionosphere on the duskside, and up from the ionosphere on the dawnside. The other major prediction implied by Figure 10.29 is that the electric field is small earthward of the inner edge of the plasma sheet, which, mapped to the ionosphere, corresponds to the equatorward edge of the diffuse auroral precipitation. It is in fact true that the intensity of the ionospheric electric field usually drops sharply near the equatorward edge of the auroral zone. This tendency is clear in the observational data shown in Figure 10.17.

The tendency of the inner edge of the plasma sheet to shield the region earthward of it from the convection electric field was not included in the simple electric-field model employed in Section 10.5.7. That major oversimplification of the assumed electric-field configuration causes the quantitative predictions of that model to be unreliable (as suggested, for example, by the model's prediction of a larger plasmaspheric bulge than is actually observed). However, more sophisticated calculations based on a realistic shielded electric field confirm the basic qualitative tendencies of the simple model.

It should also be remarked that shielding by the inner edge of the plasma sheet is frequently ineffective. Time variations in the applied convection potential or other magnetospheric parameters can cause temporary penetration of the shielding. It is also possible that the plasma sheet is sometimes incapable of effective shielding for extended periods. Furthermore, neutral winds generated by magnetospheric activity can cause significant electric fields at low and middle latitudes in the ionosphere; see Spiro, Wolf, and Fejer (1988) for a discussion.

10.7 IONOSPHERIC CURRENTS

In the ionosphere, between 100 and 500 km altitude, the conductivity along the magnetic-field lines is expected to be very good, but there is also conductivity perpendicular to **B**, as a result of collisions between the charged particles and the neutral atmosphere. The current is of two types: the Pedersen current, which is in the direction of $\mathbf{E}_\perp$, and the Hall current, which is perpendicular to $\mathbf{E}_\perp$, in the direction of $-\mathbf{E}\times\mathbf{B}$. The Pedersen current results from the acceleration of the ions and electrons in the direction of $\pm \mathbf{E}_\perp$ after each collision. The Hall current results from the fact that both ions and electrons try to $\mathbf{E}\times\mathbf{B}$-drift in response to an electric field applied perpendicular to $\mathbf{B}$. However, the two species are not equally successful in $\mathbf{E}\times\mathbf{B}$ drifting. Because the electrons have much higher cyclotron frequencies and are tied much more tightly to the field lines than are the ions, collisions do not resist the $\mathbf{E}\times\mathbf{B}$ motions of electrons as efficiently as they resist ion $\mathbf{E}\times\mathbf{B}$ drift. Because the electrons are also negative, this implies a current in the $-\mathbf{E}\times\mathbf{B}$ direction. If we assume that the field lines are directed down into the ionosphere, as is approximately true in the northern auroral region, the height-integrated horizontal current $\mathbf{J}_\perp$ carried by the ionosphere is given by an expression of the form

$$\mathbf{J}_\perp = \Sigma_P \mathbf{E}_\perp + \Sigma_H \hat{\mathbf{e}}_B \times \mathbf{E} \tag{10.36}$$

where $\hat{\mathbf{e}}_B$ is a unit vector in the direction of the magnetic field. The conductances Σ_P and Σ_H (the height-integrated Pedersen and Hall conductivities, respectively) are typically of the same order of magnitude, but they vary substantially with local time and latitude. They typically are of the order of 10 mho (1 mho = 1 siemens) on the dayside at low and middle latitudes, and also in the auroral zone in conditions of substantial electron precipitation. They drop down to less than 1 mho at low latitudes at night, when neither sunlight nor auroral precipitation is available to enhance the level of ionization. For more details about ionospheric conductance, see the discussion in Chapter 7.

Because the Pedersen and Hall conductances tend to be comparable, ionospheric currents thus tend to be at an angle of about 45° to $\mathbf{E}_\perp$, as illustrated in Figure 10.18*b,c*. For example, on the dawnside, $\mathbf{E}_\perp$ is approximately equatorward, while the current is equatorward and westward.

10.8 MAGNETIC-FIELD-ALIGNED POTENTIAL DROPS

In our simplified discussion of magnetosphere–ionosphere coupling, we treated magnetic-field lines as perfect conductors. However, significant potential drops are observed to occur along auroral-zone magnetic-field lines. These potential drops usually occur in regions of strong upward field-aligned current.

Downward field-aligned current can be carried relatively easily by upward flow of ionospheric electrons, which are available in great numbers and are light and easy for electric fields to move. However, upward field-aligned current requires either an upward flow of ionospheric ions (which is somewhat difficult because of their relatively large mass) or a downward flow of electrons from the plasma sheet (which can be difficult if the electrodynamically required upward current exceeds what can be supplied by electrons in the loss cone). If the field line has difficulty carrying the required amount of upward current, an electric field is created upward along the field line. Observed potential drops of as much as several kilovolts are frequently observed along auroral field lines. The most dramatic effects of these potential drops is acceleration of downcoming plasma-sheet electrons to energies above those characteristic of the normal plasma sheet, and such electrons are responsible for most bright auroral forms. Field-aligned potential drops also effect a partial electrical decoupling of the magnetosphere and ionosphere.

10.9 LOSS OF MAGNETOSPHERIC PARTICLES INTO THE EARTH'S ATMOSPHERE

Two major processes cause magnetospheric particles to be lost in the earth's atmosphere, namely, pitch-angle scattering and charge exchange. A particle that conserves its first adiabatic invariant is likely to be lost to the atmosphere if its mirror magnetic field B_m exceeds the magnetic field at a few hundred kilometers altitude. In view of equation (10.10), this condition can be written

$$\mu < \frac{W}{B_i} \tag{10.37}$$

where B_i is the field at the ionospheric end of the flux tube, nominally at 100 km above the surface. The equatorial pitch angle α_e then satisfies the inequality

$$\mu B_e = W \sin^2 \alpha_e < W\left(\frac{B_e}{B_i}\right)$$

or

$$\sin \alpha_e < \sqrt{\frac{B_e}{B_i}} \tag{10.38}$$

where B_e is the minimum field magnitude on the flux tube (i.e., the equatorial field). Particles that satisfy condition (10.38) are said to be in the "loss cone." Because $B_e \sim 50$ nT in the inner plasma sheet, and $B_i \sim 50{,}000$ nT, condition (10.38) requires that the particle velocity be within a few degrees of the magnetic-field line. The loss cone represents a small solid angle, and only a tiny fraction of plasma-sheet particles would precipitate if the first adiabatic invariant were always conserved.

TABLE 10.3. Lifetimes (Hours) against Charge Exchange for H$^+$ with 45° Equatorial Pitch Angle as a Function of Energy and L Value

Height (R_E)	3	3.5	4	4.5	5
n_H (cm^{-3})	800	470	300	210	150
3 keV	2.2	3.7	5.7	8.2	11.5
10 keV	1.7	2.9	4.6	6.5	9.1
30 keV	3.2	5.4	8.5	12.0	16.8
50 keV	7.6	12.9	20	29	40
100 keV	40	69	110	153	215

Source: From Tinsley (1976).

Auroral-particle precipitation proceeds primarily because plasma waves change particle pitch angles in violation of the first adiabatic invariant. Observations suggest that pitch-angle scattering keeps the loss cone about half full for electrons in the inner plasma sheet, and somewhat less than half full in the case of ions. Estimated loss lifetimes for electrons from the inner plasma sheet typically are a few hours, which is comparable to the convection time. Thus, an electron has a substantial probability of being lost by precipitation in traversing the inner plasma sheet. Loss lifetimes for plasma-sheet ions typically are more than an order of magnitude longer. An ion takes much longer to complete a bounce than does an electron of the same energy. If each mirroring process is regarded as a "chance to precipitate," then an electron has more chances per unit time than an ion does.

Pitch-angle scattering occurs much more slowly in the radiation belts than in the plasma sheet, and radiation-belt particles can orbit the earth for days, weeks, or months, depending on their species, energy, and location.

Magnetospheric ions are subject to loss by an additional mechanism, namely, charge-exchange reactions of the form

$$X^+ + Y \rightarrow X + Y^+ \tag{10.39}$$

where X^+ is an energetic magnetospheric ion (typically kilovolts or more), and Y is a low-energy neutral atom from the earth's exosphere, the uppermost part of the neutral atmosphere. The result of the reaction is a magnetospheric ion of very low energy, which contributes negligibly to the energy and current, plus an energetic neutral, which flies off in a straight line, unaffected by the magnetosphere's electric and magnetic fields. The neutral may escape from the earth's neighborhood, or it may plunge into the atmosphere. Charge-exchange rates can be calculated from a model of the exosphere and the relevant reaction cross sections. Some estimates made by Tinsley (1976) are given in Table 10.3.

10.10 CONCLUDING COMMENT

In accord with usual textbook practice, this description of the inner and middle magnetosphere has emphasized aspects that we know and understand. The first half of the chapter illustrates the fact that more than 30 yr into the space age, we know quite a bit about the interior of the magnetosphere. The second half of the chapter illustrates that many of the basic observational features can be understood on the basis of simple (and sometimes elegant) theory.

By emphasizing what is known and understood, we have neglected exciting aspects of modern research in magnetospheric plasma physics. Many aspects of the behavior of the magnetospheric interior are, in fact, not understood. For example, we do not understand how plasma-sheet plasma convects earthward from the tail. The process apparently involves much more than simple adiabatic drift; see Spence and Kivelson (1990) and references therein. We do not understand the physics of the pitch-angle scattering that causes plasma-sheet particles to drizzle into the auroral zone to cause the diffuse aurora. The physics of bright auroral arcs is understood only partially. There is still heated debate about the essential physics of the magnetospheric substorm, which is the most basic dynamic process that occurs in the interior of the magnetosphere (see discussion in Chapter 13). Finally, as this book was being written the analysis of data from the CRRES spacecraft was causing major changes in our ideas about particle transport in the radiation belts.

Magnetospheric physics therefore remains an active research area, even after more than 30 yr of space exploration. Most of the physical phenomena that we still do not understand involve coupling of small-scale, intermediate-scale, and large-scale plasma processes. Untangling these complex multiscale processes will require increasingly delicate and comprehensive observations combined with the development of increasingly sophisticated plasma theory and computer simulations.

ADDITIONAL READING

Carovillano, R. L., and J. M. Forbes (eds.). 1983. *Solar Terrestrial Physics*. Dordrecht: Reidel.[1]

Kamide, Y. 1988. *Electrodynamic Processes in the Earth's Ionosphere and Magnetosphere*. Kyoto: Kyoto Sangyo University Press.[2]

Parks, G. K. 1991. *Physics of Space Plasma*. Reading, MA: Addison-Wesley.

Schulz, M., and L. J. Lanzerotti. 1974. *Particle Diffusion in the Radiation Belts*. Berlin: Springer-Verlag.

[1] Most of the theoretical topics in the preceding chapter are covered in more detail in the chapter by Wolf in this book.

[2] Contains a particularly complete observational discussion of ionosphere–magnetosphere coupling.

PROBLEMS

10.1. Consider the bounce motion of a charged particle on a dipole magnetic-field line that extends to a maximum distance LR_E from the dipole. Consider specifically a particle that mirrors close to the equatorial plane.

(a) Applying the equation for the strength of the dipole magnetic field, show that for points near the equatorial plane, the magnetic-field magnitude is given by an expression of the form

$$B(L, s) \approx \frac{B_0}{L^3}\left[1 + \frac{\zeta}{2}\left(\frac{s}{LR_E}\right)^2\right]$$

where s is the distance from the equatorial plane, and ζ is a numerical constant. Find the value of ζ.

(b) By substituting this result in equation (10.9), show that for $|s| \ll LR_E$, the conservation-of-energy expression for motion along the field line is like that of a harmonic oscillator. Find an expression for the frequency of the oscillation in terms of particle energy W, shell parameter L, and atomic weight A. Compare with Figure 10.21.

10.2. Derive equation (10.17).

10.3. Derive equation (10.22).

10.4. Consider the integral $\int ds/B$, where the integral is taken along a magnetic-field line from the southern ionosphere to the northern ionosphere. Because $1/B$ is the area corresponding to one unit of magnetic flux, the integral $\int ds/B$ represents the volume of a magnetospheric magnetic-flux tube that contains one unit of magnetic flux.

(a) Show that for a dipole field and $L \gg 1$,

$$\int \frac{ds}{B} \approx \frac{32 R_E L^4}{35 B_0}$$

For $L \gg 1$, only a small fraction of the flux-tube volume lies below the ionosphere. Therefore, one can extend the integration all the way to the point dipole, rather than stopping at the ionosphere. Hint: Remember that $ds^2 = r^2 d\theta^2 + dr^2$.

(b) Consider an isotropic distribution of particles in the flux tube, each particle having velocity v. The number density of these particles is n. Show that the number of particles hitting a unit area at the end of the tube is $nv/4$. Hint: Consider a cylindrical tube intersecting a unit circle, such that the angle between the axis of the cylinder and the normal to the unit circle is θ. Over all azimuthal angles, integrate the number of particles hitting the circle as θ goes from 0° to 90°.

(c) Show that the loss rate $1/\tau$ from a dipole flux tube, specifically the number of particles lost from the flux tube per unit time divided by the number of particles in the tube, is given by

$$\frac{1}{\tau} = \frac{35\,v}{128 L^4 R_E}$$

This is the loss rate for the limit of strong pitch-angle scattering (i.e., pitch-angle scattering that is so rapid that the loss cone remains full). Plasma-sheet electrons frequently run close to this limit.

(d) Evaluate the loss rate numerically for a 1-keV plasma-sheet electron at $L = 10$.

11 PULSATIONS AND MAGNETOHYDRODYNAMIC WAVES

M. G. Kivelson

11.1 INTRODUCTION

WHEN PHYSICAL SYSTEMS experience perturbations, it is common for them to respond by emitting waves. For example, a sound wave in a gas like the atmosphere is produced by a change in pressure at the source of the wave, whether it is a hi-fi speaker system or a dynamite blast. The pressure perturbation then travels through the atmosphere. By knowing the properties of the atmosphere, one can predict the speed at which the signal will propagate, the local speed of sound: $c_s = (\gamma p/\rho)^{\frac{1}{2}}$, where γ is the ratio of the specific heat at constant pressure to the specific heat at constant volume in the gas, p is the gas pressure, and ρ is the gas density. An electromagnetic wave in a vacuum or in a dielectric medium can be established by driving a time-varying current in an antenna. Here, too, it is possible to predict the speed at which the signal will propagate, provided that one can characterize the medium. Conversely, various properties of a system can be probed by measuring the properties of waves found within it, such as frequency, wavelength, and polarization. For example, in the relation $f\lambda = $ constant, between the frequency f and the wavelength λ of an electromagnetic wave in an isotropic medium, the value of the constant provides information on the dielectric properties of the system. The wave polarization (for an electromagnetic wave specified by the direction of the varying electric field of the wave) is related to the wave's propagation direction. For an electromagnetic wave in a vacuum, the plane of the electric and magnetic perturbations is always normal to the direction of propagation. Sound waves, on the other hand, are polarized along the direction of propagation, the polarization direction being that of the gradient of fluctuating pressure. For a closed system, the oscillations normally are combinations of standing waves whose frequencies are governed by the size of the system, as well as by its material properties.

In a plasma, as in a gas, we might expect to find waves that are similar to sound waves, but a plasma is composed principally of charged particles that carry currents. Thus, its electromagnetic properties are of paramount importance, but plasma density and pressure are also relevant. As a consequence, plasma waves differ from both sound waves and electromagnetic waves.

TABLE 11.1. Ranges of Periods and Frequencies in Different Pulsation Classes

	Pc-1	Pc-2	Pc-3	Pc-4	Pc-5	Pi-1	Pi-2
T (s)	0.2–5	5–10	10–45	45–150	150–600	1–40	40–150
f	0.2–5 Hz	0.1–0.2 Hz	22–100 mHz	7–22 mHz	2–7 mHz	0.025–1 Hz	2–25 mHz

In this chapter, we discuss the nature of the lowest-frequency waves that occur in plasmas. A "low" frequency must be lower than the natural frequencies of the plasma, such as the plasma frequency f_p and the ion gyrofrequency f_{ci} that were introduced in Chapter 2. Such waves are referred to as ultra-low-frequency waves. Higher-frequency waves will be treated in Chapter 12. We shall show that the combination of mechanical forces (present because of the gaslike properties of the plasma) and electromagnetic forces (present because the particles are charged) creates unique types of waves: magnetohydrodynamic (MHD) waves. We shall point out how they differ from the waves found in neutral dielectric media.

The equations for conducting fluids, basically expressions of Newton's laws of motion and Maxwell's equations (Maxwell, 1873), were known to physicists for more than half a century before it was recognized that electromagnetic waves can propagate in conducting fluids even though they cannot propagate in rigid conductors. The MHD wave solutions were eventually derived by Hannes Alfvén (1942), but direct confirmation of the existence of the waves was difficult to obtain, as they decay rapidly in most laboratory situations. One can show that the decay rate can be small only if the spatial scale of the system is sufficiently large. Thus it was that the study of Alfvén's predicted waves became principally the task of space physicists.

The first observations of ultra-low-frequency (ULF) fluctuations (with periods ranging from seconds to minutes) of magnetic fields were made on the ground (Stewart, 1861), almost a century before their links to plasmas in near-earth space were established. Early studies of the magnetic pulsations measured by ground-based observers noted that waves could be grouped into classes that appeared to differ in fundamental ways. Some were *continuous pulsations,* quasi-sinusoidal in form, and each with a well-defined spectral peak. These were called Pc pulsations, and they were broken into subgroups on the basis of their periods (starting with Pc-1 in the 0.2–5-Hz band and ending with Pc-5 in the 1.7–6.7-mHz band). Other pulsations in the same frequency band contained power at many different frequencies. Such waves were called Pi for *irregular pulsations.* The names assigned to different frequency bands (Jacobs et al., 1964) are shown in Table 11.1. Typical magnetic-pulsation signatures are illustrated in Figure 11.1, which includes both ground-based observations from a chain of stations at different latitudes and

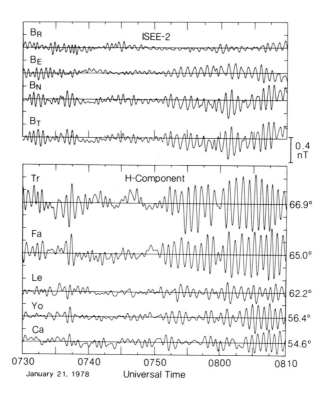

FIG. 11.1 Examples of approximately 1-min waves in the magnetosphere (upper panels) and on the ground (lower panels) at stations whose latitudes are indicated to the right.

simultaneous measurements from a spacecraft in the near-equatorial magnetosphere.

Dungey (1954a,b) was the first to suggest that MHD waves in the outer atmosphere were the sources of the oscillating or pulsating magnetic fields observed on the surface. In particular, the distinct periods of Pc pulsations suggested a resonant process, and Dungey proposed that the pulsations were caused by waves standing along magnetic-field lines and reflected at the ionospheres at the two ends. That idea has been further developed and is generally supported by studies of both ground-based and spacecraft data. We shall return to it after developing the theory and deriving some of the properties of MHD waves in a conducting fluid.

11.2 BASIC EQUATIONS

MHD waves are found as solutions to the equations introduced in Chapter 2 to express the conservation laws and Maxwell's equations. They are repeated here for convenience. Equation (2.29b) guarantees that mass is conserved as the fluid moves:

$$\frac{\partial \rho}{\partial t} + \nabla \cdot \rho \mathbf{u} = 0 \quad \text{(continuity equation)} \tag{11.1}$$

11.2 BASIC EQUATIONS

Momentum conservation is assured by equation (2.32), in which we set $\mathbf{F}_g = 0$ and assume neither sources nor losses:

$$\rho\left(\frac{\partial \mathbf{u}}{\partial t} + \mathbf{u}\cdot\nabla\mathbf{u}\right) = -\nabla p + \mathbf{j}\times\mathbf{B} \tag{11.2}$$

Maxwell's equations in the low-frequency limit [equations (2.35), (2.36b), and (2.37)] will be needed. These equations are

$$\frac{\partial \mathbf{B}}{\partial t} = -\nabla\times\mathbf{E} \quad \text{(Faraday's law)} \tag{11.3}$$

$$\nabla\times\mathbf{B} = \mu_0\mathbf{j} \quad \text{(Ampère's law)} \tag{11.4}$$

and the requirement that $\mathbf{B}$ be divergenceless:

$$\nabla\cdot\mathbf{B} = 0 \tag{11.5}$$

We add Ohm's law in the form [equation (2.46)]

$$\mathbf{E} + \mathbf{u}\times\mathbf{B} = 0 \tag{11.6}$$

and an equation that states that the specific entropy (entropy per unit volume) is conserved in the convecting magnetized plasma:

$$\left(\frac{\partial}{\partial t} + \mathbf{u}\cdot\nabla\right)\left(\frac{p}{\rho^\gamma}\right) = 0 \tag{11.7}$$

Here, as in earlier chapters, p is the pressure, ρ is the mass density, $\mathbf{u}$ is the flow velocity, $\mathbf{j}$ is the electric-current density, $\mathbf{B}$ is the magnetic field (magnetic induction), μ_0 is the magnetic permeability of free space, $\mathbf{E}$ is the electric field, and γ is the ratio of the specific heat at constant pressure to the specific heat at constant volume; γ is frequently referred to as the polytropic index.

As the derivative acting on the expression in parentheses on the right in (11.7) is just the time rate of change in a frame that follows the plasma as it flows through the system, the equation requires that the plasma obey an adiabatic equation of state. In most space plasmas, γ is $\frac{5}{3}$.

We can express the current in terms of the magnetic field, and the electric field in terms of $\mathbf{u}$ and $\mathbf{B}$, by using (11.4) and (11.6). Equation (11.2) becomes

$$\rho\left(\frac{\partial \mathbf{u}}{\partial t} + \mathbf{u}\cdot\nabla\mathbf{u}\right) = -\nabla p + (\nabla\times\mathbf{B})\times\mathbf{B}/\mu_0 \tag{11.8}$$

Suppose that the variations in a system are only in the x-direction and that $\mathbf{B} = B\hat{\mathbf{z}}$. Then the right-hand side of (11.8) can be written as

$$-\hat{\mathbf{x}}\left(\frac{\partial p}{\partial x} + \frac{B}{\mu_0}\frac{\partial B}{\partial x}\right)$$

and the x-component of the momentum equation takes the form

$$\rho\left(\frac{\partial u_x}{\partial t} + u_x \frac{\partial u_x}{\partial x}\right) = -\frac{\partial}{\partial x}[p + (B^2/2\mu_0)] \tag{11.9}$$

As discussed in relation to equation (2.48), the quantity $B^2/2\mu_0$ is the magnetic pressure. The fluid momentum responds to gradients of both the magnetic and thermal pressures.

The set of equations (11.1)–(11.7) or the modified forms that follow must be simultaneously satisfied. Plasma structures that remain at rest in the moving fluid compose a particularly simple subset of solutions of this collection of equations. Such spatially varying plasma properties appear as time variations to an observer not moving with the flow. Using the density ρ as an example, one finds that if $\rho = \rho(\mathbf{x})$ in the plasma rest frame, it depends on space and time in the form

$$\rho(\mathbf{x}, t) = \rho\left(\mathbf{x} - \int_{t_0}^{t} dt' \, \mathbf{u}(\mathbf{x}, t')\right) \tag{11.10}$$

in the observer's frame. Here, t_0 is the time when the fluid element is at $\mathbf{x} = 0$. This form automatically satisfies equations (11.1) and (11.7), as the sum of the two derivatives vanishes independently of the form of $\rho(\mathbf{x})$ in the plasma rest frame. The additional equations are satisfied if $\mathbf{u}$, p, and $\mathbf{B}$ are constant. This solution is referred to as an entropy "wave." The reference to entropy reflects the fact that if ρ varies at constant pressure, then the specific entropy p/ρ^γ varies.

There is another nonpropagating solution with $\mathbf{k}$ perpendicular to $\mathbf{B}$ for which the total pressure $p + B^2/2\mu_0$ is constant across planar surfaces that convect with the flow. In order to satisfy all of the required equations, the component of $\mathbf{B}$ normal to the surface must vanish, and the normal component of $\mathbf{u}$ must not change across the surface. The components of both $\mathbf{B}$ and $\mathbf{u}$ tangential to the surface may vary. This solution is the limit of the slow mode for $\mathbf{k} \perp \mathbf{B}$. In the nonlinear regime, the entropy wave relates to the contact discontinuity, and the slow-mode wave relates to the tangential discontinuity.

11.3 EQUATIONS FOR LINEAR WAVES

The preceding section described a convecting perturbation of arbitrary amplitude that satisfies the set of equations. This section describes a set of waves that propagate relative to the fluid. For simplicity, we assume that the perturbations carried by the waves are small.

Let us assume that the plasma is initially at rest, which means that there are neither flows nor electric fields, and also assume that no currents are flowing. The wave perturbations introduce finite but small $\mathbf{E}$, $\mathbf{u}$, and $\mathbf{j}$. The magnetic field, mass density, and pressure also change, so that

$$\mathbf{B} \to \mathbf{B}+\mathbf{b}, \qquad \rho \to \rho+\delta\rho, \qquad p \to p+\delta p$$

All of the perturbed quantities, $\mathbf{b}$, $\delta\rho$, δp, $\mathbf{u}$, $\mathbf{E} = -\mathbf{u} \times \mathbf{B}$, and $\mathbf{j} = \nabla \times \mathbf{b}/\mu_0$, are assumed to be small enough that only terms linear in any of them need be retained. This means that squares or high powers and cross products will be dropped. The perturbed quantities then must satisfy the equations

$$\frac{\partial \delta\rho}{\partial t} + \rho \nabla \cdot \mathbf{u} = 0 \tag{11.1'}$$

$$\rho \frac{\partial \mathbf{u}}{\partial t} = -\nabla \delta p + (\nabla \times \mathbf{b}) \times \mathbf{B}/\mu_0 \tag{11.8'}$$

$$\frac{\partial \mathbf{b}}{\partial t} = \nabla \times (\mathbf{u} \times \mathbf{B}) \tag{11.3'}$$

We must have $\nabla \cdot \mathbf{b} = 0$ to satisfy equation (11.5). If this condition holds initially, the divergence of (11.3') $[(\partial/\partial t)(\nabla \cdot \mathbf{b} = 0)]$ shows that the condition is automatically satisfied at all times.

The adiabatic requirement also becomes an initial condition, because

$$\frac{\partial \delta p}{\partial t} = \frac{\gamma p}{\rho} \frac{\partial \delta \rho}{\partial t} = c_s^2 \frac{\partial \delta \rho}{\partial t}$$

becomes

$$\frac{\partial}{\partial t}\left(\frac{\delta p}{c_s^2 \delta \rho}\right) = 0$$

and the constant value of the ratio of δp to $\delta \rho$ is set by the initial conditions. Substitution of δp in terms of $\delta \rho$ leaves us with seven unknowns that describe the wave perturbations: $\delta \rho$, $\mathbf{u}$, and $\mathbf{b}$, and seven equations: (11.1'), (11.8'), and (11.3'), once again counting each component of a vector equation separately. In the following sections, we shall solve for the wave properties, making various simplifying assumptions.

11.4 WAVES IN COLD PLASMAS

The simplest system in which MHD waves exist is a cold magnetized plasma. The concept of "cold" needs to be defined, because the temperature need not be zero. All that is meant is that the plasma pressure [which is given by equation (2.33)] is unimportant. Equation (11.9) shows that if the ratio of the plasma pressure to the magnetic pressure is small [i.e., $\beta \ll 1$, where β is defined in equation (2.49)], then the plasma pressure is not important.

In describing the properties of waves, it is convenient to introduce the exponential notation

$$e^{ix} = \cos x + i \sin x \tag{11.11}$$

Expressing the solution of a differential equation in complex exponential form simplifies the derivation, because the derivative of an exponential is just a multiple of the original exponential. However, any equation written in terms of the complex exponential must be satisfied separately by the terms proportional to i and the terms independent of i, and this requirement is equivalent to solving the equation in terms of sines and cosines.

For a plane wave propagating in the x-direction, with wavelength λ and frequency f, the oscillating quantities can be taken as proportional to

$$e^{ikx}e^{-i\omega t} = e^{i(kx-\omega t)} \tag{11.12}$$

where $k = 2\pi/\lambda$ and $\omega = 2\pi f$. If the proportionality factors are complex, the different perturbed quantities may have arbitrary relative phase, δ, where $\tan \delta$ is the ratio of the imaginary part to the real part of the amplitude factor. The solution is, in any case, oscillatory. At a fixed spatial location, the solution oscillates in time with frequency f, and at a fixed time, the solution oscillates spatially with a wavelength λ. Notice that in the second form of (11.12) the argument of the exponential is constant if $x = x_0 + (\omega/k)t$. This means that the solutions are constant at a position that moves along the x-axis with a velocity $dx/dt = \omega/k$, that is, at the wave phase velocity [equation (2.50a)]:

$$v_{ph} = \omega/k \tag{11.13}$$

We use equations (11.1'), (11.8'), and (11.3'), writing them in the forms that apply to the cold-plasma limit ($p = 0$). In doing so, we shall also assume that the background magnetic field and the plasma density are constant (i.e., $\partial\rho/\partial x = 0$, $\partial \mathbf{B}/\partial x = 0$, $\partial\rho/\partial t = 0$, and $\partial \mathbf{B}/\partial t = 0$) and that the wave is moving along the x-direction, meaning that only the derivatives in x and t need be retained:

$$\frac{\partial \delta\rho}{\partial t} + \rho \frac{\partial u_x}{\partial x} = 0 \tag{11.14}$$

$$\rho \frac{\partial \mathbf{u}}{\partial t} = -\hat{\mathbf{x}} \frac{\partial(\mathbf{b}\cdot\mathbf{B}/\mu_0)}{\partial x} + \left(\frac{B_x}{\mu_0}\right)\frac{\partial \mathbf{b}}{\partial x} \tag{11.15}$$

$$\frac{\partial \mathbf{b}}{\partial t} = B_x \frac{\partial \mathbf{u}}{\partial x} - \left(\frac{\partial u_x}{\partial x}\mathbf{B}\right) \tag{11.16}$$

The assumed exponential dependence on x and t of the wave properties ($\delta\rho$, $\mathbf{u}$, and $\mathbf{b}$) implies that time and x derivatives can be replaced by $-i\omega$ and $-ik$, respectively. Then these equations become

$$i(\omega\delta\rho - k\rho u_x) = 0 \tag{11.14'}$$
$$i[\omega\rho\mathbf{u} - k(\hat{\mathbf{x}}(\mathbf{b}\cdot\mathbf{B}) - B_x\mathbf{b})/\mu_0] = 0 \tag{11.15'}$$
$$i[\omega\mathbf{b} + k(B_x\mathbf{u} - u_x\mathbf{B})] = 0 \tag{11.16'}$$

The equations are now just a set of algebraic equations, which are easy to deal with. One need only eliminate variables by substitution. Without loss of generality, we can assume that **B** lies in the x–z plane, and we have already assumed that $\mathbf{k} = k\hat{\mathbf{x}}$, and so $\mathbf{B} = (B\cos\theta, 0, B\sin\theta)$. Here θ is the angle between **B** and **k**. By elimination, we can obtain

$$[(\omega/k)^2 - v_A^2 \sin^2\theta]u_x + v_A^2 \sin\theta\cos\theta\, u_z = 0 \tag{11.17a}$$

$$[(\omega/k)^2 - v_A^2 \cos^2\theta]u_y = 0 \tag{11.17b}$$

$$[(\omega/k)^2 - v_A^2 \cos^2\theta]u_z + v_A^2 \sin\theta\cos\theta\, u_x = 0 \tag{11.17c}$$

$$v_A = (B^2/\mu_0\rho)^{\frac{1}{2}}$$

The coefficients in (11.17) are squares of velocities, with ω/k the wave phase velocity and v_A the *Alfvén velocity* [see equation (2.51)]. Equations (11.17a–c) are homogeneous equations that have solutions only if the determinant of their coefficients vanishes. This requirement gives an equation for $v_{ph} = \omega/k$ that is called a *dispersion relation*. The roots of the dispersion relation give the values of the phase velocity for MHD waves in the cold plasma:

$$(\omega/k)^2 = v_A^2 \cos^2\theta \tag{11.18a}$$

$$(\omega/k)^2 = v_A^2 \tag{11.18b}$$

Dispersion relations such as (11.18a) and (11.18b) impose relations between the frequency and the vector wave number that must be satisfied in order for a wave to exist in the plasma. The requirements that electromagnetic waves in free space propagate at the speed of light ($\omega/k = c$, or the equivalent form in terms of frequency and wavelength) or that sound waves in a neutral gas propagate at the speed of sound ($\omega/k = c_s$) are familiar examples of dispersion relations. Equations (11.18a) and (11.18b) show that the dispersion relations of waves in magnetized plasmas depend on the magnitude of the magnetic field, the density of the plasma, and, under some conditions, the direction of wave propagation. Except for $\theta = 0$, the two equations cannot both be satisfied for the same k and ω, and so the dispersion relation implies two independent solutions.

It is important to remember that a solution is valid only if all three of the equations (11.17a–c) are satisfied. Consider first the wave that satisfies the dispersion relation (11.18a), referred to as the *shear Alfvén wave*. By substituting (11.18a) into (11.17a–c), we see that (11.17b) is satisfied for any value of u_y, but (11.17a) and (11.17c) are satisfied only if $u_x = u_z = 0$. Thus the shear Alfvén wave propagates with the phase velocity $v_A\cos\theta$ and sets the fluid into motion in the direction perpendicular to the plane containing the propagation vector **k** and the background field. Other parameters of the wave, such as its electric and magnetic perturbations, can be expressed in terms of u_y by using (11.6), (11.16′), and so forth, and setting $u_x = 0$ and $u_z = 0$. Equation (11.14′) shows that this type of wave does not change the density of the fluid (because $u_x = 0$). The relative orientations of the perturbation vectors in this type of wave are illustrated in Figure 11.2a. Because the perturbation of the

FIG. 11.2. Schematic of wave polarizations for (a) the Alfvén wave and (b) the fast compressional wave. Displacements of the field lines (thick curves) at maximum displacement for (c) the Alfvén wave and (d) the fast compressional wave. The thin lines represent the unperturbed field. Plasma-pressure and magnetic-pressure perturbations versus time for (e) the slow compressional wave and (f) the fast compressional wave.

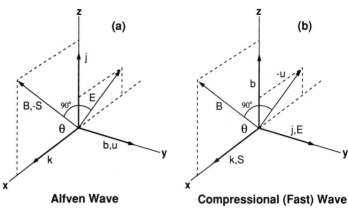

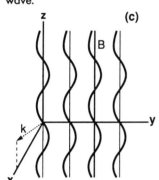

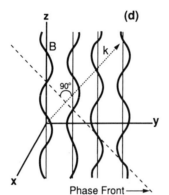

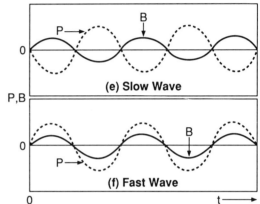

magnetic field is perpendicular to the background field, the field magnitude is constant (to linear order in the perturbation field) even in the presence of the wave:

$$|\mathbf{B}+\mathbf{b}|^2 \approx B^2 + 2\mathbf{B}\cdot\mathbf{b} = B^2$$

As the foregoing identity implies that the magnetic pressure (which is the only relevant pressure in a cold plasma) is constant, the wave is noncompressional.

The second dispersion relation (11.18b) automatically satisfies (11.17a) and (11.17c) for any value of u_x and u_z, but the only way to satisfy (11.17b) is to set $u_y = 0$. This means that a second type of wave can exist, one that sets the fluid into motion within the plane containing **k** and **B**. As u_x does not vanish, (11.14′) implies that this type of wave changes the fluid density. As well, nonvanishing perturbations of the field magnitude are produced. This can be seen as follows:

$$|\mathbf{B}+\mathbf{b}|^2 \approx B^2 + 2\mathbf{B}\cdot\mathbf{b} = B^2 + 2ku_x B^2/\omega$$

where (11.15′) and (11.16′) have been used to obtain the final form. As u_x does not vanish, the magnitude of the field and the magnetic pressure will fluctuate when the wave is present. That is why this type of wave is often referred to as a *compressional wave*. Figure 11.2b shows the polarization of the wave schematically. The wave energy propagates along the direction of the Poynting flux vector:

$$\mathbf{S} = \frac{1}{\mu_0} \mathbf{E} \times \mathbf{b}$$

This direction is along $\pm\mathbf{B}$ in the shear Alfvén wave, but at an arbitrary angle (parallel to k) relative to $\mathbf{B}$ in the compressional wave. This is an important aspect that distinguishes the two wave modes. Figures 11.2c and 11.2d show schematics of the displaced field lines in shear Alfvén waves and fast waves, respectively. In the shear Alfvén waves, the perturbations are all perpendicular to $\mathbf{B}$, and the distance between the perturbed field lines is constant. In the fast wave, the perturbations are oblique to $\mathbf{B}$, and the oblique phase fronts result in varying separations between the perturbed field lines.

If the wave propagates in directions other than along the background magnetic field (i.e., if $\theta \neq 0$), the phase velocity (11.18b) is larger than (11.18a). For this reason, the compressional mode is also referred to as the *fast mode*.

An additional property of interest in a wave analysis is what is referred to as the *group velocity*, $\mathbf{v}_g$. It describes the velocity of energy or information transfer by a physically realistic wave packet that is not strictly monochromatic and may contain a spread of wave vectors. Such a wave packet can be described as a superposition of monochromatic waves using Fourier analysis. Examples have been worked out by Jackson (1962) for the case of electromagnetic waves in dielectric media, but the approach applies to plasma waves as well. The analysis shows that if the spreads about the mean are sufficiently small, an expansion

$$\omega = \omega_0 + (\mathbf{k} - \mathbf{k}_0) \cdot \nabla_k \omega$$

can be introduced. The pulse then retains its shape and propagates with the (vector) velocity $\mathbf{v}_g = \nabla_k \omega$ [see equation (2.50b)]. Note that this is a vector velocity, which is found by expressing ω as a function of the vector $\mathbf{k}$ and taking derivatives with respect to its components. For the two types of waves we have discussed, we find

$$\mathbf{v}_g = v_A \hat{\mathbf{B}} \quad \text{(for the shear Alfvén wave)} \tag{11.19a}$$

$$\mathbf{v}_g = v_A \hat{\mathbf{k}} \quad \text{(for the fast-mode wave)} \tag{11.19b}$$

where $\hat{\mathbf{B}}$ and $\hat{\mathbf{k}}$ are unit vectors along $\mathbf{B}$ and $\mathbf{k}$. These equations tell us that the fast-mode wave can carry energy and information in any direction, whereas the energy and information content of a shear Alfvén wave are strictly guided along the background field, even if phase fronts are oriented arbitrarily.

11.5 WAVES IN WARM PLASMAS

In a warm plasma, the plasma-pressure terms cannot be dropped from the equations (i.e., β is no longer small compared with unity). The pressure-gradient term must be considered in the momentum equation, and (11.7) is also needed. For small-amplitude perturbations, the linear-analysis approach is once again appropriate. We end up with a set of equations analogous to (11.17), and the requirement that the determinant of the coefficients vanish gives a dispersion relation

$$(\omega^2 - \cos^2\theta \, k^2 v_A^2)[\omega^4 - \omega^2 k^2(c_s^2 + v_A^2) + \cos^2\theta \, k^4 v_A^2 c_s^2] = 0 \qquad (11.20)$$

which has three solutions:

$$\omega^2 = v_A^2 \cos^2\theta \, k^2 \qquad (11.20a)$$

$$\omega^2/k^2 = \tfrac{1}{2}\{c_s^2 + v_A^2 \pm [(c_s^2 + v_A^2)^2 - 4c_s^2 v_A^2 \cos^2\theta]^{\frac{1}{2}}\} \qquad (11.20b)$$

Comparing the dispersion relations (11.18) and (11.20), we find that the finite temperature of the plasma has introduced an additional wave mode and changed the properties of the fast mode previously discussed. The shear Alfvén mode appears again as the solution of (11.20a). Its phase velocity still depends only on the Alfvén velocity. All of its properties (e.g., polarization perpendicular to both **B** and **k**, and no change of density or field magnitude) remain the same as in the cold-plasma case.

The roots of (11.20b) depend not only on the Alfvén velocity but also on the sound speed. These roots apply to compressional wave modes (i.e., waves that do change the density and the field magnitude). The two solutions are referred to as the fast (positive sign) and the slow (negative sign) waves and are also called magnetoacoustic wave modes. The electric, magnetic, and current polarizations of the fast and slow waves are shown in Figure 11.2. However, when the thermal pressure varies along **B**, the flow velocity can have a parallel component, and **u** will not be perpendicular to **B**. The thermal-pressure perturbations that are a feature of waves in a warm plasma are in phase with the magnetic-pressure perturbations in the fast wave, but are out of phase with them in the slow wave. Figures 11.2e and 11.2f show schematically the phase relations of the field and pressure perturbations in fast and slow waves.

The fast mode is produced when the total pressure of the plasma (the sum of particle pressure and field pressure) changes locally in the system. For example, if the solar-wind pressure on the dayside of the magnetosphere increases suddenly, a gradient of total pressure, positive toward the dayside boundary, will develop. The pressure perturbation serves as a source of compressional waves [see equation (11.8'), where the pressure-gradient term is a source of plasma motion]. The waves have p and p_B in phase and are therefore fast-mode waves. As they radiate away from the boundary source, they carry away the excess

11.5 WAVES IN WARM PLASMAS

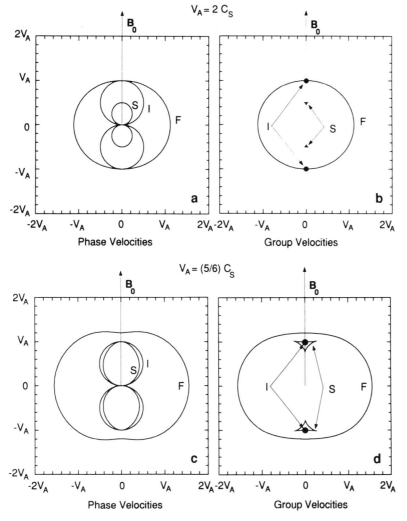

FIG. 11.3. Friedrichs diagrams for (a) the phase velocity and (b) the group velocity for $v_A = 2c_s$ and (c) the phase velocity and (d) the group velocity for $v_A = \frac{5}{6}c_s$. Wave modes are fast (F), intermediate (I), and slow (S).

pressure. Ultimately, these waves are able to reduce total pressure gradients. This wave mode propagates almost isotropically.

A convenient way in which to represent the phase velocity of a wave is in a polar plot, referred to as a Friedrichs diagram, with one axis aligned with the background field. The angle relative to that axis is the angle between **k** and **B**, and the distance from the origin represents that phase velocity. This type of plot is given for the case $c_s^2 < v_A^2$ in Figure 11.3a. For the fast wave, both the phase velocity and the group velocity are largest for propagation perpendicular to **B**. For the slow wave and the intermediate wave, the phase velocity vanishes for $\mathbf{k} \perp \mathbf{B}$.

The group velocity can also be represented in a polar plot, this time with the angle relative to the **B**-direction representing the angle between **B** and the group velocity; the length of the vector represents the magnitude of the group velocity. The plot is shown in Figure 11.3b for the case $c_s^2 < v_A^2$. The fast-mode group velocity is finite in all directions and largest perpendicular to **B**. As pointed out in the discussion of Poynting flux, these waves can carry energy in any direction, as $\mathbf{v}_g$ remains finite at all angles. The intermediate wave has a group velocity that is along $\pm\mathbf{B}$, with amplitude v_A for all $\mathbf{k}$, and so it appears as a pair of points on the plot. This is consistent with the directions of $\mathbf{S}$ in Figure 11.2. The slow-mode group velocity is $\pm c_s$ for $\mathbf{k}$ along $\mathbf{B}$. As the angle between $\mathbf{k}$ and $\mathbf{B}$ increases, the group velocity increases slightly and rotates a bit away from $\mathbf{B}$. As the angle continues to increase, the group velocity decreases and aligns more closely with $\mathbf{B}$. This accounts for the peculiar quasi-triangular curves in Figure 11.3b. Independent of $\mathbf{k}$, the slow-mode group velocity remains nearly aligned with $\mathbf{B}$. It carries energy only over a relatively narrow range of angles and is referred to as "field-guided."

The slow-mode wave is different from the fast-mode wave in several ways. For the slow mode, total pressure (i.e., the sum of particle pressure and magnetic pressure) is approximately constant across the background field. As described earlier, slow waves carry energy predominantly along the background field. Field-aligned gradients of the total pressure drive slow-mode waves. In particular, when the sound speed is much smaller than the Alfvén speed, the slow mode propagates along $\mathbf{B}$ at the sound speed and reduces plasma-pressure gradients. Figures 11.3c and 11.3d show the phase and group velocities, respectively, for the case $c_s^2 > v_A^2$. Qualitatively, the features of wave propagation are unchanged relative to the previous case, but for field-aligned propagation, the slow and intermediate modes adopt the Alfvén speed.

If fast- and slow-mode waves reduce pressure gradients, what does the shear Alfvén wave do? It acts to reduce the bending of the magnetic field. Plasma flow across the field can increase the bending of the field. The associated field perturbations will create currents that act to reduce the additional curvature of the field line. The closure of the currents that flow through the plasma is, in part, along the magnetic field, so that the shear Alfvén wave introduces field-aligned currents. Figure 11.2 shows that the perturbation current in the Alfvén wave (but not in the two compressional waves) has a nonvanishing component along $\mathbf{B}$.

This discussion of MHD waves in warm plasmas links closely with the earlier discussion of shocks in Chapter 5. The shock develops when the information required to slow and divert a flow cannot propagate upstream fast enough. The pileup of waves unable to propagate upstream leads to nonlinear conditions that establish shocks in the flow. The shock front found farthest upstream of an obstacle is linked to the fast magnetoacoustic wave, the wave that propagates fastest, and in all directions, and serves to reduce pressure gradients and to slow and divert the flow. The intermediate wave propagates more slowly and nonisotropically. It serves to rotate the field. Only under special conditions can it develop into a shock. Most rotations observed in space plasmas appear to be unrelated to shocks. Slow-mode waves, because of their nonisotropic propagation, can play at most a limited role in

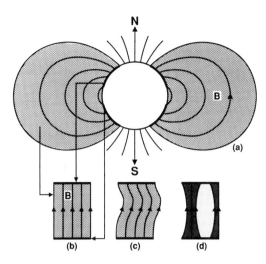

FIG. 11.4. Schematic (a, b) of the dipole field and its relation to a box model of the magnetosphere. Dipole field lines are straightened and bounded by the off-equatorial ionosphere at the top and bottom. Perturbations (c, d) of field and plasma in a shear Alfvén wave and in a fast compressional wave. The density of shading illustrates increases and decreases in the plasma density.

reducing pressure gradients, but slow shocks have been observed in space plasmas.

11.6 IONOSPHERIC BOUNDARY CONDITIONS

The frequencies of ULF waves that can be excited in a plasma depend not only on the wave modes but also on the boundary conditions. For the magnetosphere, the boundaries are the magnetopause and the ionosphere. Here we consider the conditions that must be satisfied at the ionospheric boundary of a flux tube. The ionosphere both reflects and transmits to the ground the ULF signals incident from above. We treat the ionosphere in a qualitative manner by representing it as a thin conducting sheet. The ionosphere lies above a neutral atmosphere that is in turn bounded by the earth. Figure 11.4 shows schematically a dipolar magnetosphere containing plasma (shown by stippling). In this figure, the high-latitude ionosphere forms boundaries at the ends of (most) field lines. The near-equatorial ionosphere serves as an inner boundary, and the magnetopause as an outer boundary. In Figure 11.4b, the field lines have been straightened to form a "box model" of the magnetosphere. If the conductivity of the ionosphere is very high, both the electric field and the wave displacement must vanish at the ionospheric ends of the field lines (as well as on the left side of the box). This means that any wave incident on the ionosphere will be reflected back toward the other ionosphere. (At lower conductivities, the waves are only partially reflected.) Like waves on a string, the Alfvén waves can satisfy the reflection condition only for certain selected wavelengths. If the length of the field line between the two ionospheres is l, the allowed wavelengths along the field direction $\lambda_\parallel$ are

$$\lambda_\parallel = 2l/n$$

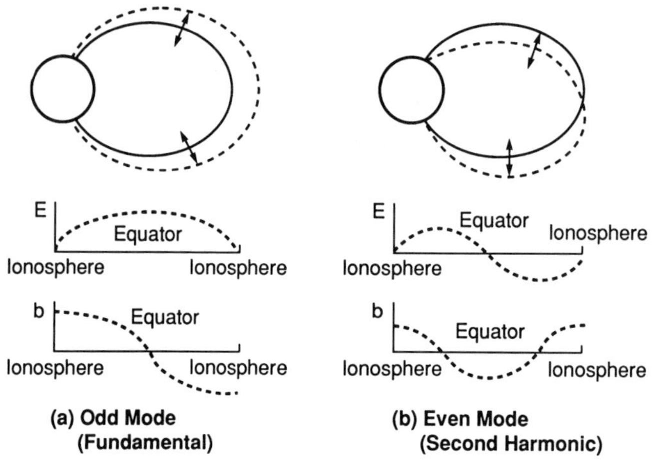

FIG. 11.5. Standing oscillations in a dipole field. Top: Schematic illustrations of the field displacements in a fundamental and second harmonic of the field-line resonances. Dashed lines are the displaced field lines. Bottom: Plots of the perturbation electric and magnetic fields versus distance along the field line from one ionosphere to the other.

(a) Odd Mode (Fundamental)

(b) Even Mode (Second Harmonic)

where n is an integer. Recalling that for a shear Alfvén wave with $k_\| = k_\| \cos \theta = 2\pi/\lambda_\|$ representing the component of **k** along the background field,

$$\omega = v_A k_\| = v_A 2\pi/\lambda_\|$$

it follows that the allowed frequencies of these waves standing on field lines are

$$f = nv_A/(2l) = nB/(2l\sqrt{\mu_0 \rho}) \qquad (11.21)$$

Thus, only certain resonant frequencies can be established. These frequencies are controlled by the length of the field lines between the ionospheres, the strength of the magnetic field, and the plasma density. If the field geometry is known, it is possible to infer the plasma density by measuring the frequencies of shear Alfvén waves present in a magnetic cavity bounded by the northern and southern ionospheres. This is just the point that Dungey (1954a,b) made in his early papers on waves in the magnetosphere. Figures 11.4c and 11.4d show how field and plasma might be deformed if standing Alfvén waves or compressional waves, respectively, perturbed the magnetosphere. For the former, the density remains constant. For the latter, the density changes as the flux-tube volume changes. The structures of the wave perturbations along the magnetic-field line are illustrated for the two lowest harmonics in Figure 11.5. The upper part of the diagram illustrates the displacement of the flux tube in a dipolar field for the fundamental ($n=1$) and second harmonic ($n=2$) of the standing waves. The lower diagrams show how the electric (**E**) and magnetic (**b**) perturbations vary with distance along the background field. Even and odd modes are identified by the symmetry of the transverse magnetic perturbations about the equator, where $\mathbf{E}=0$, $\mathbf{u}_\perp=0$, and field lines do not move.

11.7 MHD WAVES IN A DIPOLAR MAGNETIC FIELD

The foregoing discussion has concentrated on waves in a uniform background magnetic field. A model of the background magnetic field that is slightly more complicated, but considerably more realistic for a planetary magnetosphere, is a dipole field. For a cold plasma in a dipole field, the MHD waves are very similar to those discussed for a uniform plasma.

Consider first a perturbation that compresses the system at the last field line on the right in the model illustrated in Figure 11.4a. The motion may not be uniform along the field, and so the boundary field lines will bend and will move closer to the shell of field lines just inside the boundary, thereby increasing the magnetic pressure. The pressure perturbation propagates into the system, producing changes in the components of the field in the $\hat{\nu}$ and $\hat{s}$ directions (the coordinate system is shown in Figure 11.6). This type of perturbation can be identified as a fast-mode wave.

A perturbation that sets an entire shell of plasma into azimuthal motion creates a wave perturbation in the $\hat{\varphi}$ direction (Figure 11.6) that bends the field without changing its magnitude. Such a wave is a shear Alfvén wave.

Under most circumstances, the two wave modes are coupled, meaning that it may not be possible to set up a compressional wave without setting up a shear Alfvén wave somewhere in the system. If the compressional wave is monochromatic (i.e., has a single frequency, say f_{fast}), the coupling will be strongest on a field line for which f_{fast} is a resonant frequency, which means that it matches the frequency of a shear Alfvén wave that can stand on that field line. This is not unexpected, because any oscillating system responds strongly to a driving force that contains a signal at its natural or resonant frequencies. Both the Alfvén wave and the driving compressional wave have local maxima at the resonant field line.

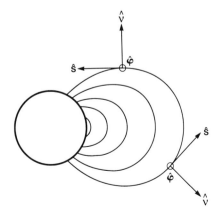

FIG. 11.6. Illustration of the unit vectors in a local dipole coordinate system at different latitudes.

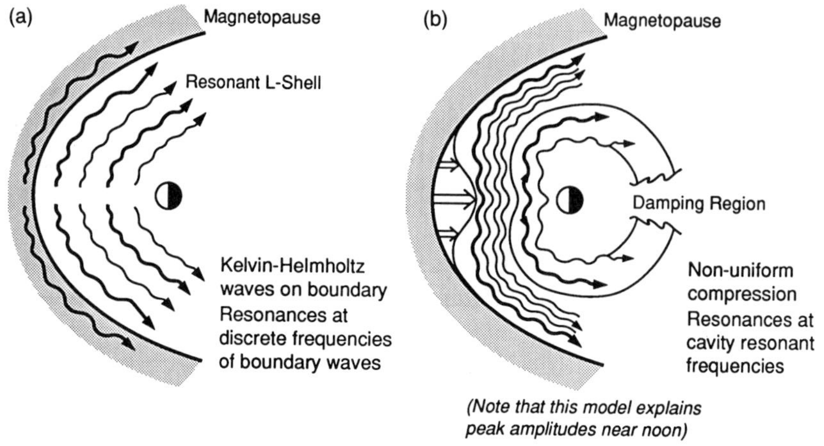

FIG. 11.7. Representations of wave perturbations throughout the dayside magnetosphere produced by (a) Kelvin-Helmholtz waves on the surface and (b) a compression of the nose of the magnetosphere.

(a) Magnetopause; Resonant L-Shell; Kelvin-Helmholtz waves on boundary Resonances at discrete frequencies of boundary waves

(b) Magnetopause; Damping Region; Non-uniform compression Resonances at cavity resonant frequencies

(Note that this model explains peak amplitudes near noon)

Theories have been developed to describe how wave disturbances at the magnetopause boundary pump energy into the magnetospheric cavity and deposit it near magnetic shells where the conditions for the transverse resonances are satisfied (Southwood, 1974; Chen and Hasegawa, 1974; Kivelson and Southwood, 1986). A schematic illustration of this process, the field-line resonance theory of magnetospheric ULF waves, is shown in Figure 11.7a, where the waves are shown as wiggly lines moving away from local noon. The line thickness represents wave amplitude, which decreases inward but peaks locally at the resonant L shell. Eddy motions are induced within the magnetosphere by wave perturbations. The eddy flows reverse sense across amplitude extrema. Wave magnetic perturbations are proportional to flow perturbations, and this means that wave polarization also varies with location in the equatorial plane of the magnetosphere. As the wave must carry energy across the magnetic field, the wave mode coupling the boundary to the resonant field lines must be a compressional mode. The model assumes that the waves on the magnetopause are surface waves whose amplitude decays away from the surface. The polarization patterns shown can be mapped down to the ionosphere along magnetic-field lines. For Pc-5 waves (periods of 2.5–10 min), distributions of wave polarizations consistent with this model have been reported from ground observations (Samson and Rostoker, 1972). ULF waves have also been investigated by spacecraft using instrumentation to measure electric and magnetic fields and plasma flow velocities, and the theoretical picture of resonant field lines given here has received ample confirmation (Perraut et al., 1978; Takahashi and McPherron, 1982; Takahashi, McPherron, and Hughes, 1984).

Recent work has focused on the response of the magnetosphere to impulsive perturbations on the boundary, such as those produced when the solar-wind dynamic pressure incident on the magnetopause changes abruptly. In connection with the response of the magnetosphere to an

11.7 MHD WAVES IN A DIPOLAR MAGNETIC FIELD

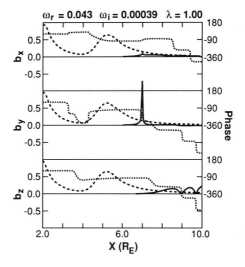

FIG. 11.8. The amplitude (solid traces) and phase (long dashes) of a global-mode wave versus equatorial distance in a box model of the magnetosphere with a spatially varying Alfvén speed (short dashes); z is field-aligned; x is the direction of the gradient of the Alfvén velocity. (From Zhu and Kivelson, 1989.)

impulsive source, the idea of the magnetosphere as a resonant cavity has recently been receiving renewed attention (Figure 11.7b). If the near-equatorial ionosphere and the magnetopause can serve to reflect signals propagating across the field, much as the northern- and southern-hemisphere ionospheres serve to confine signals propagating along the field, the compressional wave frequencies will be quantized just as the shear Alfvén frequencies are quantized. Figure 11.8 shows characteristic wave amplitudes that would be established following an interval of transient impulsive disturbance in a box model of a magnetosphere with reflecting boundaries at the equatorial ionosphere and the magnetopause. The amplitude of the field-aligned component b_z, oscillates within an envelope of decreasing field strength. The b_y component (azimuthal in a realistic magnetosphere) is vanishingly small except in the immediate vicinity of the resonant field lines, at which the compressional and transverse frequencies match. The b_x component (radial in a realistic magnetosphere) shares qualitative features of the structures of b_y and b_z, with smaller amplitude.

The transient response that develops immediately following the impulsive disturbance of the boundary is itself of interest. Figure 11.9 shows an example of the types of signals that can be observed at the ground immediately following an impulsive disturbance in the solar wind. A sudden increase of solar-wind density, evident in the left-hand panels, sets up waves within the magnetosphere. A chain of ground stations recorded waves that were most intense and long-lasting near the middle of the chain. MHD wave theory for the magnetospheric cavity can provide an interpretation of these observations. The right-hand panel of Figure 11.9 shows schematically that cavity resonances produce waves of different characters on different field lines, with peak power and long duration at intermediate latitudes.

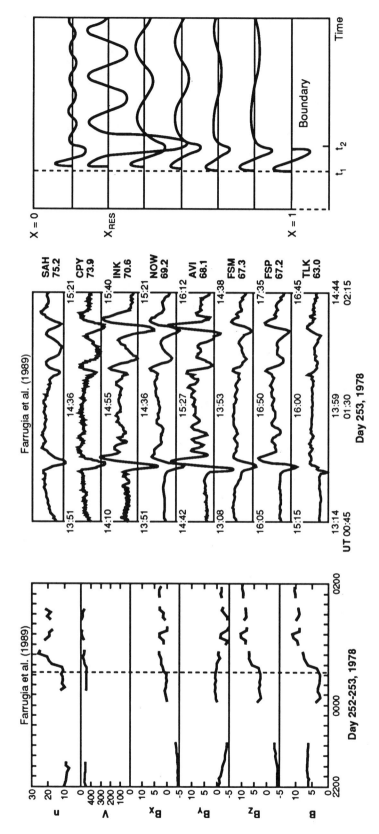

FIG. 11.9. An example of waves set up in the magnetosphere following an increase in solar-wind dynamic pressure. Left: Solar-wind data, with the dashed line indicating the time of the jump in dynamic pressure. Middle: Waves observed along a latitudinal chain of ground magnetometers. Right: Predicted waves from a cavity model of the magnetosphere. (Data from Farrugia et al., 1989.)

11.8 SOURCES OF WAVE ENERGY

In describing the properties of ULF waves, we have not addressed the questions of wave generation. Diverse processes can excite waves, and several different driving mechanisms are important in generating the waves observed in the magnetosphere or the solar wind. Any process that modifies the equilibrium of the plasma and the field can serve as an energy source for waves. As for most magnetospheric phenomena, the energy comes principally from the solar wind, but other energy sources in the ionosphere or internal to the magnetosphere can be important.

The departure from equilibrium that drives the waves is often related to large-scale convective flows. In particular, the shear in the flow across the magnetopause can produce surface waves, of the sort mentioned in the preceding section, through what is called the Kelvin-Helmholtz instability. The process is closely related to the one that produces waves on the surface of a lake when a strong wind is blowing. These surface waves compress the magnetosphere, and the perturbations generate compressional waves that decay or propagate across field lines. Field-line resonances, described in the preceding section, couple the energy of compressional waves into shear Alfvén waves. Other compressional perturbations of the magnetopause can serve as sources of wave energy. Examples are the displacements of the magnetopause that occur when a solar-wind shock passes by, or those that are produced by time-varying dayside reconnection. Waves in the solar wind can be convected through the bow shock and under certain circumstances can introduce wave power into the magnetosphere.

Steady convective flows need not generate waves, but time-varying flows typically drive MHD waves. The time-varying convective flows can be generated within the magnetosphere (e.g., at the time of substorm onset), or they can be generated in the ionosphere (e.g., in regions locally heated by precipitation of energetic particles). In either case, when the motion of one end of a flux tube changes, waves grow and bounce back and forth along the flux tube until the entire flux tube begins to move as a whole.

Convection is not the only source of wave energy. Waves can grow when the velocity-space distribution of the plasma is not in an equilibrium configuration, either because it is anisotropic or because the particle energy distribution is anomalous. Unstable velocity-space distributions develop in the ring-current region when particles are injected by enhanced convection during substorms and storms. Then it is possible to find groups of particles that are in resonance with the waves; particles that can bounce and drift in phase with ULF waves, in some cases causing the wave power to grow. An example of particles in resonance with waves nominally in the Pc-3 to Pc-5 band is shown in Figure 11.10, which again uses the box model. Electric-field intensity is shown by the

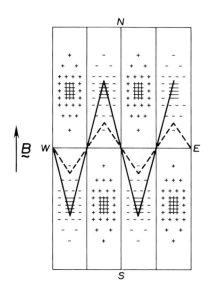

FIG. 11.10. Schematic of two bouncing particles with different equatorial pitch angles and therefore different mirror fields drifting relative to a ULF wave that stands between the northern and southern ionospheres in a box model of the magnetosphere. The plus and minus signs represent the sign of a wave electric field, and their density indicates the amplitude of that field. The particle along the solid trajectory will experience greater acceleration or deceleration as it spends time in field regions of very strong perturbation.

density of symbols. The loci of the guiding centers of bouncing, drifting particles are shown as diagonal lines. The orbits have been chosen so that the particles drift through exactly one wavelength in each full bounce; they are resonant with the wave. The lines remain in regions of negative E and the ions on these paths will lose energy. Ions on the dashed-line orbit will gain energy, as they always experience $E > 0$. Nonresonant particles will move through both positive and negative E and therefore will not change energy, as viewed over multiple bounces.

The higher-frequency wave classes of Table 11.1 (Pc-1, some Pc-2 waves, and Pi-1) arise from a local interaction with ions in motion along the field. Particles can resonate with higher-frequency waves that, in the particle rest frame, match their gyrofrequency. This means that the particle gyrofrequency must equal the wave frequency Doppler-shifted to account for the velocity of the particle's motion along the field. Waves produced in this way are also found upstream of the bow shock, where ions that have been reflected back upstream from the bow shock produce ion-cyclotron instabilities in the solar wind. Resonances that fall into the Pc-1 and Pc-2 classes are ion-cyclotron resonances, which will be discussed in connection with other "kinetic" wave processes in Chapter 12.

11.9 INSTABILITIES

In Section 11.3, we linearized the equations that govern the plasma and fields, thereby obtaining wave solutions whose average amplitudes are constant in time. This implies that even over many wave cycles, the plasma neither loses energy to the waves nor gains energy from the waves. If, on the other hand, energy and momentum are transferred, either from the waves to the plasma or from the plasma to the waves, the average wave amplitude will change with time.

Wave growth requires a source of free energy in the plasma. Plasma conditions that lead to nonlinear growth are referred to as *instabilities*. The plasma conditions that can lead to wave growth include *beams*, in which directed particle fluxes are superimposed on a plasma at rest, *anisotropic distributions* of particle pitch angles, and *nonequilibrium spatial distributions* of plasma. Thus, the departures from equilibrium that can lead to wave growth can be present either in the phase-space or in the configuration-space distribution.

If waves described by the time dependence of equation (11.12) are to grow, ω must have a positive imaginary part (i.e., $\omega = \omega_0 + i\gamma$, where both ω_0 and γ are real, and $\gamma > 0$). Equation (11.12) shows that for positive γ, the amplitude grows exponentially with time. Therefore, γ is called the *growth rate*. Notice that if γ is negative, the wave decays.

An exponentially growing wave satisfies the linear approximation only for times short compared with $1/\gamma$; at longer times, the waves become *nonlinear*, and the mathematical formulation that we have presented in this chapter is no longer applicable.

Some of the instabilities that can develop in a uniform background magnetic field are closely related to the linear waves introduced in this chapter. We shall identify only one example, the *mirror instability*. This instability of an anisotropic plasma requires that the perpendicular plasma pressure exceed the parallel pressure. The condition for wave growth of the mirror instability is

$$1 + \beta_\perp (1 - \beta_\perp/\beta_\parallel) < 0$$

where $\beta_\perp$ ($\beta_\parallel$) is the ratio of $p_\perp$ ($p_\parallel$) to the magnetic pressure. If this inequality is satisfied, the uniform field develops bubbles of low field strength separated by regions of enhanced field strength. Where the field is stronger than the unperturbed field, the particle mirror points shift in such a way that the plasma density decreases. Where the field is weaker than the unperturbed field, the plasma density increases. This field configuration reduces the anisotropy of the plasma and lowers the energy of the system.

The mirror instability is a purely growing wave with $\omega_0 = 0$. Notice that the phase relations between plasma and magnetic pressure are the same as in the slow mode, but this is a nonpropagating wave.

Wave instabilities make an important contribution to the configuration and transport properties of a magnetosphere. For example, changes in the curvature of the field can be produced by instabilities referred to as *ballooning* and *firehose* instabilities. Plasma stirring or even steady transport can, under appropriate circumstances, be produced by the *interchange instability*. In all cases, the perturbations grow because the plasma distribution is not in a minimum-energy state, and the nonlinear waves act to bring it closer to a minimum-energy configuration.

11.10 WAVES IN PLANETARY MAGNETOSPHERES AND ELSEWHERE

Although the discussion has focused on MHD waves present in the terrestrial magnetosphere, such waves are present wherever magnetized plasmas are subject to forces that introduce perturbations on appropriately long time scales (i.e., long with respect to the ion gyroperiod). Time-varying patterns of magnetic structure or of plasma flow are not normally imposed over the entire system simultaneously. Thus, non-equilibrium pressure gradients or flow patterns develop, and ULF waves can act to restore equilibrium. For example, if plasma is set into motion on one part of a magnetic-flux tube, the plasma elsewhere on the flux tube must respond to the changes. This requires signals to propagate along the flux tube. Signals that carry field-aligned current from one part of the flux tube (such as the equatorial magnetosphere or the solar corona) to another (say the ionosphere or the solar photosphere) are essential for getting the entire flux tube to move as an entity. Such signals must be carried by shear Alfvén waves.

MHD waves are observed in the solar wind; special forms of such waves are observed upstream of planetary bow shocks. The characteristic wave periods change linearly with the magnitude of the solar-wind magnetic field. The solar-wind flow convects these waves toward the magnetosphere, and they introduce wave power into the magnetospheric cavity. By monitoring the power in Pc-3 and Pc-4 waves (periods from tens of seconds to minutes) on the ground, the magnitude of the interplanetary magnetic field can be estimated.

The study of MHD waves in planetary magnetospheres is not yet complete, but one example will serve to illustrate the value of studying them. Figure 11.11 presents *Voyager 2* data from Jupiter. Plotted are the perturbations of magnetic and particle pressures measured by the spacecraft instruments. The fluctuations are proportional to the fluctuations of the field-aligned component of the magnetic field. As the particle detector is not able to determine the mass of the ions that are detected, there is an uncertainty in the particle pressure. In this plot it has been assumed that the ions are protons. However, heavy ions like sulfur and oxygen are also known to be present in the Jovian magnetosphere. If the pressure were calculated assuming singly charged oxygen ions, the particle-pressure fluctuations would be larger by a factor of 10. Notice the anticorrelated changes between the two perturbation pressures. Recalling that the slow mode maintains the total pressure approximately constant and therefore corresponds to perturbations with anticorrelated particle and field pressures, it is natural to expect that these fluctuations represent a slow-mode type of disturbance. Yet, although these two fluctuating pressures are strongly anticorrelated, their sum is not constant. The amplitude of the particle-pressure fluctuations is about one-

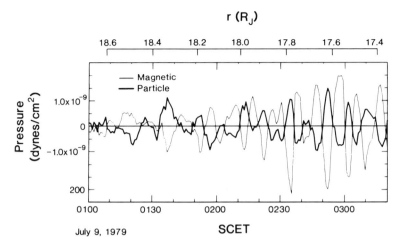

FIG. 11.11. Example of MHD waves observed in the Jovian magnetosphere by Khurana and Kivelson (1989). The particle and field pressures are in antiphase, as in slow-mode or mirror-mode waves.

third that of the magnetic-pressure fluctuations. However, if the pressure were recalculated assuming that the plasma contains about 30 percent singly ionized oxygen ions, the total pressure fluctuations would become negligibly small. That possibly was the best available method for determining the composition of the intermediate-energy plasma population in the Jovian magnetosphere with measurements available from the *Voyager 1* and *2* spacecraft, but new data from the *Ulysses* flyby (February 1992) and ultimately from the *Galileo* spacecraft (1995) will provide direct composition measurements. Then we shall be able to evaluate the accuracy of our estimates based on an understanding of the properties of ULF waves.

ADDITIONAL READING

Below are listed some good review articles on ULF waves:

Southwood, D. J., and W. J. Hughes. 1983. Theory of hydromagnetic waves in the magnetosphere. *Space Sci. Rev.* 35:301.

Hughes, W. J. 1983. Hydromagnetic waves in the magnetosphere. In *Solar Terrestrial Physics*, ed. R. L. Carovillano and J. M. Forbes (p. 453). Dordrecht: Reidel.

Pilipenko, V. A. 1990. ULF waves on the ground and in space. *J. Atmos. Terr. Phys.* 52:1193.

Samson, J. C. 1991. Geomagnetic pulsations and plasma waves in the Earth's magnetosphere. In *Geomagnetism*, vol. 4 (p. 481). New York: Academic Press.

PROBLEMS

11.1. In regions of low plasma β in a dipole field, which represents much of the dayside magnetosphere, the cold-plasma approximation is appropriate.

(a) Use your knowledge of the properties of a dipole magnetic field ($B_{eq} \propto L^{-3}$, length of field line proportional to L, volume of flux tube proportional to L^4, equation of field line $r = LR_E\cos^2\lambda$) to explain why the fundamental excitations of field lines at large L occur at lower frequencies than do the fundamental excitations of field lines at small L. Assume that the density is uniform (1 electron per 1 cm³) throughout the dipolar region of the magnetosphere and that at 6.6 R_E the fundamental frequency is 14 mHz. Make a rough plot showing how the fundamental frequency varies with L.

(b) Actually, the magnetospheric-plasma density often varies inversely with the flux-tube volume over large parts of the outer magnetosphere. Make a rough plot of the fundamental frequency of field-line excitations normalized to 14 mHz at $6.6R_E$ in a dipole field assuming this type of variation for the density.

(c) Although the density variation used in part (b) is a good approximation, the magnetospheric density actually drops by a factor of 100 or more across the plasmapause. Allow for a plasmapause at $L = 4$, and assume that the density jumps by 100 inside the plasmapause. Again provide a rough plot of the fundamental frequency versus L.

(d) Where on the surface of the earth would you expect to find pulsations of 50 mHz for the assumed conditions of part (c)?

11.2. Suppose that a standing Alfvén wave is established on a field line at $L = 5$, where the magnetosphere is approximately cylindrically symmetric. The ambient particle population is taken to include both energetic and cold plasma. Near the equator, the density is ρ in kilograms per cubic meter. Locally near the equator the uniform-field approximation is valid; the ambient field is B_0/L^3 and is oriented along the z-direction. The standing wave is a superposition of waves with **k** parallel and antiparallel to $\hat{z}$.

(a) Assume that the magnetic perturbation **b** is radial. Determine the wave electric-field and the fluid-velocity perturbations as functions of **b**, ρ, and L. Pay attention to the vector character of the perturbations to determine their directions. Identify the direction of plasma displacement.

(b) The wave oscillations displace the plasma. The rate of displacement is slow enough that the plasma responds adiabatically. Show why this is true, using nominal dipole field values.

(c) Explain why you must consider the variations of particle flux with both L and W (particle energy) if you wish to determine how the particle flux measured at a spacecraft is modulated by a wave.

(d) Assume that only cold electrons and ions are present ($W \approx 0$). Show that the magnitude of the density variations takes the form

$$\delta n = \frac{b(\partial n/\partial L)}{R_E \omega \sqrt{\mu_0 \rho}}$$

11.3

(a) Plasma boundaries in the magnetosphere sometimes are described as standing fronts that can be thought of as waves propagating against a flow. Consider the high-latitude magnetopause in the noon–midnight meridian of the magnetotail for a strictly southward-oriented interplanetary magnetic field. Sketch the change of the field across the boundary. Carefully consider the changes that must occur across the boundary. What MHD wave mode produces these changes?

(b) Waves are detected in the solar wind as well as in the magnetosphere. Assume that the solar-wind magnetic field is oriented along the spiral angle. Wave perturbations with magnetic polarization perpendicular to the ecliptic plane are observed. What MHD wave mode is relevant to such perturbations?

(c) Waves generated at the magnetopause can be observed behind the earth's bow shock. Only one of the MHD wave modes can travel from the magnetopause to the nose of the bow shock. Which wave mode is it, and how do you reach this conclusion?

12 PLASMA WAVES

C. K. Goertz and R. J. Strangeway

12.1 INTRODUCTION

WAVES ARE IMPORTANT because they propagate energy from one point to another. Some waves transmit information out of a plasma, so that an observer can deduce what is going on inside the plasma. Waves can grow or decay and can either take energy out of a plasma or increase its energy. Because a plasma consists of at least two fluids, the number of possible waves is larger than in a normal fluid, where sound waves are the only waves possible. In this chapter we shall discuss a number of simple plasma waves whose properties can be deduced from the plasma fluid equations.

12.2 WAVES IN A TWO-FLUID PLASMA

Waves in a plasma are generally electromagnetic in nature, and as such the wave electric and magnetic fields are governed by Maxwell's equations:

$$\nabla \cdot \mathbf{E}(\mathbf{x},t) = \rho_q(\mathbf{x},t)/\epsilon_0 \tag{12.1}$$

$$\nabla \cdot \mathbf{B}(\mathbf{x},t) = 0 \tag{12.2}$$

$$\nabla \times \mathbf{E}(\mathbf{x},t) = -\frac{\partial \mathbf{B}(\mathbf{x},t)}{\partial t} \tag{12.3}$$

$$\nabla \times \mathbf{B}(\mathbf{x},t) = \mu_0 \mathbf{j}(\mathbf{x},t) + \mu_0 \epsilon_0 \frac{\partial}{\partial t} \mathbf{E}(\mathbf{x},t) \tag{12.4}$$

These equations were introduced in Chapter 2, where they were identified as Poisson's equation (2.34), the divergence-free condition on **B** (2.37), Faraday's law (2.35), and Ampère's law (2.36a).

The charge density and current density in (12.1) and (12.4) are the terms that relate the wave fields to the response of the plasma. Before discussing how we calculate the current and charge densities, we should point out that there is some redundancy in the set of equations. As noted in Chapter 2, equation (12.2) does not provide any additional information beyond (12.3). Similarly, the time derivative of (12.1) and the divergence of (12.4) can be added to give $\partial \rho/\partial t + \nabla \cdot \mathbf{j} = 0$, which is simply the requirement that charge be conserved. Consequently, we usually shall not use equations (12.1) and (12.2), although in certain circumstances (when the wave is electrostatic, as we discuss later) Poisson's equation can be used instead of Ampère's law. There are six equations in (12.3) and (12.4), but nine unknowns (three electric-field, three magnetic field, and three current components). If the current density were solely dependent on the wave fields, then that relationship would provide an additional three equations, and we would have a complete set of equations for all the unknowns. A simple example is where the current is related to the wave electric fields through a scalar conductivity. In general, the conductivity is a tensor that can depend on many parameters of the plasma. In principle, it is the conductivity tensor that allows us to characterize the different types of waves supported by a plasma, although in practice we can determine the plasma response without explicitly deriving the conductivity tensor.

In order to determine the current density, we must specify how the wave fields affect the plasma. When the plasma is sufficiently cold that the thermal velocities of all species in the plasma can be neglected, the current density can be derived from the Lorentz-force law. Although this law specifies how an electric field and a magnetic field will affect the motion of a single particle, in a cold plasma all particles of a species at a particular position and time will respond in exactly the same way, and we can derive macrophysical quantities, such as the current density, simply by summing over all the particles in a volume element. At the other extreme, we can use a kinetic treatment to specify how a distribution of particles responds to a wave. In this case we use the Vlasov (or collisionless Boltzmann) equation to specify how a distribution function of particles is affected by the wave fields. In between these two extremes there are other types of equations that specify how a plasma will respond to waves. For example, as discussed in Chapter 11, we can use magnetohydrodynamics (MHD), which is derived by taking moments of the Vlasov equation, or we can use the equations for a two-fluid plasma, as we do here. The advantage of these latter two approaches is the relative simplicity; the fluid equations have three spatial dimensions and time, rather than the seven-dimensional phase space of Vlasov theory. The disadvantage is that some important effects, such as Landau damping, which is caused by a resonance with particles moving at the phase speed of the wave, cannot be obtained from a fluid equation.

For a two-fluid plasma, where the subscript s identifies electrons and ions, the relevant equations are

$$\frac{\partial}{\partial t} n_s + \nabla \cdot (n_s \mathbf{u}_s) = 0 \tag{12.5}$$

$$\frac{\partial}{\partial t}\mathbf{u}_s + \mathbf{u}_s \cdot \nabla \mathbf{u}_s - \frac{q_s}{m_s}(\mathbf{E} + \mathbf{u}_s \times \mathbf{B}) + \frac{\nabla p_s}{n_s m_s} = \frac{\mathbf{F}'}{n_s m_s} \tag{12.6}$$

$$\mathbf{j} = \sum_s n_s q_s \mathbf{u}_s \tag{12.7}$$

$$\rho_q = \sum_s q_s n_s \tag{12.8}$$

In these equations, all variables depend on **x** and t. These equations were introduced in Chapter 2 as the continuity equation (2.29a) and the momentum equation (2.31b), to which have been added the definitions of the current density and the charge density. The contributions of sources $S(\mathbf{x}, t)$ and losses $L(\mathbf{x}, t)$ are small or zero in many space-physics applications, and so they have been dropped. In the ionosphere, however, they can become important. Here, $\mathbf{F}'$ represents the forces per unit volume, other than pressure gradients and electromagnetic forces, acting on a fluid element at position **x** and time t. Examples include gravity, friction with neutrals, viscous forces, and so forth. In many space-physics applications, they can be neglected.

Equations (12.5), (12.6), and (12.7) constitute a set of 11 equations. However, in addition to the wave electric field and magnetic fields, there are two densities, two pressures, and two velocities in (12.5) and (12.6), which together with the current density in (12.7) give 13 unknowns. The complete set of equations (12.3)–(12.8) consists of 18 equations with 20 unknowns, remembering that (12.1) and (12.2) can be derived from the other equations. Thus, as for the single-fluid equations discussed in Chapter 2, the system cannot be solved as it stands. To provide a complete description of the plasma, we must add two equations. These equations relate the fluid pressure p_s to the other fluid variables n_s and $\mathbf{u}_s$. We could add an energy equation, but only at the expense of adding another variable, the heat flux. This can be related to still higher-order terms, and so forth. A commonly used method for closing the set of fluid equations is a prescription for the fluid pressure in terms of a polytropic law:

$$p_s = \text{constant} \times (n_s)^{\gamma_s} = n_s T_s \tag{12.9}$$

where γ_s is the "ratio of specific heats." For a three-dimensional adiabatic change in pressure, $\gamma_s = \frac{5}{3}$. For an isothermal change (constant temperature), $\gamma_s = 1$. For an isobaric change (constant pressure), $\gamma_s = 0$. T_s is the average thermal energy of a plasma particle, which is related to the temperature by $T_s = k_B T$, where k_B is the Boltzmann constant.

There are, of course, many nontrivial solutions to these equations. In this chapter we shall deal with wave solutions to the plasma equations. Consider an initially uniform system subjected to periodic perturbations that vary spatially with a wave number k or periodically in time with a frequency ω. For example, we can put an antenna in a plasma and charge it, whereupon the plasma will no longer be charge-neutral, and an electric field will appear. This will cause the electrons and ions to move. The motion will tend to neutralize the charge (i.e., ions will move toward the electron-density enhancement, and electrons away from it).

12.2 WAVES IN A TWO-FLUID PLASMA

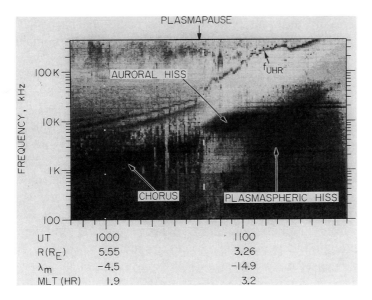

FIG. 12.1. Example of a spectrogram of electric-field perturbations from the *ISEE 1* spacecraft on a near-equatorial pass through the inner magnetosphere of the earth. Shading indicates signal intensity (dark is high, light is low) as a function of frequency (ordinate) and time (abscissa). The latter axis is labeled with universal time (UT) in hours, geocentric radial distance, magnetic latitude (degrees), and magnetic local time (MLT); f_{UHR} is the upper hybrid resonance frequency given in equation (12.69). Other characteristic wave emissions are labeled. Kurth and Gurnett (1991) have presented an overview of magnetospheric plasma waves. (Courtesy of D. A. Gurnett.)

Will the plasma respond by smoothly returning to the initially unperturbed (charge-neutral) case? Will it oscillate (i.e., will the charge density vary periodically in time)? Will the perturbation propagate away from the point where it was introduced? This situation is analogous to a quiet pond perturbed by a stone thrown into it. Where the stone hits the water, the water is momentarily displaced. We know that this displacement propagates away from the point of impact as a wave. Similar phenomena occur in a plasma, and we would like to predict the properties of the waves generated.

The waves in a plasma produce perturbations in the electric and magnetic fields that can be measured by detectors (e.g., dipole antennas or search coils) on board spacecraft or on the ground. Measurements of the electric and magnetic fields as functions of time can be processed to give Fourier spectra. Often the results are displayed in the form of dynamic spectrograms (plots versus frequency and time), of which Figure 12.1 is an example. Such spectrograms are generated separately for electric and magnetic fluctuations. The gray scale (or, in many cases, color scale) indicates the power at a certain frequency. At times the power is distributed fairly uniformly over a broad range of frequencies; at other times the spectrum peaks at certain frequencies. For quantitative work, the power spectra that characterize the waves at one particular time can also be useful. Such spectra can be thought of as vertical cuts through Figure 12.1. From these spectra, important properties of the plasma, such as its density and chemical composition, can be inferred. In some frequency ranges, only electric fields have detectable amplitudes. This is an indication that the waves are electrostatic. From the ratio of the electric-field and magnetic-field amplitudes, the wave propagation speed can be determined; the polarization of the fields yields information about the direction of the wave propagation.

Many different types of waves occur in space plasmas. They are named according to different schemes, and the nomenclature is often confusing. Waves may be named by the region in which they appear. Examples include auroral hiss, plasmaspheric hiss, and auroral kilometric radiation. They may also be named according to the frequency at which they are observed when the frequency is close to a natural frequency of the plasma, such as lower hybrid waves, ion-cyclotron waves, electron-cyclotron waves, and upper hybrid resonance noise. Some waves are called electrostatic to indicate that they carry no magnetic-field perturbation. An example is the electrostatic ion-cyclotron wave. Radio waves, which are electromagnetic plasma waves that can escape from a plasma, are often named according to their wavelengths (e.g., auroral kilometric radiation, Jovian decametric radiation, and decimetric radiation). Finally, some waves are named by the way they sound when processed as audio signals and played through a speaker: whistlers, chorus, lion roar, hiss, and so forth.

12.3 WAVES IN AN UNMAGNETIZED PLASMA

Although space plasmas are almost always magnetized, some plasmas do not have a magnetic field in them. Even when they do, we can quite often neglect the magnetic field in discussing plasma responses (the conditions will become clear later). The reduced complexity of the governing equations can be further simplified by approximations.

12.3.1 Langmuir Waves

We know that ions are much more massive than electrons ($m_i/m_e \geq$ 1,836). As a first approximation, let us assume that the ions are infinitely massive (i.e., they don't move at all). They are uniformly distributed in space, with a fixed density n_0. We shall also neglect all magnetic fields [$\mathbf{B}(\mathbf{x}, t) = 0$]. Then we are left with a set of only four equations: the continuity equation and the force equation for the electrons, Poisson's equation, and equation (12.8) for the charge density. We make another simplifying assumption, namely, that any variation occurs only in one dimension (the $\hat{\mathbf{x}}$ direction). This may seem unrealistic, but there are many cases where one expects such a situation. For example, the solar wind varies mainly in the radial direction away from the sun. Sometimes we have beams of electrons or ions propagating in a fixed direction through a plasma. Waves also propagate in a fixed direction in a uniform plasma, and they cause the plasma properties to vary only along the direction of propagation.

With these simplifications, we are left with three equations [using equation (12.8) for the right-hand side of equation (12.1)]:

$$\frac{\partial}{\partial t}n_e + \frac{\partial}{\partial x}(n_e u_e) = 0 \tag{12.10}$$

$$m_e n_e \left(\frac{\partial}{\partial t}u_e + u_e \frac{\partial u_e}{\partial x}\right) = -\frac{\partial p_e}{\partial x} - e n_e E_x \tag{12.11}$$

$$\frac{\partial E_x}{\partial x} = e(n_0 - n_e)/\epsilon_0 \tag{12.12}$$

This simple-looking set is still complex, because the equations are not linear in the variables. For example, in equation (12.10), the product of two variables (n_e and u_e) appears. As surprising as it may seem, we cannot solve these equations exactly, except for very special cases. For example, $n_e = n_0$, $u_e = 0$, and $p_e =$ constant, and $E_x = 0$ would be a solution, although a rather trivial one. This solution corresponds to a stationary electron fluid of uniform density (the quiet pond).

What happens if we perturb this state a little? For example, we could imagine that the electron density is not everywhere equal to n_0, but differs from it by a small amount, $n_1(x, t)$; that is,

$$n_e(x, t) = n_0 + n_1(x, t) \tag{12.13}$$

Clearly, in this case the electric field cannot be zero everywhere, because the right-hand side of equation (12.12) is nonzero:

$$\frac{\partial E_x}{\partial x} = -e n_1(x, t)/\epsilon_0 \tag{12.14}$$

Let us denote this field as E_1 to indicate that it is related to the small perturbation $n_1(x, t)$. We also see from equation (12.11) that the electron fluid will begin to move when the electric field is nonzero. Thus, we write

$$E_x = E_1(x, t) \tag{12.15}$$
$$u_e = u_1(x, t)$$

assuming that the perturbation flow speed is also small. We have assumed here that in the unperturbed plasma, the electrons are at rest ($u_0 = 0$). The case where this is no longer true will be treated later. All quantities with a subscript 1 are small quantities. Their products are even smaller, and we can safely neglect them. This procedure is called *linearization*. Linearizing all three equations (12.10)–(12.12) yields the following set:

$$\frac{\partial n_1}{\partial t} + n_0 \frac{\partial u_1}{\partial x} = 0 \tag{12.16}$$

$$m_e n_0 \frac{\partial u_1}{\partial t} = -e n_0 E_1 - \frac{\partial p_1}{\partial x} \tag{12.17}$$

$$\frac{\partial E_1}{\partial x} = -e n_1/\epsilon_0 \tag{12.18}$$

where we have used $\partial \rho_0/\partial x = 0$. These equations are easy to solve because they are all linear (no products of the variables occur). Consider first the special case of cold electrons ($p = 0$). We eliminate u_1 and E_1 from the set of equations by taking the time derivative of (12.16) and inserting the spatial derivatives of (12.17) and (12.18) to obtain an equation

$$\frac{\partial^2 n_1}{\partial t^2} + \left(\frac{n_0 e^2}{\epsilon_0 m_e}\right) n_1 = 0 \tag{12.19}$$

This is a harmonic-oscillator equation whose frequency is $\omega = \pm \omega_{pe}$ the *electron plasma frequency* [see equation (2.28)]. That such oscillation should occur is easy to understand. To produce an initial perturbation n_1, we must displace some electrons and hence create a positive-charge density at the position where they started. This positive-charge perturbation attracts the electrons, which will tend to move back to their original position, but will overshoot it, come back, overshoot it, and so on. Without any damping, the energy put into the plasma to create the perturbation will remain in the plasma, and the oscillation will continue forever.

Another way (other than elimination of variables) to obtain a solution is to guess it. We expect an initial perturbation to propagate through the plasma as a wave. Thus, a wave solution of the form

$$E_1(x, t) = \tilde{E}_1 \exp(-i\omega t + ikx) \tag{12.20}$$

$$n_1(x, t) = \tilde{n}_1 \exp(-i\omega t + ikx)$$

$$u_1(x, t) = \tilde{u}_1 \exp(-i\omega t + ikx)$$

should satisfy the equations, as discussed in Chapter 11, equation (11.12). The quantities with tildes are the complex amplitudes. The (complex) amplitudes obey the algebraic equations

$$-i\omega \tilde{n}_1 + ikn_0 \tilde{u}_1 = 0 \tag{12.21}$$

$$-i\omega \tilde{u}_1 + \frac{e}{m_e} \tilde{E}_1 = 0$$

$$+ e\tilde{n}_1/\epsilon_0 + ik\tilde{E}_1 = 0$$

To have a nontrivial solution, the coefficient-matrix determinant must be zero:

$$\begin{vmatrix} -i\omega & ikn_0 & 0 \\ 0 & -i\omega & \dfrac{e}{m_e} \\ e/\epsilon_0 & 0 & ik \end{vmatrix} = 0 \tag{12.22}$$

As we found before, the solution is

12.3 WAVES IN AN UNMAGNETIZED PLASMA

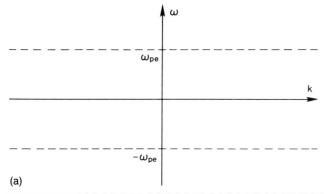

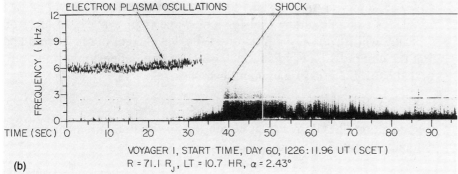

FIG. 12.2. (a) Dispersion diagram (dashed lines) for plasma waves in a cold, unmagnetized plasma ($p_e = p_i = 0$). The frequency is independent of the vector wave, and the group velocity is zero. (b) Example of plasma waves observed by *Voyager 1* upstream (to the left) and downstream of the Jovian bow shock. The electron plasma oscillations near 6 Hz are believed to be created by 1–10-keV electrons streaming away from the shock (Scarf et al., 1979b). (Courtesy of D. A. Gurnett.)

$$\omega^2 = \frac{n_0 e^2}{m_e \epsilon_0} = \omega_{pe}^2 \tag{12.23}$$

$$\omega = \pm \omega_{pe}$$

Thus, the wave frequency is independent of k and equal to the plasma frequency.

The expressions (12.22) and (12.23) are examples of what is called a dispersion relation, which represents the relation between the wave's frequency ω and its wave vector k. In this case, ω is a constant that does not depend on k. Nevertheless, it is useful to plot the dispersion relation of this "electron plasma wave" (Figure 12.2) in the form of a dispersion diagram. The group velocity of a wave, that is, the velocity with which a wave packet or an initial spatially localized perturbation propagates, is given by

$$v_g = \frac{\partial \omega}{\partial k} \tag{12.24}$$

in a one-dimensional case [see equation (2.50b)]. In this case, the group velocity is zero. Thus, a spatially localized perturbation in a cold plasma will not propagate at all, but will oscillate at the plasma frequency ω_{pe}. There is no equivalent to this oscillation in an ordinary fluid, for which there is no electric (restoring) force. An example of electron plasma oscillations observed upstream of the Jovian bow shock is shown in the bottom panel of Figure 12.2. The spectrum is narrow-band and peaked at the electron plasma frequency, which according to the simultaneous measurement of the plasma density in the solar wind was 6 kHz, corresponding to a density of $n_0 = 0.4$ cm^{-3}. Such oscillations are seen at all bow shocks, including that of the earth. It is believed that these plasma oscillations are generated by an instability due to the existence of mildly relativistic electrons streaming away from the shock. A similar situation will be treated later.

We now drop the assumption of zero pressure (cold plasma). In the linear approximation, we then must include the perturbation of the pressure gradient, $\nabla p_1 = n_0 \nabla T_1 + T_0 \nabla n_1$, in equation (12.17). We must now use some physical insight to relate T_1 to n_1 and vice versa. We could, for example, argue that for long-wavelength waves, an electron typically will travel only a fraction of the wavelength in one period and that the compression is adiabatic ($\gamma = 3$ for a one-dimensional situation). In that case, the pressure gradient becomes

$$\nabla p_1 = 3 T_0 \nabla n_1 \tag{12.25}$$

[see equation (2.44)]. We then obtain the algebraic relations among the complex amplitudes:

$$-i\omega \tilde{n}_1 + ikn_0 \tilde{u}_1 = 0 \tag{12.26}$$
$$+3ikT_0 \tilde{n}_1 - i\omega m_e n_0 \tilde{u}_1 + en_0 \tilde{E}_1 = 0$$
$$e\tilde{n}_1/\epsilon_0 + ik\tilde{E}_1 = 0$$

The coefficient matrix is then

$$\begin{vmatrix} -i\omega & ikn_0 & 0 \\ \dfrac{3ikT_0}{m_e n_0} & -i\omega & \dfrac{e}{m_e} \\ e/\epsilon_0 & 0 & ik \end{vmatrix} = 0 \tag{12.27}$$

The solution is now

$$\omega^2 = \omega_{pe}^2 + 3k^2 T_0/m_e = \omega_{pe}^2 + \tfrac{3}{2} k^2 v_e^2$$

where the thermal velocity of the electrons is $v_e = (2T_0/m_e)^{\frac{1}{2}}$ or

$$\omega = \pm \omega_e (1 + 3k^2 \lambda_D^2)^{\frac{1}{2}} \tag{12.28}$$

and $\lambda_D = v_e/\sqrt{2}\,\omega_{pe}$ is the electron Debye length, defined in equation (2.27). This dispersion relation is shown in Figure 12.3. A plasma wave that obeys this dispersion relation is called a *Langmuir wave*. Because we have used the assumption that $v_e \ll \omega_{pe}/k \approx \omega/k$, a good approximation to the group velocity [see equation (12.24)] is

$$v_g = \frac{\partial \omega}{\partial k} = 3(k\lambda_D) v_e/\sqrt{2} \tag{12.29}$$

12.3.2 Plasma Waves with Ion Effects Included

In the preceding section we discovered the Langmuir wave. From the dispersion relation, we see that these are high-frequency waves ($\omega > \omega_{pe}$) that do not involve ions (which were assumed to be infinitely massive and thus immobile). When the ions are allowed to move, the properties of the wave change. Ion contributions are important only for slow variations or low-frequency waves. Ions cannot react quickly enough to affect waves of very high frequency. In this section we shall show that this is indeed the case. Linearizing the continuity equations, the force equations for electrons and ions, as well as Poisson's equation, and assuming that the ions and electrons both obey an equation of state as given by (12.25), we arrive at the following five equations:

$$\frac{\partial n_{e1}}{\partial t} + n_0 \frac{\partial u_{e1}}{\partial x} = 0 \tag{12.30}$$

$$\frac{\partial n_{i1}}{\partial t} + n_0 \frac{\partial u_{i1}}{\partial x} = 0 \tag{12.31}$$

$$m_e n_0 \frac{\partial u_{e1}}{\partial t} + \gamma T_e \frac{\partial n_{e1}}{\partial x} + e n_0 E_{x1} = 0 \tag{12.32}$$

$$m_i n_0 \frac{\partial u_{i1}}{\partial t} + \gamma T_i \frac{\partial n_{i1}}{\partial x} - e n_0 E_{x1} = 0 \tag{12.33}$$

$$\frac{\partial E_{x1}}{\partial x} = e(n_{i1} - n_{e1})/\epsilon_0 \tag{12.34}$$

where we have assumed that the charge of the ion is $q_i = e$ and that variations occur only in the x-direction. We again use the complex exponential wave solution to reduce these to algebraic equations and eliminate u_{e1} and u_{i1} to obtain

$$-m_e \omega^2 \tilde{n}_{e1} + \gamma T_e k^2 \tilde{n}_{e1} - i k e n_0 \tilde{E}_{x1} = 0 \tag{12.35}$$
$$-m_i \omega^2 \tilde{n}_{i1} + \gamma T_i k^2 \tilde{n}_{i1} + i k e n_0 \tilde{E}_{x1} = 0 \tag{12.36}$$

Poisson's equation is then

$$ik\tilde{E}_{x1} = e(\tilde{n}_{i1} - \tilde{n}_{e1})/\epsilon_0 \tag{12.37}$$

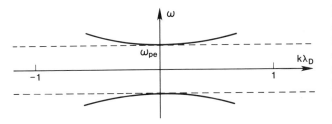

FIG. 12.3. Dispersion diagram (solid lines) for Langmuir waves in a warm, unmagnetized plasma. The ion mass is assumed to be infinitely large. Dashed lines are drawn at $\omega = \pm \omega_{pe}$.

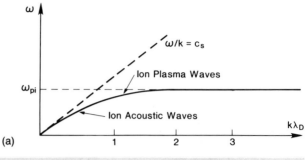

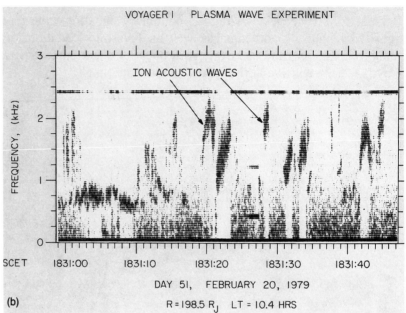

FIG. 12.4. (a) Dispersion diagram for ion-acoustic waves in a warm, unmagnetized plasma with a finite mass of ions. Dashed lines indicate the asymptomic high and low limits for k. (b) Spectrogram as in Figure 12.1. An example of ion-acoustic waves observed by *Voyager 1* upstream of the Jovian bow shock (Scarf et al., 1979a). (Courtesy of D. A. Gurnett.)

The determinant of the coefficient matrix of these three equations must vanish, which means

$$\begin{vmatrix} -m_e\omega^2 + \gamma_e T_e k^2 & 0 & -iken_0 \\ 0 & -m_i\omega^2 + \gamma_i T_i k^2 & iken_0 \\ +e/\epsilon_0 & -e/\epsilon_0 & ik \end{vmatrix} = 0 \quad (12.38)$$

Because $m_i/m_e \gg 1$ (although not infinite), we can neglect $m_e\omega^2$ as compared with $m_i\omega^2$. The dispersion relation is then

$$\omega^2 = k^2 \frac{\gamma_i T_i}{m_i} + \frac{k^2 \gamma_e T_e/m_i}{1 + \gamma_e k^2 \lambda_D^2} \quad (12.39)$$

This dispersion relation is shown in Figure 12.4. In the limit of small $k\lambda_D$,

$$\omega^2 = k^2 \left(\frac{\gamma_e T_e + \gamma_i T_i}{m_i} \right) = k^2 c_s^2 \quad (12.40)$$

The group velocity of this so-called ion-acoustic wave is independent of k:

$$v_g = c_s = \left(\frac{\gamma_e T_e + \gamma_i T_i}{m_i}\right)^{\frac{1}{2}} \qquad (12.41)$$

Another important limit is obtained for cold ions ($T_i \to 0$) and $k\lambda_D \gg 1$ (i.e., short wavelengths). In this case, ions oscillate with a frequency

$$\omega^2 = \pm \omega_{pi} \qquad (12.42)$$

Here $\omega_{pi} = n_0 e^2/\epsilon_0 m_i$ is the ion plasma frequency.

An example of ion acoustic waves observed by *Voyager 1* upstream of the Jovian bow shock is shown in the bottom panel of Figure 12.4, where the wave frequency is of the order of 1–2 kHz. However, the solar-wind density is around 0.5 cm^{-3} at Jupiter, and the ion plasma frequency is only about 0.15 kHz, which is much lower than the observed frequencies. The difference in frequency is caused by the Doppler shift. The phase speed of ion-acoustic waves is less than or equal to the sound speed (12.40), which is about 15 km·s^{-1} for typical electron temperatures ($\sim$3 eV). The solar-wind velocity is about 400 km·s^{-1}, and because of the solar-wind flow relative to the spacecraft the waves can be Doppler-shifted by at least an order of magnitude in frequency, up to the observed frequencies. A more accurate estimate of the Doppler shift requires knowledge of the wavelength of the ion-acoustic waves, which are electrostatic and have a short wavelength. However, measurement of a wave's wavelength is difficult, although under some circumstances it can be done by one spinning spacecraft when the antenna length is longer than the wavelength.

12.3.3 Two-Stream Instability

Thus far we have dealt with a plasma in which both species are initially at rest ($u_{0s} = 0$). There are many examples in space physics in which one species (e.g., electrons) moves relative to another: Solar flares produce a significant flux of energetic electrons moving through the ions in interplanetary space; energetic electrons move away from a planetary shock through the solar wind; aurorae are produced by energetic electrons moving relative to ions along auroral magnetic-field lines; electron beams can be injected from spacecraft. Consider a cold-electron beam traveling through a stationary ion fluid at a speed u_0. The linearized fluid equations are then

$$\frac{\partial n_{e1}}{\partial t} + n_0 \frac{\partial u_{e1}}{\partial x} + u_0 \frac{\partial n_{e1}}{\partial x} = 0$$

$$m_e n_0 \left(\frac{\partial u_{e1}}{\partial t} + u_0 \frac{\partial u_{e1}}{\partial x}\right) + e n_0 E_1 = 0$$

$$\frac{\partial n_{i1}}{\partial t} + n_0 \frac{\partial u_{i1}}{\partial x} = 0$$

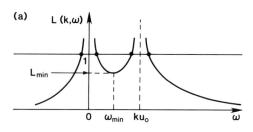
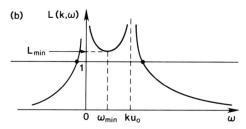

FIG. 12.5. Left-hand side of equation (12.46). (a) The case for $L_{min}<1$. The solution yields four real roots. (b) The case for $L_{min}>1$. The solution yields two real and two complex roots.

$$m_i n_0 \frac{\partial u_{i1}}{\partial t} - e n_0 E_1 = 0$$

$$\frac{\partial E_1}{\partial x} = e(n_{i1} - n_{e1})/\epsilon_0 \tag{12.43}$$

Using the wave solution again, as in the preceding section, we find, after some algebra, the dispersion relation

$$\omega^2 = \omega_{pi}^2 + \omega_{pe}^2 \frac{\omega^2}{(\omega - ku_0)^2} \tag{12.44}$$

For infinitely massive ions ($m_i \to \infty$, $\omega_{pi} \to 0$), we have

$$\omega = ku_0 \pm \omega_{pe} \tag{12.45}$$

The plasma oscillation frequency is simply Doppler-shifted by ku_0. When m_i is finite, the dispersion relation is a quartic equation in ω with four solutions. One instructive way to look at equation (12.44) is to rewrite it in the following form:

$$\frac{\omega_{pi}^2}{\omega^2} + \frac{\omega_{pe}^2}{(\omega - ku_0)^2} = L(\omega, k) = 1 \tag{12.46}$$

The left-hand side of (12.46) is shown in Figure 12.5. We see that $L(\omega, k)$ has a minimum value for $0 < \omega < ku_0$ (noting that $m_e \ll m_i$):

$$L_{min} = L\left(\left(\omega = \left(\frac{m_e}{m_i}\right)^{\frac{1}{3}} ku_0\right), k\right) \approx \frac{\omega_{pi}^2}{(m_e/m_i)^{\frac{2}{3}} k^2 u_0^2} + \frac{\omega_{pe}^2}{k^2 u_0^2} \tag{12.47}$$

If $L_{min} < 1$ (top panel), there are four real solutions of equation (12.46). If, however, $L_{min} > 1$ (bottom panel), there are only two solutions for real k. The two other solutions must be complex: $\omega = \omega_r \pm i\omega_i$. One of these ($\omega = \omega_r + i\omega_i$) leads to an exponentially increasing perturbation, because

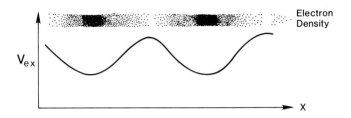

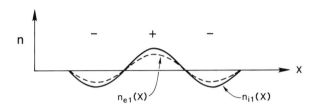

$$\exp(-i(\omega_r + i\omega_i)t) = \exp(-i\omega_r t)\exp(\omega_i t) \tag{12.48}$$

FIG. 12.6. Motion of electrons in the potential created by the charge distribution of a Langmuir wave. The charge density of ions is lower in shaded regions at the top, and the electron velocity is plotted as a curve. Bottom: Density perturbations of ions (solid line) and electrons (dashed line). Plus and minus signs correspond to net local charge density.

The other solution is an exponentially decaying solution.

Is the exponentially growing solution realistic? If so, where does the energy come from for increasing E, n, and u? The answers can be found by looking at the physics of the situation. Let us consider a wave with a large value of k and low frequency ω, so that the phase velocity ω/k is less than u_0. The ion density is perturbed, as shown in Figure 12.6. The electrons in the beam move across this ion-density perturbation (they move faster than the wave). As the electrons move toward a region where the ion density is enhanced and the local charge density is positive, they speed up. As their speed increases, the average distance between the electrons increases (much as the average distance between cars increases when the speed of the cars increases after a traffic jam), and the electron density decreases. Thus, the charge imbalance becomes even greater (more positive net charge). As the electrons move toward a region of ion depletion and hence negative charge density, they are slowed down, their density increases, and the charge imbalance becomes greater (more negative net charge). In other words, the original perturbation of ρ_q increases. The energy comes from the energy of the electron beam. The condition for instability is $L_{\min} > 1$ or

$$k^2 u_0^2 < \omega_{pe}^2 + \frac{\omega_{pi}^2}{(m_e/m_i)^{\frac{2}{3}}} = \omega_{pe}^2 \left(1 + \frac{(\omega_{pi}/\omega_{pe})^2}{(m_e/m_i)^{\frac{2}{3}}} \right) \tag{12.49}$$

Because $\omega_{pi}^2/\omega_{pe}^2 = m_e/m_i$, the second term in the parentheses on the right in equation (12.49) is small, and the condition for instability or wave growth is

$$|ku_0| \leq \omega_{pe} \tag{12.50}$$

The real part of the frequency (i.e., the wave frequency) is given roughly by

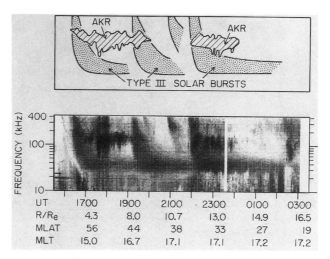

FIG. 12.7. Example of type III burst of solar radio-emission recorded by *ISEE 1* satellite. The spectrogram at the bottom is of the form of Figure 12.1. Top: Schematic in which shading is used to distinguish contributions of type III radio bursts (stippled) from contributions of auroral kilometric radiation (AKR). In the type III solar bursts, the frequency drifts downward because the electrons emitted from the sun move outward through a medium (the solar wind) whose density decreases with radial distance from the sun. Thus, the plasma frequency also decreases. (Adapted from Calvert, 1981.)

$$\omega \sim \left(\frac{m_e}{m_i}\right)^{\frac{1}{3}} k u_0 \tag{12.51}$$

We can show that for $\omega_{pi} \ll |\omega| \ll \omega_{pe}$, the solution to equation (12.46) is given by

$$\omega = \omega_{pe}\left(\frac{m_e}{m_i}\right)^{\frac{1}{3}}\left(\frac{1}{2} \pm i\frac{\sqrt{3}}{2}\right) \tag{12.52}$$

Recalling that the imaginary part of the frequency gives the coefficient of t in a positive exponential, we see that the growth of the waves can occur quite rapidly, and the electrons can lose their energy quickly. Thus, electron beams should not propagate over long distances through a plasma.

This conclusion is at the heart of a famous puzzle in space physics. Solar flares generate electron beams that move through interplanetary space at high velocities ($\sim 0.3c$). According to the foregoing discussion, they should lose their energy quickly to plasma waves. And, in fact, they do generate waves called type III solar radio emissions that are observed in space near the earth or even on the ground (Figure 12.7). The link between the electron beams and the observed waves is a little more complex than it seems, because it involves the transformation of the electrostatic plasma oscillations with a frequency near ω_{pe} into electromagnetic waves at the same frequency. Only the electromagnetic waves can propagate to the observer. The plasma oscillations do not propagate and can be observed only if the spacecraft happens to be in the region of space where they are created. By the time the electrons reach the earth, they should have lost almost all their energy. This, however, does not seem to occur. Solar energetic electrons are seen far beyond the orbit of the earth, with high energies. Several solutions to this dilemma have been proposed, involving rather subtle arguments. Although we cannot discuss them in detail, we can qualitatively understand one aspect of the problem quite easily.

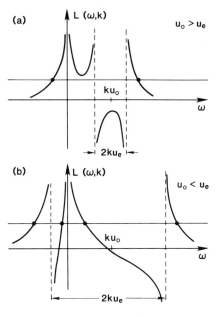

FIG. 12.8. Left-hand side of equation (12.53), the two-stream instability in a warm plasma. (a) The case for $u_0 > u_e$. In this case, the solution may yield two complex roots and two real ones. Instability is possible. (b) The case for $u_0 < u_e$. In this case, the solution yields four real roots, and no instability can occur.

In any realistic situation, the electrons in the beam are not cold, but have a thermal spread (i.e., not all electrons move exactly at the speed u_0). The dispersion relation (12.46) is modified to the following form:

$$\frac{\omega_{pi}^2}{\omega^2} + \frac{\omega_{pe}^2}{(\omega - ku_0)^2 - k^2 u_e^2} = 1 \tag{12.53}$$

where $u_e^2 = v_e^2 \gamma_e / 2$, assuming $\gamma_e = 3$ [cf. (12.25)].

The left-hand side of this equation for $u_0 > u_e$ is plotted in Figure 12.8a. We now see that, in general, there are four solutions for ω [equation (12.53) is of fourth order in ω]. Again, we find that there is a minimum between $\omega = 0$ and $\omega = ku_0$ if $u_0 > u_e$, and an instability may occur. But if $u_0 < u_e$, Figure 12.8b shows that all four roots of equation (12.53) are real, and no instability occurs. The thermal spread of the electrons reduces the strength of the instability. Thus, we expect that an initially cold electron beam ($u_0 > u_e$) will cause a plasma wave to grow. But as the electric field in the plasma grows, the electrons are slowed and "heated." By this we mean that, on the average, the electrons will lose energy, but some may gain energy, and the thermal spread (i.e., v_e) will increase. This effect depends on the fact that E_1, n_1, and u_1 are not so small anymore, so that nonlinear terms (i.e., terms that are proportional to the perturbation amplitudes) become important. The evolution of this two-stream instability has been extensively studied by numerical simulations.

The theory of this nonlinear interaction is complicated, but the result should be clear. As u_0 decreases and v_e increases, the instability weakens (ω_i becomes smaller) and eventually stops. The spreading and slowing of the beam in velocity space cannot be described by fluid equations.

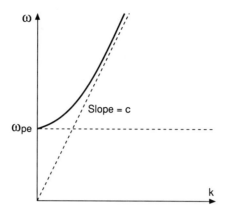

FIG. 12.9. Dispersion diagram for electromagnetic waves in a cold, unmagnetized plasma. For small values of k, the group velocity (i.e., the slope of the solid curve) goes to zero, whereas for large values of k (short wavelength), the group velocity and phase velocity approach the speed of light (slope follows dashed line).

The process is often referred to as "quasi-linear diffusion." At the end, we expect that the electron beam has slowed and spread in velocity space to such a degree that waves do not grow anymore, but actually decay. A stable situation can occur in which $\omega_i \simeq 0$ and a "warm" electron beam can propagate through the plasma without losing any more energy.

12.3.4 Electromagnetic Waves in an Unmagnetized Plasma

In this section we discuss waves that carry not only an electric field but also a magnetic field. We still assume that the unperturbed plasma has no magnetic field [$\mathbf{B} = \mathbf{B}_1(\mathbf{x}, t)$]. We shall first discuss transverse waves, so that $\mathbf{k} \cdot \mathbf{E}_1 = 0$ and $\mathbf{k} \cdot \mathbf{B}_1 = 0$. The last equality is, of course, imposed by equation (12.2) and is always true. In this case, we don't need Poisson's equation. The remaining equations in their linearized form are

$$\nabla \times \mathbf{E}_1 = -\frac{\partial \mathbf{B}_1}{\partial t} \tag{12.54}$$

$$\frac{1}{\mu_0}\nabla \times \mathbf{B}_1 = \mathbf{j}_1 + \epsilon_0 \frac{\partial \mathbf{E}_1}{\partial t} \tag{12.55}$$

$$m_e n_0 \frac{\partial \mathbf{u}_{e1}}{\partial t} = -\nabla p_{e1} - en_0 \mathbf{E}_1 \tag{12.56}$$

We neglect the ion motion, which is justified for high-frequency waves, and so $\mathbf{j}_1 = -en_0 \mathbf{u}_{e1}$. Note that the Lorentz force does not appear in equation (12.56) because it is of second order (proportional to $\mathbf{u}_{e1} \times \mathbf{B}_1$). Furthermore, $\mathbf{E}_1$ and $\mathbf{u}_{e1}$ are perpendicular to $\mathbf{k}$, and thus $p_{e1} = 0$. After assuming the exponential form of the solution and using some algebra, we find the dispersion relation for electromagnetic waves:

$$\omega^2 = \omega_{pe}^2 + k^2 c^2 \tag{12.57}$$

Here we have used $c^2 = (\epsilon_0 \mu_0)^{-1}$, where c is the speed of light in a vacuum.

12.3 WAVES IN AN UNMAGNETIZED PLASMA

If the plasma frequency is much smaller than the wave frequency, the wave becomes a free-space light wave, with $\omega = kc$. The dispersion relation (12.57) is plotted in Figure 12.9. It is sometimes useful to introduce an index of refraction:

$$n = c/v_{ph} = ck/\omega \tag{12.58}$$

For these waves,

$$n = \sqrt{1 - \omega_{pe}^2/\omega^2} \tag{12.59}$$

We see that for $\omega < \omega_{pe}$, the index of refraction becomes imaginary. For a real frequency $\omega < \omega_{pe}$, k is purely imaginary, and such a wave would decay in space and not propagate. Because there are two conjugate solutions for k ($k = \pm ik_i$), we may ask why $k = -ik_i$ does not correspond to wave growth. The answer is a bit complicated, but essentially involves the physical argument that a cold, stationary plasma does not have free energy to support a growing mode, although it can absorb a wave launched by, for example, an antenna or a spacecraft.

The fact that electromagnetic waves propagate or decay, depending on the relation between ω and ω_{pe}, can be used to diagnose the plasma. Consider two antennas separated in space. One antenna acts as a transmitter, and the other as a receiver. If the transmitter emits waves with $\omega > \omega_{pe}$, the receiver will detect a signal. If $\omega < \omega_{pe}$, the wave is evanescent, and the receiver will detect a reduced level of intensity. By sweeping through the frequency from high to low values (or vice versa), we can measure the value of the plasma frequency (ω_{pe}) and hence infer the plasma density between the two antennas.

There is another important application of the fact that the index of refraction $n \to 0$ as $\omega \to \omega_{pe}$. A wave moving through a medium of changing ω_{pe} is reflected when $\omega = \omega_{pe}$ and $n = 0$. This principle allows us to measure remotely the plasma density in the ionosphere. Imagine a transmitter on the ground emitting a short pulse of radiation (at a frequency ω) upward. The wave will propagate through the atmosphere into the ionosphere and may reach a height where the plasma frequency ω_{pe} will equal ω. At that height, the wave will be reflected, and after a certain time delay Δt, it will return to the ground, where it can be detected. Because the time delay is equal to twice the wave travel time from the ground to the reflection height h, we can calculate the height h where the electron density is $n_e = (\epsilon_0 m_e \omega^2/e^2)^{\frac{1}{2}}$. By using different frequencies ω, we can deduce a height profile for the ionospheric electron density $n_e(h)$. The device of transmitter and receiver is called an ionosonde. An example of the output of such an ionosonde, called an ionogram, is shown in Figure 12.10. Because the ionospheric plasma is magnetized (i.e., the background magnetic field is not zero), there are several different electromagnetic waves, all propagating at slightly different velocities. A single transmitter can emit a mixture of these waves that will return to the receiver after different time delays. Hence, an

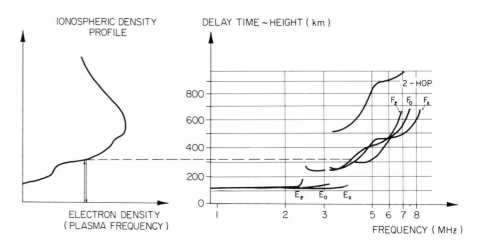

FIG. 12.10. Left: Schematic profile of electron density (increasing to the right) versus altitude. Right: Example of an ionogram. The ordinate values show the time it takes a pulse of waves at a certain frequency to propagate to the reflection layer (where $\omega = \omega_{pe}$) and back and thus are proportional to height. Because different waves can exist in the magnetized ionosphere, there are several traces, labeled by the polarization of the electric-field vector. Budden (1985), among others, discussed ionograms and ionospheric sounding.

ionogram often contains multiple traces that track each other relatively closely, as we discuss more completely in Section 12.4.3.

Another important application of equation (12.59) is in ionospheric modification. At the height where the electron plasma frequency is equal to the wave frequency (where the density is equal to the *critical density*), the wave group velocity

$$v_g = \frac{\partial \omega}{\partial k} = \frac{kc^2}{\omega} = nc \tag{12.60}$$

goes to zero. The wave amplitude becomes very large there because it cannot propagate. The large wave electric field can accelerate electrons and drive currents in the plasma and heat it (i.e., the wave modifies the background plasma). If the power from the transmitter is large enough, the heating can be quite significant. Naturally, as the wave loses energy to heating the plasma, the amplitude of the reflected waves becomes lower than the incident amplitude. This method of plasma heating is also used in laser fusion.

12.3.5 Summary of Waves in an Unmagnetized Plasma

In the foregoing sections, we have discussed three distinct types of waves. Two waves (the Langmuir waves and the ion plasma waves) do not have a wave magnetic field. They are called electrostatic waves ($\partial \mathbf{B}/\partial t = 0$, or $\nabla \times \mathbf{E} = 0$). The high-frequency Langmuir waves do not involve ion motion; the ion plasma waves are low-frequency waves and are affected by ion motion. The electromagnetic wave, which does have a magnetic field, can propagate only if its frequency is above the plasma frequency. In the next section, we shall see that when there is a stationary magnetic field in the plasma, the wave properties become more complex. In particular, we shall find that electromagnetic waves with $\omega < \omega_{pe}$ can propagate through a magnetized plasma.

12.4 WAVES IN A MAGNETIZED PLASMA

When a magnetic field exists in the unperturbed plasma, the motion of charged particles is restricted. Particles can move freely along the magnetic field, but not perpendicular to the magnetic field. The Lorentz force $q(\mathbf{v} \times \mathbf{B})$ acts perpendicular to the magnetic field and causes the charged particles to gyrate about the magnetic field at their respective gyrofrequencies $\Omega_s = q_s B/m_s$, so that $\Omega_{ce} < 0$ and $\Omega_{ci} > 0$.

We are still dealing with linear waves, which involve the first-order electric field $\mathbf{E}_1$ and the first-order magnetic field $\mathbf{B}_1$. In this section, however, we must also include the zeroth-order (unperturbed) magnetic field $\mathbf{B}_0$. Thus, the Lorentz force contains a first-order term $(\mathbf{u}_1 \times \mathbf{B}_0)$ that did not appear in an unmagnetized plasma. Because of the presence of $\mathbf{B}_0$, there is now also a unique direction, namely, the direction of this field. Waves, of course, can propagate in all directions. It is convenient to discuss special cases characterized by the direction of the wave vector $\mathbf{k}$ relative to the magnetic field $\mathbf{B}_0$ and the perturbation electric field $\mathbf{E}_1$. We can distinguish among six different cases:

1. *Parallel propagating waves* have

    $\mathbf{k} \times \mathbf{B}_0 = 0$

2. *Perpendicular waves* have

    $\mathbf{k} \cdot \mathbf{B}_0 = 0$

We can also distinguish waves by the direction of the wave electric field relative to the propagation vector:

3. *Longitudinal waves* have

    $\mathbf{k} \times \mathbf{E}_1 = 0$

4. *Transverse waves* have

    $\mathbf{k} \cdot \mathbf{E}_1 = 0$

Any wave with $\mathbf{E}_1$ neither exactly parallel to $\mathbf{k}$ nor perpendicular to $\mathbf{k}$ can be constructed as a superposition of these two waves. There are two distinct physical modes distinguished by whether or not they have a wave magnetic field:

5. *Electrostatic waves* have no magnetic-field perturbation:

    $\mathbf{B}_1 = 0$

6. *Electromagnetic waves* involve a magnetic-field perturbation:

    $\mathbf{B}_1 \neq 0$

Waves can be classified according to this scheme as, for example, parallel propagating electrostatic waves, perpendicular electromagnetic waves, and so forth. From Faraday's law, we see that if $\mathbf{k} \times \mathbf{E}_1 = 0$ (longitudinal waves), $\partial \mathbf{B}/\partial t = 0$, and hence $\mathbf{B}_1 = 0$. Thus, longitudinal waves are electrostatic, and vice versa. All transverse waves are electromagnetic waves, but not all electromagnetic waves are transverse waves.

12.4.1 Upper Hybrid Waves

Upper hybrid waves are electrostatic (these are usually the simplest waves, and that is why we deal with them first). We shall first consider a cold magnetized plasma ($T_e = T_i = 0$) with infinitely massive ions ($m_i \to \infty$), just as we did in the unmagnetized-plasma case. Thus, the ions form a uniform background fluid of density n_0 and positive charge. The relevant equations are

$$\frac{\partial n_e}{\partial t} + \nabla \cdot (n_e \mathbf{u}_e) = 0 \tag{12.61}$$

$$m_e n_e \left(\frac{\partial \mathbf{u}_e}{\partial t} + \mathbf{u}_e \cdot \nabla \mathbf{u}_e \right) + e n_e \mathbf{E} + e n_e \mathbf{u}_e \times \mathbf{B} = 0$$

$$\nabla \cdot \mathbf{E} = e(n_0 - n_e)/\epsilon_0$$

It can be shown that the other Maxwell equations are not relevant.

The background magnetic field $\mathbf{B}_0$ is in the $\hat{\mathbf{z}}$-direction. We assume that the wave is propagating in the $\hat{\mathbf{x}}$-direction ($\mathbf{k} = k\hat{\mathbf{x}}$): that is, we are dealing with a perpendicular wave with $\mathbf{k} \cdot \mathbf{B}_0 = 0$. Linearizing the set of equations (12.61) yields the algebraic equations

$$-i\omega \tilde{n}_e + ikn_0 \tilde{u}_{ex} = 0 \tag{12.62}$$
$$-i\omega m_e \tilde{u}_{ex} + e\tilde{E}_x + e\tilde{u}_{ey} B_0 = 0$$
$$-i\omega m_e \tilde{u}_{ey} - e\tilde{u}_{ex} B_0 = 0$$
$$ik\tilde{E}_x + e\tilde{n}_e/\epsilon_0 = 0$$

where, for convenience, we have dropped the subscript 1 on the first order wave perturbations. We now have four independent variables ($\tilde{n}_e$, $\tilde{u}_{ex}$, $\tilde{u}_{ey}$, $\tilde{E}_x$). For (12.62) to have a nontrivial solution, the determinant of the coefficient matrix must be zero:

$$\begin{vmatrix} -i\omega & ikn_0 & 0 & 0 \\ 0 & -i\omega m_e & eB_0 & e \\ 0 & eB_0 & i\omega m_e & 0 \\ e/\epsilon_0 & 0 & 0 & ik \end{vmatrix} = 0 \tag{12.63}$$

After some algebra, we obtain the dispersion relation

$$\omega^2 = \omega_{pe}^2 + \Omega_{ce}^2 = \omega_{UH}^2 \tag{12.64}$$

where $\Omega_{ce} = -eB_0/m_e$ is the electron gyrofrequency. The plasma response to a perturbation is an oscillation whose frequency is $\omega_{UH} = (\omega_{pe}^2 + \Omega_{ce}^2)^{\frac{1}{2}}$. This frequency is called the *upper hybrid frequency*. If $B_0 \to 0$, we recover the cold-plasma oscillations.

The reason for the appearance of Ω_{ce} is easy to see. An electric field in the x-direction will accelerate the electrons in the negative x-direction. As the electrons pick up speed, the Lorentz force will become larger and turn the electrons toward the positive y-direction. Eventually, the electrons are turned around and move against the electric field, losing energy. Because the Lorentz force acts like an additional restoring force, the frequency of the oscillation is increased.

12.4.2 Electrostatic Ion Waves

The upper hybrid wave is a high-frequency wave with ω much greater than both ω_{pi} and Ω_{ei}. We now look at low-frequency waves, which are influenced by ion motion. We are restricting ourselves to electrostatic waves (i.e., $\mathbf{k} \times \mathbf{E}_1 = 0$, $\mathbf{B}_1 = 0$). The only Maxwell equation we need is Poisson's equation. The linearized equations of continuity and momentum yield

$$m_e n_0 \frac{\partial \mathbf{u}_{e1}}{\partial t} + \gamma T_e \nabla n_{e1} + e n_0 \mathbf{E}_1 + e n_0 \mathbf{u}_{e1} \times \mathbf{B}_0 = 0 \qquad (12.65)$$

$$\frac{\partial n_{e1}}{\partial t} + n_0 \nabla \cdot \mathbf{u}_{e1} = 0$$

$$m_i n_0 \frac{\partial \mathbf{u}_{i1}}{\partial t} + \gamma T_i \nabla n_{i1} - e n_0 \mathbf{E}_1 - \frac{e n_0}{c} \mathbf{u}_i \times \mathbf{B}_0 = 0$$

$$\frac{\partial n_{i1}}{\partial t} + n_0 \nabla \cdot \mathbf{u}_{i1} = 0$$

Without loss of generality, we can take

$$\mathbf{k} = k(\sin \theta, 0, \cos \theta) \qquad (12.66)$$

where $\mathbf{B}_0 = B_0 \hat{z}$ defines the $\hat{z}$-direction, and θ is the angle between $\mathbf{k}$ and $\mathbf{B}_0$. Now the solution of the set of equations (12.65), with Poisson's equation, gets a little messy, and we shall not show the steps here. The result for the dispersion relation of *electrostatic ion waves* is

$$1 - \frac{k^2 c_s^2}{\omega^2} + \frac{\Omega_{ci}}{\omega} \tan^2 \theta \times \qquad (12.67)$$

$$\left[\frac{1}{\frac{\omega}{\Omega_{ce}}\tan^2\theta - \frac{\Omega_{ce}}{\omega}\left(1 - \frac{\omega^2}{\Omega_{ce}^2}\right)} - \frac{1}{\frac{\omega}{\Omega_{ci}}\tan^2\theta - \frac{\Omega_{ci}}{\omega}\left(1 - \frac{\omega^2}{\Omega_{ci}^2}\right)} \right] = 0$$

We have assumed $k^2 \lambda_{de}^2 \ll 1$, $k^2 \lambda_{di}^2 \ll 1$, $\omega \ll \omega_{pe}$, $|\Omega_{ce}| \ll \omega_{pe}$, where λ_{de} and λ_{di} are the electron and ion Debye lengths, respectively. This is quite a horrendous-looking equation, and it is rarely used in its full glory. Let us consider some special cases:

Parallel-propagating Electrostatic Ion Waves ($\theta \to 0$). In this case, $\theta \to 0$, $k = k_z$, and the third term of (12.67) is zero provided that $\omega^2 \neq \Omega_{ce}^2$ or Ω_{ci}^2. Then

$$1 - k_z^2 c_s^2/\omega^2 = 0 \tag{12.68}$$
$$\omega = \pm k_z c_s$$

This is just the ion-acoustic wave. Because the wave propagates along $\mathbf{B}_0$, and $\mathbf{E}_1$ is then also along $\mathbf{B}_0$, we could have guessed this answer, because for parallel propagation the magnetic field has no influence.

What if $\omega^2 = \Omega_{ce}^2$ or Ω_{ci}^2? In this case, the denominators can equal zero for parallel propagation, and (12.67) is indeterminate. Instead, let us first take one limit

$$\omega \to \Omega_{ci} \tag{12.69}$$

As $\theta \to 0$, the first denominator is finite, while the second denominator approaches zero, and the first (electron) term in the bracket in equation (12.67) can be neglected. The other must be kept. This yields

$$1 - \frac{k^2 c_s^2}{\Omega_{ci}^2} - \frac{1}{1 - \cot^2\theta\left(1 - \frac{\omega^2}{\Omega_{ci}^2}\right)} = 0 \tag{12.70}$$

This is the dispersion relation for an *ion-cyclotron wave*. For $\omega \to |\Omega_{ce}|$, we obtain the dispersion relation for an *electron-cyclotron wave*:

$$1 - \frac{k^2 c_s^2}{\Omega_{ce}^2} - \frac{m_e/m_i}{1 - \cot^2\theta\left(1 - \frac{\omega^2}{\Omega_{ce}^2}\right)} = 0 \tag{12.71}$$

Perpendicular Electrostatic Ion Waves ($\theta \to \pi/2$). In this case, we find

$$1 - \frac{k^2 c_s^2}{\omega^2} + \frac{\Omega_{ci}\Omega_{ce}}{\omega^2} - \frac{\Omega_{ci}^2}{\omega^2} = 0 \tag{12.72}$$

Clearly, the last term is much smaller than the third term ($\Omega_{ci} \ll |\Omega_{ce}|$) and can be neglected. We thus obtain the dispersion relation for *lower hybrid waves*:

$$\omega^2 = k^2 c_s^2 + |\Omega_{ci}\Omega_{ce}| \tag{12.73}$$

For long-wavelength waves ($k \to 0$), we obtain an oscillating solution at

$$\omega = \sqrt{|\Omega_{ce}\Omega_{ci}|} = \omega_{LH} \tag{12.74}$$

This frequency is called the *lower hybrid frequency*. These waves are of great significance in the auroral regions, where they are often associated with field-aligned currents and may be responsible for ion heating in the auroral region.

Why this frequency appears is easy to understand physically. The wave electric field is perpendicular to $\mathbf{B}_0$. The very massive ions move along $\mathbf{E}_1$; the Lorentz force on them is small. Of course, their displacement in $\hat{x}$ (along $\mathbf{E}_1$) is limited because the electric field oscillates in time. The electrons gyrate about the magnetic field and $\mathbf{E} \times \mathbf{B}$-drift in the $\hat{y}$-direction. Their displacement in the $\hat{x}$-direction is roughly given by

$$\Delta x_e = \frac{E_1}{B_0 \Omega_{ce}}$$

The ion displacement in the $\hat{x}$-direction is roughly given by

$$\Delta x_i = \frac{eE_1}{m_i \omega^2}$$

We see that $\Delta x_e = \Delta x_i$ only if $\omega = \omega_{LH}$. The two displacements must be equal to avoid a buildup of charge.

The Case of k_z Small, but not Zero, and $\tan \theta \gg 1$. We are looking for low-frequency waves, and we thus neglect ω/Ω_{ce} with respect to unity. If

$$1 \ll \tan \theta \ll \left(\frac{m_i}{m_e}\right)^{\frac{1}{2}}$$

$|\omega \tan^2 \theta / \Omega_{ce}| \ll 1$ for $\omega \sim \Omega_{ci}$. The electron term in (12.67) can be neglected, and we obtain

$$\omega^2 = k^2 c_s^2 + \Omega_{ci}^2 \qquad (12.75)$$

These waves are called *electrostatic ion-cyclotron waves* (EIC waves), which propagate with $k_z/k_x \gg (m_e/m_i)^{\frac{1}{2}}$. Thus, the angle of propagation must be more than 2° off the direction perpendicular to the magnetic field $\mathbf{B}_0$.

Summary of Electrostatic Ion Waves in a Magnetized Plasma. For parallel-propagating waves (i.e., $\theta \to 0$), we found three solutions:

$\omega^2 = k^2 c_s^2$ (ion-acoustic waves)
$\omega^2 = \Omega_{ci}^2$ (ion-cyclotron waves)
$\omega^2 = \Omega_{ce}^2$ (electron-cyclotron waves)

For nearly perpendicular propagation [i.e., $1 \gg k_z/k_x \gg (m_e/m_i)^{\frac{1}{2}}$], we found one solution:

$\omega^2 = k^2 c_s^2 + \Omega_{ci}^2$ (EIC waves)

For perpendicular propagation (i.e., $\theta \to \pi/2$), we found two solutions (one trivial solution was not discussed):

$\omega^2 = k^2 c_s^2 + |\Omega_{ci} \Omega_{ce}|$ (lower hybrid waves)
$\omega^2 = 0$

12.4.3 Electromagnetic Waves in a Magnetized Plasma

12.4.3.1 PERPENDICULAR PROPAGATION
Let us first look at a *perpendicular wave,* because it is the easiest. In this case, $\mathbf{k} = k\hat{\mathbf{x}}$, without loss of generality. We assume a cold plasma, and thus the relevant equations are in linearized form:

$$m_e n_0 \frac{\partial \mathbf{u}_{e1}}{\partial t} + e n_0 \mathbf{E}_1 + e n_0 \mathbf{u}_{e1} \times \mathbf{B}_0 = 0 \tag{12.76}$$

$$\nabla \times \mathbf{E}_1 + \frac{\partial \mathbf{B}_1}{\partial t} = 0 \tag{12.77}$$

$$\nabla \times \mathbf{B}_1 = \mu_0 \mathbf{j}_1 + \frac{1}{c^2} \frac{\partial \mathbf{E}_1}{\partial t} \tag{12.78}$$

For high-frequency waves (which the wave in this case turns out to be), we can assume the ions to be immobile, and hence

$$\mathbf{j} = -e n_0 \mathbf{u}_{e1} \tag{12.79}$$

This set is complete, and we do not need Poisson's equation or the continuity equation. First, we consider a wave with $\mathbf{E}_1$ along $\mathbf{B}_0$ (see Figure 12.11). Equation (12.77) then shows that $\mathbf{B}_1$ is perpendicular to $\mathbf{B}_0$, and hence in (12.78), $\mathbf{j}_1$ is parallel to $\mathbf{B}_0$. Consequently, we need only consider the z component of (12.76), which is identical with equation (12.56) of the unmagnetized plasma if the electron pressure vanishes. Because the other equations are also identical with equations (12.54) and (12.55) for the unmagnetized plasma, we can, at once, write down the dispersion relation in the form (12.57):

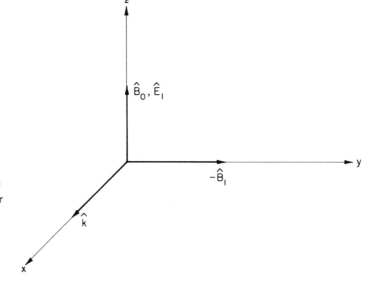

FIG. 12.11. Vector orientations of the electric- and magnetic-field perturbations for the ordinary electromagnetic wave in a cold, magnetized plasma. The wave propagates perpendicular to the background magnetic field.

12.4 WAVES IN A MAGNETIZED PLASMA

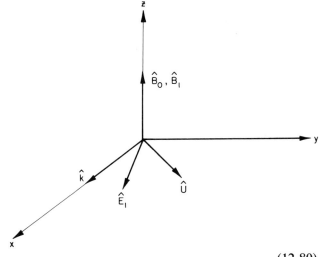

FIG. 12.12. Vector orientations of the electric- and magnetic-field perturbations for the extraordinary electromagnetic wave in a cold, magnetized plasma. The wave propagates perpendicular to the background magnetic field.

$$\omega^2 = \omega_{pe}^2 + k^2c^2 \tag{12.80}$$

This describes the *ordinary* O *wave*, which propagates as if the magnetic field $\mathbf{B}_0$ did not exist.

We now consider a wave with $\mathbf{E}_1$ in the $\hat{\mathbf{x}}$–$\hat{\mathbf{y}}$ plane (Figure 12.12). Of course, any wave may be thought of as a superposition of this wave and the wave with $\mathbf{E}_1$ along $\mathbf{B}_0$. The wave with $\mathbf{E}_1$ perpendicular to $\mathbf{B}_0$ is called an *extraordinary* (X) *wave*. For a wave solution, we have the following algebraic equation for the five unknowns $(\tilde{u}_{ex}, \tilde{u}_{ey}, \tilde{E}_x, \tilde{E}_y, \tilde{B}_z)$:

$$-i\omega m_e \tilde{u}_{ex} + e\tilde{E}_x + e\tilde{u}_{ey}B_0 = 0 \tag{12.81}$$
$$-i\omega m_e \tilde{u}_{ey} + e\tilde{E}_y - e\tilde{u}_{ex}B_0 = 0$$
$$k\tilde{E}_y - \omega \tilde{B}_z = 0$$
$$\mu_0 e n_0 \tilde{u}_{ey} + \frac{i\omega}{c^2}\tilde{E}_y - ik\tilde{B}_z = 0$$
$$\mu_0 e n_0 \tilde{u}_{ex} + \frac{i\omega}{c^2}\tilde{E}_x = 0$$

where again $c^2 = (\mu_0 \epsilon_0)^{-1}$ is used, and the subscript 1 has been dropped for harmonic perturbations. We can reduce this to four equations by using

$$\tilde{B}_z = \frac{k}{\omega}\tilde{E}_y \tag{12.82}$$

The last two remaining equations in (12.81) become

$$\tilde{u}_{ex} = -\frac{i\epsilon_0 \omega}{n_0 e}\tilde{E}_x \tag{12.83}$$

$$\tilde{u}_{ey} = \frac{i\epsilon_0}{n_0 e \omega}(k^2c^2 - \omega^2)\tilde{E}_y \tag{12.84}$$

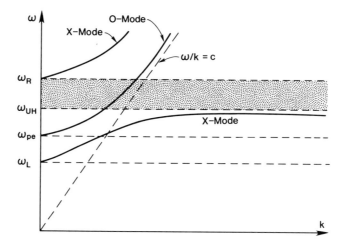

FIG. 12.13. Dispersion diagram for the perpendicular-propagating extraordinary (X) and ordinary (O) electromagnetic waves in a cold, magnetized plasma. The stop band separating the two branches of the X mode is stippled.

and the dispersion relation is

$$\begin{vmatrix} 1 - \dfrac{\omega^2}{\omega_{pe}^2} & \dfrac{i\Omega_{ce}}{\omega\omega_{pe}^2}(k^2c^2 - \omega^2) \\ \dfrac{i\Omega_{ce}\omega}{\omega_{pe}^2} & 1 + \dfrac{k^2c^2 - \omega^2}{\omega_{pe}^2} \end{vmatrix} = 0 \qquad (12.85)$$

It is instructive to write this in terms of the index of refraction:

$$n^2 = \frac{k^2c^2}{\omega^2} = 1 - \frac{\omega_{pe}^2}{\omega^2}\frac{\omega^2 - \omega_{pe}^2}{\omega^2 - \omega_{UH}^2} \qquad (12.86)$$

This dispersion relation depends on $\mathbf{B}_0$ through ω_{UH} [see equation (12.64)]. Waves whose dispersion relations depend on $\mathbf{B}_0$ are referred to as extraordinary (X) waves to distinguish them from O-mode waves, whose dispersion relations do not depend on $\mathbf{B}_0$. Figure 12.13 shows the dispersion diagram for the extraordinary (X) and ordinary (O) modes propagating perpendicular to $\mathbf{B}_0$.

Whereas the ordinary mode is linearly polarized ($\mathbf{E}_1$ only along $\mathbf{B}_0$), the X mode is elliptically polarized. From equation (12.85), we see that

$$\tilde{E}_x/\tilde{E}_y = \frac{i(\Omega_{ce}/\omega\omega_{pe}^2)(k^2c^2 - \omega^2)}{1 - \omega^2/\omega_{pe}^2} \qquad (12.87)$$

The fact that this ratio is imaginary expresses the fact that the two fields are 90° out of phase. Thus, the electric-field vector rotates in the $\hat{x}-\hat{y}$ plane, with frequency ω. Generally the magnitude of the ratio is not equal to unity, and the wave is elliptically polarized.

The index of refraction for the X mode can be zero when

$$\omega_{(L,R)}^X = \pm\frac{\Omega_{ce}}{2} + \sqrt{\omega_{pe}^2 + \Omega_{ce}^2/4} \qquad (12.88)$$

These two frequencies are called *cutoff frequencies*. At frequencies below the cutoff frequencies, the index of refraction is imaginary, and

12.4 WAVES IN A MAGNETIZED PLASMA

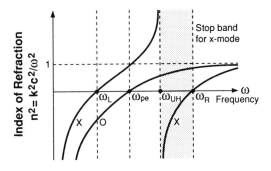

FIG. 12.14. Square of the index of refraction for the two perpendicular-propagating waves in a cold, magnetized plasma. The stop band (stippled) is the range of frequencies for which the index is purely imaginary ($n^2 < 0$) for X-mode waves, and the wave is evanescent.

the wave is evanescent. The cutoff frequency for the ordinary mode is the plasma frequency $\omega^O = \omega_{pe}$. The index of refraction can also go to infinity. This means that the wave vector becomes very large (i.e., the wavelength is very short). The phase velocity of the wave goes to zero. The frequency at which this happens is called a *resonance frequency*. The resonance frequency for the X mode is $\omega = \omega_{UH}$, the upper hybrid frequency. The ordinary mode has no resonance when propagating perpendicular to $\mathbf{B}_0$. The band of frequencies where $n^2 < 0$ is called a *stop band*. The waves in this frequency band cannot propagate; they are stopped, hence the name "stop band." For example, no X-mode waves can propagate between ω_{UH} and ω_R. The index of refraction as a function of frequency for these two waves is shown in Figure 12.14. The stop band is indicated by shading.

The existence of such a stop band is of great significance for the propagation of radio waves. All magnetized planets are emitters of radio waves. These waves are generated near the planets, where both the electron gyrofrequency and plasma frequency are high. As the waves propagate outward, these frequencies decrease. If the waves are originally created with a frequency below the upper hybrid frequency in the extraordinary mode, they will encounter a stop zone as they propagate toward a region where ω_{UH} becomes smaller than the wave frequency ω. Thus, such a wave cannot be observed beyond the stop zone, or, to put it differently, a radio wave observed at a great distance from a planet cannot have been created as an X-mode wave below the local upper hybrid frequency without mode conversion. This conclusion limits the number of possible generation mechanisms.

12.4.3.2 PARALLEL PROPAGATION Using the wave solution for the set of equations (12.76)–(12.78), with $\mathbf{k} = k\hat{\mathbf{z}}$ (parallel to $\mathbf{B}_0$), we obtain

$$-i\omega m_e \tilde{\mathbf{u}}_e + e\tilde{\mathbf{E}} + e\tilde{\mathbf{u}}_e \times \mathbf{B}_0 = 0 \qquad (12.89)$$

$$\mathbf{k} \times \tilde{\mathbf{E}} - \omega \tilde{\mathbf{B}} = 0 \qquad (12.90)$$

$$i\mathbf{k} \times \tilde{\mathbf{B}} + \mu_0 n_0 e \tilde{\mathbf{u}}_e + \frac{i\omega}{c^2}\tilde{\mathbf{E}} = 0 \qquad (12.91)$$

We see that a solution must have $\tilde{\mathbf{u}}_e$, $\tilde{\mathbf{E}}$, and $\tilde{\mathbf{B}}$ all in the $\hat{\mathbf{x}}$–$\hat{\mathbf{y}}$ plane (perpendicular to $\mathbf{B}_0$). The relations among the six unknowns ($\tilde{u}_{ex}$, $\tilde{u}_{ey}$, $\tilde{E}_x$, $\tilde{E}_y$, $\tilde{B}_x$, $\tilde{B}_y$) are given by

$$k\tilde{E}_y = -\omega \tilde{B}_x \tag{12.92}$$

$$k\tilde{E}_x = \omega \tilde{B}_y$$

$$ik\tilde{B}_y = \mu_0 n_0 e \tilde{u}_{ex} + \frac{i\omega}{c^2}\tilde{E}_x$$

$$ik\tilde{B}_x = -\mu_0 n_0 e \tilde{u}_{ey} - \frac{i\omega}{c^2}\tilde{E}_y$$

$$i\omega m_e \tilde{u}_{ex} = e\tilde{E}_x + eB_0 \tilde{u}_{ey}$$

$$i\omega m_e \tilde{u}_{ey} = e\tilde{E}_y - eB_0 \tilde{u}_{ex}$$

We can reduce these to two equations for the electric field by expressing $\tilde{\mathbf{B}}$ in terms of $\tilde{\mathbf{E}}$ and using

$$\tilde{u}_{ex} = -i\frac{\epsilon_0 \omega}{n_0 e}\left(1 - \frac{k^2 c^2}{\omega^2}\right)\tilde{E}_x \tag{12.93}$$

$$\tilde{u}_{ey} = -i\frac{\epsilon_0 \omega}{n_0 e}\left(1 - \frac{k^2 c^2}{\omega^2}\right)\tilde{E}_y \tag{12.94}$$

The factor i indicates that the velocities are 90° out of phase with the electric fields. We are now left with two equations:

$$\begin{pmatrix} 1 - \dfrac{(\omega^2 - k^2 c^2)}{\omega_{pe}^2} & i\dfrac{(\omega^2 - k^2 c^2)\Omega_{ce}}{\omega_{pe}^2} \\ -i\dfrac{(\omega^2 - k^2 c^2)\Omega_{ce}}{\omega_{pe}^2} & 1 - \dfrac{(\omega^2 - k^2 c^2)}{\omega_{pe}^2} \end{pmatrix} \begin{pmatrix} \tilde{E}_x \\ \tilde{E}_y \end{pmatrix} = 0 \tag{12.95}$$

From this, we obtain the dispersion relation

$$\frac{\omega^2 - k^2 c^2}{\omega_{pe}^2}\left(1 \pm \frac{\Omega_{ce}}{\omega}\right) = 1 \tag{12.96}$$

or, for the index of refraction,

$$n^2 = \frac{c^2 k^2}{\omega^2} = 1 - \frac{\omega_e^2/\omega^2}{1 \pm \Omega_{ce}/\omega} \tag{12.97}$$

The top sign in equations (12.96) and (12.97) refers to a wave propagating along $\mathbf{B}_0$, with the electric-field vector rotating in a right-handed sense (R wave). By this, we mean that the fingers of the right hand curl in the direction of the $\mathbf{E}$ field rotation when the thumb points along $\mathbf{k}$. The bottom sign refers to a left-handed wave (L wave). The R wave rotates in the same sense as the electrons gyrate around $\mathbf{B}_0$; the L wave rotates in the sense of the ions. Because $\Omega_{ce} < 0$ (in our convention), the

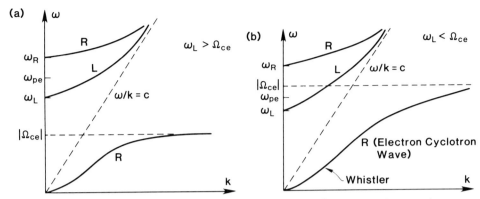

FIG. 12.15. Dispersion diagrams for parallel-propagating electromagnetic waves in a cold, magnetized plasma for two different cases. (a) The left-hand cutoff frequency ω_L is larger than the electron gyrofrequency. (b) The left-hand cutoff frequency ω_L is smaller than the electron gyrofrequency. The R wave has two "pass bands" separated by a stop band, $\Omega_{ce} < \omega < \omega_R$. The L wave exists only for $\omega > \omega_L$.

R wave has a resonance when $\omega = |\Omega_{ce}|$. In that case, an electron gyrating about the field $\mathbf{B}_0$ will experience a constant electric field that can accelerate (or decelerate) it continuously. The electrons are then in resonance with the electric field of the wave. The L mode has no resonance at these high frequencies. It has a resonance at $\omega = \Omega_{ci}$ when the ion motion is included.

The cutoff frequencies ($n \to 0$) are

$$\omega_{(R,L)}^{\text{cutoff}} = \pm \frac{\Omega_{ce}}{2} \pm \sqrt{\omega_{pe}^2 + (\Omega_{pe}^2/4)} \tag{12.98}$$

which are the same cutoff frequencies as for the X mode. The dispersion diagrams for parallel propagation are shown in Figure 12.15. The R waves with frequencies below Ω_{ce} are often called whistlers. This name derives from the nature of the dispersion relation. Whistlers are caused by lightning discharges, which are short-duration sources. The source generates a broad spectrum, but the propagation time to a distant observer will depend on the wave's group velocity (i.e., the slope of the dispersion curve), which will depend on the frequency. We can see that near the low-frequency end of the whistler branch, the high-frequency waves will propagate faster and therefore will arrive at a distant observer before the lowest-frequency waves. As time goes on, the frequency of the wave reaching the observer from the lightning source will decrease in a way reminiscent of a whistling sound with decreasing pitch. The detection of whistlers is a strong indication of lightning activity, and the discovery of whistlers in the Jovian magnetosphere (Figure 12.16) was obviously of great significance. The rate at which the frequency decreases gives information about the plasma density along the whistler propagation path from the source to the observer.

When the source is continuously active or when the observer is close

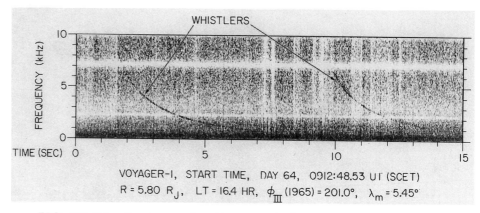

FIG. 12.16. Spectrogram from *Voyager 1* (the form is analagous to Figure 12.1) showing descending frequency signals identified as whistler waves (Scarf et al. 1981). (Courtesy of D. A. Gurnett.)

to it, the whistler has a broadband spectrum. It is then often called hiss, examples of which are shown in Figure 12.1.

12.4.3.3 THE APPLETON-HARTREE DISPERSION RELATION

In the preceding sections we have discussed wave propagation for purely parallel and perpendicular wave vectors. However, we can use equations (12.76)–(12.79) to derive a dispersion relation for waves in a cold plasma for arbitrary propagation angles. This dispersion relation is often referred to as the Appleton-Hartree dispersion relation.

In deriving this dispersion relation, we shall assume that $\mathbf{B}_0$ is along the z-axis of a cartesian coordinate system, similar to that shown in Figures 12.11 and 12.12. We shall further assume that the wave vector $\mathbf{k}$ is in the x–y plane, with θ being the angle between the wave vector and the ambient magnetic field. As in Section 12.4.3.2, we derive a set of simultaneous equations for the wave electric field $\tilde{\mathbf{E}}$ and the electron-flow velocity $\tilde{\mathbf{u}}_e$.

Assuming a first-order harmonic wave perturbation, (12.76) becomes

$$-i\omega m_e \tilde{\mathbf{u}}_e + \tilde{\mathbf{E}} + e\tilde{\mathbf{u}}_e \times \mathbf{B}_0 = 0 \tag{12.99}$$

while (12.77)–(12.79) become

$$-i\omega n_0 e \mu_0 \tilde{\mathbf{u}}_e = \left(k^2 - \frac{\omega^2}{c^2}\right)\tilde{\mathbf{E}} - \mathbf{k}(\mathbf{k} \cdot \tilde{\mathbf{E}}) \tag{12.100}$$

We shall define $R = 1 - \omega_{pe}^2/\omega(\omega + \Omega_{ce})$, $L = 1 - \omega_{pe}^2/\omega(\omega - \Omega_{ce})$, and $P = 1 - \omega_{pe}^2/\omega^2$. On further defining $S = \frac{1}{2}(R+L)$ and $D = \frac{1}{2}(R-L)$, (12.99) and (12.100) yield three simultaneous equations for the wave electric field:

$$\begin{pmatrix} n^2\cos^2\theta - S & iD & -n^2\cos\theta\sin\theta \\ -iD & n^2 - S & 0 \\ -n^2\cos\theta\sin\theta & 0 & n^2\sin^2\theta - P \end{pmatrix} \begin{pmatrix} \tilde{E}_x \\ \tilde{E}_y \\ \tilde{E}_z \end{pmatrix} = 0 \tag{12.101}$$

where we have substituted $k_x = k \sin\theta$, $k_z = k \cos\theta$, and the refractive index $n = kc/\omega$. Equation (12.101) reduces to a biquadratic equation of the general form $An^4 + Bn^2 + C = 0$ (at first sight, this should be bicubic, but the n^6 term vanishes). However, before discussing the solutions of (12.101) for arbitrary propagation angles, we shall first show that the results discussed in Sections 12.4.3.1 and 12.4.3.2 are contained within (12.101).

First, if we consider either parallel or perpendicular propagation, then the term $n^2 \cos\theta \sin\theta$ vanishes, and the solution involving $\tilde{E}_z$ is separable from the other electric-field terms. The ordinary-mode dispersion relation (12.80) is found for perpendicular propagation, on substituting the definition of P into the equation for $\tilde{E}_z$. After some algebra, the two simultaneous equations in $\tilde{E}_x$ and $\tilde{E}_y$ also yield the X-mode dispersion relation (12.86) for perpendicular propagation.

For parallel propagation, the parallel electric-field term results in plasma oscillations. For the transverse field components, (12.101) reduces to

$$n^2 = S \pm D = R \text{ or } L \tag{12.102}$$

which is the same as (12.97)

For general angles of propagation, we could simply determine the roots of the biquadratic equation given earlier. However, a more generally used form is known as the Appleton-Hartree equation (neglecting ions and collisions). In giving this form of the cold-plasma dispersion relation, we shall introduce another common notation, where we define $X = \omega_{pe}^2/\omega^2$ and $Y = |\Omega_{ce}|/\omega$ (note that this is not the X that means "extraordinary"):

$$n^2 = 1 - \frac{X}{1 - \frac{\frac{1}{2}Y^2 \sin^2\theta}{1-X} \pm \left\{\left(\frac{\frac{1}{2}Y^2 \sin^2\theta}{1-X}\right)^2 + Y^2 \cos^2\theta\right\}^{\frac{1}{2}}} \tag{12.103}$$

Figure 12.17 shows solutions of the Appleton-Hartree dispersion relation in a form similar to that of Figures 12.13 and 12.15. In this figure we have combined both parallel and perpendicular propagation and indicated by shading those regions where propagation at oblique angles is allowed. The left panel shows the dispersion curves when the electron plasma frequency is greater than the electron gyrofrequency, which is normally the case in the terrestrial magnetosphere. The right panel shows the dispersion for regions of strong magnetic field and low density, where the plasma frequency is less than the gyrofrequency. In each panel we have labeled the modes by their most common names. The superluminous (faster-than-light) modes are identified by the dispersion relation for both parallel (R or L) and perpendicular (X or O) propagation, because the wave dispersion for arbitrary angle lies between these two limiting curves. For example, the two modes that can escape from

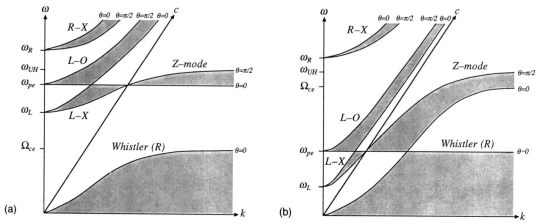

FIG. 12.17. Dispersion diagrams derived from the Appleton-Hartree dispersion relation. Shading denotes regions of oblique propagation. (a) Dispersion when the electron plasma frequency is greater than the electron gyrofrequency. (b) Dispersion when the electron plasma frequency is less than the electron gyrofrequency.

a magnetosphere are the *R*-X and *L*-O modes. These modes can escape because there is no upper frequency limit to the dispersion curves. The *L*-X mode, on the other hand, is trapped, because this mode cannot propagate above the plasma frequency as a superluminous mode. However, a wave in the *L*-X mode could convert to either the *L*-O mode or the subluminous *Z* mode. Above the speed of light, the *Z* mode becomes the *L*-X mode; the dispersion curves for these two modes all pass through $n^2 = 1$ for $\omega = \omega_{pe}$. The *Z* mode is so named because mode conversion of an *L*-O-mode wave to the superluminous branch of the *Z* mode is thought to be responsible for the third trace (labeled F_z) in Figure 12.10, which is often referred to as the *Z*-trace.

There are two waves that propagate below the speed of light in a cold plasma: the *Z* mode and the whistler mode. For low gyrofrequencies, the *Z* mode lies between the plasma frequency and the upper hybrid frequency, while the whistler mode lies below the electron gyrofrequency. When the gyrofrequency is greater than the plasma frequency, the whistler mode lies below the plasma frequency, while the *Z* mode covers a much broader range in frequency than before.

Figure 12.17 also shows why an escaping *R*-X-mode wave cannot be generated below the upper hybrid frequency. However, a wave generated in the *Z* mode could escape from a magnetosphere as an *L*-O mode. This is how continuum radiation is thought to be generated. Waves generated at the local upper hybrid frequency through some form of Landau resonance or beam instability (where the wave phase speed is comparable to the electron-drift speed) propagate as *Z*-mode waves until they encounter a region of enhanced plasma density, such that the plasma frequency is greater than the wave frequency. At this point some of the waves may be reflected as *L*-O-mode waves, and some waves may propagate into the increased-density region as *L*-X-mode waves. As the density increases, these *L*-X waves are also reflected. As the *L*-X waves finally propagate into regions of reduced density, they convert to *L*-O-mode waves, which can escape from the plasma. Other waves that escape from planetary magnetospheres, such as auroral kilometric radiation at the earth, or Jovian decametric radiation, are observed to be in the *R*-X mode. Simple linear mode conversion, as discussed earlier, cannot couple into this mode, and other instabilities that allow particles to interact directly with the superluminous *R*-X mode must be considered.

One last point from Figure 12.17 concerns the three traces shown in Figure 12.10. Figure 12.17 shows that there are three cutoffs for the superluminous waves, at ω_R, ω_{pe}, and ω_L. As an electromagnetic wave propagates into the ionosphere from below, the plasma density increases, and these characteristic frequencies increase. For a particular wave frequency, then, as this wave propagates to higher altitudes, the wave will first encounter the *R*-X-mode cutoff, and that portion of the signal propagating in this mode will be reflected (the X-trace). The *L*-O mode will continue to higher altitudes, until this mode also reflects (the O-trace). However, some of the *L*-O mode may propagate to even higher altitudes as an *L*-X mode, where reflection will occur at the *L*-X-mode cutoff (the Z-trace).

We have discussed at some length the superluminous waves. However, much of the research on plasma waves in the magnetosphere has been concerned with whistler-mode waves, whose phase speeds are less than *c*. There is a useful approximation to the Appleton-Hartree dispersion relation that is used for whistler-mode waves, known as the quasi-parallel or quasi-longitudinal approximation (in this context, "quasi-longitudinal" refers to the wave-vector direction, not the wave electric field). The quasi-parallel approximation applies when the first term in the square-root brackets in equation (12.103) is much less than the second term. This approximation is usually valid in the earth's magnetosphere when considering whistler-mode waves, because the wave frequency is less than the electron plasma frequency, and so $X \gg 1$. For whistler-mode waves, then,

$$n^2 = 1 - \frac{X}{(1 - Y \cos \theta)} = 1 - \frac{\omega_{pe}^2}{\omega(\omega - |\Omega_{ce}| \cos \theta)} \quad (12.104)$$

for frequencies much greater than the lower hybrid frequency. We must include ions in the dispersion relation if the wave frequency is close to the lower hybrid frequency. The dispersion relation (12.104) has a resonance when $\omega = |\Omega_{ce}| \cos \theta_R$. The angle is known as the resonance-cone angle, and whistlers can propagate only within a cone where $\theta \leq \theta_R$ about the magnetic field. Figure 12.18 shows a typical dispersion surface for the whistler mode.

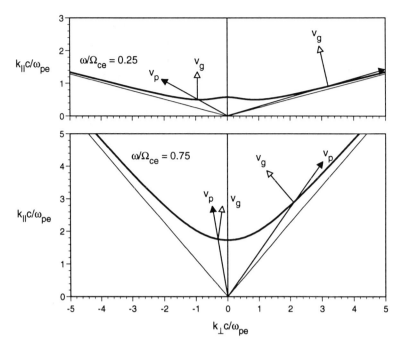

FIG. 12.18. Typical dispersion surface for the whistler mode. Top: Frequencies below half the electron gyrofrequency. Bottom: Frequencies above half the electron gyrofrequency.

In Figure 12.18 we plot the magnitude of the wave number **k** as a function of the propagation direction θ for a fixed value of frequency. Plotting such a curve is useful for understanding the relationship between the wave-propagation direction and the ray or group-velocity direction. The group velocity $\partial\omega/\partial\mathbf{k}$ lies in a direction given by the normal to the wave-vector surface. This is because the curve can be considered to be a curve of constant frequency plotted in wave-vector space. As an analogy, consider how an electric field is determined from the contours of constant electrostatic potential in configuration space.

12.4.4 Alfvénic Waves

In the preceding sections we have neglected ion motion. This should be a good approximation for high-frequency waves, but we may also have to deal with low-frequency perturbations involving a magnetic field. Those waves will be affected by ion motion. To simplify matters, we shall neglect the electron-inertia term $[m_e(\partial \mathbf{u}_e/\partial t)]$. For slow variations, it should be small. We then have

$$\mathbf{k} \times \tilde{\mathbf{E}} - \omega \tilde{\mathbf{B}} = 0 \tag{12.105}$$

$$i(c^2 \mathbf{k} \times \tilde{\mathbf{B}} + \omega \tilde{\mathbf{E}}) = \frac{n_0 e}{\epsilon_0} (\tilde{\mathbf{u}}_i - \tilde{\mathbf{u}}_e) \tag{12.106}$$

$$0 = e n_0 \tilde{\mathbf{E}} + e n_0 \tilde{\mathbf{u}}_e \times \mathbf{B}_0 \tag{12.107}$$

$$-i\omega m_i n_0 \tilde{\mathbf{u}}_i = e n_0 \tilde{\mathbf{E}} + e n_0 \tilde{\mathbf{u}}_i \times \mathbf{B}_0 \tag{12.108}$$

These can be written as algebraic equations once the direction of **k** is specified. We first look at $\mathbf{k} = k\hat{z}$ (i.e., a wave propagating along $\mathbf{B}_0$). We can show that for $\omega \ll \Omega_{ci}$, the wave polarization is linear (i.e., the electric field is not rotating), and $\tilde{\mathbf{u}}_i - \tilde{\mathbf{u}}_e$ is parallel to $\tilde{\mathbf{E}}$. It is then relatively straightforward to show that for $\omega \ll \Omega_{ci}$, the dispersion relation is

$$\omega^2 = \frac{k^2 c^2}{1 + \omega_{pi}^2/\Omega_{ci}^2} = \frac{k^2 c^2}{1 + n_0 m_i / B_0^2 \epsilon_0} \quad (12.109)$$

We recognize a familiar quantity, namely, the Alfvén speed,

$$v_A = \mathbf{B}_0 / \sqrt{\mu_0 n_0 m_i} \quad (12.110)$$

and remembering that $c^2 = 1/(\mu_0 \epsilon_0)$, we rewrite equation (12.109) as

$$\omega^2 = \frac{k^2 v_A^2}{1 + v_A^2/c^2} \quad (12.111)$$

This is the dispersion relation for *Alfvén waves*. The physical interpretation of an Alfvén wave was discussed in Chapter 11, where the approximation $v_A^2 \ll c^2$ was used.

When the wave vector is perpendicular to the magnetic field, we can have two transverse waves, one in which the electric-field vector is along $\mathbf{B}_0$ (we have already looked at that mode), and one in which it is perpendicular to it. This wave is called a *magnetosonic wave*. It is easy to show that the dispersion relation is identical with that for the parallel-propagating Alfvén wave.

12.4.5 Summary of Electromagnetic Waves in a Magnetized Plasma

For perpendicular propagation, we found two high-frequency waves: the ordinary and extraordinary waves. The ordinary wave has a dispersion relation that does not depend on the background magnetic field; it propagates very much like the electromagnetic wave in an unmagnetized plasma. This wave has a cutoff (reflection point) as $\omega = \omega_{pe}$, but no resonance frequency. The extraordinary wave has cutoff frequencies both above and below the plasma frequency. Its resonance frequency is the upper hybrid frequency.

For oblique propagation, the wave modes are mixtures of the modes found for parallel and perpendicular propagation. Superluminous wave modes are the *R*-X-, *L*-O-, and *L*-X-mode waves. Waves propagating below the speed of light are the *Z* mode (which has a resonance for perpendicular propagation at the hybrid resonance frequency) and the whistler mode. The whistler is a right-hand polarized wave for parallel propagation and has a resonance at the gyrofrequency. For oblique propagation, there is a resonance that defines a cone about the magnetic field. The whistler cannot propagate outside of this resonance cone.

At low frequencies ($\omega < \Omega_{ci}$), the electromagnetic waves are Alfvén waves or magnetosonic waves. These waves were discussed in Chapter 11.

12.5 KINETIC THEORY AND WAVE INSTABILITIES

In the preceding sections we have considered wave propagation for a plasma in which the different species can be considered to be moving as a fluid, characterized by density, flow velocity, and temperature (in the case of warm plasma). Instabilities in such a plasma are driven by the relative streaming between the different plasma populations, such as the two-stream instability discussed in Section 12.3.3. Rather than using macrophysical parameters to describe the plasma, we can use a kinetic formulation, where the dynamics of the individual particle is taken into account. In order to characterize the whole plasma using individual-particle dynamics, we use the phase-space density or distribution function $f(\mathbf{r}, \mathbf{v}, t)$ introduced in Chapter 2.

12.5.1 The Vlasov Equation

The principle of particle conservation states that the time rate of change for the number of particles in a particular volume is given by the flux of particles across the surface of the volume, in the absence of either sources or sinks for the particles. Collisions can introduce discontinuities in the motion of a particle through velocity space, thus producing effective sources and sinks, but in the absence of collisions, phase-space density obeys a continuity equation in six-dimensional space. This can be expressed as a differential equation:

$$\frac{\partial f}{\partial t} + \mathbf{\nabla} \cdot (\mathbf{v}f) + \mathbf{\nabla}_v \cdot (\mathbf{a}f) = 0 \tag{12.112}$$

where $\mathbf{\nabla}_v$ is the gradient with respect to velocity, and $\mathbf{a}$ is the acceleration, or time rate of change of velocity ($\partial \mathbf{v}/\partial t$). Any position in phase space is specified by $\mathbf{r}$ and $\mathbf{v}$, and so $\mathbf{v}$ is independent of $\mathbf{r}$. Furthermore, if $\mathbf{a}$ is determined only by forces that do not depend on velocity, that is, forces like the Lorentz force, for which $\mathbf{\nabla}_v \cdot \mathbf{a} = 0$, then (12.112) becomes

$$\frac{\partial f}{\partial t} + \mathbf{v} \cdot \mathbf{\nabla} f + \mathbf{a} \cdot \mathbf{\nabla}_v f = 0 \tag{12.113}$$

which is known as the Vlasov or collisionless Boltzmann equation. If collisions occur, then the right-hand side of (12.113) no longer vanishes. It should be noted that the term "collisions" covers any interaction process that causes particles to be correlated, apart from the "external" forces included in $\mathbf{a}$. In writing (12.113) we have not addressed the more rigorous aspects of kinetic theory, in which interparticle correlation is considered.

Two very useful theorems can be derived from (12.113). The first is Liouville's theorem, which states that phase-space density is constant along a particle trajectory. The second is Jeans's theorem, which states that any distribution function that is a function of the constants of the motion satisfies (12.113). For Liouville's theorem, we note that the derivative on the left of (12.113) is similar to the advective derivative $D/Dt = \partial/\partial t + \mathbf{u} \cdot \nabla$, often used in fluid flows, that is, the total time derivative following a fluid element. In phase space we define the Liouville operator $L = \partial/\partial t + \mathbf{v} \cdot \nabla + \mathbf{a} \cdot \nabla_v$, and (12.113) can be rewritten as $Lf = 0$. The Liouville operator is just the total time derivative following a particle's motion in phase space: $\delta \mathbf{v} = \mathbf{a} \delta t$ and $\delta \mathbf{r} = \mathbf{v} \delta t$.

Jeans's theorem can be understood by noting that the equation of motion for a particle trajectory in n-dimensional phase space is a set of n differential equations, which on integration give n constants of the motion, α_i, $i = 1, 2, \ldots, n$. If we assume that the distribution function is a function of the constants of the motion, $f(\alpha_i)$, we can rewrite the left-hand side of (12.113) as

$$Lf = \sum_{i=1}^{n} (L\alpha_i) \partial f / \partial \alpha_i \qquad (12.114)$$

Because α_i are constants of the motion, $L\alpha_i = 0$. Hence, the right-hand side of (12.114) is zero, and $f(\alpha_i)$ satisfies the Vlasov equation: This is Jeans's theorem.

12.5.2 Landau Resonance

The Vlasov equation (12.113) is the basis of any plasma-wave-dispersion relation derived from kinetic theory. Here we shall outline the methods used in obtaining a plasma dispersion relation, but we shall not present a complete derivation of a kinetic dispersion relation, as this is usually rather complicated. As discussed in Section 12.3.1, the first step is to linearize the governing equations. Equation (12.113) becomes

$$\frac{\partial f_1}{\partial t} + \mathbf{v} \cdot \nabla f_1 + \frac{q}{m} (\mathbf{v} \times \mathbf{B}_0) \cdot \nabla_v f_1 = -\frac{q}{m} (\mathbf{E}_1 + \mathbf{v} \times \mathbf{B}_1) \cdot \nabla_v f_0 \qquad (12.115)$$

where f_0 is the unperturbed distribution function of the plasma, which we assume to be uniform, and f_1 is the perturbation of f. We further assume that the only zero-order force is the Lorentz force, with $\mathbf{E}_0 = 0$. The first-order forces are due to the wave electric ($\mathbf{E}_1$) and magnetic ($\mathbf{B}_1$) fields.

The left-hand side of (12.115) is the time derivative of the first-order distribution function following an unperturbed-particle trajectory. Formally, the solution of (12.115) is

$$f_1(\mathbf{r}, \mathbf{v}, t) = -\int_{-\infty}^{t} dt' \, \frac{q}{m}(\mathbf{E}_1 + \mathbf{v} \times \mathbf{B}_1) \cdot \nabla_v f_0 \qquad (12.116)$$

where the integration follows an unperturbed-particle trajectory to the point $(\mathbf{r}, \mathbf{v})$ in phase space. From this we can determine a perturbation charge density (ρ_{q1}) for substitution into Poisson's law for electrostatic waves, or a perturbation current density ($\mathbf{j}_1$) for substitution into Ampère's law for electromagnetic waves:

$$\rho_{q1} = \sum_{\text{species}} q \int f_1 \, d\mathbf{v} \qquad (12.117)$$

$$\mathbf{j}_1 = \sum_{\text{species}} q \int \mathbf{v} f_1 \, d\mathbf{v} \qquad (12.118)$$

In practice, solving (12.116) is fairly complicated. Rather than present the full derivation of the dispersion relation, we shall make several simplifying assumptions to demonstrate some of the features of kinetic theory. First, as in Section 12.3.1, we assume a harmonic perturbation varying as $\exp[-i(\omega t - \mathbf{k} \cdot \mathbf{r})]$, in which case (12.115) becomes

$$i(\omega - \mathbf{k} \cdot \mathbf{v})\tilde{f} - \frac{q}{m}(\mathbf{v} \times \mathbf{B}_0) \cdot \nabla_v \tilde{f} = \frac{q}{m}[\tilde{\mathbf{E}}(1 - \mathbf{k} \cdot \mathbf{v}/\omega) + \mathbf{k}(\mathbf{v} \cdot \tilde{\mathbf{E}}/\omega)] \cdot \nabla_v f_0$$
$$(12.119)$$

where we have used Faraday's law to substitute for the wave magnetic field, and ~ indicates a first-order harmonic perturbation.

In (12.119), the velocity $\mathbf{v}$ is the velocity along an unperturbed-particle trajectory. In a uniform magnetic field, the constants of a particle's motion are the parallel velocity ($\mathbf{v}_\parallel$) and the magnitude of the perpendicular velocity ($\mathbf{v}_\perp$). However, the particle gyrates about the ambient magnetic field, and $\mathbf{v}_\perp$ is not time-invariant. We shall therefore assume that the waves are propagating parallel to the ambient magnetic field. This removes any gyrational terms in the scalar product $\mathbf{k} \cdot \mathbf{v}$ that would introduce gyroharmonics into the wave-dispersion relation.

As is evident from (12.117) and (12.118), f_1 enters the wave-dispersion relation through integration over velocity space. We shall assume, as usual, that the ambient magnetic field defines the z-axis of a cartesian system. In this case, the second term on the left-hand side of (12.119) contains only derivatives of the form $v_y \, \partial f_1/\partial v_x$ and $v_x \, \partial f_1/\partial v_y$. Because the charge density does not introduce any additional velocity components in the integrals, and because $f_1(\mathbf{v} \to \infty) = 0$, as required for convergence of the velocity components in the integrals over velocity space, the second term of (12.119) vanishes on integration over perpendicular velocity ($\int dv_x \int dv_y$) when calculating the charge density. We require the charge density for electrostatic waves, which have only a parallel electric field under the assumption of parallel propagation. In this case, the right-hand side (12.119) reduces to $(q/m)\tilde{E}_\parallel \partial f_0/\partial \mathbf{v}_\parallel$. For parallel-propagating electrostatic waves, we find that the perturbation charge density per species is given by

$$\tilde{\rho}_s = -i\frac{q^2}{m}\tilde{E}_\parallel \int dv_\parallel \frac{\partial f_0/\partial v_\parallel}{\omega - k_\parallel v_\parallel} \tag{12.120}$$

There is a resonance in the integration over parallel velocity in (12.120). This is the Landau resonance, which occurs when the particle velocity equals the parallel phase velocity of the wave. A physical picture of Landau resonance can be obtained by considering any particle that is moving at or near the phase velocity of the wave. Such a particle will see an electric field that is roughly constant in time and hence will be accelerated by the wave field. Particles moving much slower or faster than the wave will tend to see an electric field that averages to zero. Consequently, only those particles with velocities close to the phase velocity will resonate with the wave. Particles in resonance moving slightly faster than the wave will lose energy, while those moving slightly slower will gain energy. In a Maxwellian plasma there are more particles at low velocity than at high velocity, and so the plasma gains energy at the expense of the waves. This is Landau damping, which mathematically arises out of the fact that a Maxwellian distribution is stable. On the other hand, if a distribution function has more particles at higher velocity than at lower velocity in some part of velocity space, then that distribution will be unstable to waves that are in resonance with the particles. This is the "bump-on-tail" instability.

12.5.3 Gyroresonance

In addition to the Landau resonance, other resonances arise out of kinetic theory. These are the gyroresonances, which occur when the Doppler-shifted frequency (as observed by a particle moving with the parallel phase speed of the wave) is some integral multiple of the particle gyrofrequency. The integer can be positive or negative, but the strongest interaction usually occurs when the Doppler-shifted frequency exactly matches the particle gyrofrequency. Because the resonance is associated with particle gyration, it is reasonable to assume that the interaction is associated with perpendicular wave fields. We shall demonstrate that more clearly here.

As for Landau resonance, we shall use (12.119) as a basis, again assuming parallel propagation. This time, however, we shall further assume that the wave electric field and hence the wave magnetic field are perpendicular to the ambient magnetic field. We shall also make use of notation known as polarized coordinates, where we define $\tilde{E}_l = (\tilde{E}_x + i\tilde{E}_y)/\sqrt{2}$ and $\tilde{E}_r = (\tilde{E}_x - i\tilde{E}_y)/\sqrt{2}$. Under the assumption of a harmonic perturbation varying as $\exp[-i(\omega t - \mathbf{k}\cdot\mathbf{r})]$, it can be shown that if

$\tilde{E}_r = 0$, $\tilde{E}_x$ leads $\tilde{E}_y$ by a quarter of a wave period, and the wave is right-hand circularly polarized.

If we multiply (12.119) by $v_r = (v_x - iv_y)/\sqrt{2}$ and integrate over perpendicular velocity, we find that the left side can be written

$$\int_{-\infty}^{\infty} dv_x \int_{-\infty}^{\infty} dv_y \frac{(v_x - iv_y)}{\sqrt{2}} \int \left(i(\omega - \mathbf{k} \cdot \mathbf{v}) \tilde{f} - \frac{q}{m} (\mathbf{v} \times \mathbf{B}_0) \cdot \nabla_v \tilde{f} \right) \qquad (12.121)$$

$$= \int_{-\infty}^{\infty} dv_x \int_{-\infty}^{\infty} dv_y \, v_r i \int (\omega - k_\| v_\| + \Omega_c) \tilde{f}$$

In deriving (12.121), we have used the convergence of the integrals at infinity and integration by parts to simplify the $\partial f_1/\partial \mathbf{v}$ terms.

On integrating the right-hand side of (12.119), we make use of the fact that $f_0 = f_0(v_\|, v_\perp)$, as required by Jeans's theorem, where $v_\perp = (v_x^2 + v_y^2)^{\frac{1}{2}}$, and that f_0 is a symmetric function of $v_\perp$. In this case,

$$\int_{-\infty}^{\infty} dv_x \int_{-\infty}^{\infty} dv_y \, v_x f_0 = \int_{-\infty}^{\infty} dv_x \int_{-\infty}^{\infty} dv_y \, v_y f_0 = 0 \qquad (12.122)$$

$$\int_{-\infty}^{\infty} dv_x \int_{-\infty}^{\infty} dv_y \, v_x^2 f_0 = \int_{-\infty}^{\infty} dv_x \int_{-\infty}^{\infty} dv_y \, v_y^2 f_0 = \frac{1}{2} \int_{-\infty}^{\infty} dv_x \int_{-\infty}^{\infty} dv_y \, v_\perp^2 f_0 \qquad (12.123)$$

and

$$\int_{-\infty}^{\infty} dv_x \int_{-\infty}^{\infty} dv_y \, v_x \frac{\partial f_0}{\partial v_x} = \int_{-\infty}^{\infty} dv_x \int_{-\infty}^{\infty} dv_y \, v_y \frac{\partial f_0}{\partial v_y} = \frac{1}{2} \int_{-\infty}^{\infty} dv_x \int_{-\infty}^{\infty} dv_y \, v_\perp \frac{\partial f_0}{\partial v_\perp} \qquad (12.124)$$

Using these identities, we find

$$\int_{-\infty}^{\infty} dv_x \int_{-\infty}^{\infty} dv_y \frac{(v_x - iv_y)}{\sqrt{2}} \left(\left\{ \tilde{\mathbf{E}} \left(1 - \frac{\mathbf{k} \cdot \mathbf{v}}{\omega} \right) + \mathbf{k} \left(\frac{\mathbf{v} \cdot \tilde{\mathbf{E}}}{\omega} \right) \right\} \cdot \nabla_v f_0 \right) \qquad (12.125)$$

$$= \frac{1}{2} \int_{-\infty}^{\infty} dv_x \int_{-\infty}^{\infty} dv_y \, v_\perp \left(\left(1 - \frac{k_\| v_\|}{\omega} \right) \frac{\partial f_0}{\partial v_\perp} + \frac{k_\| v_\perp}{\omega} \frac{\partial f_0}{\partial v_\|} \right) \tilde{E}_r$$

Combining (12.121) and (12.125), and integrating over v_z, we find the current density per species:

$$\tilde{j}_{sr} = -i \frac{q^2}{m} \tilde{E}_r \int d\mathbf{v} \, v_\perp \frac{\left(\left(1 - \frac{k_\| v_\|}{\omega} \right) \frac{\partial f_0}{\partial v_\perp} + \frac{k_\| v_\perp}{\omega} \frac{\partial f_0}{\partial v_\|} \right)}{(\omega - k_\| v_\| + \Omega_c)} \qquad (12.126)$$

In (12.126), the resonance condition is the gyroresonance, and it is associated with perpendicular wave fields. This resonance is important for generating waves such as the whistler mode, which is polarized predominantly perpendicular to the ambient magnetic field, as discussed in Section 12.4. It can be shown that for a purely transverse wave in which both the wave electric field and magnetic field are perpendicular to the wave vector, the net energy change for a particle moving with the phase speed of the wave is zero. This can also be seen by transforming the numerator in (12.126) to the wave frame, that is, to a frame in which the parallel phase velocity of the wave is zero. In this frame the gradient in velocity space is a gradient with respect to pitch angle. The effect of the gyroresonance is hence to cause particles to change pitch angle in the wave frame. If the wave frequency is much less than the particle gyrofrequency, then the plasma frame and wave frame are essentially the same, and the main effect of gyroresonance is to cause pitch-angle diffusion in the plasma frame, rather than energy diffusion. This should be contrasted with Landau resonance, where the diffusion is in parallel velocity (due to the $\partial f_0/\partial v_\parallel$ term) and hence mainly in energy, rather than pitch angle.

For a wave to grow from gyroresonance, there should be a net decrease in particle energy as the particle diffuses down the phase-space density gradient defined by the numerator in (12.126). For whistler-mode waves, the wave frequency is less than the electron gyrofrequency. Resonant electrons and the wave must consequently travel in opposite directions (i.e., $\mathbf{k} \cdot \mathbf{v}_\parallel < 0$), so as to increase the Doppler-shifted wave frequency to the gyrofrequency. In this case, an electron that moves from a high to a low pitch angle in the wave frame loses energy in the plasma frame, because a contour of constant energy is simply a circle centered on the origin. For this reason, electrons can generate whistler-mode waves if there is a loss cone in the distribution, or if the perpendicular temperature is higher than the parallel temperature.

Ions can also generate whistler-mode waves, but in this case the resonant particles must be traveling faster than the wave, so that $\mathbf{k} \cdot \mathbf{v}_\parallel > \omega$. In this case, the effect of the Doppler shift is to change the sense of rotation of the wave electric field in the resonant-particle frame from right-handed to left-handed. This can be seen from our definition of polarized coordinates, where we assumed a temporal variation given by $\exp(-i\omega t)$. If we change the sign of the frequency, then $\bar{E}_l = 0$ corresponds to a left-hand circularly polarized wave. Ions gyrate in a left-handed sense about the ambient field and thus will interact strongly with a left-hand polarized wave. In this case, where the resonant particle is moving in the same direction as the wave, diffusion from a low pitch angle to a high pitch angle in the wave frame corresponds to a decrease in energy in the plasma frame. Hence, ions can generate whistler-mode waves provided the parallel temperature is sufficiently large with respect to the perpendicular temperature.

12.5.4 Summary of Kinetic Theory

Kinetic theory introduces resonance into the wave-dispersion relation. For parallel propagation, we find that Landau resonance is associated with parallel electric fields. For perpendicular electric fields, particles and fields can be in gyroresonance. Landau resonance diffuses particles parallel to the magnetic field, whereas gyroresonance causes diffusion in pitch angle. As such, then, Landau-resonant instabilities are often driven by "beamlike" distributions, whereas gyroresonant instabilities are driven by pitch-angle anisotropy.

ADDITIONAL READING

Introductory Material

Boyd, T. J. M., and J. J. Sanderson. 1969. *Plasma Dynamics*. London: Thomas Nelson & Sons.

More Formal Discussion of Plasma-Wave Theory

Clemmow, P. C., and J. P. Dougherty. 1969. *Electrodynamics of Particles and Plasmas*. Reading, MA: Addison-Wesley (reissued 1990).
Melrose, D. B. 1986. *Instabilities in Space and Laboratory Plasmas*. Cambridge University Press.
Stix, T. H. 1962. *The Theory of Plasma Waves*. New York: McGraw-Hill.
Stix, T. H. 1992. *Waves in Plasmas*. New York: American Institute of Physics [revised and extended version of Stix (1962)].

PROBLEMS

12.1. Derive equation (12.19).

12.2. Show that if the finite mass of the ions is included, the frequency of Langmuir waves in cold plasma is given by $\omega^2 = \omega_{pe}^2 + \omega_{pi}^2$, where $\omega_{pi}^2 = e^2 n_0 / m_i \epsilon_0$.

12.3. Derive equation (12.39) from equation (12.38).

12.4. Derive equation (12.44).

12.5. Derive equation (12.52).

12.6. Sketch the group velocity $\partial \omega / \partial k$ and the phase speed ω/k versus ω of the whistler mode, and discuss how this may explain the rate at which the frequency decreases with time for a lightning-generated whistler.

12.7 Show that $\omega_L = |\Omega_{ce}|$ when $\omega_{pe} = \sqrt{2} |\Omega_{ce}|$.

12.8. Show that the dispersion relation given by (12.101) gives the X-mode dispersion relation (12.86) for perpendicular propagation.

12.9. Derive the Appleton-Hartree dispersion relation as given by (12.103). Hint: Use $n^2 = 1 - X/(1-\lambda)$ and solve for λ. Also, prove and make use of the identity $P - S = PS - S^2 + D^2$.

12.10. When an elliptically polarized electromagnetic wave propagates through a magnetized plasma, the orientation of the polarization rotates as it travels. This is known as Faraday rotation. Why does it occur? (You need treat only the case of parallel propagation.)

12.11. Show through geometric construction why gyroresonant electrons lose energy (in the plasma frame) on moving to a lower pitch angle (in the wave frame) when in resonance with whistler-mode wave and hence require either a loss cone or $T_\perp/T_\parallel > 1$. Similarly, show why ions require $T_\perp/T_\parallel < 1$.

13 MAGNETOSPHERIC DYNAMICS

R. L. McPherron

13.1 INTRODUCTION

THE MAGNETIC FIELD of the terrestrial magnetosphere is produced by superposition of magnetic fields from a variety of sources. At the earth's surface, the most important source is the main field produced by currents within the earth's liquid core. Remanent magnetization and magnetization induced by the main field are also important. At ionospheric heights of about 120 km, electric currents, including the solar dynamo, the equatorial electrojet, the convection electrojets, and the substorm electrojet, are other important sources of magnetic field. Field-aligned currents link these ionospheric currents to the outer parts of the earth's magnetosphere. These currents, which transmit stresses from the outer magnetosphere to the ionosphere, produce effects on the earth indistinguishable from those produced by current systems flowing solely in the ionosphere. At about $3-5R_E$, particles drifting in the Van Allen radiation belts produce a westward ring current that, depending on the total energy of particles, reduces the horizontal component of the surface field [see equation (10.21)]. On the dayside of the earth, at distances of $10-15R_E$ at the boundary between the solar wind and the main field, there is a thin sheet of current, the magnetopause current. The magnetopause current almost entirely confines the earth's field to the magnetosphere. This current increases the magnetic field everywhere inside the magnetopause. Tangential stress between the solar wind and the earth's field drags the earth's field lines and plasma antisunward, forming a long magnetic tail behind the earth. This process produces another current called the tail current, which flows across the midnight meridian in the same direction as the ring current. Its main effect at the surface of the earth is to reduce the total field.

All of the currents above the ionosphere are controlled by the solar wind. The two most important controlling parameters are dynamic pressure, which depends on the solar-wind velocity and density, and the dawn–dusk component of electric field, which depends on the velocity and the north–south magnetic field. When any of these variables change, there are corresponding changes in the strength, location, and distribution of currents. The changes in currents are reflected by changes in the magnetic field at the earth's surface. Such changes are called *geomag-*

netic activity. The field of study that examines the relation between the solar wind and geomagnetic activity (i.e, solar-wind–magnetosphere coupling) is called solar-terrestrial physics. Geomagnetic activity can affect us in many practical ways. Radio communications, radar observations, electrical utilities, long-distance pipelines, and synchronous spacecraft are among the systems affected when geomagnetic activity reaches peak values.

The study of solar-wind coupling has a long history, beginning with studies of the relationship of geomagnetic activity to events on the sun. These studies utilize measurements of the earth's field made by instruments called magnetometers. Various techniques for measuring magnetic fields have been developed; some of the most common are described in Appendix 13A. Originally, these depended on magnets and photographic recording. Today, proton precession and fluxgate magnetometers produce digital signals that are recorded on magnetic media such as computer tapes. The original records from many locations spread over the entire globe are too complex and too voluminous to be manipulated easily. Magnetic indices have been developed to replace these data with crude measures of the strength of the various sources. Most studies of solar-wind coupling are performed with these indices. Descriptions of the most commonly used indices can be found in Appendix 13B.

Magnetic-activity indices are well correlated with events on the sun. Magnetic activity rises and falls, as does the number of sunspots during the 11-yr cycle of sunspot activity. Magnetic activity is also modulated by the location of the earth in its orbit around the sun. Annual and semiannual variations can easily be seen in magnetic indices. The rotation of the sun about its axis approximately every 27 days (as seen from the moving earth) is also an important factor. Activity indices display a very strong tendency to repeat every solar rotation.

Although the sun is the original cause of almost all geomagnetic activity, the primary means by which it exerts its influence is the solar wind. As discussed in Chapter 4, at the earth the solar wind blows radially outward at velocities of 300–$1{,}000$ km·s^{-1}, carrying particles at densities of 1–50 cm^{-3}. Embedded in the solar wind is a magnetic field that normally lies in the ecliptic plane, with a strength of 3–30 nT. It has been found that some activity indices increase as the square of the solar-wind velocity. However, almost no magnetic activity is seen unless the interplanetary magnetic field (IMF) embedded in the solar wind has a southward component, antiparallel to the earth's magnetic field near the subsolar point on the dayside magnetopause. This dependence is quantified in the half-wave rectifier model. According to this model, which is an approximation to the best available models, whenever the IMF has a southward component, activity is proportional to the cosine of the clock angle of the IMF around the earth–sun line. There is no activity predicted for northward IMF. Even when the IMF is along the expected Archimedean spiral, a southward component can occur because of the orientation of the earth's magnetic axis. Solar flares, coronal mass ejections, and the perturbed solar wind found at the interfaces between high and low-speed streams, referred to as corotating interaction regions, can also generate substantial north–south components of the IMF.

Attempts have been made to define precise solar-wind coupling parameters that can be used to predict the strength of magnetic activity. The procedures assume that the magnetosphere responds to the solar wind as if it were a deterministic system driven by the solar wind. Various techniques have been used to determine the best energy-coupling parameters and their most general linear relationships with geomagnetic activity. Such studies reveal that at time scales of less than about 3 h, only about half of the variance in magnetic-activity indices is predictable. The unpredictable residual is related to discrete events in which energy stored within the magnetosphere by the interaction of the solar wind with the magnetosphere is suddenly released. These events are known as magnetospheric substorms.

In the following sections of this chapter we shall discuss in greater detail the topics summarized here. In particular, we shall discuss the evidence for solar-wind control of the energy input that drives geomagnetic activity and identify the properties of the solar wind that are most closely coupled to the magnetosphere. A subsequent section extends the discussion to the topics of substorms and storms; this discussion focuses on the electric currents responsible for geomagnetic activity.

13.2 TYPES OF MAGNETIC ACTIVITY

13.2.1 Definitions

Electric and magnetic fields are produced by a fundamental property of matter: electric charge. Electric fields can be produced by charges at rest or moving relative to an observer, or by time-varying magnetic fields. Magnetic fields require moving charges. The two fields are different aspects of the electromagnetic force. The electric field at a point near a distribution of charges is defined as the force per unit charge when a positive test charge is placed at the observation point. It might seem that the magnetic induction **B** could be defined in a manner similar to **E**, as proportional to the force per unit pole strength when a test magnetic pole is brought close to a source of magnetization. However, no isolated magnetic pole has ever been observed. The consequence of this is that magnetic-field lines always close on themselves, rather than terminating on a pole. It is therefore common to define the magnetic field by the Lorentz-force equation, $\mathbf{F} = q(\mathbf{v} \times \mathbf{B})$, for the force on a particle of charge q with a velocity $\mathbf{v}$ [see equation (2.1)].

13.2 TYPES OF MAGNETIC ACTIVITY

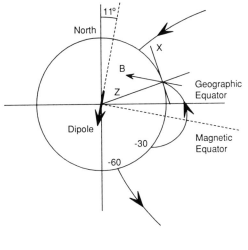

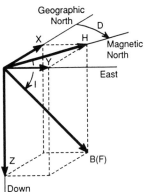

FIG. 13.1. Geometry of the earth's dipole magnetic field and the tangential plane with respect to which the earth's magnetic-field vector is measured.

FIG. 13.2. Components of the geomagnetic field as measured in different coordinate systems.

The relationships of geomagnetic coordinate systems to the earth and its dipole magnetic field are illustrated in Figure 13.1. The figure shows an observatory located at 30° magnetic latitude on the geomagnetic prime meridian, which cuts through the eastern seaboard of the United States. Reference directions are provided by the horizontal plane tangential to the earth at that station and by a radial vector directed toward the center of the earth. In this meridian, the earth's dipole axis is tilted 11° toward the geographic equator.

The nomenclature used for the various components of the vector field in geomagnetism is summarized in Figure 13.2. **B** is the vector magnetic field; F is the magnitude or length of **B**, originally referred to as the total force; X, Y, and Z are the three cartesian components of the field, measured with respect to a geographic coordinate system; X is northward, Y is eastward, and, completing a right-handed system, Z is vertically down toward the center of the earth. The magnitude of the field projected in the horizontal plane is called H. This projection makes an angle D (for declination) measured positive from north to the east. The dip angle I (for inclination) is the angle that the total-field vector makes

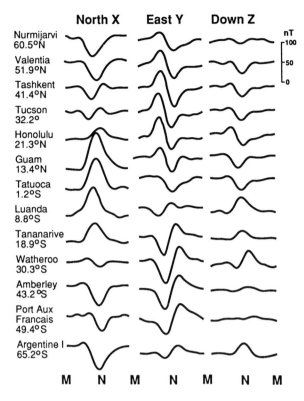

FIG. 13.3. Average quiet-day traces of the geographic north (X), east (Y), and vertical (Z) components of the surface magnetic field at stations distributed from north to south latitudes and plotted versus time. The times when the stations cross midnight (M) and noon (N) are indicated. (From Parkinson, 1983.)

with respect to the horizontal plane and is positive for vectors below the plane. It is the complement of the usual polar angle of spherical coordinates.

13.2.2 The Solar-Quiet Variation

The most thoroughly studied and best understood of the various types of geomagnetic activity is the diurnal variation S_q. On quiet days, every midlatitude observatory records a systematic variation in each field component (X, Y, Z), as illustrated in Figure 13.3. Stations at the same magnetic latitude, but separated in longitude, record similar patterns, but delayed in time by the earth's rotation. The pattern has considerable symmetry with respect to the magnetic equator and local noon, suggesting that the diurnal variation is produced by an ionospheric current system fixed with respect to the sun, under which the stations rotate. The form of this current system at equinox is shown in Figure 13.4. Two cells of current circulate around foci located at about 30° magnetic latitude. At the equator, both currents flow from dawn to dusk.

Fourier analysis of the diurnal variation at a single station reveals that it is primarily solar diurnal, with a smaller semidiurnal component. In addition, there is a weak, lunar semidiurnal (12 h 25 min) component. The primary cause of the diurnal variation is a dynamo created by

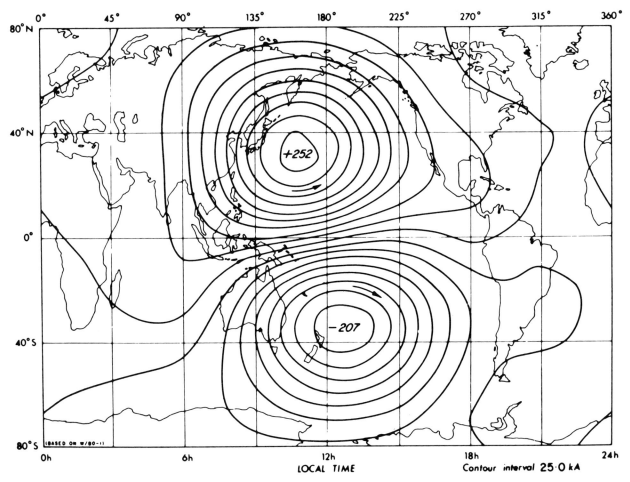

FIG. 13.4. Ionospheric current system responsible for the magnetic variations known as S_q. (From Parkinson, 1983.)

motion of electric charges in the ionosphere across the earth's magnetic-field lines. This motion is driven by winds in the ionosphere. These winds are driven by solar heating and lunar and solar tides. The dominance of the diurnal component suggests that solar heating is the primary force driving the dynamo.

13.2.3 Magnetospheric Substorms

The most frequent type of geomagnetic activity is referred to as a *magnetospheric substorm*. A substorm is the ordered sequence of events that occurs in the magnetosphere and ionosphere when the IMF turns southward and increased energy flows from the solar wind into the magnetosphere (McPherron, 1979, 1991; Akasofu, 1979; Rostoker et al., 1980). The most obvious manifestation of a substorm is the aurora, discussed in detail in Chapter 14. During a substorm, quiet auroral arcs suddenly explode into brilliance. They become intensely active and

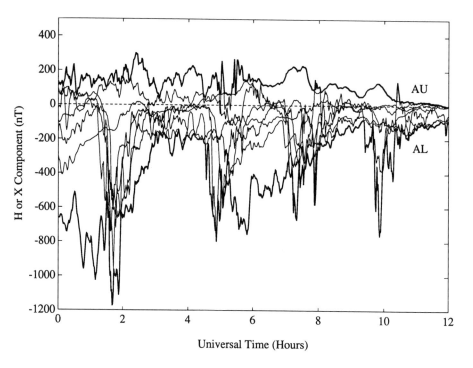

FIG. 13.5. Magnetic perturbations in the *H* component observed by auroral-zone observatories during a sequence of substorms on 3 May 1986. Positive perturbations are produced by a concentrated current or electrojet flowing eastward. Negative perturbations are produced by a westward electrojet.

colored. Over a period of about an hour, they develop through an ordered sequence that depends on time and location. Magnetic disturbances also accompany the aurora. On the surface beneath the aurora, a magnetometer will record intense disturbances caused by electric currents in the ionosphere. Figure 13.5 illustrates the form of these magnetic disturbances as a superposition of the perturbations recorded in the *H* component at a number of auroral-zone observatories. Stations located in the afternoon-to-evening sector record positive disturbances, whereas stations near and past midnight record negative disturbances relative to the field measured on quiet days. Applying the right-hand rule to currents assumed to be overhead leads to the conclusion that the currents are respectively eastward and westward toward midnight. These currents, later to be discussed further, are called electrojets, because the currents flow in concentrated channels of high conductivity produced at 120 km by the particles that generate the auroral light. Typical disturbances have amplitudes in the range of 200–2,000 nT, and durations of 1–3 h.

13.2.4 Magnetic Storms

When the coupling of the solar wind to the magnetosphere becomes strong and prolonged and geomagnetic activity becomes intense, as illustrated in Figure 13.5, a magnetic storm will develop. During a magnetic storm, auroral currents are almost continuously disturbed, such as seen on 3 May 1986 in Figure 13.5. The development of the storm is

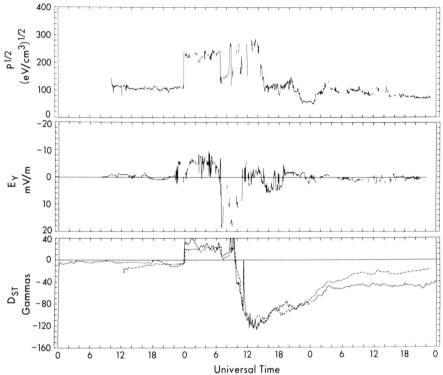

FIG. 13.6. Measurements of solar wind and magnetic field on the surface of the earth on 15–17 February 1967. Top: Solar-wind dynamic pressure. Middle: Dusk-to-dawn component of solar-wind electric field ($E_y = -uB \cos\theta$). Note that negative values of E_y are plotted upwards. Bottom: Effects of a magnetic storm as recorded in the D_{st} index (see Figure 13B.4). (From Burton et al., 1975.)

best identified at midlatitudes in the D_{st} index. D_{st} is defined as the instantaneous worldwide average of the equatorial H disturbance (see discussion in Appendix 13B). Figure 13.6 shows a classic storm signature (Burton, McPherron, and Russell, 1975) as recorded in D_{st}. A storm often begins with a sudden increase in magnetic field that may last for many hours. This initial phase is followed by a rapid and sometimes highly disturbed decrease in D_{st}. This is the storm main phase. Subsequently, D_{st} begins a rapid recovery, the first stage of the recovery phase. Eventually, a stage of long, slow recovery ensues. Typical storm durations are 1–5 days. The initial phase may be of any length, from zero to more than 25 h. Main phases last about 1 day. The recovery phase lasts many days. The distribution of storm magnitudes obeys a power law. Storms with D_{st} of order 50–150 nT occur almost every month. Several times per year the disturbance may reach 150–300 nT. Only a few times per solar cycle does one exceed 500 nT.

Studies of the spatial distribution of the storm disturbance field during the main phase indicate that it is nearly uniform over the entire earth and is directed parallel to the earth's dipole axis (i.e., southward). Such a disturbance would be produced by a current encircling the earth called a *ring current*. This ring current is created by particles drifting in the Van Allen radiation belts at distances of 3–5R_E. The initial phase is produced by the magnetopause current. An increase in solar-wind dynamic pressure forces this current closer to the earth and increases its strength. The perturbation magnetic field at the surface is northward. Recovery is caused by loss of particles from the radiation belts, leading to a decrease in the intensity of the ring current. In Figure 13.6, the correlation between dynamic pressure (top panel) and the initial phase of elevated D_{st} is obvious. Similarly, it can be seen that the main-phase decrease is a delayed response to an interval of strongly southward IMF (negative E_y).

13.2.5 ULF Waves

Magnetic activity can also take the form of periodic disturbances of the magnetic field, with periods of 1–1,000 s. These disturbances are called magnetic pulsations. They were originally discovered through microscopic examination of the minute fluctuations of the tip of a large compass needle; hence they are also referred to as magnetic micropulsations. Spacecraft observations have led to an understanding of their causal mechanisms. The subject of magnetic pulsations is discussed in Chapter 11, and examples are shown there.

13.3 MEASURES OF MAGNETIC ACTIVITY: GEOMAGNETIC INDICES

Magnetic activity at the earth's surface is produced by electric currents in the magnetosphere and ionosphere. Magnetic measurements at many locations provide one means of remotely sensing these currents and recording how they change with time. Because it is relatively inexpensive to make magnetic measurements, and because they are unaffected by weather, they provide an ideal way of routinely monitoring these currents. Over the past two centuries, a large network of more than 200 permanent magnetic observatories has been established. Data from these observatories and from temporary stations are frequently used to study magnetospheric phenomena, often in conjunction with in situ observations by spacecraft. However, the interpretation of raw magnetograms requires experience, and it is essential to accumulate data from many stations for a study. To simplify the task, several organizations routinely generate indices of magnetic activity. Ideally, an index is simple to generate, but useful, as it varies monotonically with some meaningful physical quantity, such as the total current flowing in a system.

Originally, magnetic indices were defined without much understanding of the processes that caused the disturbances. As understanding developed, new indices were defined, and older ones abandoned. Some indices are continued simply for historical reasons. They provide an ever-lengthening sequence of measurements that can be used for long-

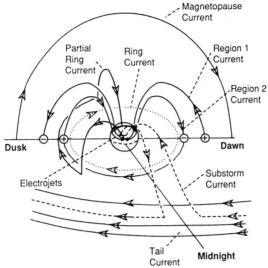

FIG. 13.7. Highly schematic representation of the various current systems linking magnetospheric and ionospheric currents and ultimately responsible for magnetic activity.

term studies of such phenomena as solar-cycle effects. Useful reviews of magnetic indices include those by Davis and Sugiura (1966), Baumjohann and Glassmeier (1986), Lincoln (1967), Mayaud (1980), Rostoker (1972), Troshichev et al. (1988), and Menvielle and Berthelier (1991).

Today, most of the current systems responsible for magnetic activity have been identified and studied. Figure 13.7 summarizes these in a highly schematic manner. They include various currents described either in this chapter or elsewhere in this book, including the magnetopause current, the tail current, and the ring current. Additional currents illustrated include the *partial ring current,* which flows near the equatorial plane principally near dusk, closing through the ionosphere, to which it is linked by field-aligned currents, and the *substorm current wedge,* a diversion of the tail current that also links into the ionosphere through field-aligned currents. The ionospheric portions of these current systems flow in enhanced-conductivity channels at high latitudes and are called the *auroral electrojets.* Sheets of field-aligned currents centered near dawn and dusk are referred to as *region-1* and *region-2 currents.* The higher-latitude region-1 currents flow into the ionosphere from the dawn sector and out at dusk. The lower-latitude region-2 currents have the opposite polarities. Not shown are polar-cusp currents, polar-cap closure of the electrojets, and currents associated with IMF-B_y effects. The primary sources of ground magnetic disturbances during substorms, such as those illustrated in Figure 13.5, are the two electrojets and the substorm current wedge. The sources of the midlatitude storm time variations shown by D_{st} in Figure 13.6 are the magnetopause current, the ring current, and the partial ring current.

13.4 SOLAR-WIND CONTROL OF GEOMAGNETIC ACTIVITY

Before the space age, it was well known that geomagnetic activity was related to events on the sun. How that influence was transmitted was not understood. Early theories due to Chapman and Ferraro (1930, 1931, 1932) suggested that solar flares ejected clouds of particles that traveled through the vacuum of space to interact with the earth's field. Despite obvious successes in correlating magnetic storms with solar flares, the flare concept did not account for perplexing storms that repeated at every 27-day solar rotation. Those storms were attributed to "magnetically effective" regions or M regions on the sun, even though nothing could be seen there to cause them (Chapman and Bartels, 1962, p. 410). It gradually became clear from studies of comet tails and other phenomena that there must be charged particles traveling outward from the sun at all times. With the launch of the first interplanetary probe, that speculation was quickly confirmed (Snyder, Neugebauer, and Rao, 1963), and a theory developed to explain what came to be called the solar wind (Parker, 1958). From that time on, it was obvious that the solar wind was the agent that transmitted the sun's influence to the earth. We begin our study of solar-wind–magnetosphere coupling with a brief review of three different variations apparently originating in changes at the sun: the solar-cycle variation, the annual variation, and the 27-day recurrence tendency.

13.4.1 Variations Related to the Sun

13.4.1.1 SOLAR-CYCLE VARIATIONS For several centuries, telescopic observations of the sun have shown that the number of spots on its surface oscillates, with a period of roughly 11 yr (see Chapters 1 and 3). More recent observations have demonstrated that there is a related oscillation in the pattern of the sun's main magnetic field. At every solar cycle, the main field reverses, so that in two spot cycles or 22 yr it returns to its initial orientation. Geomagnetic activity reflects the numbers of sunspots, as illustrated in Figure 13.8. The top panel shows a time-series plot of yearly means for the magnetic activity index AA (the first-difference time series of daily mean H at midlatitudes). Immediately below is the yearly sunspot number. High geomagnetic activity is obviously correlated with a high sunspot number. The same relation can be seen with monthly means in the bottom panel.

13.4.1.2 ANNUAL VARIATION The rotation axis of the earth is tipped at 23° to the ecliptic plane, causing a strong annual variation in the amount of sunlight that strikes the atmosphere. Because the dayside ionosphere is created primarily by photoionization, it might be expected

13.4 SOLAR-WIND CONTROL OF GEOMAGNETIC ACTIVITY

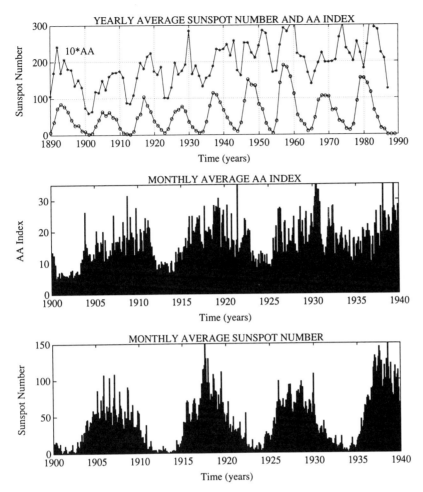

FIG. 13.8. The number of sunspots and the level of magnetic activity change together in both the annual and monthly means of sunspot and activity indices. (From Chapman and Bartels, 1962.)

that the conductivity of the ionosphere in a particular hemisphere would have an annual peak at its summer solstice. Therefore, any geomagnetic-activity index calculated from stations in that hemisphere should also show an annual modulation, peaking at the summer solstice. For indices that equally weight activity in both the summer and winter hemispheres, we might expect a semiannual variation, with peaks at each solstice. For indices independent of ionospheric conductivity, such as the storm time index D_{st}, it is not obvious what to expect. In fact, the ring current and other indices show very strong semiannual variations, as illustrated in Figure 13.9. Rather than one or two peaks per year at the solstices, there are two peaks near the equinoxes. At an equinox (March 22 or September 22), the sun crosses the earth's geographic equator. On March 5 and September 5 the earth crosses the sun's rotational equator. The phase of the activity peaks is not sufficiently well determined to decide whether the modulation in activity is related to the crossing of the sun's equator or to the crossing of the geographic equator.

The explanation of the semiannual variation put forward by Russell

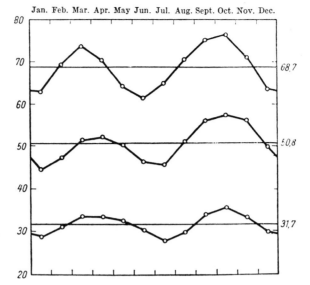

FIG. 13.9. The annual variation of magnetic activity as measured by monthly means of the u_1 index has two peaks near the equinoxes. The u_1 index is defined as the absolute difference between successive daily means in the H component of a station normalized so that the index has the same distribution as sunspot number. (From Chapman and Bartels, 1962.)

and McPherron (1973) is now generally accepted. Their argument is based on the idea that magnetic activity occurs preferentially when the IMF is southward relative to the dipole axis and that it increases with the size of the southward component, a matter that will be discussed later in this chapter. Russell and McPherron proposed that the variation in activity is controlled by the projection of the cross-flow component of the solar-wind magnetic field onto the cross-flow component of the earth's magnetic-dipole axis. At the spring equinox, the earth's dipole axis makes the largest possible angle, 35°, to the ecliptic normal at 2240 UT; at the fall equinox, this happens at 1040 UT. Depending on whether the solar-wind magnetic field **B** points inward toward or outward from the sun, the component of the ideal Archimedean spiral magnetic field, B, perpendicular to the flow will have a component $B \sin 35°$ that is either northward or southward relative to the earth's field at the subsolar point. If the projected field is southward there will be significant coupling. At other times of day, the north–south component of a field in the ecliptic plane is smaller, and the level of activity diminishes.

13.4.1.3 SOLAR-ROTATION VARIATIONS A third periodicity in geomagnetic variations is associated with the equatorial rotation rate of the sun as viewed from the moving earth. This is illustrated by the standard plot of the 3-h range index K_p presented in Figure 13.10. In this diagram, K_p values for successive solar rotations (27 days) are plotted one below another. The heights of the individual bars are proportional to the K_p index for a given interval. Intervals of intense activity are wrapped around in this display and are shown as superposed black bars. The sudden commencement of a magnetic storm is denoted by a dark triangle beneath the time line. A tendency for strong activity and weak activity to repeat at roughly the same time in each rotation is evident.

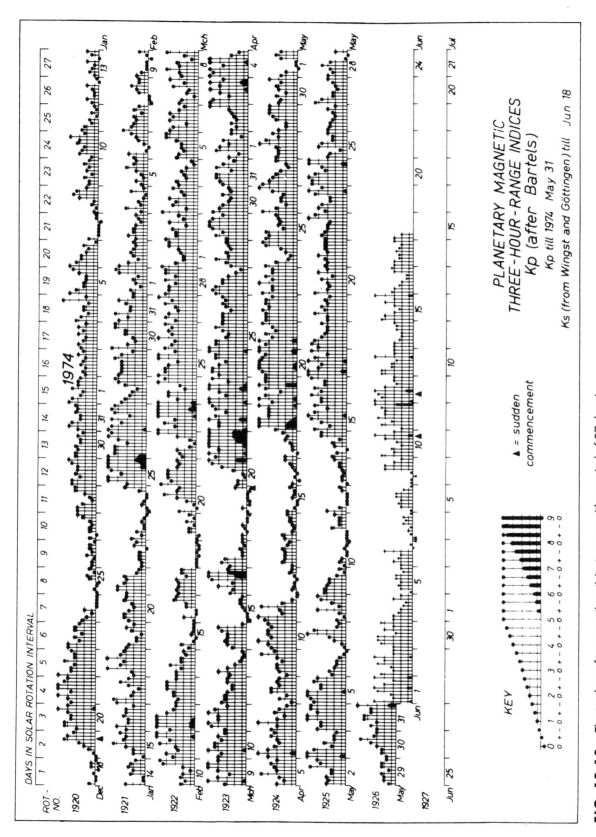

FIG. 13.10. The tendency for magnetic activity to recur with a period of 27 days is illustrated by charts of the planetary range index K_p.

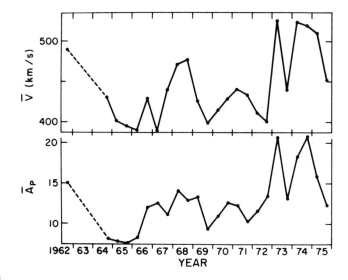

FIG. 13.11. Yearly averages of the A_p index are closely related to the solar-wind velocity, as shown by time-series plots. (From Crooker et al., 1977.)

The primary cause of the solar-rotation periodicity (27 days) in geomagnetic activity is that streams of high-speed solar wind recur at each rotation. As discussed in Chapter 4, large unipolar regions develop on the sun during the declining phase of the solar cycle. These regions produce coronal holes that become the sources of high-speed solar-wind streams. During the declining phase, the current sheet separating the north and south polarities becomes highly inclined to the solar equatorial plane. Thus, as these fast streams propagate toward the earth, they overtake slower plasma emitted from angular sectors of the sun ahead of the unipolar region, including the active streamer belt. An interaction region develops at the interface, and the magnetic field is compressed and tilted out of the plane of the ecliptic. As discussed in the next section, the combination of high solar-wind velocity and a southward IMF is particularly favorable for the development of geomagnetic activity.

13.4.2 Variations Related to the Solar Wind

13.4.2.1 DEPENDENCE ON VELOCITY One of the first major findings in the space era was the discovery of the solar wind. Measurements of its velocity quickly showed that long-term averages of geomagnetic activity were closely related to averages of the solar-wind velocity (Snyder et al., 1963). Figure 13.11 illustrates a result obtained after data on one complete solar cycle became available. The yearly means for the solar-wind speed v and the planetary index A_p closely track one another (Crooker, Feynman, and Gosling, 1977) (see Appendix 13B).

13.4.2.2 DEPENDENCE ON THE IMF Comparison of the waveforms of high-time-resolution magnetic indices and the solar-wind velocity contradicts the foregoing finding! There is virtually no relation between their joint variations. This perplexing result was not understood until it became clear that the solar-wind magnetic field was also important in controlling geomagnetic activity. That the IMF might be important was first suggested by Dungey (1961). He noted that if the IMF were antiparallel to the earth's field at the subsolar point of the magnetopause, then the two fields might be forced together and merge (see Chapter 9). He viewed this process as one in which the field lines of the earth and the IMF are first cut and then reconnected with a new topology. This view of the process leads to the name "magnetic reconnection." The amount of magnetic flux that reconnects will depend on the rate at which magnetic flux is transported to the subsolar point. One might therefore expect that geomagnetic activity would depend on the quantity uB_z, which is the rate per unit length in the equatorial plane at which the solar wind transports flux to the earth. Here, B_z is the magnitude of the z_{GSM} component. It is also the magnitude of the solar-wind electric field, according to the relation $\mathbf{E} = -\mathbf{u} \times \mathbf{B}$. However, it is expected that not all flux that flows to the magnetopause will merge. Much of it will slide around the magnetosphere. If we suppose that the fraction that merges increases with velocity, then the dependence on u^2 that is observed is not unreasonable.

These considerations led a number of investigators to examine the relationships of magnetic-activity indices to the solar-wind magnetic field. Schatten and Wilcox (1967) were among the first to show that scatter plots of K_p versus B_z revealed a linear dependence of K_p on B_z when B_z was negative (southward). Later, Arnoldy (1971) used correlation analysis between the auroral-electrojet index AE and various solar-wind parameters to obtain the result shown in Figure 13.12a. As shown by the legend, five different parameters, including u, n, B, and hourly integrals of B_n and B_s, were considered. Here n is the density of the solar wind, and B_n and B_s are, respectively, the magnitudes of the z_{GSM} component when it is greater than zero and less than zero. One parameter stood out: the integral of B_s over the hour preceding the hour in which activity was measured. Arnoldy also confirmed the result reported earlier by Hirshberg and Colburn (1969) that the correlation is highest when B_s is calculated in the geocentric solar magnetosphere (GSM) coordinate system. GSM coordinates have an x-axis pointing directly at the sun and an x–z plane that contains the earth's dipole axis (Figure 13.12b). In subsequent work, Murayama and Hakamada (1975) demonstrated that an optimum correlation is obtained with the function $AL = Cu^2 B (\cos \theta) U(\theta - \pi/2)$, where θ is the clock angle of the IMF about the x-axis (zero northward), U is the unit step function, and C is a constant of proportionality (see Appendix 13B for the substorm index AL). Figure 13.13 shows that there is almost no dependence of AL/u^2 on B_n, whereas there seems to be a linear increase in this quantity as B_s increases.

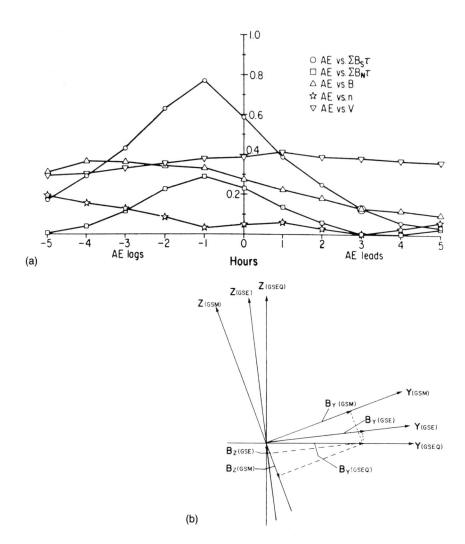

FIG. 13.12. (a) Linear-correlation functions demonstrate that hourly averages of AE are most closely related to the integral of B_s in the preceding hour. (From Arnoldy, 1971.) (b) Relationships between the GSM coordinate system, which maximizes the correlation between IMF B_s and geomagnetic activity, and the solar equatorial and solar ecliptic coordinate systems, which are less effective. The tilts as illustrated would apply on September 21 at 1040 UT, a time of maximum dipole tilt toward negative y.

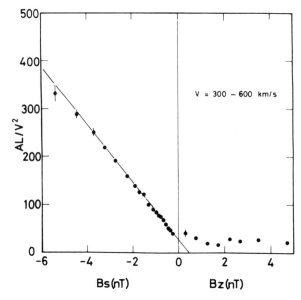

FIG. 13.13. Dependence of substorm magnetic activity, as measured by the AL index, on solar-wind velocity and the north–south component of the IMF. The plot shows the half-wave rectifier response of activity measured by AL and normalized by the square of the solar-wind velocity. The abscissa shows the hourly average B_z component when $B_z > 0$. For negative values, the observations are weighted by their durations to give B_s. (Murayama et al., 1980.)

13.4.3 Solar-Wind Coupling Parameters

13.4.3.1 EMPIRICAL STUDIES The results presented in Figure 13.13 are typical of those obtained in solar-wind–magnetosphere coupling studies. These studies make use of the long history of observations of the solar wind and geomagnetic activity to derive empirical relations. The standard approach uses statistical techniques to define relevant variables, determine the functional dependence on these variables, and then combine the variables into coupling parameters. The analysis assumes that the relation between the solar wind and the magnetic index is linear, or that it can be transformed to a linear relation piecewise by an appropriate choice of the solar-wind coupling function, as shown in Figure 13.13. This assumption allows us to treat the various processes as an electronic "black box" characterized by a transfer function. As discussed later, if this assumption is valid, then it provides a straightforward procedure for predicting geomagnetic activity.

13.4.3.2 DIMENSIONAL ANALYSIS The good correlation between magnetic-activity indices and the properties of the solar wind implies that the solar wind is the source of energy that drives the various magnetospheric processes. That observation led Akasofu (1980) to propose that both the coupling function and the activity indices should be expressed in units of power. Vasyliunas et al. (1982) studied the implications of this constraint and concluded that the principle of dimensional similitude provides a guide for choosing the variables. This principle states that the input function must be the product of (1) an energy flux relevant to the process, (2) the area on which it is incident, and (3) an arbitrary function of dimensionless variables important to the process that transfers energy to the magnetosphere. For magnetic reconnection, the relevant parameters were identified as the solar-wind kinetic-energy flux, the magnetopause cross-sectional area, the clock angle of the interplanetary magnetic field, and the solar-wind Alfvén Mach number M_A. For simplicity, the dimensionless function was taken as a separable function of Mach number and clock angle, so that the power input to the magnetosphere P is given by

$$P = ku(\rho u^2) l_c^2 (M_A^{-2})^\alpha g(\theta)$$

where $u(\rho u^2)$ is the flux of solar-wind kinetic-energy density, l_c^2 is an area equal to the square of the standoff distance to the subsolar magnetopause, $(M_A^{-2})^\alpha$ is the first term in a power-series expansion of a general function of inverse Mach number taken as a small parameter, $g(\theta)$ is the angular gating function for reconnection, and k is a dimensionless constant of proportionality corresponding to the fraction of the total flux incident on l_c^2 that reconnects.

The foregoing assumptions lead to predictions of the functional form of the solar-wind power coupling parameter. Any choice of the coupling exponent α produces a quantity that is expressed in units of power. For example, $\alpha = 0.5$ corresponds to a coupling parameter of the form

$$P = \rho u^3 l_c^2 M_A^{-1} g(\theta) = \text{constant} \cdot p_{SW}^{1/6} E_{SW} g(\theta)$$

where p_{SW} and E_{SW} are, respectively, the solar-wind dynamic pressure and electric field. This expression is close to the quantity uB_s used in many statistical studies. Kan and Akasofu (1982) carried out the analytical procedure recommended by Vasyliunas et al. (1982) and concluded that $\alpha = 1.0$. However, Bargatze et al. (1985) found $\alpha = 0.5$. The discrepancy is explained by the fact that Kan and Akasofu did not allow the size of the magnetosphere to scale with dynamic pressure, as required by the principle of dimensional similitude. Thus, it seems that the best power coupling parameter is more nearly proportional to the solar-wind electric field.

Actually, there seems to be no need to require that the input coupling parameter have energy (or power) units. Most of the available indices of magnetospheric activity, such as AE, do not have these units. These indices will almost certainly not be linearly proportional to energy input. Empirical studies, such as those summarized by Maezawa and Murayama (1986), have demonstrated that the AL index is best predicted by a function proportional to $u^2 B$, which does not have dimensions of power.

13.4.3.3 LINEAR PREDICTION FILTERING Most empirical studies of solar-wind coupling have found input parameters that are approximately linearly related to a given activity index. If the relation is truly linear and the system is time-invariant, then the technique of linear prediction filtering can be used to describe the system. This technique treats the solar-wind–magnetosphere system as a black box. Time histories of the solar-wind input and magnetic-activity-index output can be used to determine a transfer function. The Fourier transform of the transfer function is the impulse response of the system. Fourier-transform theory shows that the impulse response determined from this analysis can be convolved with any input at a later time to predict the output of the system; hence the name "linear prediction filter." As discussed by McPherron et al. (1988), this technique works well for the D_{st} index, but predicts less than half the variance in the AL index. Much of the unpredictable variance is correlated with the onset of the substorm expansion phase, suggesting that this phase of the substorm is not directly driven by the solar wind.

13.4.3.4 SOLAR-WIND TRIGGERING OF SUBSTORMS In most geomagnetic activity it is impossible to determine what causes the sudden increase in dissipation that accompanies the magnetospheric reconfiguration that is known as a substorm expansion. Akasofu (1979) attributed this to an increase in an energy-input parameter above some critical threshold. He speculated that increases in energy input cause

13.4 SOLAR-WIND CONTROL OF GEOMAGNETIC ACTIVITY

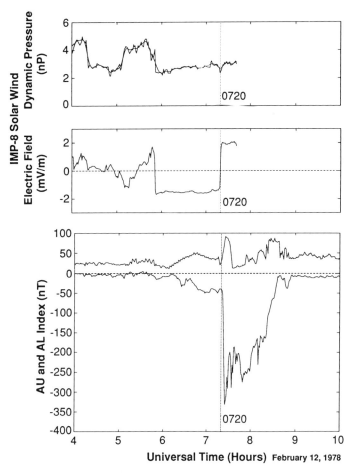

FIG. 13.14. Sudden changes in the solar-wind dynamic pressure and, more frequently, northward turnings of the IMF can trigger the sudden unloading of stored energy, as evidenced by changes in the AE-related indices. A sudden decrease in the AL index is correlated with a northward turning of the IMF, which here appears as an increase of the electric field E. Other events show that either phenomenon alone can cause the same effect in AE.

increases in the intensities of field-aligned currents, until they reach a threshold where they become unstable and generate field-aligned potential drops (Akasofu, 1980). These potentials accelerate electrons into the ionosphere, creating discrete arcs and enhanced conductivity. This enhancement draws more current, leading to larger potential drops, greater conductivity, and so forth.

Two processes are known to cause the sudden onset of dissipation. These are sudden solar-wind dynamic-pressure pulses (Burch, 1972; Kokubun, McPherron, and Russell, 1977) and sudden northward turnings of the IMF, which are equivalent to decreases in the dawn–dusk component of the solar-wind electric field (Caan, McPherron, and Russell, 1977; Rostoker, 1983; McPherron et al., 1988). Figure 13.14 shows an example of a sudden change in dissipation apparently triggered by a sudden northward turning. For both types of changes, it has been found that the IMF must have been southward for about an hour prior to the triggering event for an onset to occur. This strongly suggests that there is some threshold of stored energy or strain in the magnetospheric configuration required to produce the explosive phase of a substorm. In the case of northward turnings, the solar-wind input parameter decreases immediately, and the field-aligned currents driven by the solar wind decrease as well. We would thus expect dissipation to decrease, not explosively increase as is often observed. For such cases there must be some reservoir of stored energy feeding the enhanced dissipation. As we shall show later, this reservoir is the tail magnetic field.

13.5 MAGNETOSPHERIC CONTROL OF GEOMAGNETIC ACTIVITY

In the preceding sections we have examined magnetic activity as if the magnetosphere were a passive electric circuit that transforms the solar-wind input into a magnetic-activity index. In fact, the magnetosphere is a highly dynamic system that undergoes a more or less predictable sequence of changes each time the IMF turns southward. This sequence is called a *magnetospheric substorm*. When the IMF remains southward for an extended interval, auroral currents become continually disturbed, and the ring current grows with time. The ring current causes a strong decrease in the surface magnetic field, a signature that is known as a *magnetic storm*. In this section we shall describe the phenomena associated with substorms and storms in greater detail and then present a physical model of what causes them.

13.5.1 The Auroral and Polar Magnetic Substorms

13.5.1.1 THE AURORAL OVAL The disturbances known as substorms are most clearly seen in the auroral ovals. These are two oval-shaped bands roughly centered about the north and south magnetic poles, within which bright, active aurorae and strong magnetic disturbances are observed. The auroral oval as seen from space by the *DE-1* auroral imager is shown in Figure 14.1d. If the location of this band of auroral luminosity is plotted in the invariant magnetic-latitude–magnetic-local-time coordinate system (Wallis et al., 1982), it becomes a circle of about 20° radius centered about a point located 4° tailward of the instantaneous location of the dipole axis (Holzworth and Meng, 1975). In geographic coordinates, however, it is oval-shaped; hence its name. The region poleward of the oval is called the polar cap. Field lines in this region are open to the solar wind and are connected to the lobes of the geomagnetic tail (Paulikas, 1974).

13.5.1.2 DISCRETE AND DIFFUSE AURORAE Two distinct classes of aurorae are observed in the nightside auroral oval. Near the equatorward edge is the diffuse aurora notable for its lack of structure. Farther north and predominantly in the evening sector is the discrete aurora. Discrete aurorae are made up of long east–west bands of luminosity, with very small north–south extensions. Generally, those extensions are less than the 50–100-km resolution of current imagers, so that they cannot be resolved in images obtained with high-altitude spacecraft. The ionospheric signatures of auroral activity are discussed more fully in Chapter 14.

13.5.1.3 THE AURORAL SUBSTORM For many years it was thought that the type of aurora observed depended only on the latitude and local time of the observer and that changes in aurorae resulted primarily from the rotation of the earth. During the International Geophysical Year (1957), arrays of all-sky cameras were placed around the auroral oval to record what actually occurred. Analysis of the pictures established that the auroral type at a given location depended on universal time (UT), as well as on local time (LT) and latitude. A phenomenological model describing this development was called the auroral substorm (Akasofu, 1964). Figure 13.15 schematically illustrates the main features of this model.

In the model of the auroral substorm, activity begins from a quiet state consisting of multiple arcs drifting equatorward (panel A). The first disturbance is a sudden brightening of a portion of the most equatorward arc somewhere in the premidnight sector (panel B). This event is called the *onset* of the auroral substorm. The brightening expands rapidly westward and poleward (panel C). Within a short time a bright bulge of auroral disturbance forms in a broad region spanning the midnight sector close to where the aurora originally brightened. Within the bulge the aurora is very dynamic. Arcs appear and disappear; patches form and pulsate. Most arcs develop drapery-like folds that rapidly move along the arc. Lower borders of the arcs may become intensely colored. The interval of time during which the disturbed region is growing is called the *expansion phase* of the substorm. Eventually the auroral bulge develops a sharp kink at its westward edge, where it joins with the bright arc extending farther westward. This kink often appears to move westward, becoming more pronounced with time; hence it is called the *westward-traveling surge* (panels C and D). At the eastern edge of the bulge, torchlike auroral forms appear, extending poleward from the diffuse aurora and drifting eastward (panel D). These forms are called *omega bands,* from the shape of the dark regions defining their poleward borders. At the equatorward edge of this eastern region, dim pulsating patches of aurora appear, drifting eastward (panel E). After about 30–50 min, the auroral activity ceases to expand poleward, and the expansive (expansion) phase of the substorm has ended (panel E).

With the end of the expansion phase, auroral activity begins to dim at lower latitudes in the oval, and quiet arcs reappear. To the west, the westward-traveling surge degenerates, and a westward-drifting loop re-

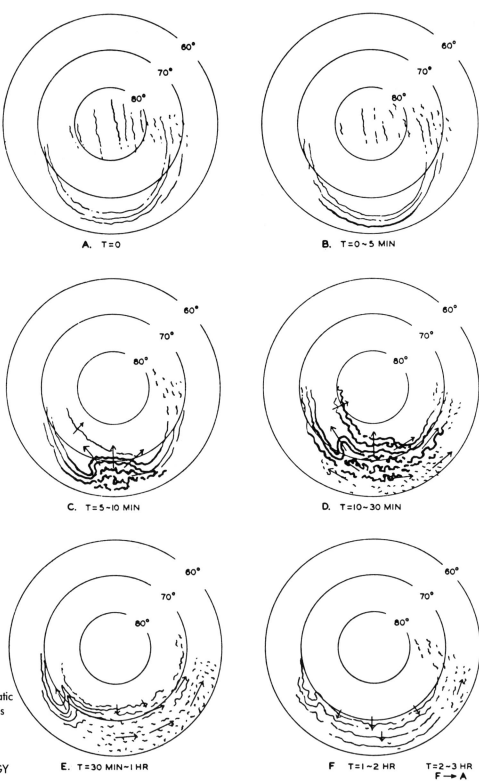

FIG. 13.15. Schematic representation of six stages in the development of an auroral substorm, as determined from all-sky camera data during the IGY (1957). (From Akasofu, 1964.)

13.5 MAGNETOSPHERIC CONTROL OF GEOMAGNETIC ACTIVITY 423

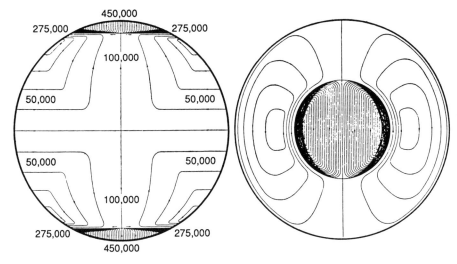

FIG. 13.16. An early view of the form of the ionospheric current system responsible for positive and negative bays in auroral-zone ground magnetograms. (From Chapman and Bartels, 1962.)

places it (panel E). In the morning sector, a pulsating aurora proceeds for some time. This phase of a substorm lasts for about 90 min and is called the *recovery phase*.

More recent observations with spacecraft imagers have modified this picture to some extent. Most significant is the observation that typical substorm expansions comprise multiple intensifications of the aurora, rather than one continuous expansion (Rostoker et al., 1987b). Each of these intensifications produces a new westward-traveling surge. However, the satellite observations demonstrate that frequently the surges do not travel far from where they form, and they do not proceed continuously, but rather in steps (Kidd and Rostoker, 1991). It appears that the original concept was incorrect because of spatial and time aliasing by the limited network of all-sky cameras. A more correct picture is that successive surges tend to form progressively farther west and farther poleward, giving the appearance of continuous expansion. In addition, it now appears that intensifications continue to occur on the poleward edge of the auroral bulge long after recovery has begun at lower latitudes (Hones et al., 1987).

13.5.1.4 THE CONVECTION ELECTROJETS Magnetic activity is associated with the auroral disturbances described by the auroral-substorm model. These disturbances were first extensively studied by Birkeland (1913), who called them elementary polar magnetic storms. Magnetogram recordings of these disturbances became known as geomagnetic bays, because their waveforms in a relative-amplitude-versus-time plot resembled the bays on a coastline. Positive bays in the *H* component usually are recorded by stations in the afternoon-to-premidnight sector, and negative bays by stations in the midnight-to-morning hours. A typical isolated bay disturbance lasts about 3 h.

Early studies of bay disturbances by a limited network of stations were statistical. Average disturbances seen by a single station at each hour of local time were calculated. Using stations distributed at different latitudes, a complete pattern of average disturbances occurring in polar and auroral regions was constructed. Figure 13.16 illustrates the result of such averaging, as reported in an early study. Contour lines show the ionospheric current, which could produce the average pattern of magnetic activity. This equivalent current system, called the *solar daily variation* (SD), is nearly symmetric about the earth–sun line, with two cells centered near the dawn and dusk terminators. The current flows sunward across the polar cap as a sheet of current. But in the auroral oval it is concentrated by high conductivity into the *eastward electrojet* (dusk) and *westward electrojet* (dawn). The right-hand rule is easily used to determine the magnetic perturbations that the two electrojets produce on the ground.

After the International Geophysical Year (1957), enough magnetic observatories existed at high latitudes to determine the instantaneous pattern of magnetic activity. It was found that the two-cell pattern apparent in the averages was also observed instantaneously, at least in the early stages of an auroral substorm (Iijima and Nagata, 1972; Kokubun, 1972). The instantaneous pattern was named DP-2 (disturbance polar of the second type). The axis of symmetry for DP-2 was found to differ from the pattern shown in Figure 13.16. It is tilted along a line from late morning to late evening, rather than along the earth–sun line. An example of this current pattern, as determined by more than 50 stations, is shown in the left panel of Figure 13.17 (Clauer and Kamide, 1985).

The current producing the ground disturbance is a Hall current, meaning that it flows at right angles to the ionospheric electric field. This current is produced by the drift of ionospheric charges in the presence of orthogonal electric and magnetic fields. In the absence of collisions, positive and negative charges would drift at the same rate, and there would be no current. In the weakly collisional ionosphere, ions drift more slowly than electrons, producing a current in a direction opposite to their mutual drift. The existence of the DP-2 current system implies that ionospheric plasma is moving in a circulation system referred to as *ionospheric convection*. The direction of this flow is approximately from noon to midnight across the poles and then back to the dayside through the auroral ovals; thus the two electrojets that establish and maintain the flow frequently are called the *convection electrojets*. The form of this convection pattern provided the initial insight into the mechanism by which the solar wind causes magnetic activity.

13.5.1.5 THE SUBSTORM ELECTROJET AND THE CURRENT WEDGE Auroral-zone magnetic activity is strongest during the expansion phase of an auroral substorm. In fact, once the concept of an auroral substorm (visible excitation of light in the ionosphere) had been

13.5 MAGNETOSPHERIC CONTROL OF GEOMAGNETIC ACTIVITY

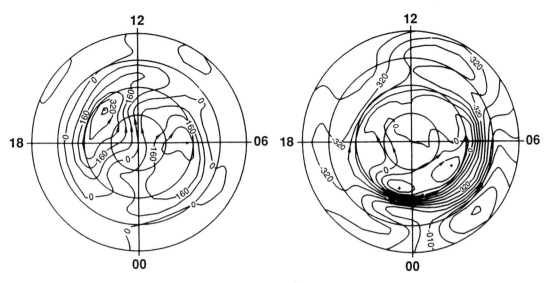

FIG. 13.17. Results from a recent determination of the patterns of ionospheric currents during magnetic disturbance. Closed contours show the flow lines for an equivalent ionospheric current that produces the observed ground magnetic perturbations. Left: The two-cell DP-2 current system present in the substorm growth phase. Right: The single-cell DP-1 current system that dominates during the substorm expansion phase. (From Clauer and Kamide, 1985.)

developed, it became clear that there was an accompanying polar magnetic substorm (current flow and convection). The main characteristic of this phenomenon is the sudden enhancement of the westward electrojet across the midnight sector. According to Akasofu, Chapman, and Meng (1965), as soon as the expansion phase begins, a westward current begins to flow along the newly brightened arc. As the auroral bulge expands poleward, and the surge moves westward, the region occupied by the current expands as well. The poleward edge of the bulge and the westward surge define the limits of the current. As the surge moves toward dusk, the westward end of the current intrudes along the poleward edge of the eastward electrojet. Because this new current is stronger than the eastward current, the associated magnetic perturbations that were initially positive become negative. Eventually, when this current dies away, the eastward current again dominates, leaving the magnetograms with an H-component waveform called an indented positive bay. Near midnight, the poleward expansion of the westward current causes the z component to reverse sign as the center of the current passes over a station. Eventually the current begins to decrease. This happens first at the equatorward edge of the auroral oval, and later at higher latitudes. This apparent poleward motion of the equatorward edge of the westward current is the recovery phase as seen in magnetic disturbances.

Initially there was considerable debate concerning the appropriate equivalent-current pattern for the expansion-phase current. The term "equivalent current" is used to describe a current system confined to the ionosphere that produces the magnetic signatures observed by a network of ground magnetometers. Ground observations cannot determine whether or not the currents are confined to the ionosphere. If the ionospheric conductivity is uniform, the same ground signatures can be produced by current systems that include field-aligned currents. Akasofu et al. (1965) believed that the pattern was primarily a single cell centered at midnight, with strong currents across the auroral bulge and weaker return currents at both higher and lower latitudes. It was suggested that the twin-cell system was really an artifact of averaging the temporal development of the single-cell system. That was not correct. The two different current systems are characteristic of different phases of the substorm. The expansion-phase current is now referred to as the DP-1 current (disturbance polar of the first type) and also as the *substorm electrojet*. The right panel in Figure 13.17 illustrates the form of this current for the same substorm characterized by DP-2 at an earlier time. Note the strong current in the sector between dawn and 2200 LT (closely spaced contours) that corresponds to the westward electrojet.

The difficulty in interpreting the early observations of the substorm electrojet can be attributed to the assumption that the current was confined to the ionosphere. Birkeland (1913) had believed that field-aligned currents were also present, but Chapman and others argued that they were confined to the ionosphere. Several workers, including Bostrom (1964), Atkinson (1967), and later Akasofu and Meng (1969), suggested that DP-1 was part of a three-dimensional system. Final proof that this was true came from a combination of ground data and synchronous-satellite data. The proof lies in the observation that the east-component magnetic perturbations at midlatitude stations and at the synchronous satellite *ATS1* have the same sign. This is possible only if the causative currents are above the spacecraft. McPherron, Russell, and Aubry (1973) pictured the actual currents, as shown in Figure 13.18. The actual three-dimensional current system that gives the signature observed on the ground is a wedge-shaped sector of magnetic-field lines. Eastward current across the midnight equatorial plane is diverted along field lines into the northern and southern auroral ovals. The current flows westward across midnight and then returns to space. The point at which the outward current leaves the ionosphere is the westward surge. This current system is called the substorm current wedge.

13.5.2 The Magnetospheric Substorm

13.5.2.1 PHASES OF A SUBSTORM The existence of two distinct current patterns during a substorm had no counterpart in the original model of auroral and polar magnetic substorms suggested by Akasofu (1968). Because of that, the model was amended by McPherron (1970), who noted that many phenomena preceded the onset of the

13.5 MAGNETOSPHERIC CONTROL OF GEOMAGNETIC ACTIVITY

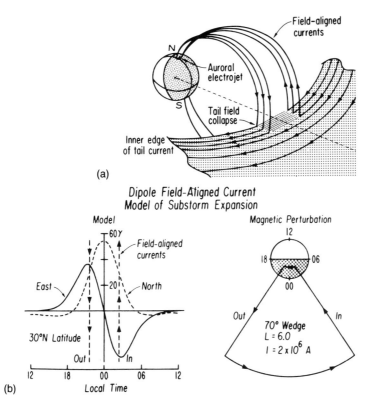

FIG. 13.18. Schematic illustration of the three-dimensional current system that is responsible for the DP-1 current system during the expansion phase of a polar magnetic substorm. (a) Diversion of the cross-tail current through the midnight ionosphere. (b) Magnetic perturbations caused by this current system along a chain of northern midlatitude magnetic observatories. The wedge shape of the projected equivalent current accounts for the name "substorm current wedge." (From Clauer and McPherron, 1974.)

expansion phase in the ionosphere. Foremost among them were weak positive and negative bay signatures similar to those seen later in the expansion phase. There also appeared to be an increasing probability of weak, short-duration intensifications of the aurora and electrojet, accompanied by bursts of ULF waves called Pi-2 bursts. Yet another feature that occurred before expansion was a gradual increase in the size of the polar cap. McPherron interpreted these phenomena as the *growth phase* of a substorm. The growth phase is an interval of time, prior to the onset of expansion, during which energy extracted from the solar wind is stored in the magnetosphere. The expansion phase corresponds to the release or unloading of that stored energy, and the recovery phase is the return of the magnetosphere to its ground state.

Figure 13.19 illustrates the three substorm phases in terms of the AU and AL indices during the substorm shown in Figure 13.17. As discussed in Appendix 13B, AU and AL are the envelopes of the superposed *H*-component traces from a worldwide chain of auroral-zone magnetometers underneath the electrojets. The beginning and end of an isolated substorm are defined by the departure and return of both those indices to background levels defined by quiet-day variations. The growth phase is then the initial interval of slowly growing AU and AL. These changes are created by the DP-2 system. Occasionally during the growth phase, a pseudobreakup will momentarily disturb the development, as if a substorm expansion with its DP-1 system were about to start. Eventually that does happen, and a full expansion phase ensues. Often, several

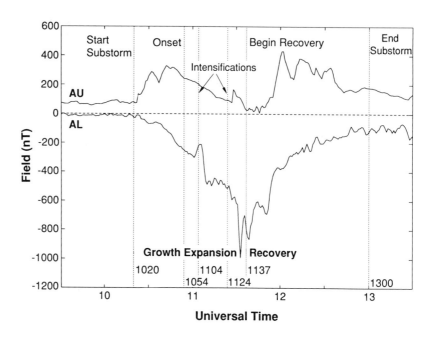

FIG. 13.19. AU and AL indices for a particularly well studied substorm. The three phases of this substorm (growth, expansion, and recovery) can be identified (see labels at bottom) by examination of the slope of the AL index. Times corresponding to the beginning and end of each phase are denoted by labels at the top. The onset of the substorm expansion (1054 UT) is often characterized by a sudden increase in the rate of decrease of AL. Subsequent increases are called intensifications. Note the indented positive bay (reduction in AU) during the middle of the substorm. Vertical lines are drawn at times when the B_z component of the tail field increased sharply. (From McPherron and Manka, 1985.)

distinct intensifications of the AL index occur during the expansion phase. These are caused by new substorm current wedges forming, each with a surge at its western end, and each closing westward through the ionosphere as a westward current filament. The initial onset and some of the intensifications usually are apparent in the AL trace as a change in slope. Eventually the AL index reaches a minimum and begins to recover. The interval of increasing AL (decreasing magnitude) is usually called the recovery phase, although some intensifications or surges continue to occur at high latitudes. As the DP-1 current system dies away, the DP-2 current system reappears, but later it too disappears. The entire sequence of events consisting of the three phases is the substorm. The rise and fall of the substorm electrojet constitute only the expansion and early recovery phase, although many people mistakenly refer to that rise-and-fall process as a substorm.

13.5.2.2 MAGNETOSPHERIC EFFECTS During a typical isolated substorm, the magnetosphere undergoes a distinct sequence of changes in its magnetic field and plasma, associated with the changes in the auroral current systems. We shall identify the key features here and expand the description in the next section. The IMF turns southward. The dayside magnetopause is eroded, and the associated magnetic flux is transported to the tail lobes. The plasma sheet thins, and the tail current moves earthward. A connected pair of X- and O-type neutral lines form in the near-earth plasma sheet. Magnetic reconnection at this X-line forms a bubble of plasma in the plasma in the plasma sheet. This bubble is disconnected and pulled out of the center of the tail. The extra flux in the lobes reconnects earthward of the bubble and convects back to the dayside. Particles energized at the X-line are injected into the inner magnetosphere and drift in the radiation belts. Eventually the near-earth portion of the X-line moves tailward, establishing a distant X-line. These events constitute the three phases of a substorm as seen in the magnetosphere, and we shall refer to the complete sequence as the *magnetospheric substorm*.

13.5.3 Magnetic Storms

13.5.3.1 PHASES OF A MAGNETIC STORM

An isolated substorm is created by a brief (30–60-min) pulse of southward IMF. When the IMF remains southward for longer times, activity becomes more complex. There is a series of overlapping auroral-zone activations, each injecting particles into the inner magnetosphere. The injected particles drift in a ring around the earth. Protons drift westward, and electrons eastward, creating a westward current called the *ring current*. Some particles from each activation are accelerated by drift across the enhanced magnetospheric electric field. The stronger the electric field, the greater their energy, and the closer the ring current is to the earth. In addition, particles are accelerated out of the ionosphere into the equatorial plane, so that heavy ions such as oxygen become important in the ring current, as discussed in Chapter 10. The ring current causes large decreases in the H component over most of the earth's surface. This effect is known as a *magnetic storm*. As long as injection of particles continues, the ring current will grow toward some asymptotic value in which the rate of injection equals the rate of loss. The time during which the ring current is growing is called the *main phase* of the magnetic storm. However, as soon as the IMF weakens, or turns northward, the ring current stops growing, and the ground perturbations begin to decrease. The ground perturbations decrease principally because particles are lost from the ring current. The loss process occurs in several steps. First, the rate of dayside reconnection decreases, and the convection boundaries move to larger radial distances. The ionosphere begins to refill flux tubes within the new boundary. As the cold ionospheric plasma encounters the ring-current plasma, ion-cyclotron waves begin to grow, and these waves scatter the ring-current protons into the loss cone. Other ring-current ions charge-exchange with the cold neutral hydrogen. Ring-current ions become energetic neutral atoms and are lost to the atmosphere or outer space. The low-energy ions that replace them contribute little current, and so the strength of the ring current decreases with time. This is the *recovery phase* of the storm. Many storm recoveries occur in at least two stages. The first stage results from the rapid loss of oxygen ions, and the second from the slower loss of protons.

Some magnetic storms are preceded by an initial phase of enhanced H component in ground magnetometer records. This effect is unrelated to the ring current and is caused by an enhancement of the magnetopause current. Many magnetic storms follow solar flares or coronal mass ejections. In either case, a high-speed parcel of the sun's atmosphere sweeps through slower solar wind, compressing and distorting the magnetic field ahead of it. That parcel of gas then encounters the earth's field and compresses it, enhancing the magnetopause currents and thereby producing positive perturbations in H at the earth's surface. This compression of the field is called a *sudden impulse*. In many storms, this phase will last for 4–16 h, as long as the IMF is northward. Eventually the IMF will turn southward, and there will be a sequence of substorms producing a magnetic storm. The H component that results from superposition of the enhanced magnetopause current (positive ΔH) and the ring-current perturbation (negative ΔH) is negative, even though the solar-wind dynamic pressure may remain elevated for some time.

13.6 PHENOMENOLOGICAL MODELS OF SUBSTORMS

13.6.1 The Ground State of the Magnetosphere

The occurrence of large-scale, systematic changes throughout the ionosphere during a substorm suggests that events are simultaneously occurring in space on the field lines connected to the aurora and electrojets. A phenomenological model has been developed to describe the associated magnetospheric phenomena that compose the magnetospheric substorm (Coroniti, McPherron, and Parks, 1968; Akasofu, 1968; Rostoker et al., 1980, 1987a). Detailed descriptions of this model have been given by McPherron et al. (1973), Akasofu (1977), McPherron (1979, 1991), Hones (1979), and Baker et al. (1984). In this section we present an abbreviated description of this model using schematic illustrations to represent the findings from many experimental studies. We begin with a description of the ground state of the magnetosphere as it is observed during relatively quiet conditions.

Figure 10.4 represents the topology of the magnetosphere in its ground state between substorms. Three types of field lines are present. There are interplanetary magnetic-field lines connected to the sun at both ends and generally excluded from the magnetosphere by the currents on the magnetopause. There are dipolelike field lines connected to the earth at both ends. These pass through the earth's equatorial plane, although they may be highly distorted by the tail and magnetopause currents. Finally, there are open field lines connected at one end to the earth and at the other to the sun. The open field lines have their footprints at latitudes poleward of the auroral oval and form the *polar caps*.

The topology of the earth's magnetic field defines many of the important regions of the magnetosphere. The polar cusp is the region of open field lines just poleward of the boundary between open and closed field lines on the dayside (see Figure 9.19). Near the magnetopause, these field lines link to a region of weak magnetic field where solar-wind pressure creates an indentation in the magnetopause. These open field lines provide a path for solar-wind plasma to reach the ionosphere. On the nightside, the boundary between open and closed field lines is connected to the distant X-line. The X-line is the separatrix between open and closed field lines. Above and below the X-line the field lines are open, connected to the polar cap and to the solar wind. Earthward in the equatorial plane they are closed and connected to the earth. Tailward they disconnect from the earth and hence are solar-wind field lines, despite their shape. The closed field lines of the tail confine the plasma sheet. This is a region of weak magnetic field filled with charged particles. Drift of these charges in the tail magnetic field produces the tail current that in turn creates the magnetic field in which they drift. The open field lines from the polar caps pass through the lobes of the tail. These regions contain few particles, because particles are easily lost either to the polar cap or to the solar wind. Close to the earth there is a rapid transition from dipole-like to tail-like field lines on the nightside. This occurs at the inner boundary of the plasma sheet. This boundary exists because charged particles of most energies in the tail are unable to drift arbitrarily close to the earth. The region earthward of the boundary is a "forbidden region" for these particles. For cold particles of either sign, the earthward boundary is the separatrix of the magnetospheric electric-potential distribution (see Chapter 10). This separatrix is symmetric around the dawn–dusk meridian, bulging outward somewhere near dusk. During steady conditions, the separatrix corresponds to the plasmapause, the boundary separating a region of high-density cold plasma originating in the ionosphere from lower-density plasma drifting earthward from the magnetotail. For higher-energy particles, the boundaries are more complex. Very roughly, the region accessible to tail particles of all energies and charge states has a horseshoe shape in the equatorial plane, with the open ends nearly meeting at local noon. Inside of this boundary lies the plasmasphere discussed in Chapter 10.

The locations of the inner edge of the plasma sheet and the convection separatrix depend on the electric field applied to the magnetosphere by the solar wind. The stronger the electric field, the closer to the earth the boundary will be. In the ground state of the magnetosphere, the boundary is typically at $10R_E$ near midnight. Fluctuations in the convection electric field enable particles from the tail to penetrate into the forbidden region, and in fact these fluctuations are the main process that populates the forbidden region with energetic particles. The energetic particles that penetrate the separatrices constitute the outer radiation belt, and their drift creates the ring current.

In the ionosphere, the nightside portion of the auroral oval contains particles lost from the plasma sheet. These particles hit the ionosphere and cause emission of light. The dayside portion of the auroral oval contains particles lost from open field lines in the polar cusp and from the closed field lines of the low-latitude boundary layer, as discussed later. This region is also horseshoe-shaped, but points tailward. Thus the oval has two main parts: one produced by particles lost from field lines with their feet on the dayside, and one produced by particles lost from closed field lines with their feet on the nightside.

Two other important regions of the magnetosphere are depicted in Figure 9.19. One is the *polar mantle*. The polar mantle exists because the solar-wind electric field is present on the open field lines of the polar cusp as well as on closed field lines. This electric field causes the particles to drift across field lines, initially poleward while they are entering and leaving the cusp. Fast particles do not drift far across the field before they leave the cusp and stream tailward along open field lines close to the magnetotail boundary. Lower-energy particles drift farther poleward and emerge from the cusp on field lines deeper in the tail lobes. As the mantle particles stream tailward, they continue drifting across the tail lobes toward the center of the tail. Eventually these particles reach the midplane of the tail. It is generally believed that the distant X-line is located at the point where particles with the characteristic energy of the cusp reach the midplane. This probably occurs at about 100–200 R_E downstream.

The plasma-sheet boundary layer is also depicted in Figure 10.4. This region forms on field lines near the upper and lower boundaries of the plasma sheet. The plasma-sheet boundary layer contains structured beams of particles streaming along closed field lines. Isotropization of the pitch angles of these particles produces the central plasma sheet, which fills the center of the tail.

A word of caution is needed here. It is not clear that either the polar mantle or the plasma-sheet boundary layer exists during quiet conditions associated with a northward IMF. The reason for this is that they are associated with open field lines in the polar cusp and with an electric field in the plasma sheet. Both of these are produced primarily by magnetic reconnection at the subsolar point and at the distant X-line, as discussed later. The mantle, as depicted in the drawing, may exist only during substorms while the IMF is southward. Similarly, the plasma-sheet boundary layer may be sharply defined only while reconnection is occurring at the distant X-line and a strong electric field exists inside the plasma sheet. We believe this is primarily during the recovery phase of a substorm.

It is important to view the magnetosphere in equatorial projection as well. Figure 13.20 portrays this view. The inner edge of the plasma sheet wraps around the earth from the nightside. The separatrix between corotating and convecting plasma (the plasmapause) is shown as a boundary closer to the earth. This drawing is not quite as described earlier. The reason is as follows: Most of the particles in the plasma

13.6 PHENOMENOLOGICAL MODELS OF SUBSTORMS

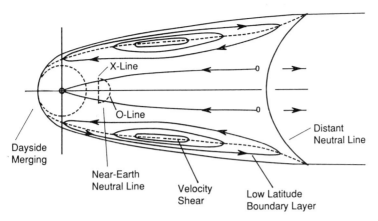

FIG. 13.20. Equatorial projection of the quiet magnetosphere. A dashed line defines the inner edge of the plasma sheet. The distant X-line should be farther from the earth at the center of the tail, and it should reach the flanks at nearly twice the distance of its center point.

sheet have finite energy and thus are subject to gradient drift and curvature drift. Because of these drifts, the forbidden boundary for electrons is located progressively farther outside of the plasmapause as energy increases. Roughly speaking, the boundary for protons is farther inside as energy increases. Thus, there is a gap between the inner edge of plasma-sheet electrons and the plasmapause, whereas the plasma-sheet ions reach and penetrate the plasmapause. These features are illustrated in Figures 10.25 and 10.26.

Data from the *ISEE 3* spacecraft suggest that the X-line in the distant tail is parabolic in shape, opening away from the earth; it may be as close as $100R_E$ at the center of the tail, but probably is $200R_E$ distant at the flanks.

On the flanks of the magnetosphere is the last major region of the magnetosphere to be defined, the low-latitude boundary layer. This region is generally believed to be on closed field lines, with plasma moving tailward at velocities that decrease away from the magnetopause. These two layers are thought to be produced by some type of viscous interaction or by reconnection with a northward IMF. The viscous interaction may be supplied by processes such as particle scattering, wave penetration, pressure-induced bumps on the boundary, or surface waves that transfer momentum from the flowing magnetosheath plasma to the outer portions of the magnetosphere. The inner edge of the boundary layer is a stagnation line at which the direction of convection reverses. Between the dawn and dusk convection reversals, plasma, and field normally flow sunward. Streamlines of flow in the equatorial plane form closed loops about the region of convection reversal. These streamlines project onto the two-cell ionospheric convection pattern and DP-2 current system mentioned earlier. In the drawing, we depict the ends of the boundary layers as merging with the ends of the distant X-line, although it is not certain that this is correct. If reconnection is occurring at the X-line, then field lines that close at the X-line flow sunward in the center of the plasma sheet, separating the return flows from the two boundary layers.

13.6.2 The Driven Model

In the ground state of the magnetosphere described in the preceding section, open field lines connect the polar caps to the solar wind. This configuration can exist only after magnetic reconnection interconnects the magnetic fields of the solar wind and the earth. Once interconnected, the interplanetary electric field is directly applied to the earth's ionosphere, driving electric currents. Wherever there are divergences in the electric field, or discontinuities in the ionospheric conductivity, there will be field-aligned currents. Changes in the ionospheric field-aligned currents and magnetospheric currents cause geomagnetic activity. The driven model of substorms was introduced by Perreault and Akasofu (1978) to explain the high correlation between the magnetic indices that are sensitive to these currents and the solar wind. To quantify the dependence on u, B, and field direction around the earth–sun line θ, Perreault and Akasofu developed the energy-coupling parameter ϵ (a power), defined as $\epsilon = l_0^2 u B^2 \sin^4(\theta/2)$, where $l_0^2 = (6R_E)^2$ is the area on the magnetopause through which magnetic energy (Poynting flux) enters the magnetosphere. In the driven model, a southward turning of the IMF (increasing θ) enhances the coupling of the solar-wind electric field to the ionosphere and hence the strength of the various currents. To explain the sudden enhancement of the aurora characteristic of the expansion phase, Akasofu (1979, 1980, 1981) postulated that an outward field-aligned current near midnight (located in the Harang discontinuity) develops a field-aligned potential drop when its density exceeds a threshold value. This potential accelerates electrons downward, increasing the current, as required by the increased solar-wind coupling, and simultaneously increasing the ionospheric conductivity. The higher conductivity demands more current for a fixed solar-wind electric field, further increasing the field-aligned current and its potential drop. This instability and its subsequent saturation correspond to the substorm expansion phase.

The driven model is implicitly a reconnection model, although the details of magnetic reconnection are never discussed. In particular, the existence of two or more X-lines on the dayside and nightside is not mentioned. No significance is attributed to delays between the onset of dayside and nightside reconnection, nor to the changes in magnetospheric configuration brought about by these delays.

13.6.3 The Boundary-Layer-Dynamics Model

Another reconnection model for substorms is the boundary-layer-dynamics model (BLD) developed by Rostoker and Eastman (1987). This model was developed to explain why fast flows in the magnetotail during substorm expansions are most frequently directed earthward (Eastman, Frank, and Huang, 1985). The authors of the model thought this meant that the source of the substorm expansion was distant from the earth. The model used a magnetic-field mapping in which the high-latitude ionospheric convection reversal at dusk and the Harang discontinuity near midnight were mapped to the same topological feature of the magnetosphere. Because the convection reversal at dusk clearly mapped to the inner edge of the low-latitude boundary layer on the duskside of the plasma sheet, continuity of mapping forced the authors to conclude that the Harang discontinuity at midnight mapped to a location close to the distant X-line. They therefore explained the expansion phase as a consequence of the sudden onset of reconnection at this distant X-line. Plasma energized by reconnection will jet earthward, enhancing the velocity shears at the inner edge of the low-latitude boundary layer. These shears become unstable to the growth of the Kelvin-Helmholtz instability, wrapping portions of the interface into vortices, preferentially on the duskside. Field-aligned currents flow out of the velocity shear on the dawnside of the plasma sheet, and into the center of vortices on the duskside. The three-dimensional current system produced by this process is the substorm current wedge. Inside the wedge, the currents cause an increase in the vertical component of the magnetic field (dipolarization), and outside the wedge they cause a decrease (more taillike field). As in the driven model, the authors postulate that the outward field-aligned current requires a field-aligned potential drop that accelerates electrons downward, creating the westward-traveling surge. It also accelerates heavy ionospheric ions outward. The authors attribute the occasional observations of tailward plasma flows, threaded by a southward magnetic field in the plasma sheet, to the effects of these ions in a region just west (duskward) of the outward field-aligned current.

The BLD model is based on a magnetic-field mapping that is known to be incorrect (e.g., Fairfield and Mead, 1975). The Harang discontinuity near midnight maps close to synchronous orbit, not the distant X-line. Furthermore, observations of tailward flow accompanied by a southward field in the plasma sheet occur predominantly at midnight, which is the center of typical substorm current wedges, not at earlier local times west of the westward-traveling surge.

13.6.4 The Thermal-Catastrophe Model

Another model that invokes reconnection to drive convection, but otherwise attributes to it no essential role, is the *thermal-catastrophe model* of Smith, Goertz, and Grossman (1986) and Goertz and Smith (1989). In this model, energy enters the magnetotail from surface-wave perturbations on the tail magnetopause. The wave energy propagates into the plasma-sheet boundary layer, where it is resonantly absorbed at field lines for which the natural frequencies match the incident frequencies. The absorbed energy heats the plasma, which convects to the central plasma sheet and eventually earthward. In this model the substorm growth phase begins when the intensity of compressional waves crossing the lobes increases. Power input to the boundary layer increases, and for a given level of convection, more heat remains in the local plasma, causing the temperature to rise. However, it is a characteristic of the resonant-wave absorption process that the opacity of the boundary layer maximizes at a fixed temperature dependent on the properties of the plasma. At this temperature the boundary layer becomes totally opaque, and all incident wave energy is absorbed. The plasma temperature rises explosively, as convection is unable to maintain an equilibrium. The plasma sheet then dynamically adjusts to a geometry appropriate to the elevated temperature. At the higher temperature, the opacity decreases, and convection is again able to carry off the energy deposited by waves. The expansion phase is attributed to the adjustment of the state of the boundary-layer plasma from one temperature to another at constant energy. This is the so-called thermal catastrophe. There have been attempts to correlate wave power in the lobes with substorm activity, but the evidence of a connection remains questionable.

13.6.5 Magnetosphere–Ionosphere Coupling Models

The models reviewed earlier emphasize the magnetospheric aspects of substorms. The members of another class of models focus on the ionosphere and its coupling to the magnetosphere. Kan (1990) and Rothwell et al. (1988) are proponents of such models, which are referred to as magnetosphere–ionosphere coupling (MIC) models. These models emphasize the positive feedback that changes in ionospheric conductivity can have on the sources of field-aligned current in the magnetotail. In particular, these models appear to provide possible explanations for the dynamic development of various auroral features, such as the surge, Pi-2 pulsation burst, and poleward bulge. The Kan (1990) model considers the consequences of instantaneously imposing an enhanced two-cell convection pattern on the magnetosphere. Alfvén waves reverberate between the ionosphere and the magnetotail, creating the distribution of field-aligned currents. This model is basically the driven model mentioned earlier. It makes no attempt to consider the phenomenology of the magnetotail and does not ascribe any particular importance to reconnection in the tail, other than as a source of convection electric field.

The Rothwell et al. (1984, 1989) model postulates an existing substorm current wedge and considers its ionospheric consequences. This model is very similar to one developed earlier for auroral arcs by Sato (1978) and Miura and Sato (1980). It invokes a mechanism called the feedback instability, which is a consequence of the finite delays imposed on the system by the inductance of field-aligned currents. It allows the ionospheric and magnetospheric developments of currents and fields to become uncoupled to some extent and results in poleward and westward expansion of the ionospheric portion of the substorm current wedge. A unique feature of this model is the development of a sheet of outward field-aligned current on the poleward edge of the auroral bulge that goes unstable in the same way as the outward current at the western edge of the current wedge. This potential drop becomes so strong that it alters plasma flow in the plasma sheet.

13.6.6 The Near-Earth Neutral-Line Model

The best-developed model of substorms is the near-earth neutral-line (NENL) model (McPherron et al., 1973; Russell and McPherron, 1973; McPherron, 1991). This model attempts to provide an internally consistent explanation for most magnetospheric phenomena, but does not attempt to explain many ionospheric observations. The unique feature of this model is the formation of a *plasmoid,* or bubble of closed field lines, that is ejected from the plasma sheet during the expansion and recovery phase of a substorm. The formation of the plasmoid is a direct consequence of the substorm growth phase that initiates the substorm.

In the NENL model, the substorm begins when a southward turning of the IMF activates dayside reconnection. This occurs in a ground-state magnetosphere (see Figure 10.4) that has a distant X-line separating the closed field lines of the plasma sheet (auroral oval) from the open field lines of the tail lobe (polar cap). This X-line is assumed to be inactive, with little or no reconnection occurring. Dayside magnetic flux from the earth connects to the IMF and is transported over the polar caps by the solar wind, where it is added to the outer portions of the tail lobes. The removal of this flux from the dayside initiates a convective flow of plasma toward the reconnection region, but the flow is retarded by the finite conductivity of the ionosphere at the foot of the field lines. Because of this, the magnetospheric return flow is unable to balance the rate at which flux reconnects, and the dayside magnetopause erodes earthward. The onset of convective flow propagates around the earth into the plasma sheet as a rarefaction wave, initiating earthward flow. Dayside erosion increases magnetopause flaring, thus increasing the dynamic pressure on the boundary. The dynamic pressure then reduces the flaring angle, compressing the tail until a corresponding increase in tail-lobe magnetic pressure balances the external pressure. This enhanced pressure is applied to the plasma sheet, and in combination with rarefaction brought about by flow toward the dayside, it causes the nightside plasma sheet and the tail current to thin. Simultaneously, the increased drag on the magnetotail caused by the newly opened field lines passing through the magnetotail boundary is balanced by the tail current moving earthward, closer to the earth's dipole moment. These changes increase the lobe field intensity, cause the plasma sheet to thin, and its

438 MAGNETOSPHERIC DYNAMICS

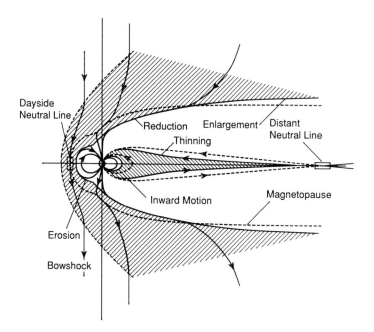

FIG. 13.21. Schematic illustration of the changes in magnetic field and plasma sheet expected in the situation where the reconnection rate on the dayside exceeds that on the nightside. Increased magnetopause flaring, plasma-sheet thinning, and earthward motion of the tail current are the main effects.

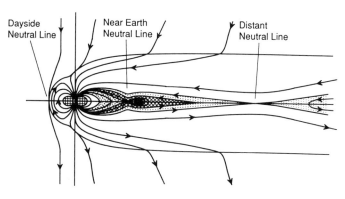

FIG. 13.22. Initial stage of the substorm expansion phase illustrating the first step in the formation of a substorm plasmoid.

inner edge to move earthward. These changes are summarized in Figure 13.21 and have been described quantitatively by the model of Coroniti and Kennel (1972).

At some time during the late growth phase, the vertical component of magnetic field across the plasma sheet becomes sufficiently small that ions in the cross-tail current no longer behave adiabatically. Coroniti (1985) and Baker and McPherron (1990) postulate that at that point, magnetic reconnection begins in the central plasma sheet. Reconnection proceeds slowly at first, cutting closed field lines of the plasma sheet at a new X-line. As successive field lines are cut, they form closed loops within the plasma sheet, centered about an O-line located tailward of the X-line, as illustrated in Figure 13.22. Azimuthal localization of the X-line requires that the X-0 and O-type lines connect at their end points (Russell and McPherron, 1973). As time progresses, the current sheet becomes thinner, and the rate of reconnection increases until it becomes explosive. If reconnection severs the last closed field lines at the edge of

13.6 PHENOMENOLOGICAL MODELS OF SUBSTORMS

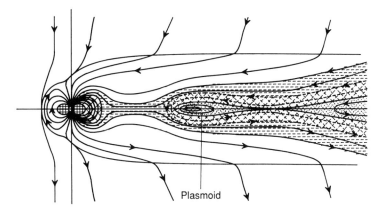

FIG. 13.23. Intermediate stage in the substorm expansion phase, showing a plasmoid moving away from the earth as a consequence of severance of the last closed field line originally connected to the distant neutral line.

the plasma sheet, a full-fledged substorm expansion takes place. Otherwise the disturbance is quenched, and the disturbance is called a pseudobreakup.

Once the open field lines of the tail lobe reconnect, they wrap around the plasmoid formed in the earlier stage and begin to pull it down the tail. The combination of magnetic tension and pressure from the plasma flowing tailward from the X-line accelerates the plasmoid away from the earth. As it leaves, the plasma sheet behind the X-line collapses to a thin sheet, as illustrated in Figure 13.23. Subsequently, open flux of the tail lobe reconnects at the X-line, forming closed field lines earthward of the X-line and open interplanetary magnetic-field lines tailward of the X-line. This process continues for some minutes until the balance of forces in the plasma sheet changes and the X-line begins to move down the tail. Earthward of the X-line, the plasma sheet thickens, and strong earthward flows are observed. As the X-line moves toward its distant location, the currents and aurora begin to die at the lower edge of the auroral bulge, although sporadic intensifications are observed on the poleward edge. This is the beginning of the recovery phase. With time, all disturbances die away, the substorm is over, and the magnetosphere returns to its ground state.

13.6.7 The Current-Sheet-Disruption Model

In 1984, the AMPTE *CCE* spacecraft was launched into an orbit with apogee of $8.8R_E$ and an inclination of 4.8°. That mission began to explore a region of the magnetosphere just beyond synchronous orbit that had been little studied by previous spacecraft. Substorm onsets observed in the regions near apogee in the midnight sector revealed signatures that had not been anticipated. Takahashi et al. (1987) reported on an event in which the magnetic field rotated slowly from nearly northward to quite taillike. Then, in a 3-min interval, oscillations with periods of about 13 s appeared suddenly in all components of the field. The amplitudes of individual fluctuations exceeded 40 nT, with some cycles of the wave producing field orientations near 80° southward. Ion fluxes in the range 100 keV to 1 MeV increased at *CCE* by an order of magnitude, and a nearby spacecraft reported injection of energetic electrons. Following that burst, the field orientation was nearly vertical, and within another 15 min its magnitude reached 60 nT, close to the dipole value at the spacecraft location. In another event, when *CCE* was very close to midnight and almost exactly in the magnetic equator, Lui et al. (1988) reported that large fluctuations produced southward fields, and in 4 min the field magnitude increased from 10 nT to the local dipole value of 40 nT. Energetic ions appeared, and their distribution showed strong asymmetries, with enhancements alternately earthward and tailward of the spacecraft. For the first event, Takahashi et al. (1987) offered an interpretation based on the assumption that the spacecraft observed radial oscillations of the position of a nearby neutral line, an interpretation that accounted for the reversal of the sign of the magnetic field. Lui et al. (1988) argued that their observations were not consistent with an X-type neutral-line geometry because of inconsistencies of particle and field signatures. They suggested, instead, that instabilities that occur in a thin current sheet can produce the signatures observed. This suggestion has been developed into a new model of the substorm, the *current-sheet-disruption model*.

In the model as described by Lui (1991a, b), a thin current sheet develops in the inner magnetosphere during the substorm growth phase, for the same reasons as in the near-earth neutral-line model. As the current sheet thins, the ions become nonadiabatic and begin to stream across the current sheet in serpentine orbits. The streaming ions interact with adiabatic electrons drifting in the opposite direction (the kinetic cross-field-streaming instability), producing lower hybrid waves (Lui et al., 1990). At the same time, the density gradient on the boundary of the plasma sheet drives the lower-hybrid-drift instability. The combination of the two types of waves produces an anomalous resistance in the plasma sheet that disrupts the cross-tail current. However, because of the high inductance of the tail circuit, the current must continue to flow. It accomplishes this by diversion along field lines, particularly those of the substorm current wedge.

According to observations reported by Lopez and co-workers (Lopez et al., 1988a–c; Lopez and Lui, 1990), the current-sheet disruption begins close to synchronous orbit and expands radially outward into the tail. This view is supported by two spacecraft observations of delays in substorm-related changes in the tail-lobe field. Jacquey, Sauvaud, and Dandouras (1991) and Ohtani, Kokubun, and Russell (1992) have both modeled these variations as the results of tailward propagation of current-sheet disruption. Recent Viking spacecraft auroral observations are also consistent with an origin very near the earth for the substorm expansion (Elphinstone et al., 1991). Using average magnetic-field models, the auroral oval and expansion onsets are found to project to distances just outside synchronous orbit, presumably the region in which the current-sheet disruptions take place. Furthermore, Murphree et al. (1991) and Murphree and Cogger (1992) have presented evidence that the expansion onset, or at least its initial stage, takes place on closed field lines. Such observations have led Lui, Lopez, Murphree, Ohtani, and others to conclude that the substorm expansion does not involve reconnection on the open field lines of the tail lobe. It has been suggested, instead, that the current disruption launches a rarefaction wave down the tail that induces plasma-sheet thinning and reduction in B_z. At some point down the tail, late in the expansion phase, or perhaps at the beginning of the recovery phase, these effects initiate reconnection and the subsequent generation of a plasmoid.

13.7 CONCLUSIONS

In this chapter we have described the important role of the solar wind in the generation of geomagnetic activity. We have continued the discussion showing that the energy coupled to the earth's magnetic field is transported and controlled by internal processes that produce the variety of magnetic variations known as magnetic activity. Some of this energy eventually reaches the earth's atmosphere, while the rest returns to the solar wind. The primary phenomenon organizing this energy transport is the magnetospheric substorm. The substorm is created by the superposition of two types of processes. One type is directly driven by the solar wind, and the other type corresponds to the unloading of energy stored in the magnetosphere by the driven processes. Isolated substorms are caused by short intervals of a southward IMF (30–60 min). Isolated substorms have identifiable and repeatable structures consisting of three phases called growth, expansion, and recovery. It is generally accepted that the growth phase is produced by changes in the configuration of the earth's field caused by unbalanced reconnection. Until nightside reconnection begins and returns magnetic flux to the dayside, the configuration must evolve as the amounts of open and closed flux in various regions change. Eventually this leads to a flux catastrophe that we believe is the expansion phase.

At the present time, different models posit quite different mechanisms for the expansion and recovery phases of substorms. The driven model discusses neither nightside reconnection nor its effects. The boundary-layer-dynamics model starts reconnection at the distant X-line at expansion onset. The thermal-catastrophe model starts it at the distant X-line soon after it begins on the dayside. The substorm expansion is then an explosion of the plasma-sheet boundary layer due to absorption of ULF waves originally generated on the high-latitude boundaries of the magnetotail. The near-earth neutral-line model starts reconnection close to the earth on closed field lines in the late growth phase. In this model, expansion onset is the time at which severance of the last closed field line occurs. The current-sheet-disruption model argues that the expansion phase is caused by an instability of the near-earth current sheet that diverts current through the ionosphere. This disruption apparently does not alter open field lines until the recovery phase begins. The magnetosphere–ionosphere coupling models suggest that the ionosphere is the cause of cross-tail current disruption and, like the disruption model, suggest that the reduction in cross-tail current then leads to a formation of an X-line and reconnection. Regardless of which picture of the onset is ultimately found to be appropriate, the onset of the expansion appears to be an internal process seldom directly driven by the solar wind.

From the wide variety of models postulated for the expansion phase, it becomes apparent that we do not yet understand substorms. Kan (1990) has suggested that greater effort be made to develop a global model of substorms that will incorporate all of the well-established observations. In this regard, we have noted a distressing tendency for many researchers, both young and old, to ignore observations that are inconsistent with their particular models. For example, the boundary-layer-dynamics model and current-disruption model tend to ignore observations in the central plasma sheet of tailward flow threaded by a southward field within a minute or two of expansion onset. These models also ignore the fact that the magnitude of the lobe field decreases significantly during the expansion phase, indicating that a reduction in the amount of open flux is occurring in this phase, not just during the recovery phase. Furthermore, observations have established that plasmoid release must occur before the beginning of substorm recovery, not at the beginning of recovery as in the disruption model. On the other hand, the near-earth neutral-line model does not provide an explanation for the auroral-arc breakup at low latitudes that maps to the current sheet just outside synchronous orbit. Neither can it explain how an entire expansion phase can occur on closed field lines, as suggested by some of the recent Viking auroral images. But if this interpretation of the images is correct, how is open flux of the lobe converted to closed flux on the dayside without obvious manifestations?

The conclusion of our discussion is that there are still problems in our models of substorms. Several models are useful in systematizing a selected subset of observations, but no model has yet proved capable of accounting for all of the observations. In the next decade we can expect to obtain additional data from spatial regions that have not been fully investigated. New instruments will improve our knowledge of particle responses to substorm dynamics. Missions involving several spacecraft will give us insight into the three-dimensional structure of the current and field perturbations. We can be optimistic that these new measurements will enable us to reject some proposed models and improve others. We can be certain that research on the complex phenomenon that we call a substorm will continue to be challenging and rewarding.

Appendix 13A INSTRUMENTS FOR MEASURING MAGNETIC FIELDS

13A.1 INTRODUCTION

MAGNETIC VARIATIONS MEASURED by instruments on the ground and in space provide information about the solar wind and the earth's magnetic field. These fields are produced by the superposition of the effects of a variety of different current systems. The processes that produce these currents are the subjects of interest to space physicists. A number of different instruments have been developed to measure magnetic fields. Originally, those instruments were basically variations of the compass. More recently, such instruments have begun to be replaced with more sophisticated devices based on other principles, including magnetic hysteresis, proton precession, and the Zeeman effect. In this appendix we describe some of the most common magnetic instruments in use today.

13A.2 MAGNETIC VARIOMETERS AND STANDARD OBSERVATORIES

Magnetic fields can be measured in a variety of ways. The simplest measurement device still in use today is the compass. A compass consists of a permanently magnetized needle, balanced to pivot in the horizontal plane. In the presence of a magnetic field and the absence of gravity, a magnetized needle would align itself exactly along the magnetic-field vector. When balanced on a pivot in the presence of gravity, the needle aligns with a component of the field. In the familiar compass, this is the horizontal component. A magnetized needle may also be pivoted and balanced about a horizontal axis. If this device (called a dip meter) is first aligned in the direction of the magnetic meridian, as defined by a compass, then the needle will line up with the total-field vector and measure the inclination angle I. Finally, it is possible to measure the magnitude of the horizontal field by use of the oscillations of the compass needle. It can be shown that the period of oscillation of a compass needle depends on properties of the needle and the strength of the field.

FIG. 13A.1. Examples of older instruments used to measure variations in the earth's magnetic field. Magnets suspended by quartz fibers reflect light beams onto photographic paper.

Magnetic observatories continuously measure and record the earth's magnetic field at a number of locations. In the older observatory instruments, magnetized needles with reflecting mirrors are suspended by quartz fibers inside of brass-and-glass housings like those shown in Figure 13A.1. Light beams reflected from the mirrors are imaged on a photographic negative mounted on a rotating drum. Variations in the field cause corresponding deflections on the negative. Typical scale factors used for such instruments correspond to 2–10 nT·mm^{-1} vertically and 20 mm·h^{-1} horizontally. A print of the developed negative is called a magnetogram. Typical magnetograms are illustrated in Figure 13A.2. The three traces in each panel are respectively the D, Z, and H components, as defined in Section 13.2.1. The bottom panel shows a storm's sudden commencement at 0826 UT on 22 March 1979 that is followed by a long chain of Pc-5 magnetic pulsations (see Chapter 11). A strong negative bay disturbance produced by a substorm began at 1054 UT. Definitions of these phenomena are given in Section 13.2.

Magnetic observatories have recorded data in this manner for well over 100 yr. Their magnetograms are photographed on microfilm and submitted to World Data Centers, where they are available for scientific or practical use. Among these uses are the creation of world magnetic maps for navigation and surveying, correction of data obtained in air,

APPENDIX 13A: INSTRUMENTS FOR MEASURING FIELDS

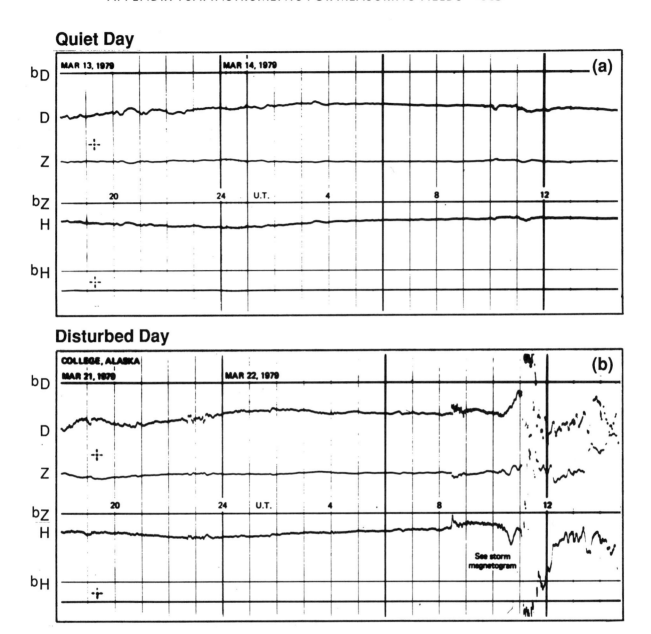

FIG. 13A.2. Example of magnetograms from a standard magnetic observatory at College, Alaska. The top panel displays a quiet day (14 March 1979), and the bottom panel a disturbed day (22 March 1979). Horizontal lines are baselines for the traces produced by the three variometers. The magnetogram spans two days of universal time, because the record at this observatory is changed at 8 A.M. local time.

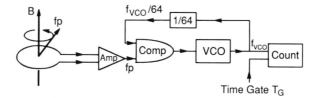

FIG. 13A.3. Schematic illustration showing how absolute measurements of the total magnetic field are made using the principle of proton precession.

land, and sea surveys for minerals and oil, and scientific studies of the interaction of the sun with the earth.

13A.3 VAPOR AND LIQUID MAGNETOMETERS

Today, other methods of measuring magnetic fields are more convenient, and the older instruments are being gradually replaced. One new type of instrument is the proton-precession magnetometer. The proton magnetometer takes advantage of the fact that the magnetic moment of a proton makes it act like a small bar magnet. Using this property, it is possible to temporarily align a proton in an external magnetic field created by passing a strong current through a coil wrapped around a container of liquid containing protons (water or kerosene). When the polarizing field is turned off, the protons try to align themselves with the earth's field. However, because they are spinning as well as magnetized, they initially behave like gyroscopes and precess around the earth's field. The precession frequency f_p is proportional to the magnitude of the earth's field B through a constant called the gyromagnetic ratio ($f_p = B/g_p$). This frequency can be measured by detecting magnetic effects of the gyrating protons with the polarizing coil. When the polarization field is suddenly released, all the protons are aligned in the same direction and begin to gyrate in phase, producing a time-varying magnetic field at the precession frequency. Very quickly the protons interact with other particles and are scattered in direction and phase. These interactions cause the signal strength to decay with time. Figure 13A.3 summarizes this technique of measurement. The exponentially damped sinusoid detected by the polarizing coil is amplified (Amp) and presented to a phase comparator (Comp). The voltage output of the comparator is input to a voltage-controlled oscillator (VCO) running at approximately 64 times the input frequency (f_{VCO}). The output of the VCO is fed back through a circuit that reduces the frequency by 64 and inputs it to the phase comparator. As long as the two inputs to the comparator have different frequencies, the output voltage of the comparator will continue to change the VCO frequency until it matches the precession frequency. The output of the VCO is also the input to a digital counter that counts the number of cycles of the VCO over a specified time interval (time gate). The various constants can be adjusted to force the output count to exactly equal the field in nanotesla, or any desired fraction thereof.

A proton-precession magnetometer is an absolute instrument that uses a fundamental property of matter to obtain an unbiased measurement of the strength of a magnetic field. To obtain the direction of the field, some other method must be devised. A method that uses the proton-precession magnetometer as a sensor consists of a coil system with three orthogonal axes surrounding a proton sensor. Accurately known currents are successively passed through each of the three coils, and the total field, consisting of the earth's field plus that of the coil, is measured. A total of four measurements, one for each coil and one without any field in the coils, is sufficient to calculate the three components of the vector field.

Instruments similar to the proton-precession magnetometer, but operating at much higher speed and with greater resolution, have been developed. These instruments are based on the Zeeman effect. In the presence of a magnetic field, electromagnetic radiation given off by an atom is split into two or more slightly different frequencies. This happens because the energy levels of the outermost atomic electron are split into a number of sublevels, depending on the orientation of the electron spin axis relative to the magnetic field. The separation depends on the strength of the magnetic field and is much less than the separation of the major levels. When an electron drops from an excited state to a lower level, the frequency of light given off depends on which sublevel the electron occupies at the beginning and the end of the transition.

A magnetic sensor that uses the Zeeman effect is constructed in the following way: Two cells containing atomic vapor (e.g., rubidium) and a photocell are aligned. The first cell is heated, causing light to be generated at all frequencies corresponding to the possible transitions between the two lowest atomic levels. Light from the first cell passes through a circularly polarizing filter and into the second cell. The polarized light is absorbed by the vapor in the second cell, raising its electrons to specific sublevels of the higher energy state. The electrons then fall to the lower level, giving off light. However, because the light is circularly polarized, transitions are allowed only between certain sublevels. After a brief interval, all electrons occupy the highest sublevel of the lowest energy state, and no more absorption is possible, because transitions from the high level are forbidden. While absorption is in progress, the amount of light falling on the photocell decreases, but when all electrons are "pumped" to the highest sublevel, it returns to normal. The pumped electrons can be redistributed to lower sublevels by applying a radio-frequency signal corresponding to the frequency difference between the sublevels of the higher state. This is done with a coil surrounding the second cell. If the frequency corresponds exactly to the Zeeman splitting, the electrons change levels, and the pumping process can be repeated. The required frequency is generated by using the output of the photocell to control the frequency of an oscillator. If proper phase shifts are introduced between the photocell and coil, the system oscillates

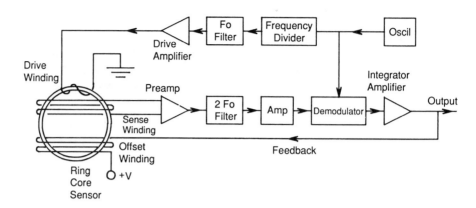

FIG. 13A.4. Schematic illustration showing the basic components of a fluxgate magnetometer.

between absorption and radiation. The field strength is measured by determining this oscillation frequency with a counter and converting it to an equivalent magnetic-field strength.

13A.4 FLUXGATE MAGNETOMETERS

Today the simplest and most common method of measuring vector magnetic fields is with a fluxgate magnetometer like that displayed as a block diagram in Figure 13A.4. The sensor in this instrument is a transformer wound around a high-permeability core (ring-core sensor). The primary winding of the transformer is excited by high-frequency current (5 kHz). The permeability of the core and the strength of the current are chosen so that the core is driven into saturation on each half cycle of excitation. The secondary winding detects a time-varying voltage that is related to the input through the hysteresis curve of the core material. For high-permeability materials, this curve is very nonlinear, and the output signal is highly distorted, containing all harmonics of the input signal. In the special case of no external magnetic field along the axis of the transformer, the hysteresis loop is executed in a symmetric manner. For this case it can be shown that only odd harmonics of the drive frequency are present in the output. If, however, an external field is present, the output reaches saturation in one-half cycle sooner than it does in the other half. This lack of symmetry introduces even harmonics into the output signal. The amplitude and phase of all even harmonics are proportional to the magnitude and direction of the field along the transformer axis.

In practice, the strength of even harmonics is very weak relative to that of odd harmonics. To amplify and detect these weak signals, the odd harmonics must first be eliminated. This is accomplished with the race-track or ring-core transformer, which is designed with two identical, parallel cores. Each core is excited by separate coils carrying equal currents in opposite directions. One secondary coil wound around both primaries is used to detect the output. For zero external field, the two coils induce exactly equal but opposite effects in the secondary, producing zero output. For nonzero field, the odd harmonics still cancel, and only even harmonics appear in the output. The second harmonic is amplified and detected, giving a voltage proportional to the field along the transformer axis. Three components of a vector field are measured by three separate sensors with their transformer axes in mutually orthogonal directions.

The electronics needed to measure one component of the magnetic field is constructed as shown in Figure 13A.4. A precision oscillator (Oscil) generates a string of pulses at a frequency $2f_0$, where f_0 is the final drive frequency. This signal is passed to the demodulator circuit as a reference signal, and to a frequency divider ($\frac{1}{2}$). The output of the divider (f_0) is passed through a narrowband filter to the drive amplifier. The string of pulses from the drive amplifier is applied to the primary (Drive) winding of the transformer. A secondary winding around the transformer (Sense) detects the total induced signal and passes it to a preamplifier (Preamp). The output of the preamp passes through a narrowband filter of frequency $2f_0$ and is further amplified. A strong signal of frequency $2f_0$ is then presented to the synchronous demodulator. This is simply an electronic double-pole, double-throw switch. Each time the input waveform starts to change sign, the switch is activated by the reference signal, making the output signal positive or negative depending on whether the input lags or leads the reference by 180°. The output waveform thus has a frequency of $2f_0$. The output of the demodulator is input to an integrator/amplifier that smooths over many cycles of the rectified waveform, producing a near-dc voltage, with amplitude proportional to the amplitude of the second harmonic component output by the sensor, and with sign depending on the phase of the second harmonic relative to the reference signal. These two quantities are, respectively, proportional to the magnitude and direction of the component of the external magnetic field along the axis of the transformer. This low-frequency signal is the magnetometer output. The output voltage is also used to supply a current to an offset winding wrapped around the transformer. The coil constant of this winding and a feedback resistor are chosen so that the current that flows in the offset winding will exactly cancel the magnetic field along the coil axis. Thus, the fluxgate magnetometer serves only as a null detector, making the entire instrument very linear over a large dynamic range.

A fluxgate magnetometer is not an absolute instrument like the proton-precession magnetometer, and so it must be calibrated against standards. For measurements of the accuracy required by magnetic observatories, this calibration is difficult and includes sensor offset, sensitivity, temperature coefficients, and alignment angles. Such calibrations require large nonmagnetic test facilities with three-axis calibration coils, proton magnetometers, and optical theodolites.

13A.5 DIGITAL MAGNETIC OBSERVATORIES

Modern magnetic observatories usually include both proton-precession and fluxgate magnetometers mounted on granite pillars in nonmagnetic, temperature-controlled rooms. The outputs from the instruments are electrical signals that are digitized and recorded on magnetic media by systems with components as summarized in Figure 13A.5. Many observatories also transmit their information almost immediately to central facilities, where it is collated with data from other locations in a large computer database.

A typical digital data-acquisition system has the components shown in Figure 13A.5. A voltage output from some detector circuit is presented to a multiplexer (MUX). This device is simply an electronic selector switch with the added feature that it captures and holds the voltage for the time it takes to convert it to a digital value. The output of the MUX is applied to an analog-to-digital converter (ADC) that produces a digital representation of the input voltage. The rate at which the ADC converts a voltage is controlled by a precision clock. This clock is also an input to a microprocessor controlling the rate at which it performs calculations. The microprocessor produces control signals that switch the MUX from one input channel to another and initiates data conversion in the ADC. The output of the ADC is sent to the microprocessor, which stores it in memory while formulating an output record. This record is written to disk memory for temporary storage. Periodically, disk memory is dumped to some other device such as a modem (modulator/demodulator) that outputs to a radio or telephone, a tape device, or a computer network connection.

Magnetic measurements are often made at locations remote from the fixed observatories. Such measurements are part of a survey to better define the earth's main field or to detect anomalies in it. They are routinely carried out by people on foot or in ships, aircraft, and spacecraft. For surveys near the earth's surface, the proton-precession magnetometer is also always used, because it need not be precisely aligned.

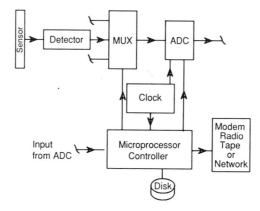

FIG. 13A.5. Block diagram showing the major components of a system to obtain and record digital data from the analog voltage output of an electronic magnetometer.

Well above the earth's surface, the main field decreases rapidly, and the requirement of precise alignment is less stringent. Thus, fluxgate magnetometers are generally used on spacecraft. Calculation of components of the vector field in a coordinate system fixed with respect to the earth requires knowledge of the location and orientation of the spacecraft.

Appendix 13B STANDARD INDICES OF GEOMAGNETIC ACTIVITY

13B.1 INTRODUCTION

MAGNETIC INDICES are widely used in studies of solar-terrestrial physics. Ideally, an index can be simply derived and is a monotonic function of some important physical parameter related to the phenomenon causing the disturbance. The original indices were simply averages of subjective impressions of the level of disturbance. With time, various components of magnetic activity were identified, and new indices were defined to isolate these specific phenomena. In this appendix we describe some of the most common indices in use today.

13B.2 THE CHARACTER FIGURES Ci AND C9

One of the simplest magnetic indices, and one with nearly the longest record, is the character figure C. This index is generated daily at a specified set of collaborating observatories. It consists of a subjective determination of the level of disturbance observed in each Greenwich day at each observatory. A number (0, 1, 2) is assigned depending on whether the record appears to be quiet, moderately disturbed, or severely disturbed. The average over all observatories is termed the international character figure Ci. A nonuniform tabular transformation of this average to a scale from 0 to 9 is called the C9 index. Figure 13B.1 shows the standard tabulation of this index for a 3-yr interval approaching solar maximum in 1980. Daily values are plotted horizontally, with data from successive solar rotations placed below those for the preceding rotation. Notice that the sizes of figures increase with their numerical size, thus allowing a ready appreciation of the intervals of disturbance. Often the effects of an active region on the sun, or a region of high dynamic pressure in the solar wind corotating with the sun, will stand out in such plots as a pronounced 27-day recurrence tendency (e.g., see the last half of 1978 near the end of a rotation interval). The daily sum K_p index defined in the next section has been calibrated against the historical

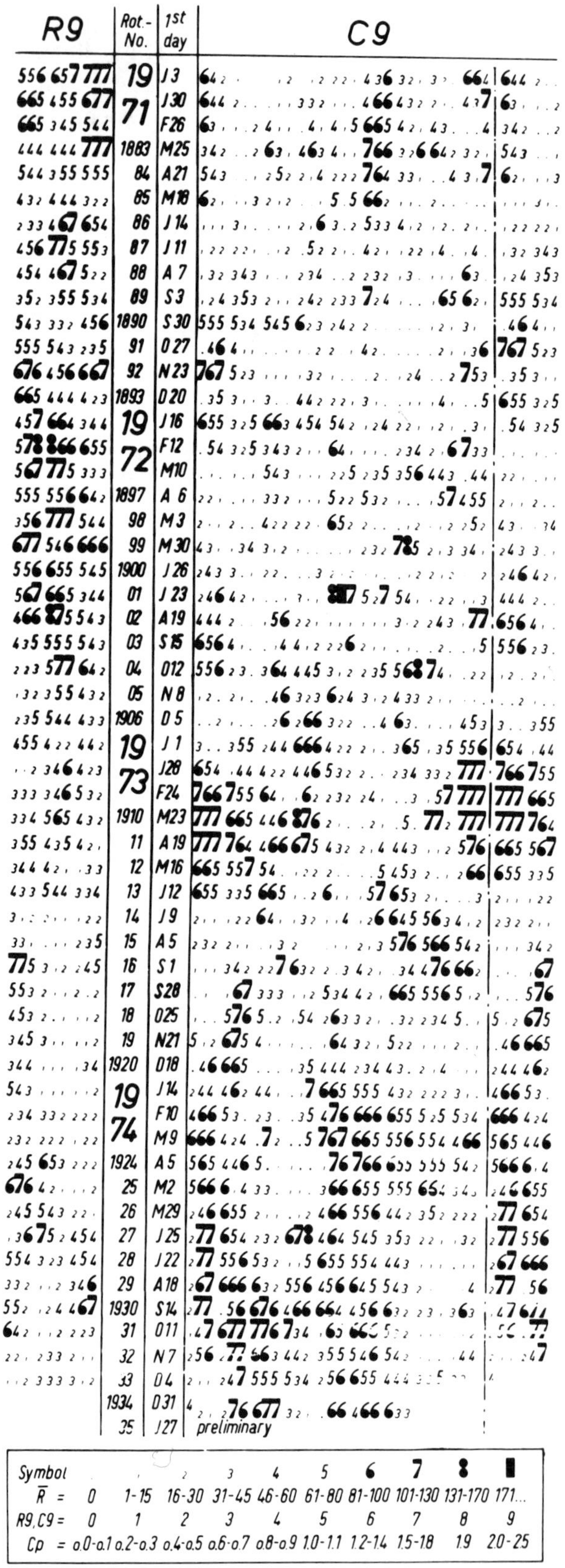

FIG. 13B.1. International character figure C9 displayed in standard form. The first major column displays the 3-day mean sunspot number, R9. The second column is the solar-rotation number or the calendar year. The third column shows the month and day of the first day of each solar rotation. The main column presents 27 consecutive days of the C9 index. The final column shows C9 for the first six days of the next rotation. The sizes of the figures are adjusted to correspond to the numerical value of the index, to emphasize disturbance.

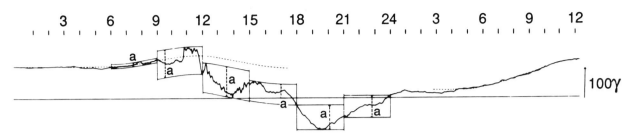

FIG. 13B.2. Illustration of the procedure by which the 3-h range index K is determined at a single station.

record of C9 and provides almost identical data. As a basis of comparison, it should be noted that the two magnetograms shown in Figure 13A.2 correspond to C9 values of 2 (quiet) and 7 (severely disturbed), respectively.

13B.3 THE RANGE INDICES K, K_p, A_k, and A_p

The K index is a range index for a field component at a given station that nominally measures the magnitude of disturbances caused by phenomena other than the diurnal variation and the long-term components of the storm time variation. It is calculated in the manner illustrated in Figure 13B.2. The Greenwich day is divided into 3-h intervals. For each interval, the diurnal variation appropriate to the season, phase of the moon, station, and component is visualized. This diurnal variation is translated vertically to the level for which it touches the minimum value of the trace in the interval to the level for which it touches the maximum value of the trace in the interval. The vertical separation between the translated diurnal variations is the range of the component in that interval. The range is then converted to a quasi-logarithmic K index by a table specific to the observatory. The first transformation table was created for a European observatory and was used to generate K values for several years. Subsequently, the range values for the year 1938 at that reference observatory were tabulated and compared to range values at other observatories in the same year. Separate tables were produced for each observatory, such that their distribution of K values was the same as the distribution at the reference observatory.

It is very useful to have a single index like C9 derived from the K values of observatories distributed over the entire earth. Because in any interval these are located at different distances from the source of disturbance, they measure quite different K values. To create a planetary K index, K_p, the station indices are first standardized to K_s indices. This is done through tables that create equal distributions of K_s values of each station in every 3-h interval of every season. These tables, specific to each observatory, translate the integral (0–9) K values into 28 fractional K_s values quantized to units of $\frac{1}{3}$ (0, $\frac{1}{3}$, $\frac{2}{3}$, . . . , 9). The K_p

index is then defined as the arithmetic average of the K_s values at 13 standard observatories. Originally, ranges were calculated for all three components, and the most disturbed component was used to define K. From 1964 onward, the vertical component was no longer used because it was shown to be sensitive to underground conductivity anomalies.

For long-term studies, it is useful to define daily average indices. Because of the logarithmic nature of the K indices, they are difficult to average. Consequently, linear range indices have also been defined. The A_k index at a station is obtained by multiplying the central value of the range used to define the integral K by a factor obtained by dividing the lower limit of the range corresponding to $K=9$ by 250. This number is typically of order 2; hence the A_k range index may be thought of as being twice the average amplitude of disturbance in the interval. A similar procedure is used with the K_p index to define a planetary A_p index with finer graduations.

13B.4 THE POLAR-CAP INDEX PC

The polar-cap index, PC, has only recently been proposed (Troshichev et al., 1988). This index is designed to be a measure of the strength of the sheet current flowing sunward across the polar cap, closing the two auroral electrojets. Thus, it provides a measure of the penetration of the solar-wind electric field into the magnetosphere. It is calculated using the following formula for ΔF:

$$\Delta F = \Delta H \sin \beta + \Delta D \cos \beta, \qquad (\beta = \lambda + \delta + \mathrm{UT} + \varphi)$$

In this relation, ΔH and ΔD are deviations of the H and D components from quiet levels, δ is the average declination of the station, λ is the geographic station longitude, UT is converted to degrees by multiplying by 360/24, and φ is the angle the average current vector makes with respect to local noon. ΔF is essentially the magnetic perturbation perpendicular to the average current direction. Figure 13B.3 illustrates this pattern when the IMF is southward, with B_y nearly zero.

The polar-cap conductivity and hence the return current are strong functions of solar illumination. Troshichev et al. (1988) have tried to determine this dependence using linear regression against various solar-wind coupling parameters. They have found the highest correlation (~0.7) for the linear relation

$$\Delta F = \alpha E_M + K = \alpha u B_T \sin^2 \theta / 2 + K$$

where α and K are constants, E_M represents the merging electric field, θ is the clock angle of the IMF about the x-axis, B_T is the magnitude of the projection of the IMF into the GSM y–z plane, and u is the solar-wind speed. Those authors propose that the PC index be defined by $\mathrm{PC} = \Delta F / \alpha$.

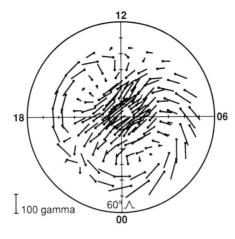

FIG. 13B.3. Polar-cap current pattern driven by a southward IMF. The strength of the ground perturbation normalized for ionospheric conductivity variations is the Pc index. (From Vennerstrom and Friis-Christensen, 1987.)

The advantages of the proposed PC index are several. First, it contains almost the same information about the solar wind as the AL index, with which it is highly correlated. Second, it can be calculated with data from only one station. Third, it is not contaminated by effects of the substorm current wedge, and therefore it appears to measure the driven response of the magnetosphere (the difference between driven and unloading processes was discussed earlier in this chapter). Its disadvantage is its sensitivity to changes in conductivity and its dependence on IMF B_y effects. It remains to be determined whether or not this index will be adopted as a worldwide standard and will produce new and interesting scientific conclusions.

13B.5 THE SUBSTORM INDICES AU, AL, AE, AND AO

The auroral-electrojet indices were defined by Davis and Sugiura (1966) to obtain a measure of the strength of the auroral electrojets relatively uncontaminated by effects of the ring current. The technique for calculating them can be understood by reference to Figure 13B.4. The third set of traces in this diagram displays the H-component traces from a worldwide chain of auroral-zone magnetic observatories. Monthly mean values are first subtracted from each station's trace to give a base value of zero. The traces are then plotted with respect to a common baseline, and upper and lower envelopes are calculated. The AU (auroral upper) index is defined at any instant of time as the maximum positive disturbance recorded by any station in the chain. Similarly, AL is defined as the minimum disturbance defined by the lower envelope. If the disturbances were caused by an infinite sheet current, then AU and AL would be proportional to the maximum overhead current density in the two electrojets. A single measure that approximates the total effect of both

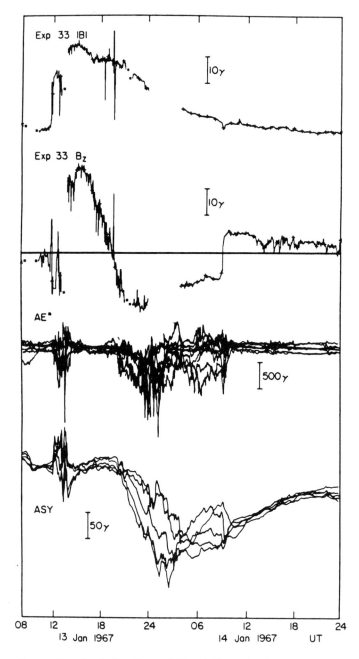

FIG. 13B.4. Illustration showing the type of data used to create the auroral-electrojet indices (AU, AL, AO, AE) and the disturbance storm time and asymmetry indices (D_{st}, A_{sym}). AU is the upper envelope of auroral-zone deviation of H (third panel, labeled AE) from a reference value; AL is the lower envelope, AO is the average, and AE is the separation of envelopes. At midlatitudes, D_{st} and A_{sym} are, respectively, the average deviation of H from a quiet day and the separation of the upper and lower envelopes (bottom panel, labeled ASY). The top two traces show that magnetic activity is produced by a strong interplanetary magnetic field pointing southward ($B_z < 0$) and parallel to the earth's dipole axis. (From Wolf et al., 1986.)

electrojets is defined as $\mathrm{AE} = \mathrm{AU} - \mathrm{AL}$. For completeness, AO is defined as the average of AU and AL: $\mathrm{AO} = (\mathrm{AU} + \mathrm{AL})/2$.

13B.6 THE STORM INDICES D_{st} AND A_{sym}

A measure of the strength of the ring current is the disturbance storm time index D_{st}. This index is calculated by a technique similar to that used in the auroral-electrojet indices, with several refinements. These are required because secular variations (long-term changes in the main field) and diurnal variations at each station can be as large as the storm time disturbance. The basic problem is to define a sequence of quiet values that can be used to define the trend and the seasonally dependent diurnal variation. Unfortunately, quiet days often occur in the recovery phase of magnetic storms. At such times the H trace is depressed, but is increasing exponentially with time. If these days are included in the determination of the secular trend, the result will be biased to low values. A similar problem arises in obtaining an average quiet day appropriate to a season. Storm recovery tends to make the values of H at the ends of days higher than at the beginnings. It is also difficult to define a quiet day. The quietest day in a month may contain some intervals of disturbance that can bias the calculation of an average variation. For the secular trend, the best that can be done is to take a sequence of midnight values (minimum diurnal variation) that occur during rare instances of no activity, well separated from magnetic storm recoveries. A polynomial fit to these values can then be subtracted from all data acquired by a station in a given year. From these data, one then selects quiet intervals identified by AE or some other index. These intervals are corrected for recovery trends and offset to zero at local midnight to remove storm bias. The data are then arranged in a two-dimensional matrix with rows for each day of the season and columns for each hour of the day. A two-dimensional Fourier analysis is performed, and only the low harmonics are retained. The trend coefficients and Fourier harmonics can then be used to predict a quiet H at any time of day and year.

After the quiet day is removed, the amplitude of the residual is adjusted by dividing by the cosine of the station's magnetic latitude. This is equivalent to the assumption that the measured H perturbation is the projection of an axial disturbance onto the tangential plane at the observatory (see Figure 13.1). The modified residuals can then be plotted against a common baseline, as illustrated in the bottom traces of Figure 13B.4. The D_{st} index is then defined as the instantaneous average around the world of the adjusted residuals (see the bottom trace). The A_{sym} index is a measure of the departure of the H perturbations from the axial symmetry expected for a ring current. It is defined as the instantaneous separation of the upper and lower envelope of H traces in the bottom set of traces. It is entirely equivalent to the AE index in the auroral zone. Similarly, D_{st} is similar to the AO index. As shown in Chapter 10, D_{st} is proportional to the total energy in the drifting particles that create the current (Dessler and Parker, 1959).

ADDITIONAL READING

Akasofu, S.-I. 1968. *Polar and Magnetospheric Substorms*. Dordrecht: Reidel.
Chapman, S., and J. Bartels. 1962. *Geomagnetism*, vol. 1. Oxford: Clarendon Press.
Coroniti, F. V., and C. F. Kennel. 1972. Changes in magnetospheric configuration during the substorm growth phase. *J. Geophys. Res.* 77:3361–70.

Kamide, Y., and J. Slavin (eds.). 1986. *Solar Wind Magnetosphere Coupling*. Tokyo: Terra.

Kan, J., T. A. Potemra, S. Kokubun, and T. Iijima. 1991. *Magnetospheric Substorms,* Geophysical Monograph 64. Washington, DC: American Geophysical Union.

McPherron, R. L. 1991. Physical processes producing magnetospheric substorms and magnetic storms. In *Geomagnetism,* vol. 4, ed. J. Jacobs (pp. 593–739). London: Academic Press.

Russell, C. T., and R. L. McPherron. 1973. The magnetotail and substorms. *Space Sci. Rev*. 11:111–22.

14 THE AURORA AND THE AURORAL IONOSPHERE

H. C. Carlson, Jr., and A. Egeland

14.1 INTRODUCTION

NUMEROUS NATURALLY OCCURRING celestial phenomena have been observed and admired since the dawn of human history, but few have stirred human imagination, curiosity, and fear as much as the aurora. The aurora (also called the northern lights and polar lights) is certainly one of the most spectacular of nature's phenomena (Figure 14.1a–d).

When we search for records of the northern lights dating from more than 1,000 yr ago, we find that most come from the Mediterranean countries, that is, from low latitudes. Yet auroral displays are seen in that area only after unusually strong solar activity. The time lapse between such large auroral events can be 50–100 yr. Furthermore, an aurora seen at such low latitudes is significantly less dramatic and colorful than those at the higher latitudes of common auroral displays (see Section 14.3). Nonetheless, the ancient low-latitude events were dramatic enough to strike fear into the hearts of those who saw them.

At much higher latitudes, the aurora borealis (i.e., in the north) and aurora australis (in the south) routinely appear in the so-called auroral zones, far from most population centers. Even today, the southern auroral zone (roughly around Antarctica) is inhabited only intermittently. The northern auroral zone, which crosses Alaska, northern Canada, northern Scandinavia, and Siberia, has always been accessible to frontiersmen (hunters and fishermen) living in the polar region. More recently, the area under the northern lights has become permanently, although sparsely, populated.

In earliest historical times, inhabitants of Greenland and the Nordic countries interpreted the northern lights as omens from the gods portending disaster, as signs from deceased relatives, as signs of a battle among the gods, or as weather signs. From more recent and more scientific Scandinavian records (*The King's Mirror,* written about 1230 A.D.), it appears that the regions of auroral activity have shifted significantly during the past 1,000 yr. For those interested in the history of the aurora, monographs by Brekke and Egeland (1994) and Eather (1980) are available.

Those who appreciate the beauty of nature may find nothing compara-

FIG. 14.1a. Auroral forms seen from the ground during the polar night.

FIG. 14.1b. View from space of the aurora from low earth orbit (~250 km) of the Space Shuttle *Discovery*, 29 April 1991. (Courtesy of NASA.)

FIG. 14.1c. Aurora seen from a Defense Meteorological Satellite Program (DMSP) satellite.

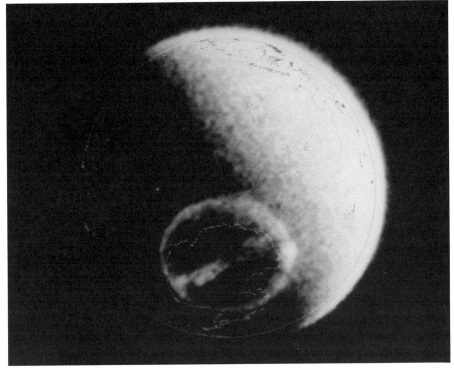

FIG. 14.1d. Aurora seen from the *Dynamics Explorer DE-1* in high earth orbit near $4R_E$ during an unusual event where the unusually circular oval has an extra transpolar sun-aligned crossbar, resembling the greek letter θ, giving this the name of "theta aurora." (From Frank et al., 1986.)

ble to a night with a magnificent auroral display. It is just as beautiful to watch today as it was in the earliest days of human history. This was given poetic expression by Tromholt (1885) in his book *Under the Rays of the Aurora Borealis:*

> Lovely celestial display! Before your fascinating mysterious play, in which enigmatic forces of Nature flood the heavens with light and color throughout the long Polar night, the golden sunsets of the Pacific Ocean, the gorgeous flora of the Tropics, the resplendent lustre of gems of Golconda, must pale. Lovely celestial display!

Less poetic, but fascinating in a different way, is the recognition that an aurora is the optical manifestation of auroral-particle precipitation and its interaction with atmospheric constituents (see Section 14.2). Auroral emissions are produced by particles, originating from the sun and the earth's atmosphere, that collide with the earth's atmosphere along streamlines modulated by electric and magnetic fields in the magnetosphere and ionosphere. The size and form of the aurora thereby reflect the forces acting on these auroral particles as they journey from their source to the earth's upper atmosphere (see Section 14.2). Auroral morphology, the study of the occurrence of the aurora in space and time, is described in Section 14.3, as is the electrodynamics of polar-cap arcs.

The auroral-substorm concept is discussed in Section 14.4. Sections 14.5 and 14.6, respectively, discuss the auroral ionosphere and its effect on radio waves. The basics of thermal balance and energy balance, plasma convection, and thermospheric responses controlled by geomagnetic activity, the interplanetary magnetic field (IMF), and local magnetic time are presented in Section 14.7. The auroral boundaries, as defined by optical and particle signatures, as well as current- and plasma-convection-reversal (i.e., electric-field) signatures, are discussed in Section 14.8.

This treatment seeks to provide an introduction to the terminology and morphology necessary to read the extensive literature on these subjects, to introduce the relevant physical processes, and to describe the relationships among upper-atmospheric boundaries. Within the constraints of available space, we trace the development of some key concepts, spiced with some of the unsolved challenges of auroral physics.

The advantage of using the aurora as a monitor of those near-earth processes that arise through the link to the magnetosphere, rather than any conceivable system of in situ measurements, is that the size of the auroral oval differs in scale from the magnetosphere by perhaps a factor of 10^6. The high spatial and temporal resolution available through ground-based observations provides another advantage of studying the aurora from below (see Section 14.8).

The first International Polar Year (1882–3) can be regarded as marking the beginning of modern auroral research. The driving force behind the effort was Kristian Birkeland, the great auroral pioneer (Birkeland, 1908, 1913). In his day, only the simplest ground instruments were available for auroral investigations. Today, auroral research is conducted mainly through the use of sophisticated instruments on board rockets and satellites, as well as advanced balloon and ground-based equipment. Even artificial aurora have been produced in the earth's atmosphere (e.g., Winckler, 1980). Some of the mysteries of the northern lights have been solved, partly or fully, but new problems have appeared, and the study of the auroras continues to engross many scientists.

14.2 AURORAL-PARTICLE PRECIPITATION: THE AURORAL SPECTRUM

The optical spectrum of an aurora consists of a great number of spectral lines and bands – from ultraviolet to infrared wavelengths. The auroral radiation is emitted by atmospheric constituents that are excited by precipitating particles. Figure 14.2 shows parts of the optical spectrum of an aurora. These emissions are primarily due to a two-step process in which precipitating energetic auroral particles (electrons and ions) collide with the atoms and molecules of the earth's upper atmosphere, converting their kinetic energy, in part, into energy stored in the chemically excited states of atmospheric species; the chemically excited states relax, giving off photons of wavelengths determined by the energy transitions in the relaxation processes. We shall summarize some of the main characteristics of the energetic particles and the auroral optical emissions they produce.

14.2.1 Scattering and Absorption of Auroral Particles

The primary auroral particles, populations of electrons and ions with energies from less than 100 eV up to small multiples of 100 keV, can be measured directly by the use of instrumented rockets and satellites. Some of these will precipitate into the atmosphere, causing atmospheric excitation and ionization, as discussed in Chapter 7. Near the earth, such particles are found mainly above 55° magnetic latitude. At greater distances above the earth, they have their sources in the plasma sheet of the geomagnetic tail and in the polar-cusp region on the dayside of the magnetosphere, as described in Chapter 10. As the high-energy tail of the energy spectrum (>30 keV for electrons; >1 MeV for protons) is not important for the auroral emissions, it will not be discussed further here.

The rate of precipitation of auroral particles into the upper atmosphere is schematically illustrated in Figure 14.3a. The dots represent mainly the higher-energy (>20 keV) auroral particles, and the triangles

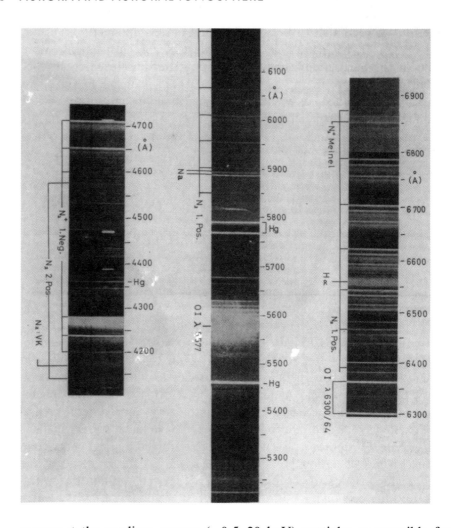

FIG. 14.2. Selected parts of the auroral spectrum in the visible range. (For details, see Vallance Jones, 1974.)

represent the medium-energy (~0.5–20 keV) particles responsible for the visual aurora. The stars mark the particles (<1 keV) that enter the magnetosphere through the polar cusp, causing the dayside oval aurora. Interpretation of Figure 14.3a, dating from 1971, illustrates the way in which straightforward ideas can advance our understanding of auroras. The most energetic particles lie on a circle of constant latitude, as would trapped particles leaking out of a loss cone as they drifted in a longitudinal ring current (see Chapter 10). The medium-energy particles lie on a circle tipped back away from the sun, as would particles accelerated downward along the earth's higher-latitude magnetic-field lines draped upward and downwind from the solar source of the solar wind. The low-energy particles are confined to the footprint of those midday magnetic-field lines that plausibly could funnel solar-wind particles directly into the earth's upper atmosphere with minimal acceleration. Far more extensive observations show overlapping of principal zones, a more gradual transition from one to another, and a much more sophisticated framework for relating particle populations to boundaries (e.g.,

14.2 AURORAL-PARTICLE PRECIPITATION

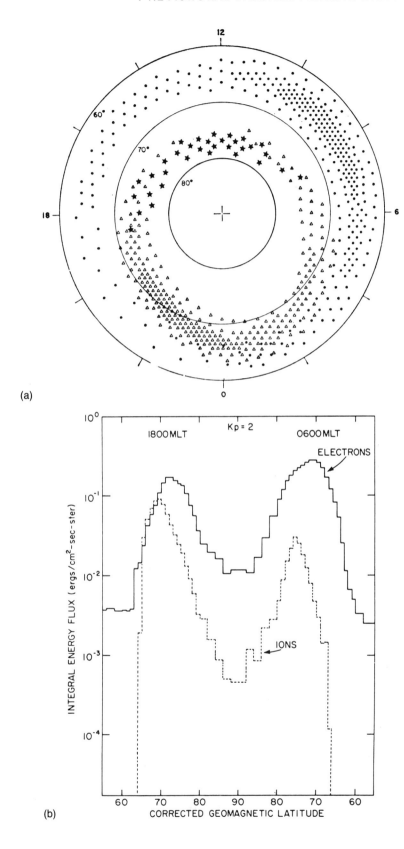

FIG. 14.3. (a) Idealized representation of a three-zone auroral-particle precipitation pattern. The auroral-oval (medium-energy) precipitation (splash type) is represented by the triangles, the auroral-zone (high-energy) precipitation (drizzle type) by the dots, and the polar-cusp (low-energy) precipitation on the dayside by the stars. The average flux is indicated approximately by the density of the symbols. The coordinates are geomagnetic latitude and geomagnetic time. (From Hartz, 1971.) (b) Integrated energy flux into the auroral ionosphere across the dawn–dusk plane as a function of geomagnetic latitude for electrons and protons.

Figure 9.18). The strong asymmetry in the location of the aurora on the dayside and nightside of the earth gives the impression of an oval-shaped band around the polar regions. This band, within which auroras are common, is referred to as the auroral oval and is discussed fully in Section 14.3.3. Precipitating ions producing auroras show a dawn–dusk asymmetry, displaced toward dusk with respect to auroral electrons, as illustrated in Figure 14.3b.

In the high-latitude region on the dayside (i.e., between 70° and 80° Λ, where Λ is magnetic latitude), the characteristics of the energetic particles are similar to those of the magnetosheath; that is, the average energies are well below 1 keV. Both electrons and protons from the magnetosheath penetrate down to the atmosphere in the cusp/cleft region, where they produce dayside auroras. The precipitation occurs in a narrow region at about 78° invariant latitude stretching from late morning to early evening magnetic local time (Section 14.3.3), referred to as a cusp or a cleft.

The proton and electron motions in near-earth space are governed by the three adiabatic invariants introduced in Chapters 2 and 10. A fraction of these particles will have their mirror points in the atmosphere, below about 200 km in altitude. Particles penetrating the atmosphere collide with atmospheric atoms and molecules and gradually lose their energy to the neutrals. The energy loss rate for a subrelativistic electron is given by the formula

$$-\frac{dW_e}{dx} = -\frac{dW_e}{Q\,ds} = \frac{2\pi e^4 Z A_0}{W_e A}\ell n(W_e/I) \tag{14.1}$$

where $dx\,(= Q\,ds)$ is the atmospheric depth, given in grams per square centimeter, A_0 is Avogadro's number, Z is the average atomic number of atmospheric atoms of atomic weight A, I is the average energy loss per ionization, Q is the mass density of scattering atoms, and ds is a differential distance along the electron trajectory. In fact, the main sink for fast, charged particles in the magnetosphere is the atmosphere (Rees, 1989).

Precipitated charged particles in the ionosphere are subject to inelastic and elastic collisions with the atmospheric constituents. They lose their energy gradually by (1) ionizing and exciting the upper atmosphere, (2) dissociating atmospheric molecules, (3) heating the upper atmosphere, and (4) producing bremsstrahlung x-rays. (This latter process, which is discussed in Chapter 7, is negligible for low-energy particles and will not be discussed further here.)

Thus, energy deposited in the upper atmosphere by precipitating particles is, in part, used to produce optical emissions (i.e., the aurora). As discussed in Chapter 7, a downcoming beam of monoenergetic particles entering the atmosphere will penetrate to about the altitude of "unity optical depth" for the particles. Most of the absorption will be within a neutral scale height of this altitude. In a realistic situation, one must sum the collision cross section s_{ij} over the different absorption processes available for each of the j atmospheric constituents present. These are weighted by the relative cross sections for the i processes and the relative number densities of the j constituents present. The cross section has an important energy dependence, with higher-energy particles penetrating more deeply. The approximate penetration depths for various proton and electron energies are shown in Figure 7.4. Because the particle penetration is governed by statistical processes, the actual penetration depths are not identical even for two particles with identical initial conditions. The values given in Figure 7.4 should therefore be considered as the average height where most of the energy is absorbed for vertical incidence. A detailed discussion of the problems of particle scattering and absorption is given in the monograph by Rees (1989).

Experimental data show that fast electrons and protons produce about one ion pair (ion–electron) per 36 eV of their initial energy. This can be written symbolically for electrons and protons, respectively, in the following equations:

$$X + e \rightarrow X^+ + e_n + (e - 36 \text{ eV}) \tag{14.2a}$$

$$X + H^+ \rightarrow X^+ + e_n + (H^+ - 36 \text{ eV}) \tag{14.2b}$$

where X is an atmospheric constituent, and e_n is a thermal electron in the ambient electron gas, rather than an energetic auroral electron. Because the ionization potential of the atoms and molecules, on average, is about 15 eV, about 40 percent of the energy goes into ionization, whereas about 60 percent goes into the motion of the product electron, which subsequently thermalizes.

14.2.2 The Auroral Spectrum

As discussed in Section 14.2.1, the kinetic energy of the auroral particles can be deposited – through collisions – into the translational, vibrational, and rotational energies of atoms and molecules, expended in impact-excitation of bound electrons from their ground state to a higher level, or spent in electron ionization by impact (Vallance Jones, 1974). The distribution of energy among these initial options sets the stage for the energy subsequently liberated in the ultraviolet (UV), visible, and infrared (IR) emissions of auroras. Thus, the auroral emissions contain atomic lines and molecular-band spectra of the primary constituents of the upper atmosphere (Figure 14.2), plus some important emissions from minor species (e.g., NO, He, and CO_2, which are efficient in the cooling of the thermosphere via strong IR emissions). The auroral emissions can therefore be considered as the "fingerprints" of the atmospheric constituents.

This section provides a brief description of the mechanisms that account for the auroral spectrum. For more thorough reviews of auroral spectroscopy, the reader is referred to Vallance Jones (1974) and Omholt (1971).

Photons may be emitted spontaneously as the excited species relax to lower energy levels and/or ground states, or further chemical reactions may take place as the energy cascades, with photons emitted along the way. The optical-emission wavelength λ, in nanometers, is related to the released energy E, in kiloelectron volts, by $E = 1.240/\lambda$.

Some energy reactions and transfer mechanisms important in auroral physics are the following:

Electron Impact

$$e + N \rightarrow N^* + e' \tag{14.3a}$$

Energy Transfer

$$X^* + N \rightarrow X + N^* \tag{14.3b}$$

Chemiluminescence Reaction

$$M + XN \rightarrow MX^* + N \tag{14.3c}$$

Cascading

$$N^{**} \rightarrow N^* + h\nu \tag{14.3d}$$

Here, N, X, and M are atmosphere neutrals, asterisks represent different levels of excitation, and $h\nu$ is an emitted photon. In each collision, the primary electron (e) is replaced by secondary electrons (e') of lower energy [equation (14.2)], which can be responsible for further excitation. The energy and altitude dependences of the absorption of auroral particles will not be discussed further here (see Chapter 7).

An additional class of energy-transfer processes involves ionization of neutrals. When ionization occurs, the ion may be left in an excited state and may radiate light immediately. The auroral spectrum can be used to detect the ionization process. This is the case with the first-negative-band system. These bands are excited directly from the ground state of N_2. Symbolically, the process involved is

$$N_2 + (e \text{ or } H^+) \rightarrow (N_2^+)^* + (e' \text{ or } H^{+\prime}) + e_n \tag{14.4a}$$

where $e' = e - 36$ eV, $H^{+\prime} = H^+ - 36$ eV. The asterisk indicates that the molecular ion is created in an excited state. This process is followed by radiation of the first negative bands:

$$(N_2^+)^* \rightarrow N_2^+ + (391.4\text{- and } 427.8\text{-nm}) \text{ aurora} \tag{14.4b}$$

The probability of process (14.4a) followed by (14.4b), compared with the probability of all possible ionization processes, is nearly independent

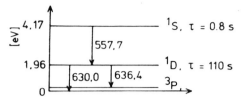

FIG. 14.4. Energy level of the oxygen atom. The terms as well as the radiative half-lives of the ^1S and ^1D levels are indicated, and the wavelengths of photons emitted in transitions between energy levels are shown.

of energy for electrons between 0.5 and 20 keV (not for electrons <0.5 keV). The intensity of these auroral bands can therefore be used to determine the net downward electron energy. Quantitative considerations show that about 25 ion pairs are produced for each photon emitted in the λ 391.4-nm band. The corresponding figure for the λ 427.8-nm band is 75 ion pairs per photon. For protons, the situation is somewhat more complicated (as will be discussed later).

The brightest visible feature of the aurora, the "green line" at 557.7 nm, is due to the transition of an electron from the ^1S excited state to the ^1D state of atomic oxygen, as illustrated in Figure 14.4. Another commonly observed line – particularly in the polar cusp and cap – is the "red line" at 630 nm as the ^1D state relaxes to the ground state. (Fine structures of the electron shells in the ground state allow 636.4-nm emission as well.) If the O(^1S)-state electron gives up its full 4 eV in a single step, instead of two nominal 2-eV steps (^1S to ^1D, and then ^1D to ^3P), it emits a photon at 297.2 nm, that is, about half the wavelength of the emissions from the smaller energy steps (Figure 14.4).

The forbidden oxygen line at 557.7 nm and the red doublet at 630 and 636.4 nm can be excited by the following process (where e' has less energy than e):

$$O(^3P) + e \rightarrow O(^1S) + e' \tag{14.5a}$$

followed by

$$O(^1S) \rightarrow O(^1D) + h\nu \quad (557.7 \text{ nm}) \tag{14.5b}$$

or

$$O(^1S) \rightarrow O(^3P) + h\nu \quad (297.2 \text{ nm})$$

For the red doublet, we have

$$O(^3P) + e \rightarrow O(^1D) + e' \tag{14.5c}$$

followed by

$$O(^1D) \rightarrow O(^3P) + h\nu \quad (630/636.4 \text{ nm}) \tag{14.5d}$$

where ^1S and ^1D have total electron spin $s=0$, while ^3P has spin $s=1$ (Vallance Jones, 1974). These excitations have a small probability for higher energies. There are also great numbers of permitted oxygen and nitrogen lines from higher excited states. Most of the permitted transitions observed in O and N have excitation potentials of about 10–13 eV, and in O^+ and N^+, 20–30 eV above the ground state of the neutrals. The sodium doublet at 589.0 and 589.6 nm (^2S–^2P) is occasionally observed in auroras. Also, the helium line at 587.6 nm (^3P–^3D) has been observed on rare occasions.

In sunlit auroras (Størmer, 1955), the spectral distribution is somewhat different from that of ordinary auroras, with enhanced intensity of the first negative bands. This is due to resonance scattering of sunlight by N_2^+ produced by the primary particles.

The statistical residence time in an excited state before emission is determined by the *Einstein transition probability*. Governed by quantum-mechanical selection rules, *allowed* or *permitted transitions* occur very rapidly (in times on the order of 10^{-7} s). Transitions that violate these selection rules are called *forbidden transitions*. They do occur, but only after a much longer time; for example, the transition for $O(^1S)$ shown in equation (14.5b) occurs in about 0.8 s, and that for $O(^1D)$ shown in equation (14.5d) requires about 110 s. The latter is so slow that much below 200 km, an $O(^1D)$ atom is likely to suffer a collision that will knock it out of the $O(^1D)$ state before it has a chance to emit. Thus, the 630-nm (OI, or neutral oxygen) emission is expected to peak above 200 km, even though the excitation to the $O(^1D)$ state is expected to peak near an altitude of 100 km. (A roman numeral following an atomic symbol refers to the ionization state, with I designating un-ionized, II designating singly ionized, etc.) This *quenching* by collision with other atmospheric constituents substantially reduces the number of 630-nm photons emitted below the number of $O(^1D)$ states excited. The 1.96 eV used by the oxygen atom in exciting the $O(^1D)$ state goes into vibrationally exciting the local atmospheric gases, instead of dissipating into the atmosphere (Figure 14.4).

Molecules also can store energy in the form of vibrational energy (along the molecular axis) and/or rotational energy (along a transverse axis). Because of the close spacing of vibrational-energy levels, auroral emissions from molecules have bandwidths of nanometers, whereas atomic-line bandwidths are on the order of 0.1 nm or less.

A few definitions and categorizations should help to put these ideas of auroral emission processes and rates in perspective. The photon-emission intensity I (cm$^{-3} \cdot$ s^{-1}) is

$$I = N^*A$$

where N^* (cm^{-3}) is the density of the excited-state emitting molecules and A (s^{-1}) is the Einstein coefficient. The density of excited molecules is the ratio P/L: the excitation rate per unit volume P (cm$^{-3} \cdot$ s^{-1}) divided by the loss rate L (s^{-1}). L must include all collisional deactivation (or quenching) processes of the excited state by each quenching species, weighted by its rate coefficient (molecule$^{-1} \cdot$ cm$^{-3} \cdot$ s^{-1}).

An important part of the optical spectrum in the visible region is shown in Figure 14.2, and the average intensities of the most prominent

TABLE 14.1. Typical Relative Auroral Intensities

Molecule	Atomic Emission (nm)	Relative Intensity	
		Night	Day
O	^1D–^1S 557.7	1	1
	^3P–^1D 630/636.4	0.1–0.5	1–100
N	^4S–^2D 519.9	0.01	
	^2D–^2P 1,040	1	
O_2	b-X atmospheric (0–0) bands	2	
	a-X IR (0–0) bands	10^2–10^3	
O_2^+	B-A, first negative bands	0.4–1	
N_2	B-A, first positive bands	5–20	
	A-X, Vegard-Kaplan bands	1	
	a-X, Lyman-Birge-Hopfield band	0.5–1	
N_2^+	B-X, first negative bands	0.5–1	0.3
	A-X, Meinel bands	7–20	

Note: Several bands of O_2, O_2^+, N_2, and N_2^+ radiation, each band extending over a range of wavelengths and containing emissions at discrete wavelengths, contribute to the auroral emission spectrum. These bands are listed in standard auroral spectroscopy notation in column 2 of this table. For an explanation of the nomenclature and other details, see Chamberlain (1961).

auroral emissions relative to the green line are listed in Table 14.1 (Omholt, 1971; Vallance Jones, 1974). The auroral intensities are highly variable, and the values listed in the table are estimated averages. Because of the difficulties inherent in accurate optical-intensity measurements, the intensities quoted have relatively large uncertainties.

Some weak but important hydrogen lines, first discovered by Vegard in 1939, exist in the auroral spectrum (Vegard, 1939). The emissions H_α at 656.3 nm and H_β at 486.1 nm result from excited hydrogen atoms that are produced when energetic protons (H^+) bombard the atmosphere. The excitation mechanism, illustrated in Figure 14.5, can be written

$$X + H^+ \rightarrow X^+ + H^* \tag{14.6a}$$

followed by the auroral emission

$$H^* \rightarrow H + h\nu \quad \text{(hydrogen/proton aurora)} \tag{14.6b}$$

The hydrogen atom again collides:

$$H + X \rightarrow H^+ + X + e_n \tag{14.6c}$$

and process (14.6a) can start again.

The hydrogen atom formed in process (14.6a) has almost the same velocity and direction as the original proton. The fast atom – after process (14.6b) – collides with an atmospheric particle (X) and may be reionized (14.6c) or excited. The latter is more likely to occur at low particle energies. An average particle goes through a great number of processes of electron capture and loss before it has lost its energy and is brought to rest in the upper atmosphere.

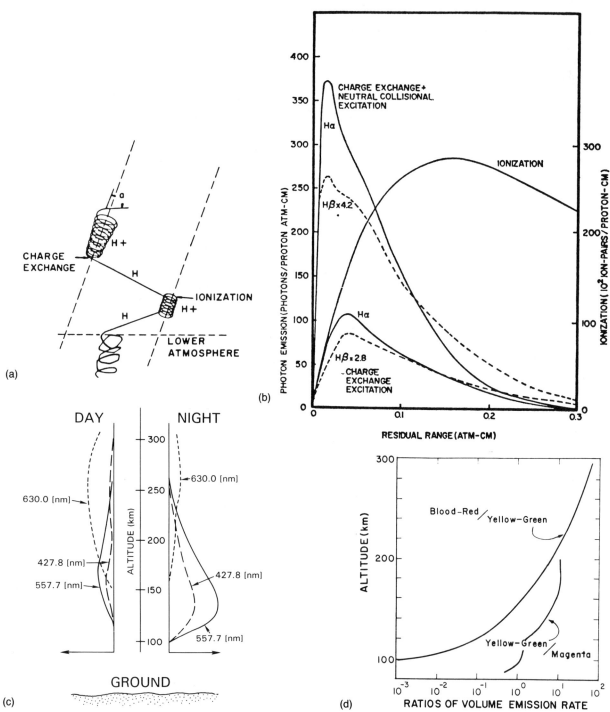

FIG. 14.5. (a) Typical path of protons entering the atmosphere. The figure is schematic and not to scale. (b) Photon-emission curves for H_α and H_β and ionization curves for protons in the air. (From Omholt, 1971.) (c) Main differences between dayside and nightside auroras versus altitude. (d) Ratios of emission intensities in various spectral ranges versus altitude. On the basis of observations of spectral ratios, the average auroral height can be determined.

As a fast particle (proton/atom) penetrates the atmosphere and slows down, it spends a large fraction of its time as a neutral atom. Therefore, its probability for emission of light increases until its energy drops below the excitation level. Figure 14.5b shows the photon emission in H_α and H_β per proton as a function of residual range. When a photon is emitted, it has a Doppler displacement that depends on the velocity of the emitting hydrogen atom and the angle between the velocity vector and the direction of the photon (Omholt, 1971). The first realistic estimates of the auroral-particle energies (made before the space age) were based on such Doppler profiles.

The protons arrive in helical paths (Figures 14.5a), spiraling around the magnetic lines of force with a given pitch angle. With a known distribution of an ensemble of protons of initial energies and pitch angles, we can estimate the total emission from hydrogen as a function of height, as well as the Doppler profile of the hydrogen light for any direction of observation. These computations are not difficult in principle, but somewhat cumbersome in practice (Omholt, 1971). As a result of charge exchange, the proton auroras are more defocused (i.e., diffuse) than the incident proton precipitation. These auroras occur in an oval displaced duskward of the electron oval (Figure 14.3b) and have a response time different from that of electron auroras (Section 14.4).

The spectral characteristics listed earlier are more generally described in terms of the colors observed in auroral forms. The 630-nm (OI) emission, which is created by auroral "soft" primaries, is seen as the red aurora; this forms the diffuse background radiation in which the discrete arcs are embedded (Figure 14.6b). "Blood-red" auroras are produced by low-energy electrons ($\ll 1$ keV). Reports of "red lower borders" (usually fast-moving) indicate the presence of particles with energies greater than 10 keV. The majority of the auroras are yellow-green, but sometimes appear gray. The gray appearance refers to observations below the color threshold of the eye and corresponds to low particle flux. Figure 14.5d shows how the ratios of the main visual emissions change relative to one another as a function of altitude. Specifically, blood-red auroras dominate in the altitude region above 200 km, whereas a magenta color predominates below approximately 100 km. Blood-red is the (OI) emission at 630–636.4 nm, yellow-green is the (OI) emission at 557.7 nm, and magenta is a combination of N^2 and O_2^+ emissions near 600 nm and N_2^+ first-negative-band emissions in the blue end of the spectrum.

The 630-nm (OI) emission can also be excited by a process called thermal excitation. The ambient electron gas will have a thermal or Maxwellian distribution of energies, with a population decreasing exponentially with increasing energy. The fraction of the electron population above a fixed energy is obviously strongly dependent on the electron temperature. For electron temperatures much above 3,000K, there may be enough electrons in the high-energy *tail* of the thermal distribution to excite detectable emission from O atoms. This follows because the

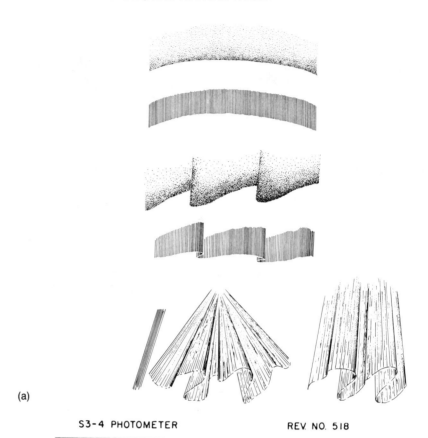

(a)

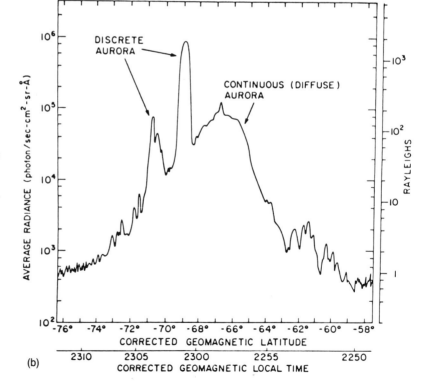

(b)

FIG. 14.6. (a) Illustrations of typical forms of the northern lights. Top to bottom: homogeneous arc; arc with ray structure; homogeneous band; band with ray structure. The three lower forms are (left to right) rays, corona, and draperies. (b) Downward-looking photometric measurements of discrete and diffuse (continuous) aurora performed by the S^3-4 satellite. Intensity plotted versus satellite position in latitude and local time.

O(^1D) level is only 1.96 eV above the ground state, as shown in Figure 14.4. Thus, thermal electrons can produce excitation of 630-nm (OI) auroras. Stable auroral red (SAR) arcs are formed on the equatorward edge of the auroral oval by this excitation process, provided that increased electron-gas heating from above and/or decreased electron-gas cooling by the ionosphere below allow the electron temperatures in the F region to rise to well over 4,000K (e.g., Kozyra et al., 1990).

14.2.3 Auroral Intensities

An aurora appears as a luminous cloud, having an apparent surface brightness. Absorption within the visible spectrum is negligible. Hence, the apparent surface brightness is proportional to the integrated emission per unit volume along the line of sight. The surface brightness is used to define the intensity of an aurora.

If the surface brightness I is measured in photons per square centimeter per second per steradian, then $4\pi I$ represents the total emission in photons per square centimeter per second integrated along the line of sight. This is defined as the intensity of an aurora. The unit adopted for $4\pi I$ is the rayleigh (R). One rayleigh is equal to an integrated emission rate of 10^6 photons per square centimeter per column per second (inclusion of "column" in the units refers to the unknown height of the column above the apparent source; it is included to show that this is a volume emission, not a true surface emission). The observed intensity of a particular auroral form depends on the direction of observation. A thin auroral layer covering a large part of the sky is most intense when viewed at low elevation angles.

The auroral intensity in rayleighs at a particular wavelength λ from an incoming electron/proton beam, with isotropic pitch-angle distributions assumed at all energies, can be evaluated from the following expression:

$$I(X(\lambda))[R] = \frac{\pi}{10^6} \int_{E=0}^{\infty} j(E) \cdot P(E) \, dE \tag{14.7a}$$

where $j(E)$ is the differential electron/proton energy spectrum, and $P(E)$ is the total number of photons for gas X at wavelengths λ (i.e., for N_2^+ at 427.8 nm or for H_β at 486.1 nm). For protons, $P(E)$ depends strongly on the energy spectrum (see Problem 14.2), and H^+ above 20 keV contributes significantly to the output light. For the N_2^+ first negative bands, Rees and Roble (1986) have given the following formula:

$$4\pi I(427.8 \text{ nm})[R] = 213 \left(\frac{E_{\text{ave}}}{2}\right)^{0.0735} \cdot \epsilon \frac{\text{erg}}{\text{cm}^2 \cdot \text{s}} \tag{14.7b}$$

where ϵ is the energy flux, and E_{ave} is the average energy of incident auroral particles.

When classifying auroral intensities, the line used for reference is the green oxygen line at 557.7 nm, which is dominant in the wavelength region near the maximum sensitivity of the human eye. Typical intensities of nightside auroral arcs and bands vary from one to a few tens of kilorayleighs. During an active period with a bright display, auroral intensity in or near the zenith may be several hundred kilorayleighs. Thus, as 1 kR near 550 nm is the visibility threshold for the dark-adapted naked eye, the optical aurora is a relatively weak but very dynamic optical phenomenon. The large variations in intensity are closely correlated with the net downward particle energy. In order for the northern lights to be visible to the eye, the particle energy input to the atmosphere must be about $1 \text{ erg} \cdot \text{cm}^{-2} \cdot \text{s}^{-1}$ or approximately $10^{-3} \text{ W} \cdot \text{m}^{-2}$. For a medium-strong northern light about 10 km wide and 1,000 km long, approximately 10^6 kW are needed, which is comparable to the power capacity of a large power plant. Because only about 1 percent of the particle input to the atmosphere is used to produce visible light, it is clear that there is an enormous quantity of energy deposited in the upper strata of the atmosphere during each auroral night.

Auroral Photography. On a clear, dark winter night, the northern lights may look bright to the dark-adapted naked eye. Still, it is not easy to take good pictures of the aurora with short exposures, though with a modern 35-mm camera with a small-focal-ratio lens, it is possible. With fast color film (e.g., 400 ASA), an exposure time of 1 s to a few seconds is normally needed, and therefore it is advisable to use a tripod for mounting the camera.

14.3 AURORAL DISTRIBUTION IN SPACE AND TIME

14.3.1 Auroral Forms

Looking at an auroral display, one normally sees a bewildering number of auroral forms and situations (Størmer, 1955). Each instantaneous auroral situation may be considered to be composed of various superimposed elementary auroral forms or structures that vary in space and time. For practical purposes, we have to consider only four such elementary forms, as illustrated in Figure 14.6: (1) the quiet homogeneous arc and band stretching along the magnetic east–west direction across the sky in a straight or curved line; (2) auroral rays and combinations of rays, which may vary considerably in length; (3) diffuse or irregular auroral clouds; (4) spirals and curls. Their intensities may vary over several orders of magnitude. For a more detailed description of the different auroral forms, the reader is referred to Størmer's excellent book *The Polar Aurora* (1955).

Another scientifically useful method of categorizing auroral forms is to identify them as discrete or diffuse forms (Figure 14.6b). However, even diffuse auroras may contain some weak discrete structures. This has been shown by the use of very sensitive optical instruments. The *diffuse aurora* is so named because its weak, small striated structures were not observable in the satellite data where it was first documented as a separate form. It is associated in the evening with the region of greatest proton energy flux and maps magnetically just inside the boundary of stable trapping (see Chapter 10). This explains its more or less circular form around the magnetic pole. A diffuse aurora in the E region is produced by higher-energy particles bouncing from one hemisphere to another and losing energy to the atmosphere as they diffuse into the *loss cone* while drifting in longitude, electrons to the east and protons to the west.

14.3.2 Height Distribution of Auroras

Generally, the light from an aurora is proportional to the deposition of energy into the atmosphere by the primary particles. As discussed in Section 7.2.2, the height distribution is related to the energy and pitch-angle distribution of the precipitating particles, as well as the atmospheric composition. The energy distribution and flux of the particles can be studied directly only by rocket and satellite techniques. Still, auroral height measurements, as well as spectral ratios (Figure 14.5d), may provide useful information on representative energies and systematic variations in the energy spectrum.

Figure 14.7 shows the distribution of 12,330 auroral points measured by Størmer (1955). Størmer's measurements demonstrated that auroral arcs and bands, homogeneous as well as rayed, lie predominantly within the height interval 95–150 km, whereas isolated rays or bundles of rays can lie significantly higher. Long rays may stretch several hundred kilometers up into the atmosphere. As Figure 14.5c illustrates, the average height of dayside auroras is significantly higher (by 100–200 km) than that for the nighttime auroras.

The detailed height distributions for individual forms have been measured by both rockets and ground-based techniques. Auroral height and the height distribution for each form are related to the average energy and the energy distribution of the precipitating particles.

14.3.3 Auroral Locations, Auroral Morphology

In the literature up to the 1960s, auroral location was described in terms of the auroral zones, which represent the regions in which nighttime auroras occur in an average statistical sense. Within the auroral zones, which are centered roughly 23° from the geomagnetic poles and are approximately 10° wide, auroras are observed by the naked eye on more than 50 percent of clear nights, even during years of low solar

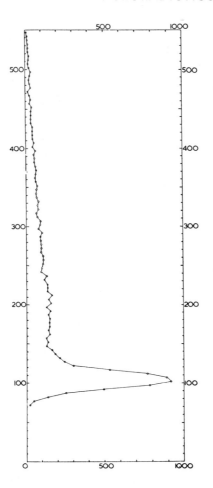

FIG. 14.7. Statistical distribution of 12,330 height measurements of the northern lights derived by Størmer and his colleagues. The vertical scale gives the height in kilometers, and the horizontal scale gives the number of measurements. Most of the northern lights are found between 90 and 150 km.

activity. The frequency of occurrence and the intensities of auroras are clearly correlated with the activity of the sun. This is borne out by the 27-day recurrence tendency, which correlates with the rotational period of the sun, as well as by the 11-yr variation that correlates with the sunspot cycle. However, as first pointed out by Størmer (1955), the auroral activity peaks 1–2 yr after a sunspot maximum. The auroral zones follow $L =$ constant to a good first approximation. The instantaneous distribution of maximum auroral activity versus magnetic time and latitude was found by Feldstein and Starkov (1967) to be given by an oval-shaped belt called the auroral oval (Figure 14.8).

The auroral ovals (one for each hemisphere) are continuous bands centered near 67° magnetic latitude at magnetic midnight and near about 77° at magnetic noon during quiet periods and periods of moderate activity (Akasofu, 1968). (Examples of auroras, as seen from both space and ground, are shown in Figure 14.1.) The latitude extremes of the oval are the average boundaries of auroral luminosity, as seen by all-sky cameras during the International Geophysical Year (IGY) 1957–8. The locations of the auroral oval for quiet, moderate, and active conditions

14.3 AURORAL DISTRIBUTION IN SPACE AND TIME

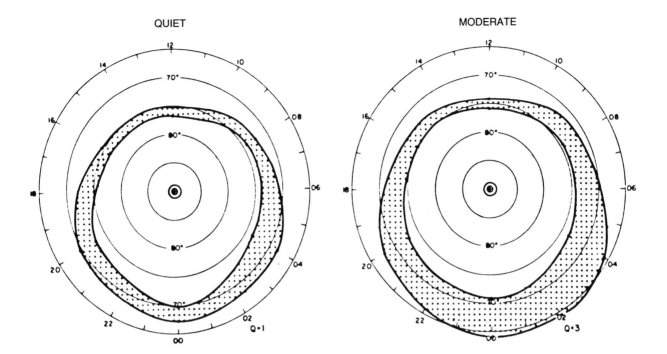

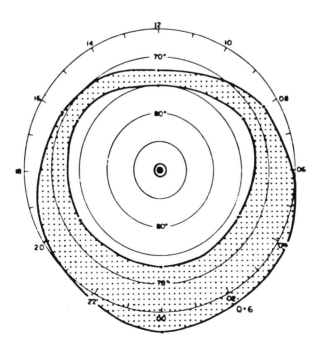

FIG. 14.8. Variation in the size of the auroral oval with activity. The shaded area represents the distribution of maximum auroral activity in the northern hemisphere. Coordinate system is corrected geomagnetic (CG) latitude and CG local time, and noon is at the top. (Adapted from Feldstein and Starkov, 1967.)

are shown in Figure 14.8 as functions of geomagnetic local time and geomagnetic latitude. Evidently, the location of an aurora is connected with disturbances in the geomagnetic field, the aurora moving mainly equatorward with increasing magnetic activity.

Auroras also appear regularly poleward of these ovals, but with less frequency and intensity. Equatorward, the occurrence rate falls rapidly. The occurrence of auroras at low latitudes is associated with enhanced magnetic activity and magnetic storms.

As an observation station rotates with the earth, its proximity to the oval (of a particular activity level) fixed in the earth–sun frame of reference, or magnetic local time (MLT), changes with a 24-h periodicity. Because the magnetic pole is displaced from the earth's rotation axis (roughly an 11° difference in the northern hemisphere, tipped toward the U.S. east coast), there is also a longitude dependence for many auroral effects. These are best ordered by universal time (UT), that is, the time counted positive westward from Greenwich with 15° per hour.

By means of careful comparisons among ground-based, rocket-borne, and satellite observations, it has been found that nightside auroras are primarily due to electrons between 1 and 15 keV, whereas dayside auroras are primarily due to electrons with energies less than 1 keV (Figure 14.3). This explains well the large differences between both the emission altitudes and spectra of nightside and dayside auroras.

Unstructured auroras, down to about 95 km in a circular band of roughly constant magnetic latitude nearly equal to that of the equatorward edge of the midnight auroral oval, are produced by electrons of approximately 5–20 keV. Diffuse and structured auroras are produced at intermediate altitudes by particles of intermediate energy (on the order of 1–10 keV) from the magnetospheric plasma sheet (see Section 14.8 and Chapter 10).

In contrast to the morphological and statistical descriptions of the auroral zones developed up to the 1960s, and the auroral oval (Figure 14.8) described thereafter, a more physical view of an instantaneous auroral state and location controlled by the solar wind, or by the interplanetary magnetic field (IMF), emerged after the 1970s. For a northward IMF, the solar wind couples poorly to the magnetosphere; the auroral oval becomes quite contracted, and the auroral intensity becomes weak. For a southward IMF, the coupling is much stronger and the oval much larger; its size expands with increasing solar-wind pressure, and the auroral intensity increases. Ground-based measurements show poleward motion of the poleward boundary of the midday aurora as the IMF changes from southward to northward, with monotonic progression of the contraction over extended periods.

Conjugacy of nightside auroral forms (mirror images, as it were) between the northern and southern hemispheres has been seen and taken as strong support for the argument that these auroras are on closed field lines.

A complication in describing and studying auroras arises from the fact that the sun usually is not in the geomagnetic equatorial plane. Because of the inclination of the earth's axis of rotation with respect to the earth's orbital plane (23°) and the angle between the dipole axis and the axis of rotation (~11°), the direction to the sun may deviate as much as ±34° from the geomagnetic equatorial plane during a year. These effects give rise to both diurnal and seasonal variations in auroral occurrences. The most important variation is probably the seasonal one, due to the angle between the geomagnetic axis and the earth–sun direction. There is a pronounced equinoctial maximum in auroral occurrences.

14.3.4 Dayside Cusp/Cleft Auroras

The increasing interest in dayside ionospheric phenomena in recent years has been motivated by general questions of the solar-wind–magnetosphere–ionosphere interactions. In this section we shall concentrate on optical ionospheric signatures on the dayside (i.e., $\sim \pm 6$ h of magnetic noon) in the cusp/cleft region, which corresponds to the auroral oval between about 70° and 80° Λ.

Some characteristic differences between dayside and nightside auroras regarding emission wavelengths and height distributions are seen in Figures 14.5c and 14.5d. Notice also that the dayside portion of the oval is located closer to the magnetic pole (i.e., centered at a magnetic latitude between 75° and 78° Λ) than are the nightside auroras, and the width of the dayside oval is approximately half of that at night (Figure 14.8).

Cusp auroras form in the daytime polar thermosphere, where atomic species are abundant (see Chapter 7). Consequently, the intensity of atomic lines in these auroras is significantly greater than the intensity of molecular bands, leading to a simplification of the dayside auroral spectrum. Even atomic lines that are not observed in nighttime auroras, partly because their wavelengths coincide with those of the more intense molecular bands, are detected in midday auroras. The relative weakness of the auroral molecular bands in midday cusp auroras also permits daytime observation of the chemiluminescent airglow OH emissions from the polar mesosphere.

The dominant dayside cusp aurora is, by definition, the diffuse band where emissions with I (630 nm) $\gg I$ (557.7 nm) are normally observed in the region 11–13 MLT. The red-line emission above 200 km is caused by the soft fluxes of magnetosheath-particle precipitation (Figure 14.4). Its intensity during quiet conditions is normally below 1 kR; that is, the dayside cusp aurora is subvisual, and the average height is 250 km or higher. The cusp aurora is diffuse in character.

The first direct evidence that the dayside aurora is related to plasma entering from the magnetosheath into the polar F region dates back no further than the 1970s. Ground-based dayside auroral measurements

TABLE 14.2. Cusp/Cleft-Aurora Electron Precipitation

Auroral Forms	I (630 nm)	I (557.7 nm)	Electron Energy	Energy Flux (erg·cm^{-2}·s^{-1} = 10^{-3} W·m^{-2})
Quiet cusp (midday gap)	<1 kR	<0.3 kR	<0.2 keV	<0.5
Active cusp (i.e., transient events)	Typically 0.5–5 kR	Typically 0.5–10 kR	<2 keV	Typically 0.5–5

were largely neglected until 1980, partly because of inaccessibility of ground sites with the necessary proximity to both the geomagnetic and geographic poles. Lack of equipment with the sensitivity necessary to study these weak, subvisual auroras was an additional factor.

Recently (e.g., Sandholt and Egeland, 1989) it has been realized that these dayside ionospheric effects are as important to solar-terrestrial research as are the nightside processes. Research related to the dayside cusp/cleft regions (based on coordinated ground, rocket, and satellite observations) has increased significantly, and important findings are being reported. The islands of Svalbard, Franz Josef Land, and part of Greenland, located near 75° Λ, are the only accessible sites in the northern hemisphere where midday auroras can be observed by optical methods (which can be used only if the sun is at a minimum of 10° below horizon near magnetic noon). Recent developments of more sensitive all-sky TV cameras and multichannel scanning photometers (sensitive primarily to 630- and 557.7-nm wavelengths) have provided high-resolution data (both in time and space) that can be studied together with coordinated particle and field measurements from polar satellites.

The region of cusp auroras is also called the "midday gap" because of the lack of discrete structures. Some typical parameters, including particle characteristics, are given in Table 14.2. Discrete, active auroral forms (often multiple, long-lived auroral bands) are observed before and after the midday gap. Transient discrete forms (often with a recurrence period of ~10 min) occur between 09 and 16 MLT (even at 12 MLT). The discrete arcs are produced by particles of higher energy than those producing the 630-nm region, but the average energy is more variable than on the nightside. Altitudes range from 150 to 200 km, and the intensities and occurrences of the arcs do not vary directly as the magnitude of the auroral substorm on the nightside. Discrete arcs are also found in the polar cap. They are also short-lived arcs of the same spectral type as the dayside arcs. At times, nightside and dayside auroras are simultaneously active. The most striking feature of simultaneous nightside and dayside auroras is the poleward expansion on the nightside, coinciding with equatorward expansion on the dayside. However, active cleft auroras are often observed without any simultaneous changes in the nightside, and vice versa. The large-scale dynamics of the cusp/cleft auroras are mainly controlled by the IMF B_z component. During IMF $B_z < 0$ conditions, the auroral intensity is particularly sensitive to the solar-wind activity, which controls the efficiency of plasma transfer. When the IMF vector turns due north, the auroral intensity decreases, and the aurora contracts poleward, indicating reduced plasma transport into the cusp region.

One type of active event (breakup arcs, poleward expansion, longitudinal motion depending on IMF B_y) can be superposed on a quiet, gaplike background arc. The majority of these midday transients, less than 10 min in duration, with intensities of a few kilorayleighs (10–20-kR events are rare), are not likely to be observed by satellite imagery, because of inadequate sensitivity and temporal/spatial resolution.

14.3.5 Polar-Cap Sun-aligned Auroras

Observations during IGY 1957–8 demonstrated that deep within the polar cap (poleward of ~75°–80° magnetic latitude) there were auroras oriented along the sun–earth direction. Seen only a small percentage of the time, they always pointed toward the sun; in the winter-night central polar cap, one could virtually tell the time by noting their direction as the earth rotated under them.

In the late 1970s, satellite and ground-based data showed that such occurrences correlated with northward IMF conditions. However, it was not until the 1980s that the improved sensitivity of intensified-imaging photometers showed these arcs (an example is shown in Figure 14.9a) to be present about half of the time, when the IMF was northward (or near zero). Those images, when combined with ground-based optical and radar measurements and satellite observations, showed that such arcs were of electrodynamic origin and determined their thermal and energetic characteristics.

A simple situation for which electrodynamic effects lead to an optical-arc signature along a boundary line is illustrated in Figure 14.9b. Consider flow near a boundary, and assume a vertical magnetic field and initially uniform conductivity across the boundary. Suppose the plasma velocity reverses (left-hand side) or that there is a velocity gradient (right-hand side) across the boundary line. This means that the electric field is discontinuous across the boundary. The horizontal electric-field differential will produce a horizontal Pedersen-current convergence at the boundary in the absence of a vertical current. Thus, a vertical (actually, magnetic-field-aligned) current flows with the magnitude required to maintain a divergence-free current state.

The difference between the left-hand-shear reversal and the right-hand-shear differential flow is merely the velocity of the frame from which the flow is viewed. It is the rest frame of the neutral gas that determines the currents. The special significance of the neutral-gas rest

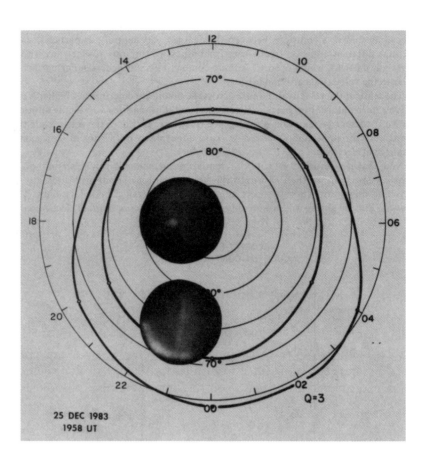

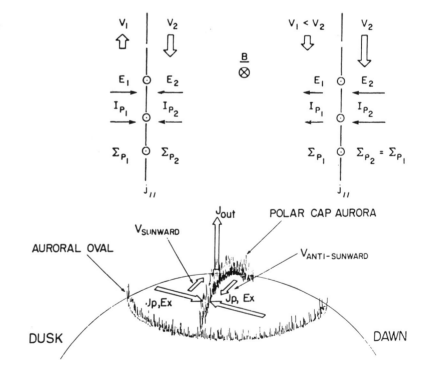

FIG. 14.9. (a) Example of a sun-aligned arc, connected to the premidnight auroral oval, extending over 2000 km sunward, as seen in a pair of 630-nm all-sky imaging photometers. (From Carlson, 1990.) (b) Two examples of a simple electrodynamic situation that would produce a sun-aligned arc. (c) Cartoon representation of the simple electrodynamics characteristic of stronger sun-aligned arcs. The plasma flow on the duskside may more generally be (for a range of arc intensities) sunward, stagnant, or antisunward, but slower than the antisunward flow on the dawnside.

frame may be understood in part by recalling that it is collisions between the charged and neutral-gas particles, particularly different collision frequencies (mobilities) of the ions and electrons with the neutral-gas particles, that produce a finite conductivity and result in Pedersen current perpendicular to the velocity-shear boundary.

The sense of the velocity differential across the boundary determines whether it drives a horizontal convergence or divergence of current, and thus requires a vertically upward or downward current to maintain a divergence-free state. In Figures 14.9b and 14.9c, the sense leads to an upward current, presumably carried into the ionosphere by a flux of downcoming suprathermal electrons.

Note that if there is a series of velocity differentials across a portion of the ionosphere, with plasma alternately speeding up and slowing down, the upward current sheets will be interspersed between downward current sheets along velocity-difference boundary lines. The downward currents presumably will be carried by upgoing thermal electrons.

Combined ground-based images and satellite data have shown that the simple arc electrodynamics are in fact the electrodynamics that pertain to sun-aligned arcs that are stable in time and extensive in sunward direction (Carlson et al., 1988). We should note that a change in conductivity across a boundary in the ionosphere can drive current sheets out of the ionosphere even in the absence of a velocity shear. We have found no examples of this effect alone creating sun-aligned arcs.

Because the current carriers of the vertical (actually, magnetic-field-aligned) current are suprathermal electrons with energies of tens to hundreds of electron volts, they excite the optical signature of the arc. By the same token, they will be expected to produce impact ionization, thereby enhancing the conductivity within the arc and modifying the distribution of currents that flow within the arc itself. This feedback effect must be allowed for in detailed self-consistent treatments of the current patterns related to the arc.

Here, our main point is that the sun-aligned arcs are visual markers of velocity-shear lines, lines of sharp velocity differentials, of a specific sense: greater antisunward velocity on the dawnside than on the duskside of a polar-cap sun-aligned arc. The shear line points toward the sun because the antisunward flow is driven by coupling to the solar wind blowing past the earth and away from the sun (as described in Chapter 10). Because sun-aligned arcs are signatures of sharp slowing (or even reversal) of an antisunward ionospheric plasma flow, they are thereby a most valuable tool for discovery and definition of the character of polar-cap convection for northward IMF conditions (i.e., the half of the time for which convection is most poorly understood). These sun-aligned arcs are also of importance to the thermal balance and energetics of the polar-cap ionosphere and thermosphere (Valladares and Carlson, 1991).

14.4 THE AURORAL SUBSTORM

An auroral substorm, which can have many different auroral forms, large variations in both color and intensity, and rapid motions, may look completely disordered. However, an experienced observer will soon notice that such large auroras follow a fixed pattern, almost as in a classical play or symphony, with four scenes or movements (Akasofu, 1968).

For a nighttime observer near the oval, the display usually starts with one or more quiet, homogeneous arcs of fairly low intensity (1–10 kR) elongated approximately in the geomagnetic east–west direction. After some time, possibly hours, the aurora starts to move equatorward, increases in intensity, and may develop ray structure and take the form of less regular bands. Then, suddenly, the whole sky explodes, and the aurora spreads over the entire sky. Simultaneously, the aurora moves rapidly, with changes in form and intensity, at times increasing to several hundred kilorayleighs, and the individual structures in the aurora may show apparent speeds in an easterly or westerly direction of several tens of kilometers per second. After a few minutes, the aurora becomes weaker and rather diffuse. This marks the beginning of what is called the recovery phase.

Substorms must be considered within a fixed earth–sun frame of reference, as is true of all auroral phenomena. A ground-based observer thus must remove the effects of the earth's rotation in tracing auroral motion in this reference frame. Whereas that seems straightforward in principle, the offset of the geomagnetic pole from the geographic pole introduces more complexity than at first meets the eye.

Each (four-scene) active period is called an auroral substorm. An illustration of the development of an auroral substorm in the dipole local-time coordinate system is shown in Figure 14.10a. An example of an auroral "breakup" is shown in Figure 14.10b, a sequence of auroral pictures taken with an imager on board the Dynamics Explorer *DE-1* spacecraft at roughly 20,000 km from the earth. The proton aurora (Section 14.2.2) moves poleward of the trapping boundary some time after the poleward expansion of the discrete arcs associated with the substorm. The most spectacular part of the display usually lasts only a few minutes (<10 min), the whole active period being perhaps 0.5–1 h. Afterward, the sky is more or less covered with a homogeneous surface of weak auroral light. The intensity is then usually between 1 and 5 kR, but this aurora may be subvisual. Because of the lack of contrast, the intensity of such weak, widespread emission surfaces, compared with more discrete and distinct forms, is often underestimated by the eye. The duration and intensity of the substorm may vary greatly. The usual time interval between two substorms in one night ranges from 0.5 to 3 h. For disturbed periods, several auroral breakups may occur in one night.

14.4 THE AURORAL SUBSTORM

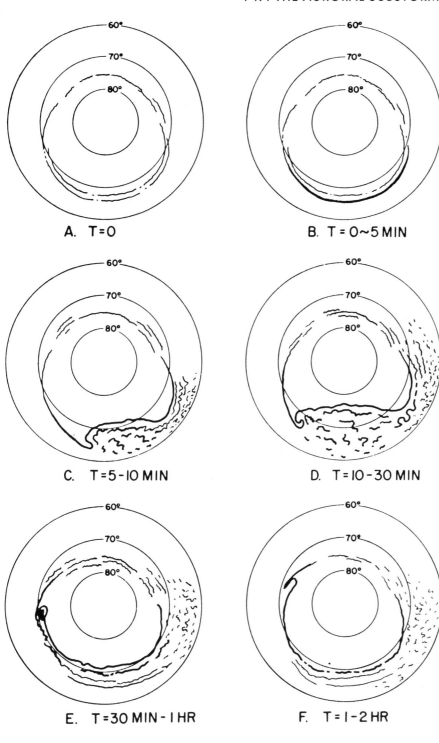

FIG. 14.10. (a) An auroral breakup event is illustrated by this sequence of drawings. The sequence of the pictures is indicated by the time labels given from $T=0$. (From Akasofu, 1968.) *(Continued)*

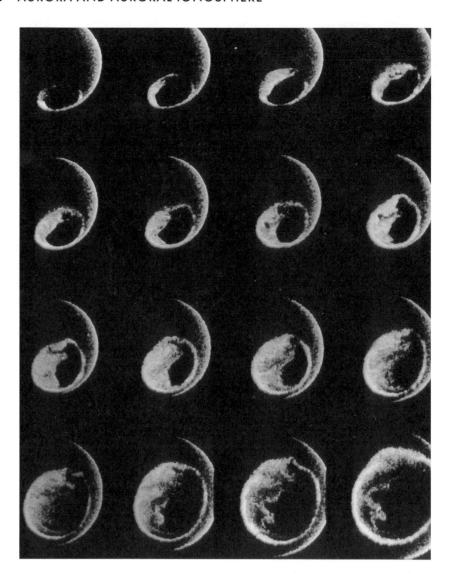

FIG. 14.10 (cont.) (b) Sequence of auroral pictures from *DE-1* during an auroral substorm. Time increases from left to right and from top to bottom. (From Frank and Craven, 1988.)

During an auroral display, several forms may appear simultaneously, partly overlapping or embedded in each other. Auroral arcs and bands frequently show ray structure. Each form usually moves and varies in intensity with time. For homogeneous, quiet forms, the velocities are small (≈ 100 m·s^{-1}), whereas particular ray structures may show rapid motion, with velocities of up to 50 km·s^{-1}. Simultaneously, there are intensity and color variations in the moving forms. Large-scale intensity variations are combined with the movements. The form of the aurora may vary with the MLT. A few hours after geomagnetic midnight, the auroral pattern is predominantly diffused, but the forms often show rapid fluctuations in intensity and pulsations. The substorm is associated with disturbance current systems, giving rise to characteristic magnetic disturbances (Akasofu, 1968; Akasofu and Chapman, 1972).

Because an aurora is the result of charged particles interacting with the earth's upper atmosphere, and the injection of these particles is due to interaction with the solar wind, the aurora depends both on the plasma streaming away from the sun and on the earth's magnetic field. The appearance of auroras in space and time depends not only on the properties of the solar wind and the geomagnetic field but also on the composition of the atmosphere. The occurrences and spatial distributions of auroras depend also on local time and the substorm phase. The physical processes that control the different phases of an auroral substorm are not yet well understood.

14.5 THE AURORAL IONOSPHERE

Before addressing the electrodynamics of the ionosphere, we shall summarize some characteristics of the ionized portion of the radiating medium: the auroral ionosphere (see also Chapter 7). The auroral ionosphere has properties significantly different from those of the ionosphere at subauroral latitudes (Rishbeth and Garriott, 1969). For the midlatitude ionosphere, the dominant ionizing source is solar radiation [extreme-ultraviolet (EUV) and UV rays and x-rays], the dominant heating source is likewise solar radiation, and transport is usually dominated by neutral winds and downward diffusion due to gravity. In disturbed auroral regions, the dominant ionizing source is particle precipitation, the dominant heating source is particle precipitation (for electrons), and neutral particles, Poynting flux, or Joule heating (for ions), and the dominant transport is driven by electric fields.

14.5.1 Ionization

Below 180–200 km in the ionosphere, primary ions in the auroral ionosphere (as at midlatitudes) are molecular ions. They are relatively short-lived, recombining rapidly with thermal electrons (dissociative recombination at a rate of 10^{-6}–10^{-7} $cm^{-3} \cdot s^{-1}$). Thus, any significant (10^4 cm^{-3}) electron density seen in the nighttime auroral region much below 200 km in altitude is a measure of production of ionization at the time and place it is seen. Because the recombination rate is the product of the number of electrons per cubic centimeter (n_e) times the number of ions per cubic centimeter (n_i) available for recombination, and because $n_e \simeq n_i$, n_e^2 is a measure of precipitating auroral particles. For a nominal altitude of maximum ionization in the auroral E region (near 100 km), n_e^{max} is a measure of auroral-particle energy deposition (see Section 14.7.1), as the characteristic energy of particles that are able to penetrate to that altitude is known. From the ground, an ionosonde can measure the maximum plasma frequency of the E region, f_0E, giving the nominal relationship $(f_0E)^2 \propto n_e^{max}$.

Molecular ions dominate in the E region; in contrast, atomic ions dominate the F region. Recombination of atomic ions with electrons is slower than recombination of molecular ions, by a factor of about 10^4. These ions are generally lost through the following two-step process:

$$O^+ + N_2 \rightarrow NO^+ + N \tag{14.8a}$$

or

$$O^+ + O_2 \rightarrow O_2^+ + O \tag{14.8b}$$

The limiting process (14.8), a chemical reaction with neutral molecules to form a molecular ion (by charge exchange), is followed rapidly by dissociative recombination (see Section 7.3). Reaction (14.8a) explains the dominance of NO^+ below the F region.

14.5.2 Motion

The earth's magnetic field greatly influences the motion of plasma in the ionosphere (this is why many ionospheric properties are better described in geomagnetic coordinates than in geographic coordinates). If collisions with neutral particles do not become very frequent, electrons and ions will move together in the presence of an electric field **E**, frozen to magnetic-field lines **B**, in the direction perpendicular to both the electric and magnetic fields at a velocity

$$\mathbf{v}_e = \mathbf{v}_i = (\mathbf{E} \times \mathbf{B})/B^2 \tag{14.9}$$

where the subscripts refer to electrons (e) and ions (i).

Because of the long lifetimes of ions at F-region heights, the transport described by equation (14.9) becomes important. Ionized particles in the F region may survive for many hours before chemically recombining. At auroral and polar latitudes, strong auroral electric fields may move ionospheric plasma at velocities of the order $1 \text{ km} \cdot \text{s}^{-1}$. Ionization may have occurred thousands of kilometers away from the location where ions are observed. As polar ionospheric convection (Heelis, 1988) is always observed in the high-latitude ionosphere around the auroral oval, the displacement of ions from their source location in the F region is a continuing process.

14.5.3 Currents

Equation (14.9) applies to the motion of the F-region ions, but does not apply to the E region. In the E region, collisions of charged particles with neutrals occur so frequently that no transport of bulk ion-density profiles can occur.

In Figure 14.11a, positive-ion trajectories are shown for different ratios between the collision frequency and the ion-cyclotron frequency Ω_i. Above about 200 km, the charged-particle trajectories are described

14.5 THE AURORAL IONOSPHERE

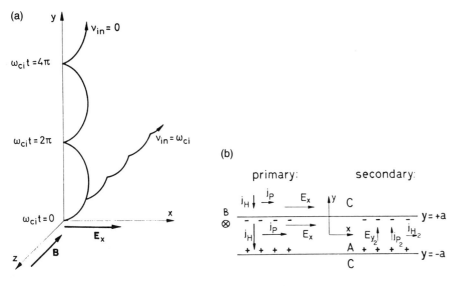

FIG. 14.11. (a) Motion of a positive ion in the absence of collisions ($v_{in} = 0$) and when the collision frequency and gyrofrequency are equal ($v_{in} = \Omega_i$). The orbit is shown for a case when there is no electric-field component E_z along the magnetic field. (b) Simple model of an auroral arc, including both primary and secondary currents. (From Egeland et al., 1973.)

in terms of bulk plasma-drift velocity $[(\mathbf{E} \times \mathbf{B})/B^2]$. However, at around 140 km, the average ion will complete only part of a gyrocycle before it collides with a neutral particle. It then will resume its motion under the influence of $\mathbf{E}$ and start on another gyroarc, until the next collision occurs. When the ion–neutral collision frequency v_{in} just equals Ω_i, the ions will drift, on average, at a 45° angle to $\mathbf{E} \times \mathbf{B}$. The subscript n refers to neutrals. When $\Omega_i \gg v_{in}$, the ions will never have a chance to move in the $\mathbf{E} \times \mathbf{B}$ direction. They will simply diffuse through the neutrals in the direction of $\mathbf{E}$ (see Problem 14.5). Because v_{in} is so much greater than the electron–neutral collision frequency v_{en}, electrons are bound to the geomagnetic-field line ($\omega_e \gg v_{en}$), and equation (14.9) applies to them down to an altitude of about 100 km. Eventually, even the electrons suffer significant collisions with the neutrals ($v_{en} \sim \omega_e$ at 80 km). Ion motion relative to electrons arising from the different effects of neutral collisions produces an electric current in the ionosphere. The mathematical treatment given in Section 7.7 applies in general.

Auroral currents (j = current density, $A \cdot m^{-2}$)

$$j = \sigma E = n_i e (v_i - v_e) \tag{14.10}$$

flow, where n_i provides current carriers, primarily above roughly 90 km; ions drift relative to electrons primarily below 140 km. Here, E and n_i lead to a current with components perpendicular to $\mathbf{B}$ and parallel to $\mathbf{E}$ (Pedersen current), perpendicular to both $\mathbf{B}$ and $\mathbf{E}$ (Hall current), and parallel to $\mathbf{B}$ (Birkeland current).

14.5.4 Secondary Currents

Auroral electrojets, intense localized currents, mainly flow from east to west, following long, narrow enhanced-conductivity channels near the midnight portion of the auroral oval during substorms (see Chapter 13). This characteristic geometry produces distinctive current patterns. In order to understand them, it is necessary to become familiar with the concept of secondary-polarization electric fields.

Consider a slab of aurorally enhanced ionospheric plasma A immersed in a background ionosphere C of lower electron density and thus conductivities as illustrated in Figure 14.11b. A primary electric field in the x-direction, E_x, is applied in both the A and C regions; it must be continuous at the interface at a consequence of $\nabla \times \mathbf{E} = 0$. The geomagnetic field is into the figure. The primary currents (J = height-integrated, $A \cdot m^{-1}$, see Section 14.7) driven by E_x include both the Pedersen current J_P along $\hat{x}$ and the Hall current J_H in the $\hat{y}$-direction. (J_P is greater in A than in C simply because of the higher conductivity in A.) However, the y current component must, in the steady state, be continuous across the A–C interface. This causes a transient accumulation of excess positive and negative charges at the A–C interface, as shown in the figure. The secondary-polarization field E_y built up in A by this process drives a secondary Pedersen current J_P that cancels the excess primary J_H across the boundary. However, this secondary polarization E_y also drives a secondary Hall current (J_H) along the arc A in the $+x$-direction. This adds to the primary J_P, making the current even more intense along the arc.

Quantitatively, the equalities just noted lead to (Egeland, Holter, and Omholt, 1973)

$$J_x = \Sigma_P E_x + \Sigma_H E_y = E_x(\Sigma_P + \Sigma_H(\Sigma_H - \Sigma_H^C)/\Sigma_P) \quad (14.11a)$$

$$= E_x(\Sigma_P + (\Sigma_H)^2/\Sigma_P) \quad \text{(for } \Sigma_H \gg \Sigma_H^C) \quad (14.11b)$$

$$J_y = -\Sigma_H^C E_x \quad (14.11c)$$

where Σ is the height-integrated conductivity, and all terms refer to the auroral arc, except Σ_H^C, which refers to background and E-region ionospheric conductivity.

This simplified argument neglects Birkeland currents ($\mathbf{J} \| \mathbf{B}$) at the arc boundaries, and when applied to auroral arcs, it gives an upper limit for the degree to which the current is enhanced. In the actual ionosphere, Birkeland currents along $\mathbf{B}$ reduce the polarization charges and thus reduce E_y. This discussion of secondary-polarization fields was applied to the auroral electrojet, but in general it applies to any enhanced-conductivity slab. [Note: Equation (14.11a) shows that the effective conductivity along the slab is enhanced. In the limiting case of a very high conductivity ratio between regions A and C, we find that the effective conductivity approaches $\Sigma_P + (\Sigma_H)^2/\Sigma_P$, as in equation (14.11b). This expression is similar in structure to the Cowling conductivity.]

14.6 AURORAL EFFECTS ON RADIO WAVES

Although it is beyond the scope of this treatment to include an exposition of the ways in which an aurora affects the radio-frequency spectrum, it would be a serious omission not to note that its effects are significant. We mention a few to convey a sense of those effects.

The large auroral and polar-cap electric fields, currents, and plasma drifts, relative to the neutral atmospheric velocity, lead to a variety of plasma instabilities. These instabilities at F-region altitudes can be reduced to two major classes. One, known for 20 yr or more, is an $\mathbf{E} \times \mathbf{B}$ or gradient-drift class of instability, somewhat modified by strong field-aligned currents. The other, discovered to be important in 1988, is an inertial or Kelvin-Helmholtz instability that develops in strong-velocity-shear regions, especially in auroral latitudes (Basu et al., 1990). These plasma instabilities cause otherwise relatively smooth plasma to develop inhomogeneous structures, **B**-aligned, over scale sizes ranging from 0.1 to 10 km, through which strong radio scintillations are produced. These auroral scintillations lead to strong amplitude fading and phase fluctuations up to gigahertz (GHz) frequencies, disrupting VHF (30–300 MHz), UHF (0.3–3 GHz), and GHz communications, and even navigation systems at lower frequencies.

Magnetic-field-aligned irregularities in the auroral ionospheric plasma, driven by instability processes, also scatter radio waves, especially at HF (3–30 MHz) and VHF ranges, much as glass rods reflect light where the geometry is similar. This auroral scatter or *clutter* can blind radar tracking, disrupt or improve HF communication, and serve as an important geophysical diagnostic tool. Its existence has been known since early in the 1930s, when amateur radio operators discovered that during an aurora it was possible to receive transmissions from an operator located to the south by directing their antennas toward the north.

Enhanced ionization at D-region altitudes leads to strong radio absorption (proportional to f^{-2}). A wave propagating through an ionized medium exchanges energy between the wave and the particles. The AC electric field alternately accelerates the electron for half a cycle and decelerates it on its next half cycle. If the electron suffers a collision with a neutral particle before it can reradiate, this ordered energy becomes disordered, heating the electron at the expense of dissipating the energy in the radio wave. Increased D-region electron densities in the auroral region (auroral absorption and/or polar-cap absorption events) seriously attenuate HF communication from earth. Measurements of attenuation of extraterrestrial radio waves passing through the attenuating D region can serve as a diagnostic for D-region ionizing sources (Rosenberg et al., 1991).

Wave–particle interactions at auroral latitudes also lead to a variety of electromagnetic emissions. The particle energy is converted to electromagnetic energy through wave–particle plasma processes, not all of which are understood (Helliwell, 1988).

14.7 ENERGY TRANSFER TO THE IONOSPHERE

Particle energy, momentum, and mass are transferred to the magnetosphere from the solar wind. Consequently, significant energy is conveyed to the ionosphere in the form of particle energy and/or as electromagnetic energy.

We have addressed the role that the auroral particles play in direct excitation of auroral optical emissions (Section 14.2), as well as in ionization of neutrals. Both mechanisms imply that energy has been deposited. However, another form of energy deposition often exceeds the rate of energy input to the auroral and polar regions by particles. The downward Poynting flux ($\propto \mathbf{E} \times \mathbf{H}$) from the magnetosphere can be dissipated in the ionosphere–thermosphere and can do mechanical work against $\mathbf{j} \times \mathbf{B}$ forces there.

We can think of this type of energy input to the ionosphere as starting with mechanical energy in the solar-wind generator that gets converted to electromagnetic energy. That, in turn, is conveyed down the geomagnetic-field lines (as Poynting flux) into the ionosphere, where it is dissipated as Joule heating. Joule heating in the ionosphere is $\mathbf{j} \cdot \mathbf{E}$ in the ionospheric rest frame. Thus, for an applied electric field, energy is dissipated by the current component parallel to $\mathbf{E}$, that is, the Pedersen current; Hall current (perpendicular to $\mathbf{E}$) is nondissipative. That leads to a Joule-heating dependence on characteristic energy, because higher-energy particles penetrate more deeply into the atmosphere, where $\Sigma_H > \Sigma_P$, and the currents tend to be nondissipative. Joule heating is then less important. Less energetic particles produce ionization at higher altitudes where Pedersen conductivities are greater than Hall conductivities and Joule heating is relatively more important. Because the current density $j_P = \sigma_P E$, where σ_P is the Pedersen conductivity, height-integrated Joule heating can be expressed as $\Sigma_P E^2$, where $\Sigma_P = \int \sigma_P \, dh$. For limited times and locations, this heating rate may be many ergs per square centimeter per second, while the energy deposited in sunlit regions by solar UV is about 0.5 erg·cm^{-2}·s^{-1} above roughly 110 km. The latter is a valuable reference quantity to use as a measure of the significance of upper-atmospheric energy input. It should be compared with the integral (over altitude) of the rate of Joule heating (erg·cm^{-2}·s^{-1}) expressed in the rest frame of the neutral atmosphere as

$$\mathbf{j} = \sigma(\mathbf{E} + \mathbf{v}_n \times \mathbf{B}) \tag{14.12}$$

where $\mathbf{v}_n$ is the neutral-wind velocity.

14.7 ENERGY TRANSFER TO THE IONOSPHERE

The total heating of the lower thermosphere by auroral energy dissipation, with heating rates transiently exceeding that of UV from an overhead sun, will change the temperature, density, composition, and winds of the local thermosphere. In fact, the global-scale thermospheric and ionospheric responses to such variable heating are subjects of ongoing research.

14.7.1 Particle-Energy Deposition

Arriving at an estimate of the particle heating affords an opportunity to bring out several important points. At E-region altitudes, recombination is proportional to a rate multiplied by the number of electrons and ions or n_e^2 (see Section 7.3). The production rate of ionization can be estimated by the energy of the particle divided by 35 eV, as discussed in Section 14.2. In a quasi-steady state, the production rate just balances the recombination rate. This leads to a quantitative estimate of Q_p, the rate of ionospheric heating by precipitating particles, as

$$Q_P = 5.6 \times 10^{-6} \int \alpha(h) n_e^2(h) \, dh \ (\text{erg} \cdot \text{cm}^{-2} \cdot \text{s}^{-1}) \tag{14.13}$$

where α is the effective-altitude recombination coefficient (roughly 2–3×10^{-7} cm$^3 \cdot$s^{-1} around 120 km), n_e is the electron density (cm^{-3}), and h is the altitude (km). Note that the particle heating rate is also proportional to n_e^2. A 3-MHz E-region plasma frequency equates roughly to an electron density of 10^5 cm^{-3}, a Q_P of 1 erg$\cdot$cm$^{-2}\cdot$s^{-1}, and 0.5 kR of 391.4-nm emission for a typical auroral layer. This is representative of diffuse auroras, which are produced by loss-cone particle precipitation, which is also responsible for an extensive, highly uniform, quasi-equilibrium layer of E-region ionization.

14.7.2 Joule Heating and Energy Deposition

The earlier discussion of Joule heating was formulated as a collective-interaction description (the ionosphere as a resistive load). It is often instructive to take both a collective view and a particle view of interaction processes. Thus, this section examines energy deposition from the viewpoint of particles. Auroral and transpolar electric fields drive plasma at a velocity $v_i = (\mathbf{E} \times \mathbf{B})/B^2$ at high latitudes. The plasma is electrically neutral, with $n_e = n_i$. An individual ion, upon collision with a neutral atmospheric particle, will exchange energy and be deflected in some new direction. Thus, the ion's ordered motion in the plasma-drift direction becomes disordered motion, that is, heat. From this particle view, it is thus easy to see that the ion-gas temperature heats up rapidly over a few ion–neutral collision times. [A useful number to recall for collision frequencies (Chapter 7, Figure 7.8) is that at about 140 km, the ion–neutral collision frequency roughly equals the ion gyrofrequency, about 200 s^{-1}]. The v_{in} decreases exponentially with altitude, in proportion to the neutral density. At F-region altitudes (~300km), the time for the ions to heat up by this process is only on the order of seconds.

The ion temperature increases by this v_{in} process at a rate (Banks and Kockarts, 1973)

$$\frac{\partial}{\partial t}T_e = v_{in}\left(T_n - T_i + \frac{m_n}{3k}(v_i - v_n)^2\right) \quad (14.14a)$$

so that for a quasi-steady state, ion temperature T_i exceeds T_n by

$$T_i - T_n = \frac{m_n}{3k}(v_i - v_n)^2 \quad (14.14b)$$

where m_n is the mean mass of the neutral atmospheric gas, and k is Boltzmann's constant.

Note that temperature enhancement varies as the square of the plasma velocity in the neutral rest frame and is negligible for midlatitude velocities of the order of 100 m·s^{-1} or less, but exceeds 1,000K for auroral and polar-cap velocities much greater than 1 km·s^{-1}.

14.7.3 Thermospheric Heating and Momentum Transfer

In sharp contrast to the ions, the neutral-particle gas takes much longer to respond and responds dramatically by a momentum change, rather than a thermal change. Heat into ions exposed to an electric field and an $\mathbf{E} \times \mathbf{B}$ force (which physically may be thought of as resistive Joule heating in a conductor, randomization of ordered ion-particle motion, or "frictional-drag heating" of the ion gas dragged through the neutral-particle gas) is ultimately lost to the neutral gas. However, the thermospheric temperature increases much less dramatically than the ionospheric temperature previously discussed, because of its greater particle and mass densities, which lead to far greater heat capacity.

For a steady electric field $\mathbf{E}$ moving the ions along an ordered trajectory, we noted that v_{in} causes the ordered motion of individual ions to become randomized in direction, resulting in heating. However, for an upper atmosphere at rest, the momentum transfer from the ions to the neutral particles is systematic in that it always has a component downstream in the direction of plasma flow. Whereas (Figure 7.8) an average F-region ion collides with a neutral particle on the order of once per second, an average neutral particle collides with an ion much less often. Thus, the thermosphere responds much more slowly. This is because, at F-region altitudes, there are more than 1,000 times as many neutrals as ions (~10^9 cm^{-3} neutral particles versus typically 10^6–10^5 cm^{-3} ions and electrons). Thus, the average ion has to experience 10^3–10^4 ion–neutral collisions, which occur with a frequency v_{in}, before the average neutral particle experiences one neutral–ion collision, occurring with a frequency v_{in}. For large ion densities, the thermosphere may eventually be brought up to the speed of the ions. Quantitatively, the "ion-drag" force on the neutrals acts over a time scale given by the neutral-gas momentum equation (neglecting gradient terms):

$$\frac{\partial v_n}{\partial t} = \frac{\rho_i}{\rho_n} \nu_{in}(v_i - v_n) \tag{14.15}$$

where we assume that the F-region neutral wind varies negligibly with altitude, and where ρ is the gas density (number density times mass) for the ions and neutrals, as indicated by the i and n subscripts.

The time for the thermosphere to respond is then given by $\rho_n \rho_i \nu_{in}$. Using typical values for n_n, m_n, and ν_{in} for atomic oxygen ions, we find that the response time for typical F-region altitudes and thermospheric temperatures/densities (250–450 km, 750–1,500K) is $0.23 \times 10^{10}/n_i$. Thus, for a typical daytime ionospheric density of $n_i = 10^6$ cm^{-3}, the thermosphere is dragged up to the ionosphere plasma drift velocity (typically ~ 1 km·s^{-1} over the polar cap for a southward IMF) in about 30 min, whereas for ionospheric ion and electron densities of 10^5 cm^{-3}, it takes 6 h. In 6 h, the earth has rotated 90°, and the thermosphere is exposed to a very different part of the ion convection pattern, which is fixed in the earth–sun reference frame for a fixed IMF condition. Thus, for ionospheric densities near and above 10^6 cm^{-3}, the upper-atmospheric winds are tightly coupled to the plasma drift (with roughly 30 min smoothing or averaging). For ionospheric densities near and below 10^5 cm^{-3}, this wind coupling is almost negligible.

14.8 RELATION TO BOUNDARIES AND PHYSICAL PROCESSES IN THE MAGNETOSPHERE–IONOSPHERE–THERMOSPHERE

The era of satellites has provided a sophisticated framework within which to order our developing understanding of the transfer of energy, momentum, and mass from the solar wind through the magnetosphere and into the ionosphere and atmosphere.

Figure 9.18 identifies magnetospheric regions, distinguished by particle population characteristics, from which particles precipitate into the upper atmosphere to produce auroras. The figure illustrates the topology of the magnetosphere (x points toward the sun, and z to the north, in the right-hand x, y, z coordinate system) and also the projection of the plasma source regions onto the northern polar upper atmosphere, with the sun again to the left.

Currents flow between the dawn and dusk regions of the magnetosphere across the tail current sheet, separating the tail lobes. Most of them flow straight across the magnetotail, but some are diverted along field lines into and across the auroral ionosphere. These currents are carried principally by electrons (which are more mobile than ions). The primary current to the auroral region enters the dawnside, crosses the polar cap, and exits the duskside. Currents into the ionosphere are carried by upgoing thermal electrons, which do not excite auroral radiation. Currents out of the ionosphere are usually carried by precipitating-electron current carriers, which do excite auroras. Hence, the primary currents produce auroras mainly premidnight.

As the conductivity is enhanced in the auroral oval (see Section 14.5), currents also flow from the inner (poleward) edge to the outer (equatorward) edge of the oval, leading to secondary currents. The outward return flow (secondary current) on the dawnside then also produces auroral emissions. The dynamo effect converts the kinetic energy of the solar wind, which is the ultimate source of the magnetospheric current system, to electrical energy. The net power involved is of the order of 10^6 MW when integrated over the entire magnetosphere.

The potential drop across the polar cap (typically 50 kV) also applies an $\mathbf{E} \times \mathbf{B}$ force driving ionospheric plasma antisunward across the polar cap. These plasma velocities typically are $1 \text{ km} \cdot \text{s}^{-1}$ to within a factor of 2. This convection dominates the character of the polar ionosphere, and often even that of the polar thermosphere (the part of the upper atmosphere above 120 km). The flow lines of antisunward ionospheric convection close on themselves by returning toward the sun equatorward of the auroral region.

The degree to which this solar-wind-driven convection occurs smoothly, versus a series of spurts, and the implications of that issue for the underlying physical processes and boundary relationships are important topics (Cowley and Lockwood, 1992). Improved time-resolution measurements, over large areas of the polar regions, are needed to help resolve these issues.

14.9 STABLE SUN-ALIGNED ARC: ENERGETICS AND THERMAL BALANCE

We have discussed the optical character and electrical character of sun-aligned arcs (Figures 14.1d and 14.9). Discussion of the energetics and thermal balance of such arcs at this point will serve not only to present the properties of a common class of arcs but also to review the material of Sections 14.5 and 14.7 and consolidate it with the material introduced in the earlier sections of this chapter.

We can anticipate much about the physical characteristics of a sun-aligned arc. Recall the right-hand side of Figure 14.9b. As the conductivity changes more slowly than $\mathbf{E}$ across the sun-aligned element of an arc, the decrease in $\mathbf{E}$ from the dawnside to the duskside means that the horizontal-current component decreases across the arc. This can happen only if current flows out of the plane of the horizontal-current slab. In the ionosphere, this happens through upward magnetic-field-aligned currents. We can say that these arcs satisfy a simple Ohm's-law relationship between **E** and the current it drives through the conducting ionosphere. Electrons are the current carriers. Because of their low mass, they are more mobile than ions. The electrons must be drawn into the ionosphere from the magnetosphere. The ultimate driver of all this is a solar-wind-driven mechanical force, generating an electric field in the magnetosphere that maps down magnetic-field lines to the ionosphere to drag plasma across the polar cap. Now, the presence of this sun-aligned sheet of incoming electrons causes the temperature of the electron gas to be higher above an arc (double or more) than elsewhere in the surrounding ionosphere. The actual increase depends on the incident flux of suprathermal electrons from the magnetosphere and on how well the lower ionosphere is thermally coupled to its thermospheric heat sink. For low electron densities, and thus low electron-gas cooling rates, the electron temperature increase may exceed 3,000K, leading the heated ambient electrons to enhance the 630-nm emissions.

As the cross-arc gradient of antisunward plasma flow velocity (cross-arc gradient of **E**) increases, the upward current out of the arc must increase to carry off the enhanced horizontal current converging on the arc. Field-aligned-current densities can grow by increasing the number and/or the speed of the current carriers. The latter effect causes precipitating-electron fluxes to become harder, as is observed. The fluxes of precipitating hard electrons increase E-region ionization. It then follows that intense arcs will have underlying E-region ionization, whereas weak arcs will not. It also follows that the particle energy flowing into intense, stable sun-aligned arcs can be estimated from measurements of either molecular optical emission (which can exceed kilorayleighs) or E-region electron densities (which can exceed 2×10^5 cm^{-3}). Recall that the global mean thermospheric EUV heating rate is 0.5 erg·cm^{-2}·s^{-1}. Electron-particle energy deposition within an intense arc exceeds this reference solar-radiation rate!

Plasma on the dawn edge of the arc is found to flow antisunward at about 1 km·s^{-1} and to decrease duskward across the arc. If thermospheric winds are light, the velocity difference between the ions and the neutrals can be used to calculate the Joule heating rate. It follows from the square-law dependence of the velocity difference that the heating will be concentrated near the high-plasma-velocity (dawn) edge of the arc. Measured ion and neutral velocities across an intense arc imply Joule heating rates of several ergs per square centimeter per second, concentrated near the dawn edge, with net energy into the arc exceeding that from the incident-electron source. This Joule heating rate has been confirmed, based on calculating the rate of ion heating required to maintain the approximately 1,000K by which ion temperatures have been observed to exceed those of the neutral gas. This energy that appears in the ionosphere in the form of heat is conveyed into the ionosphere in the form of electromagnetic energy (Poynting flux). Measurement of the Poynting flux simultaneously with Joule heating rates has demonstrated that the foregoing calculation is consistent with observation (Valladares and Carlson, 1991). These physical arguments are not confined to sun-aligned arcs.

Polar-cap convection can take different forms, depending on whether the solar-wind magnetic field is nearly parallel or nearly antiparallel to the subsolar field of the earth. When these fields are parallel (i.e., the IMF is northward), sun-aligned arcs are seen to occur. Near the central polar cap, the sun-aligned boundaries of the arc drift toward dawn (dusk) for negative (positive) IMF B_y in the northern hemisphere. The connection of the arcs and the relation of their drifts to the coupling of the magnetosphere and the solar wind are challenging topics, but beyond the scope of this treatment.

ADDITIONAL READING

Akasofu, S.-I., and S. Chapman. 1972. *Solar-Terrestrial Physics*. Oxford: Clarendon Press.
Alfvén, H., and C.-G. Fälthammar. 1963. *Cosmical Electrodynamics, Fundamental Principles*. Oxford University Press.
Banks, P. M., and G. Kockarts. 1973. *Aeronomy*. New York: Academic Press.
Brekke, A., and A. Egeland. 1994. *The Northern Lights, Their Heritage and Science*. Dreyer: Grøndahl.
Chamberlain, J. W. 1961. *Physics of the Aurora and Airglow*. New York: Academic Press.
Eather, R. H. 1980. *Majestic Lights*. Washington, DC: American Geophysical Union.
Egeland, A., O. Holter, and A. Omholt. 1973. *Cosmical Geophysics*. Oslo: Universitetsforlaget.
Jursa, A. S. (ed.). 1985. *Handbook of Geophysics and Space Environment*. Springfield, VA: National Technical Information Service.
Kelley, M. C. 1989. *The Earth's Ionosphere*. San Diego: Academic Press.
Meng, C. I., M. J. Rycroft, and L. A. Frank (eds.). 1991. *Auroral Physics*. Cambridge University Press.
Rees, M. H. 1989. *Physics and Chemistry of the Upper Atmosphere*. Cambridge University Press.
Sandholt, P. E., and A. Egeland (eds.). 1989. *Electromagnetic Coupling in the Polar Clefts and Caps*. Dordrecht: Kluwer.
Størmer, C. 1955. *The Polar Aurora*. Oxford: Clarendon Press.
Vallance Jones, A. 1974. *Aurora*. Dordrecht: Reidel.

PROBLEMS

14.1 Assume that monoenergetic electrons at 3 keV precipitate parallel to the geomagnetic field and produce an auroral arc of 10^3 km in east–west extent and north–south width of 10 km, with a homogeneous surface brightness of 5 kR in the 391.4-nm band.

(a) Estimate the flux needed to produce this arc.
(b) Calculate the corresponding net downward energy in watts per square meter for this electron flux.

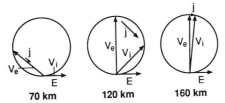

FIG. 14.12. Electric current j and the direction and magnitude of electron and ion drift ($\mathbf{v}_e$ and $\mathbf{v}_i$) relative to an ionospheric electric field $\mathbf{E}$ parallel to the x-axis (while $\mathbf{B}_E$ is along the z-axis) for three different heights.

14.2

(a) Discuss the hydrogen lines (H_α and H_β) in the auroral spectrum and explain why this aurora is diffuse in character.

(b) Discuss how the Doppler shift (see Figure 14.5b) can be used to estimate the average H^+ energy. Assume that a remote observer at an auroral latitude makes observations of the spectrum both parallel and perpendicular to the earth's magnetic field. A comparison of the spectra reveals a 1-nm shift in the wavelength of the peak intensity. Estimate the average velocity of the downflowing H^+ assuming 0° pitch angles.

(c) Assume that the H_α production versus H^+ energy can be estimated in the 0.4–100-keV energy range by a straight line on log-log coordinates, defined by $x = H_\alpha$ photon production/proton; y = energy (keV); ($x=1$, $y=1$), ($x=10$, $y=15$). Calculate the flux needed to produce 100 R at H_α, for 100, 10, and 1 keV H^+-particle energies.

(d) Discuss why the hydrogen lines are so weak in the dayside cusp and cap.

14.3 Calculate the L values for the following auroral observatories (see Section 6.2.1):

Tromsø (geomagnetic coordinates 117.54°E, 66.96°N)
Ny Ålesund (geomagnetic coordinates 131.24°E, 75.31°N)
Nord (geomagnetic coordinates 112.0°E, 80.9°N)
Qaanaaq (geomagnetic coordinates 39.9°E, 86.8°N)

14.4 Discuss how the direction of the geomagnetic-dipole axis relative to the sun varies with time (diurnal and seasonal) and how that variation will influence the auroral occurrences at the observatories listed in Problem 14.3.

14.5 The electric current j and the direction and magnitude of electron and ion drift ($\mathbf{v}_e$ and $\mathbf{v}_i$) relative to an ionospheric electric field $\mathbf{E}$ parallel to the x-axis (while $\mathbf{B}_E$ is along the z-axis) are plotted in the Figure 14.12 for three different heights. The angle a between electrons and ions relative to $\mathbf{E}$ is given, for a charged particle k, by $a_k = \tan^{-1}(\Omega_{ck}/\nu_{kn})$ (Ω and ν are gyrofrequency and collision frequency). For a height of about 180 km, $\nu_{kn} \ll (\Omega_{ck})$; electrons and ions move in the same direction, with a velocity $\mathbf{E} \times \mathbf{B}/B^2$. This drift produces no net current. Discuss the importance of collisions for two cases. In case a, assume that the electron-neutral collision frequency greatly exceeds the electron gyrofrequency (as, for example, near 70 km). In case b, assume that the electron gyrofrequency greatly exceeds the electron-neutral collision frequency (as, for example, near 180 km).

14.6 Discuss how the Biot-Savard law can be used to estimate the magnetic disturbance at the ground (in nT) for a line current at 120 km. Assume different values for the conductivity and the electric field.

14.7 Current continuity at the equatorward boundary of the arc in Figure 14.11b can be written

$$E_x^A = \frac{\Sigma_P^C}{\Sigma_P^A} E_x^C + \frac{\Sigma_H^A - \Sigma_H^C}{\Sigma_P^A} E_y^C + \frac{J_\parallel}{\Sigma_P^A}$$

where the values inside and outside the arc are indicated by A and C.

(a) Discuss the different terms.

(b) We assume that conductivity gradients play a minor role when soft particles dominate (e.g., dayside cusp aurora). The equation then reduces to

$$E_x^A = E_x^C + \frac{J_\parallel}{\Sigma_P^A}$$

where E_x^C is the polarization electric field, and $J_\parallel$ is the Birkeland current (often given approximately by $\partial E_x/\partial x \cdot \Sigma_p$, if E_x^C plays a minor role).

Discuss the importance of these different terms in the E and F regions for nightside and dayside auroras.

14.8 Explain how auroral spectrometer measurements could be used to determine the composition and height distribution of the upper atmosphere between 95 and 300 km.

15 THE MAGNETOSPHERES OF THE OUTER PLANETS

C. T. Russell and R. J. Walker

15.1 INTRODUCTION

IN THE LABORATORY, we can modify the conditions in our experiments and apparatus and record how the process changes. We would like to be able to make such modifications in studying the solar-terrestrial system, but we cannot. By and large, solar-terrestrial physics is an observational science, rather than an experimental science. There are few active experiments we can perform to determine how the system works, either in the magnetosphere or in the laboratory. We cannot create magnetospheres with quantitatively scaled parameters to see how they work. Nor can we simulate accurately all the processes with our computers. We are restricted to working with the magnetospheres that we have found in place. Fortunately, these magnetospheres differ sufficiently that comparisons among them can give some insight into the governing processes at work. Unfortunately, we have limited data on the magnetospheres of the outer planets. As shown in Figure 15.1, there have been five passes through the Jovian magnetosphere, three at Saturn, and one each at Uranus and Neptune.

As discussed in Chapter 6, Mercury provides an important contrast to the terrestrial magnetosphere because it has no significant ionosphere or atmosphere. The outer planets provide us with other comparisons. First, as we move outward in the solar wind, the properties of the solar wind change. This may affect the coupling of energy flux from the solar wind into the magnetospheres. Also, the sizes of the magnetospheres are different. Those of the outer solar system, the "gas giants," are much larger than that of Earth. The rapid rotation of all four of the gas giants leads to important centrifugal forces that far exceed those in the earth's magnetosphere. In the Jovian magnetosphere there is a significant source of plasma from the moon Io. Although the other Galilean moons provide some mass, and though there are weak mass sources in the magnetospheres of Saturn, Uranus, and Neptune, none approaches the magnitude of the Io source.

Finally, the outer planets have rings. Some of these are quite tenuous, such as Jupiter's, and have little effect on the magnetosphere. However, others, like that of Saturn, are extensive and quite thick and absorb the trapped radiation efficiently. The varying rates of absorption in the

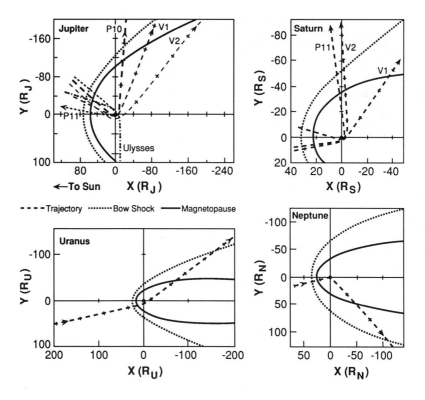

FIG. 15.1. Spacecraft trajectories used to explore the outer-planet magnetospheres. The coordinate system used has its *x*-direction oriented toward the sun and assumes that the bow shock and magnetopause are cylindrically symmetric about this direction. The trajectories are drawn in the plane containing the spacecraft and the planet. Spacecraft abbreviations: P, *Pioneer;* V, *Voyager.* Only *Voyager 2* encountered Uranus and Neptune.

different magnetospheres cause significant differences in the particle populations in the innermost parts of these magnetospheres.

15.2 THE VARIATION IN THE SOLAR-WIND PROPERTIES

As discussed in Chapter 4, the solar-wind-number density and the radial component of the interplanetary magnetic field (IMF) both vary inversely as the square of distance from the sun. The tangential component of the magnetic field, however, varies inversely as the first power. This means that the spiral angle of the IMF becomes increasingly tighter and tighter, approaching 90° to the radial direction. This change in spiral angle is not expected to have a major effect on the interaction with the outer planets. It will change the foreshock geometry somewhat, and it will change the interactions with comets by altering the strength of the draping of the IMF. However, comets outgas weakly in the outer solar system.

The electron and ion temperatures also decrease with distance from the sun, but because of heat conduction and dissipation, the radial falloff is slower than that of an adiabatic process. This has important consequences for two parameters that control aspects of the solar-wind interaction with planetary magnetospheres: the fast magnetosonic Mach

15.2 VARIATION IN THE SOLAR-WIND PROPERTIES

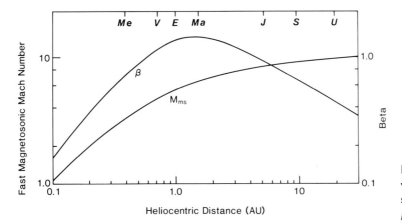

FIG. 15.2. Expected variations of the fast magnetosonic Mach number and plasma β as functions of heliocentric distance. The locations of the planetary orbits are shown at the top of the figure. (From Russell et al., 1990.)

number and the beta (β) of the plasma. The fast magnetosonic Mach number is the ratio of the velocity of the solar wind to the speed of compressional waves in the solar wind. It controls the strength of the bow shock, which in turn controls the properties of the shocked plasma in the magnetosheath that bathes the planetary magnetopause. The β of the plasma is the ratio of the thermal pressure in the plasma to the magnetic pressure. Beta also helps control the properties of the magnetosheath plasma. As illustrated in Figure 15.2, the expected magnetosonic Mach number increases from about 6 at Earth to about 10 at Saturn, whereas β maximizes at about Mars and then declines slightly in the outer solar system, under our assumptions of radial falloff. The major implication of this behavior is that, on average, the bow shocks of the outer planets are stronger than those in the inner solar system. Because of the high β values behind these strong shocks, the magnetic field in the magnetosheath is relatively weak. Figure 15.3 shows a comparison of the magnetic-field strengths measured through the bow shocks of the Earth, Jupiter, and Uranus. The large overshoot in magnetic-field strength just downstream of the shock ramp is a signature of the strengths of these shocks.

Strong shocks, in turn, are accompanied by strong fluxes of particles streaming back toward the sun. The two mechanisms that can lead to particle streaming along the interplanetary field from planetary bow shocks are leakage of the hot downstream magnetosheath particles and reflection of solar-wind particles. At the outer planets, both mechanisms should be stronger than at 1 AU, because of the size of the overshoot and the expected temperature of the magnetosheath.

The ever-tightening spiral of the IMF alters the geometry of the foreshocks of the outer planets, as illustrated in Figure 15.4 for Saturn (Orlowski, Russell, and Lepping, 1992). The tangent field line, approximately the boundary of the electron foreshock, is nearly perpendicular to the solar-wind flow, and the ion foreshock is swept back, so that the strong ULF waves associated with back-streaming ions are seen only over the terminator regions. This geometry should be compared with the

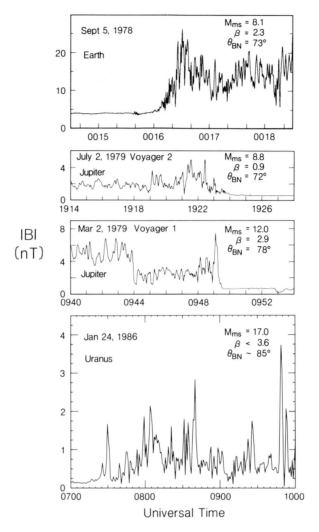

FIG. 15.3. Magnetic profile of high-Mach-number shocks at Earth, Jupiter, and Uranus. The quiet fields to the left in the top and bottom panels and to the right in the middle two panels are those of the preshock solar wind. (From Russell et al., 1990.)

terrestrial foreshock in Figure 1.14. Figure 15.5 shows ULF waves upstream of the Uranian shock. These waves are very similar to those upstream of the terrestrial bow shock and have the frequency theoretically expected for these waves. This frequency declines in proportion to the magnetic-field strength, as illustrated in Figure 15.6.

Another variation that occurs with increasing distance from the sun is the change in the gyroradius of the reflected solar-wind ions. However, because the sizes of the magnetospheres of the outer planets always are much larger than the expected size of this gyroradius, this increasing size is expected not to have any major effects on the interaction of the solar wind with the magnetospheres of the outer planets.

15.3 MAGNETOSPHERIC SIZE

As discussed in Chapter 6, the size of the magnetic cavity is proportional to the sixth root of the ratio of the square of the magnetic moment divided by the dynamic pressure of the solar wind. Because the solar-wind dynamic pressure varies inversely as the square of the heliocentric distance, the magnetospheres of the outer planets should be considerably larger than the terrestrial magnetosphere, all else being equal. However, all else is not equal. The magnetic moments of the outer planets are significantly larger than the terrestrial magnetic moment. Table 15.1 lists the relevant parameters for the earth and the four gas

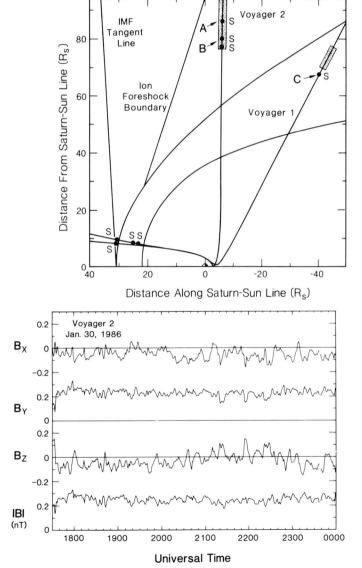

FIG. 15.4. Geometry of the Saturn foreshock, as deduced from *Voyager 1* and *2* measurements. Trajectories of the two spacecraft are shown, as are shock encounters and regions of upstream waves studied by Orlowski et al. (1992).

FIG. 15.5. Upstream magnetic fluctuations observed after the last bow-shock crossing at the *Voyager 2* encounter with Uranus. (From Russell et al., 1990.)

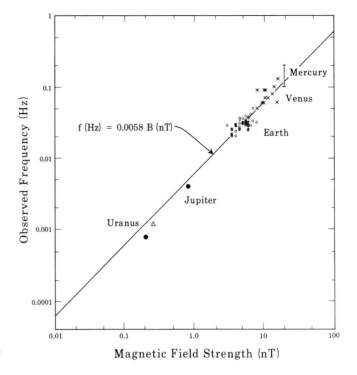

FIG. 15.6. Frequencies of waves seen upstream of the bow shocks of several of the planets as functions of the magnetic-field strength in the solar wind. (From Russell et al., 1990.)

TABLE 15.1. Magnetospheric Dimensions in the Outer Solar System

Planet	Heliocentric Distance (AU)	Magnetic Moment (M_E)	Tilt Angle	Expected Magnetopause Distance	
				Kilometers	Planetary Radii
Earth	1.0	1	10.8°	0.7×10^5	$11 R_E$
Jupiter	5.2	20,000	9.7°	30×10^5	$45 R_J$
Saturn	9.5	580	<1°	12×10^5	$21 R_S$
Uranus	19.2	49	59°	6.9×10^5	$27 R_U$
Neptune	30.1	27	47°	6.3×10^5	$26 R_N$

giants. Magnetic moments are given in terms of the terrestrial magnetic moment of 8×10^{15} T·m³. The last two columns show the expected distance of the nose or subsolar point of the magnetopause, in kilometers and planetary radii, as derived from simple pressure-balance arguments. Figure 15.7 shows the relative sizes of these magnetospheres, including that of Mercury. The expected subsolar distances are all much greater than the terrestrial distances. However, the distances expressed in planetary radii are more similar: 11 for Earth, 45 for Jupiter, and about 25 for the other three planets.

The tilts of the dipole moments to the rotation axes vary over a large range, from less than 1° for Saturn to close to 60° for Uranus. The effects of these dipole tilts are varied. The alignment of the dipole moment and those higher moments that can be measured results in an axially sym-

15.3 MAGNETOSPHERIC SIZE

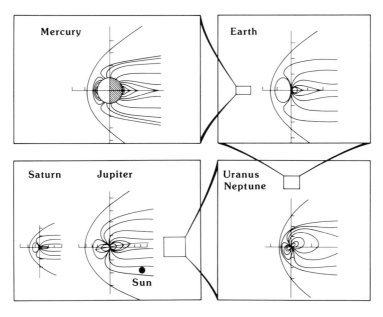

FIG. 15.7. Comparison of the sizes of planetary magnetospheres.

metric inner magnetosphere at Saturn. The roughly 10° tilt of the Jovian dipole results in a ±10° motion of Io with respect to the magnetic equator, since Io orbits Jupiter in its rotational equatorial plane. Thus, the ionized plasma torus becomes spread over ±10° in latitude. Presently at Uranus, the large tilt of the dipole results in an Earthlike magnetosphere, because the rotation axis of Uranus is pointing nearly at the sun. This situation will change drastically in one-quarter of a Uranian year. Then the angle of the dipole axis relative to the solar wind will undergo large variations, much like the present situation at Neptune.

For one planet, our theoretical expectations based on a vacuum magnetosphere are not met. The subsolar magnetopause of Jupiter is found to range from the expected 45 Jovian radii (R_J) to over $100R_J$. The reason for this discrepancy is that Io provides a significant source of mass to the Jovian magnetosphere, which in turn accelerates this mass to high velocities because of the rapid rotation of the planet. The resulting centrifugal force of the plasma pushes out against the solar wind and creates a more distant standoff distance than would be the case if the magnetopause location were determined solely by magnetic pressure. Figure 15.8 shows the magnetic field lines in the noon–midnight meridian in a model of this "magnetodisk." Though quite distorted by the centrifugal force, this magnetodisk still retains the general configuration of the terrestrial magnetosphere.

The disklike shape of the magnetosphere has an effect on the location of the bow shock. The bow shock is located at a distance from the magnetopause sufficient to allow the shocked solar wind to flow around the magnetosphere. If the magnetosphere is blunt, this distance will be larger than if the magnetosphere shape is sharp. In the analogous situa-

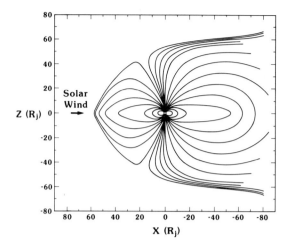

FIG. 15.8. Magnetic field lines in the noon–midnight meridian of the Jovian magnetodisk. (From Engle, 1991.)

TABLE 15.2. Characteristic Times and Velocities

Planet	Rotation Rate (Hz)	Corotation Velocity at Magnetopause (km·s^{-1})	Solar-Wind Flow Time (min)
Earth	1.26×10^{-5}	4	2
Jupiter	2.8×10^{-5}	923	200
Saturn	2.6×10^{-5}	196	45
Uranus	1.6×10^{-5}	65	25
Neptune	1.6×10^{-5}	62	23

tion in aviation, a shock wave will be detached from an aircraft according to its shape. A needle-nosed supersonic plane may have the shock attached to the nose of the plane, comparable to an Earthlike magnetosphere. The Jovian magnetosphere is sharp enough to cause a measurable reduction in this distance. At Jupiter, the nose of the bow shock is about 20 percent farther out than the magnetopause, whereas at Saturn and Earth, the nose of the bow shock is about 30 percent farther out than the magnetopause.

The enormous sizes of the outer-planet magnetospheres also have some important consequences for the convection of solar-wind plasma past the planets. At Earth, a solar-wind disturbance can travel the distance from the nose to the terminator in 2–3 min. However, at the outer planets, as shown in Table 15.2, this time is much longer, ranging from about 25 min at Uranus and Neptune to about 200 min at Jupiter.

The velocity of the plasma just inside the magnetopause is also quite different for the outer planets. If the plasma nearly corotates, then it is moving at close to 1,000 km·s^{-1} at the Jovian magnetopause. On the dawn flank of the magnetopause, this velocity should help destabilize the Kelvin-Helmholtz instability, whose growth rate is proportional to the velocity shear. On the afternoon flank, this could be a stabilizing

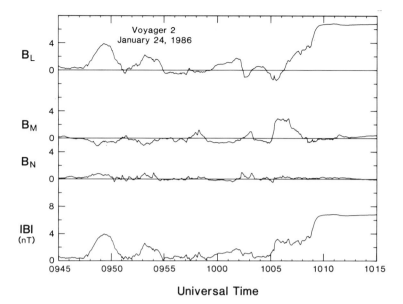

FIG. 15.9. Magnetic-field strength at the time of the *Voyager 2* crossing of the Uranian magnetopause. The last increase in the field strength into a relatively quiet region is interpreted to be the magnetopause crossing. The preceding activity is thought to be slow-mode waves excited in the high-β magnetosheath. (From Russell et al., 1989.)

effect by reducing the velocity shear. At Saturn, Uranus, and Neptune, similar but smaller effects occur.

15.4 THE ROLE OF RECONNECTION

The process known as reconnection plays a significant role in the dynamics of Earth's magnetosphere, controlling the occurrence of substorms and storms. However, in the outer solar system, the IMF may not be as important, because the magnetic-field strength in the magnetosheath is, on average, a significantly smaller fraction of the magnetospheric field than at Earth. This is illustrated in Figure 15.9, which shows *Voyager 2* magnetic-field measurements across the magnetopause at Uranus. The ratio of field strengths across the magnetopause is close to 20. Because the sum of the plasma thermal and magnetic pressures is constant across the magnetopause, and because we expect the major contribution to the pressure in the magnetosphere to be that of the magnetic field, we can conclude that the value of β in the Uranian magnetosheath is close to 400 at this time. If reconnection at the magnetopause for Uranus and Neptune is much less efficient than at Earth, we would expect a rather quiescent magnetosphere at each planet. We find that the energetic-particle flux and the ULF-wave intensity at both planets are reduced compared with the levels found closer to the sun. However, there is a suggestion of a substorm in the Uranian magnetotail data, and there are many plasma waves present in these magnetospheres.

Reconnection still has a role to play, especially at Jupiter. There have been reports of Jovian flux-transfer events on the dayside magneto-

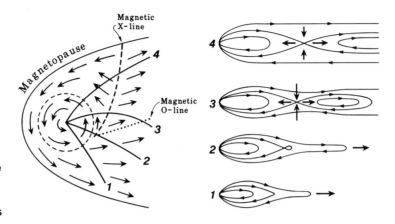

Fig. 15.10. Schematic representation of the escape of plasma down the Jovian tail via reconnection. The magnetic meridians on the right-hand side of the figure show the sequence of neutral-point formation and plasmoid formation. This process can proceed in a steady state, as envisioned here. The equatorial-plane view on the left-hand side shows the locations of the four meridian planes and the locations of the X and O neutral lines. (From Vasyliunas, 1983.)

pause, and there has been a proposal that reconnection plays a critical role in the physics of the Jovian magnetotail. Figure 15.10 shows a sketch of the proposed flow of material as it reaches the tail region after flowing around Jupiter. Reconnection here allows material to escape the closed magnetospheric field lines and flow down the magnetotail and eventually into the solar wind. In fact, energetic electrons with fluxes modulated at the synodic period of Jupiter (13 months) have been detected at 1 AU on field lines that would be expected to intersect the Jovian magnetosphere.

15.5 INTERACTION OF MOONS WITH THEIR MAGNETOSPHERES

The magnetospheres of the outer planets contain sources of mass and energy not found in the terrestrial magnetosphere: their various moons. Earth's moon generally lies well outside the main magnetospheric cavity, passing through the magnetotail once each month, and providing very little mass to the magnetosphere. At Jupiter, and to a lesser extent at Saturn, Uranus, and Neptune, the moons orbit deep inside the magnetosphere and supply significant amounts of mass to the magnetospheres. When that mass is ionized, it is rapidly accelerated to the velocity of the corotating magnetosphere, which is frozen into the ionosphere. Thus the plasma is energized at the expense of the rotational energy of the planet.

The source of mass is greatest at the Jovian moon Io, whose weak atmosphere is maintained by apparently continual volcanic eruptions. The radiation-belt particles collide with atoms and molecules in that atmosphere and with the surface of Io, knocking off atoms in a process known as sputtering, so that atoms escape from Io and form a cloud in orbit about the planet. Neutral atoms in the cloud become ionized, and the electric field of the corotating magnetosphere accelerates these particles into a torus about the planet. Corotating ions can undergo

15.5 INTERACTION OF MOONS AND MAGNETOSPHERES

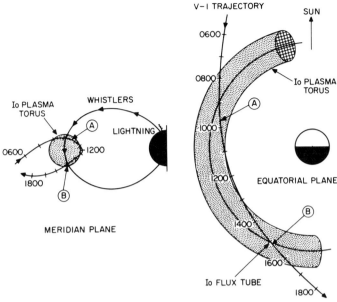

FIG. 15.11. *Voyager 1* trajectory through the Io plasma torus, illustrating the propagation of lightning-generated whistler-mode signals to the spacecraft. The arrival time of the originally broadband signal as a function of frequency provides information on the plasma density along the propagation path. (From Gurnett et al., 1979.)

charge exchange with neutral atoms. Any ionized particles that are neutralized by charge exchange will fly out of the torus (along straight lines tangent to their original circular orbits) and spread material throughout the equatorial plane of Jupiter. When this material again becomes ionized, it is accelerated to the local flow velocity and acquires significant thermal energy. Again the centrifugal force of this material stretches the field into a disklike pattern. The centrifugal force can also cause heavy magnetic-flux tubes of the plasma torus to exchange with lighter tubes outside the torus in what is called the interchange instability.

The plasma of the Io torus has been observed in many different ways. Some of its constituent ions can be seen optically from Earth. It has been probed directly by plasma instruments on *Voyager 1*, and its properties have been deduced indirectly through its effects on the propagation of lightning-generated whistler-mode waves. Figure 15.11 shows the *Voyager 1* trajectory through the Io torus and how whistlers generated in the Jovian atmosphere propagate along magnetic field lines through the torus to the spacecraft.

The force necessary to accelerate plasma to corotational velocities is supplied by a $\mathbf{j} \times \mathbf{B}$ force. The current perpendicular to $\mathbf{B}$ is linked to the ionosphere by a field-aligned current system. The perturbation in the field and plasma caused by the interaction propagates as an Alfvén wave down the field line. This Alfvénic perturbation is known as an Alfvén wing, as mentioned earlier in Chapter 8.

In the Saturnian magnetosphere, Titan is also an important source of mass, but Titan adds mass far out in the magnetosphere, close to the magnetopause. Much of the mass lost by Titan is in turn lost to the solar wind (but enough enters the magnetosphere to affect its composition). The interactions of the solar wind and magnetospheric plasma with Titan may very well resemble that of the solar wind with Venus, albeit with slightly different Mach numbers and dynamic pressures. Other moons at Saturn, and at the other planets, also supply some mass to their magnetospheres, but none to date has been found to exhibit any significant effect on these magnetospheres.

The moons of the gas giants not only provide mass but also absorb energetic radiation. Radiation-belt electrons and ions drift relative to the moons and spiral into them to be absorbed. This absorption leaves narrow gaps in the radial distribution of radiation-belt fluxes. The gaps are well defined at Saturn, where the dipole field is centered on the planet and is aligned with the spin axis, but less well defined in other magnetospheres. The speed with which the gaps are filled places constraints on the rate of radial diffusion in the magnetosphere. The rings of Saturn also are efficient absorbers of the radiation-belt particles and limit the buildup of intense fluxes of energetic particles in the inner radiation zone of Saturn.

15.6 RADIATION BELTS

With the exception of the Jovian magnetosphere, the radiation belts of the outer magnetospheres behave very much like those of the terrestrial magnetosphere. Processes such as radial diffusion and pitch-angle diffusion act to transport particles across field lines and cause the particles to precipitate into the atmosphere and be lost. At Jupiter, the mass source at Io, deep in the magnetosphere, coupled with an efficient acceleration mechanism, not yet fully understood, supplies the interior of the magnetosphere with an enormous source of energy. This energy in turn leads to an intense radiation belt that provides a hazard even to robotic spacecraft.

Figure 15.12 shows cuts through the electron radiation belts at Earth, Jupiter, Saturn, and Uranus. Neptune is similar to Uranus, with even smaller fluxes. The radiation belts of the different planets are similar. The fluxes are most intense just above the atmosphere (except at Saturn, where the fluxes maximize just outside the rings). At lowest altitudes, the spectrum is harder; that is, the flux decreases less sharply with increasing energy than at high altitudes. However, when one considers the peak fluxes, there is a great disparity. As illustrated in Table 15.3, the peak electron flux at Jupiter is about 1,000 times greater than that at Earth (>3 MeV) and the peak flux at Uranus is an order of magnitude less than that at Earth.

A similar story holds for the protons, as shown in Figure 15.13. The radiation belts look grossly similar, but the fluxes shown in Table 15.3 reveal an excess of three orders of magnitude at Jupiter and a deficit of

15.7 WAVES AND INSTABILITIES

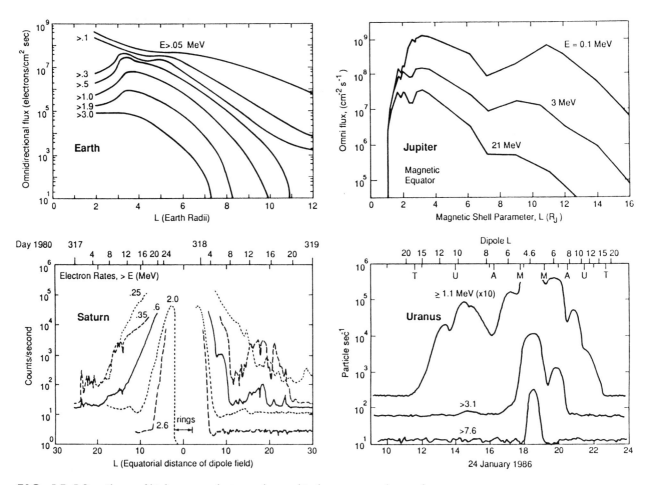

FIG. 15.12. Fluxes of high-energy electrons observed in the magnetospheres of Earth, Jupiter, Saturn, and Uranus. (Courtesy of D. J. Williams.)

three orders of magnitude at Uranus. It is clear that particles in the magnetospheres of Uranus and Neptune do not become highly energized, even when compared with the terrestrial radiation belts, which are formed in a much smaller magnetosphere. Perhaps the reason for this difference is the inefficiency of magnetopause reconnection in the outer heliosphere, concerning which we speculated earlier.

15.7 WAVES AND INSTABILITIES

On the microphysical level, the magnetospheres of the outer planets behave much as does the terrestrial magnetosphere. The distribution functions of the trapped plasma and the various beams and currents flowing in these magnetospheres are not unlike those for the terrestrial magnetosphere. Thus, these magnetospheres produce many, if not all, of the same plasma instabilities as Earth. Wave processes in the Jovian

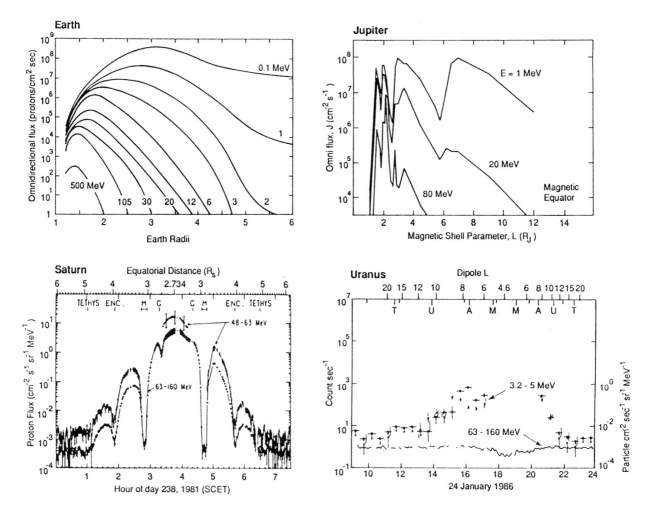

FIG. 15.13. Fluxes of high-energy protons observed in the magnetospheres of Earth, Jupiter, Saturn, and Uranus. (Courtesy of D. J. Williams.)

magnetosphere seem most intense overall. This includes wave processes both in the ULF band at periods of 1 s and longer and in the VLF band at kilohertz frequencies. ULF waves appear to be strongest in the plasma sheet and the Io torus. These waves are responsible for the diffusion of energetic particles, both radially and in pitch angle, and therefore contribute to the mass and energy transport in the magnetosphere. VLF waves are seen throughout the Jovian magnetosphere.

At ULF frequencies, Saturn does have some activity, as illustrated in Figure 15.14, which shows ion-cyclotron waves at the orbit of Dione. The presence of these waves may signal the pickup of ions from Dione by the Saturnian magnetosphere. However, at Uranus and Neptune, such ULF activity seems not to be present.

At VLF frequencies, at least some waves are present at each of the four outer planets. In particular, electron-cyclotron harmonic waves are

15.8 RADIO EMISSIONS

TABLE 15.3. Peak Energetic-Particle Fluxes

Planet	Electrons		Protons	
	Flux (cm$^{-2}\cdot$s^{-1})	Energy (MeV)	Flux (cm$^{-2}\cdot$s^{-1})	Energy (MeV)
Earth	10^5	≥ 3	10^4	≥ 105
Jupiter	10^8	≥ 3	10^7	≥ 80
Saturn	10^5	≥ 3	10^4	≥ 63
Uranus	10^4	≥ 3	<10	≥ 63

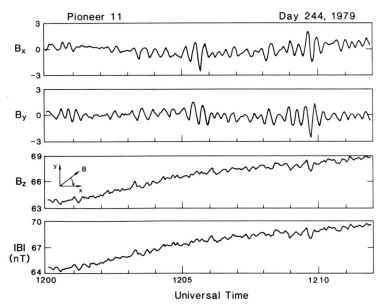

FIG. 15.14. Magnetic fluctuations observed as *Pioneer 11* crossed the L shell of Dione in Saturn's magnetosphere. (From Smith and Tsurutani, 1983.)

present at the magnetic equator. However, at Neptune, all other waves are weak. Figure 15.15 shows a spectrogram of the plasma waves seen on the *Voyager 2* passage through the magnetosphere of Uranus in 1986. Electromagnetic waves such as chorus and hiss are seen, as well as electrostatic waves such as electron-cyclotron harmonic waves and lower-hybrid-resonance emissions. The generation of these waves is thought to be due to processes that are similar to those in Earth's magnetosphere, as discussed in Chapter 12.

15.8 RADIO EMISSIONS

The outer planets announce their presence to the rest of the solar system via electromagnetic radio waves that propagate off into space, and in the case of Jupiter they can be detected more than 5 AU away. In contrast, the waves discussed in the preceding section must be measured within

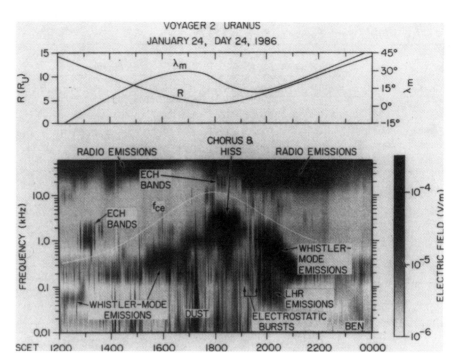

FIG. 15.15. Overview of the plasma-wave spectrum observed by *Voyager 2* in the magnetosphere of Uranus. ECH stands for electron-cyclotron harmonic radiation, and LHR for lower hybrid resonance. The top panel shows the magnetic latitude and the radial distance of the spacecraft. (From Kurth, 1990.)

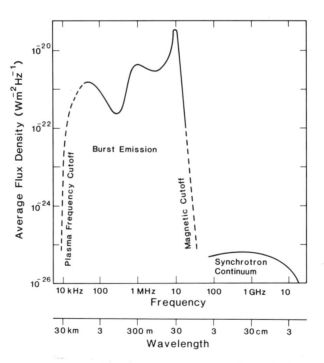

FIG. 15.16. Average power flux-density spectrum of Jupiter's nonthermal magnetospheric radio emissions. The instantaneous spectrum may be quite different. (Adapted from Carr et al., 1983.)

the magnetospheres in question. Radio emissions are generally characterized by wavelength. At the shortest wavelengths there is decimetric radiation, with wavelengths of tens of centimeters, which comes from synchrotron radiation of relativistic electrons gyrating in the inner radiation belt. Synchrotron radiation led to the discovery of the magnetosphere of Jupiter in the late 1950s, well before spacecraft were sent to the planet.

Decametric and hectometric radiations have wavelengths from 10 to 100 m. These emissions have much temporal and frequency structure and seem to be associated with instabilities in the plasma. *Voyager 1* and *2* showed that this radiation extends to even longer (kilometer) wavelengths that cannot be detected at Earth. Figure 15.16 shows a spectrum of the Jovian radio-wave flux versus frequency, illustrating the great intensity of decametric and longer-wavelength emissions.

The decametric emissions from Jupiter have been found to vary in frequency and in occurrence rate, and their variations have been analyzed extensively. Suffice it to say that the radiation received by Earth depends on the longitude on Jupiter of the Jupiter–Earth line and also on the location of Io in its orbit about Jupiter. It is suspected that instabilities in field-aligned currents connecting Io to the Jovian ionosphere are responsible for many of these waves.

15.9 CONCLUDING REMARKS

Much remains to be learned about the magnetospheres of the outer planets. Already, *Galileo* is on its way to Jupiter, scheduled to enter Jovian orbit in December 1995. The *Cassini* mission to Saturn has been approved, with injection into Saturnian orbit scheduled for early in the first decade of the twenty-first century. However, further exploration of Uranus and Neptune seems a long way off, and even though plans are being made for a mission to Pluto, the likelihood that such a mission will explore the solar-wind interaction with Pluto seems remote.

Perhaps such exploration seems moot and of little practical importance. However, each of these magnetospheres is different, and their differences enable us to identify the various possible mechanisms for their various phenomena. We cannot experiment in the usual sense, but by observing different systems we can achieve the same end. Thereby we learn the general processes by which magnetospheres operate and in turn learn more about how the terrestrial magnetosphere works.

ADDITIONAL READING

Dessler, A. J. (ed.). 1983. *Physics of the Jovian Magnetosphere*. Cambridge University Press.

PROBLEMS

15.1 If the polar cap of Jupiter can be approximated by a circular cap of extent 10° in colatitude from the dipole axis and is open, calculate the flux content of the Jovian tail. If this flux does not close across the central tail current sheet, and if the asymptotic field strength in the tail is 1 nT, calculate the tail cross-sectional area and its radius. Illustrate with a sketch.

15.2 As the solar wind flows from the sun to Neptune, how many times does the spiral magnetic field wrap around the sun? Illustrate with a sketch.

Appendix 1 NOTATION, VECTOR IDENTITIES, AND DIFFERENTIAL OPERATORS

A.1.1 NOTATION

THE COORDINATE NOTATIONS used in this text are undoubtedly familiar to most. Boldface is used for vectors; in handwritten material, vectors may be underlined or have arrows placed above them. For spatial coordinates, we may use cartesian, spherical or cylindrical coordinate systems. The following usages are common (where the first two forms are alternative usages for cartesian coordinates):

$$\mathbf{x} = (x, y, z) \text{ or } \mathbf{x} = (x_1, x_2, x_3); \quad \mathbf{r} = (r, \theta, \phi) \text{ or } \mathbf{r} = (r, \theta, z) \quad \text{(A1.1)}$$

$$\mathbf{v} = (v_x, v_y, v_z) \quad \text{or} \quad \mathbf{u} = (u_x, u_y, u_z) \quad \text{(A1.2)}$$

Integrals may be taken over a three-dimensional spatial volume (e.g., $d\mathbf{x} = d^3x = dx\, dy\, dz$) or over a three-dimensional volume in velocity space (e.g., $d\mathbf{v} = d^3v = dv_x\, dv_y\, dv_z$).

Vectors can be combined in various ways. The dot product produces a scalar from a pair of vectors

$$\mathbf{a} \cdot \mathbf{b} = a_x b_x + a_y b_y + a_z b_z \quad \text{(A1.3)}$$

The dot product is equal to the product of the length of the vectors times the cosine of the angle between them. The cross product of two vectors produces a new vector:

$$\mathbf{a} \times \mathbf{b} = (a_y b_z - a_z b_y,\ a_z b_x - a_x b_z,\ a_x b_y - a_y b_x) \quad \text{(A1.4)}$$

The magnitude of the cross product is the product of the length of the vectors times the sine of the angle between them. The time derivative at constant spatial position is written as $\partial f(x, y, z, t)/\partial t$, and the x-derivative at constant y, z, and time is written as $\partial f(x, y, z, t)/\partial x$. Vector derivatives are also very common in our work. These can be created from derivatives in different directions combined with unit vectors that identify the direction in which the derivative is taken. The resultant vector is referred to as a vector operator and is very convenient in writing complicated equations in only a few lines. The operators we shall use are defined as follows:

$$\nabla f(x, y, z, t) = \left(\frac{\partial f(x, y, z, t)}{\partial x}, \frac{\partial f(x, y, z, t)}{\partial y}, \frac{\partial f(x, y, z, t)}{\partial z} \right) \quad \text{(A1.5)}$$

which is called the gradient of f;

$$\nabla \cdot \mathbf{A} = \frac{\partial A_x(x, y, z, t)}{\partial x} + \frac{\partial A_y(x, y, z, t)}{\partial y} + \frac{\partial A_z(x, y, z, t)}{\partial z} \quad \text{(A1.6)}$$

which is called the divergence of A; and

$$\nabla \times \mathbf{A}(x, y, z, t) = \left(\frac{\partial A_z(x, y, z, t)}{\partial y} - \frac{\partial A_y(x, y, z, t)}{\partial z}, \frac{\partial A_x(x, y, z, t)}{\partial z} - \frac{\partial A_z(x, y, z, t)}{\partial x}, \frac{\partial A_y(x, y, z, t)}{\partial x} - \frac{\partial A_x(x, y, z, t)}{\partial y} \right) \quad \text{(A1.7)}$$

which is called the curl of **A**.

Both ∇f and $\nabla \times \mathbf{A}$ are vectors. The vector operator ∇ can be treated almost like any other vector provided that the order of the operators and operands in the expression is retained. You can verify this statement by examining the foregoing forms, recognizing that the divergence is a dot product of two vectors and that the curl is a cross product of two vectors. The divergence and the curl are of interest not only as mathematical constructs but also because they have physical significance. Some remarks on the physical significance of these operators follow equation (A1.42).

A.1.2 VECTOR IDENTITIES

The following formulas are taken from Book, D. L., *NRL Plasma Formulary,* 1983 revised, Naval Research Laboratory, Washington, D.C. 20375.

Notation: f and g are scalar functions; **A**, **B**, etc., are vector functions; **T** is a tensor.

$$\mathbf{A} \cdot \mathbf{B} \times \mathbf{C} = \mathbf{B} \cdot \mathbf{C} \times \mathbf{A} = \mathbf{C} \cdot \mathbf{A} \times \mathbf{B} \quad \text{(A1.8)}$$

$$\mathbf{A} \times (\mathbf{B} \times \mathbf{C}) = (\mathbf{C} \times \mathbf{B}) \times \mathbf{A} = \mathbf{B}(\mathbf{A} \cdot \mathbf{C}) - \mathbf{C}(\mathbf{A} \cdot \mathbf{B}) \quad \text{(A1.9)}$$

$$\mathbf{A} \times (\mathbf{B} \times \mathbf{C}) + \mathbf{B} \times (\mathbf{C} \times \mathbf{A}) + \mathbf{C} \times (\mathbf{A} \times \mathbf{B}) = 0 \quad \text{(A1.10)}$$

$$(\mathbf{A} \times \mathbf{B}) \cdot (\mathbf{C} \times \mathbf{D}) = (\mathbf{A} \cdot \mathbf{C})(\mathbf{B} \cdot \mathbf{D}) - (\mathbf{A} \cdot \mathbf{D})(\mathbf{B} \cdot \mathbf{C}) \quad \text{(A1.11)}$$

$$(\mathbf{A} \times \mathbf{B}) \times (\mathbf{C} \times \mathbf{D}) = (\mathbf{A} \times \mathbf{B} \cdot \mathbf{D})\mathbf{C} - (\mathbf{A} \times \mathbf{B} \cdot \mathbf{C})\mathbf{D} \quad \text{(A1.12)}$$

$$\nabla(fg) = \nabla(gf) = f\nabla g + g\nabla f \quad \text{(A1.13)}$$

$$\nabla \cdot (f\mathbf{A}) = f\nabla \cdot \mathbf{A} + \mathbf{A} \cdot \nabla f \quad \text{(A1.14)}$$

$$\nabla \times (f\mathbf{A}) = f\nabla \times \mathbf{A} + \nabla f \times \mathbf{A} \quad \text{(A1.15)}$$

$$\nabla \cdot (\mathbf{A} \times \mathbf{B}) = \mathbf{B} \cdot \nabla \times \mathbf{A} - \mathbf{A} \cdot \nabla \times \mathbf{B} \quad \text{(A1.16)}$$

A.1.2 VECTOR IDENTITIES

$$\nabla \times (\mathbf{A} \times \mathbf{B}) = \mathbf{A}(\nabla \cdot \mathbf{B}) - \mathbf{B}(\nabla \cdot \mathbf{A}) + (\mathbf{B} \cdot \nabla)\mathbf{A} - (\mathbf{A} \cdot \nabla)\mathbf{B} \quad (A1.17)$$

$$\mathbf{A} \times (\nabla \times \mathbf{B}) = (\nabla \mathbf{B}) \cdot \mathbf{A} - (\mathbf{A} \cdot \nabla)\mathbf{B} \quad (A1.18)$$

$$\nabla(\mathbf{A} \cdot \mathbf{B}) = \mathbf{A} \times (\nabla \times \mathbf{B}) + \mathbf{B} \times (\nabla \times \mathbf{A}) + (\mathbf{A} \cdot \nabla)\mathbf{B} + (\mathbf{B} \cdot \nabla)\mathbf{A} \quad (A1.19)$$

$$\nabla^2 f = \nabla \cdot \nabla f \quad (A1.20)$$

$$\nabla^2 \mathbf{A} = \nabla(\nabla \cdot \mathbf{A}) - \nabla \times \nabla \times \mathbf{A} \quad (A1.21)$$

$$\nabla \times \nabla f = 0 \quad (A1.22)$$

$$\nabla \cdot \nabla \times \mathbf{A} = 0 \quad (A1.23)$$

A second-order tensor $\mathbf{T}$ can be written in a number of different ways. It is often convenient to express such a tensor in terms of two vectors, $\mathbf{A}$ and $\mathbf{B}$, which allows one to write $\mathbf{T}$ in dyadic form:

$$\mathbf{T} = \mathbf{AB} \quad \text{or} \quad T_{ij} = A_i B_j \quad (A1.24)$$

In cartesian coordinates, the divergence of a tensor is a vector with components

$$(\nabla \cdot \mathbf{T})_i = \sum_{ij} (\partial T_{ji}/\partial x_j) \quad (A1.25)$$

$$\nabla \cdot (\mathbf{BA}) = (\mathbf{B} \cdot \nabla)\mathbf{A} + \mathbf{A}(\nabla \cdot \mathbf{B}) \quad (A1.26)$$

$$\nabla \cdot (f\mathbf{T}) = \mathbf{T} \cdot \nabla f + f \nabla \cdot \mathbf{T} \quad (A1.27)$$

Let $\mathbf{r} = \hat{\mathbf{e}}_x x + \hat{\mathbf{e}}_y y + \hat{\mathbf{e}}_z z$ be the radius vector of magnitude r from the origin to the point x, y, z. Then

$$\nabla \cdot \mathbf{r} = 3 \quad (A1.28)$$

$$\nabla \times \mathbf{r} = 0 \quad (A1.29)$$

$$\nabla r = \mathbf{r}/r \quad (A1.30)$$

$$\nabla(1/r) = -\mathbf{r}/r^3 \quad (A1.31)$$

$$\nabla \cdot (\mathbf{r}/r^3) = 4\pi \delta(\mathbf{r}) \quad (A1.32)$$

If V is a volume enclosed by a surface S, and $d\mathbf{S} = \hat{\mathbf{n}} \, dS$, where $\hat{\mathbf{n}}$ is the unit normal outward from V,

$$\int_V dV \, \nabla f = \int_S d\mathbf{S} \, f \quad (A1.33)$$

$$\int_V dV \, \nabla \cdot \mathbf{A} = \int_S d\mathbf{S} \cdot \mathbf{A} \quad (A1.34)$$

$$\int_V dV \, \nabla \cdot \mathbf{T} = \int_S d\mathbf{S} \cdot \mathbf{T} \quad (A1.35)$$

$$\int_V dV \, \nabla \times \mathbf{A} = \int_S d\mathbf{S} \times \mathbf{A} \quad (A1.36)$$

$$\int_V dV (f \nabla^2 g - g \nabla^2 f) = \int_S d\mathbf{S} \cdot (f \nabla g - g \nabla f) \quad (A1.37)$$

$$\int_V dV(\mathbf{A}\cdot\nabla\times\nabla\times\mathbf{B}-\mathbf{B}\cdot\nabla\times\nabla\times\mathbf{A})=\int_S d\mathbf{S}\cdot(\mathbf{B}\times\nabla\times\mathbf{A}-\mathbf{A}\times\nabla\times\mathbf{B}) \quad (A1.38)$$

If $\mathbf{S}$ is an open surface bounded by the contour C, of which the line element is $d\mathbf{l}$,

$$\int_S d\mathbf{S}\times\nabla f = \oint_C d\mathbf{l}\, f \quad (A1.39)$$

$$\int_S d\mathbf{S}\cdot\nabla\times\mathbf{A} = \oint_C d\mathbf{l}\cdot\mathbf{A} \quad (A1.40)$$

$$\int_S (d\mathbf{S}\times\nabla)\times\mathbf{A} = \oint_C d\mathbf{l}\times\mathbf{A} \quad (A1.41)$$

$$\int_S (d\mathbf{S}\cdot(\nabla f)\times\nabla g) = \oint_C f\, dg = -\oint_C g\, df \quad (A1.42)$$

The integral relations (A1.34) and (A1.40) are helpful in understanding the physical significance of the divergence and the curl operators. The names "curl" and "divergence" are suggestive and are particularly easy to visualize if the vector $\mathbf{A}$ is set equal to $\mathbf{v}$, the vector velocity in an incompressible fluid; $\mathbf{v}$ is assumed to be a function of position. Consider equation (A1.34) applied to the divergence of $\mathbf{v}$. From $\int_V dV\,\nabla\cdot\mathbf{v} = \int_S d\mathbf{S}\cdot\mathbf{v}$ we find that the integral of $\nabla\cdot\mathbf{v}$ over a small volume at any location in the fluid is equal to the integral of the normal component of fluid flow across the surface. The divergence of the fluid is nonvanishing at a particular point if the net fluid flow into an infinitesimal volume surrounding the point differs from the net fluid flow out of the volume, but in an incompressible fluid, that cannot happen. That means that the divergence must vanish, a condition for incompressible flow. The situation is readily visualized by drawing streamlines of the flow and noting that the same number of streamlines enter and exit the volume.

The curl can be understood in an analogous way by setting $\mathbf{A}$ equal to $\mathbf{v}$ in equation (A1.40) and taking the surface integral over a surface around a point of interest. Then $\int_S d\mathbf{S}\cdot\nabla\times\mathbf{v} = \oint_C d\mathbf{l}\cdot\mathbf{v}$, and the integral of the curl of $\mathbf{v}$ is seen to be equal to the integral of $\mathbf{v}$ around the curve bounding the surface containing the point. The integral around the closed contour C is called the circulation. If the curl of $\mathbf{v}$ is nonvanishing, there is net circulation around the point. Once again, streamlines are helpful in visualizing the situation. If some of the streamlines encircle the point of interest, the circulation and the curl of the velocity are nonvanishing.

In space physics, one often needs to consider the curl and the divergence of electromagnetic fields. In such situations, the analogues of streamlines are field lines, which are contours in space that are everywhere parallel to either the electric or magnetic field. From Maxwell's laws (see Chapter 2) we find that **E** can have a finite divergence. This means that it is possible to have more field lines that extend outward from a point than inward (or vice versa). However, if there is a net divergence of **E** at some point in space, there must be a net charge. The situation is different for magnetic fields. Magnetic field lines emerging from a spatial point must always close back on the point, which means that the magnetic field is divergence-free.

A.1.3 DIFFERENTIAL OPERATORS IN CURVILINEAR COORDINATES

Analyses of physical problems often can be greatly simplified by taking advantage of symmetry. For example, the form of the Coulomb potential is substantially more complicated in cartesian coordinates than in spherical coordinates. The axial symmetry of a dipole field is more straightforwardly represented in spherical or cylindrical coordinates than in cartesian coordinates. However, operator expressions in noncartesian (or curvilinear) coordinate systems must be treated with care. This is because one must take into account the fact that distance scales vary with position in a curvilinear coordinate system (e.g., lengths in the azimuthal direction in a spherical coordinate system depend on the colatitude θ and radial distance as $r \sin \theta \, \Delta\varphi$), and unit vectors point in different directions at different points in space (e.g., when $\theta = 90°$, the radial direction is along y). Thus, for spherical and cylindrical coordinates, the expressions for the operators do not have the simple forms given in equations (A1.5)–(A1.7). Next, we give the forms of the principal operator expressions in several important coordinate systems.

A.1.3.1 Cylindrical Coordinates

Divergence

$$\nabla \cdot \mathbf{A} = \frac{1}{r}\frac{\partial}{\partial r}(rA_r) + \frac{1}{r}\frac{\partial A_\varphi}{\partial \varphi} + \frac{\partial A_z}{\partial z} \qquad (A1.43)$$

Gradient

$$(\nabla f)_r = \frac{\partial f}{\partial r}, \qquad (\nabla f)_\varphi = \frac{1}{r}\frac{\partial f}{\partial \varphi}, \qquad (\nabla f)_z = \frac{\partial f}{\partial z} \qquad (A1.44)$$

Curl

$$(\nabla \times \mathbf{A})_r = \frac{1}{r}\frac{\partial A_z}{\partial \varphi} - \frac{\partial A_\varphi}{\partial z} \tag{A1.45}$$

$$(\nabla \times \mathbf{A})_\varphi = \frac{\partial A_r}{\partial z} - \frac{\partial A_z}{\partial r}$$

$$(\nabla \times \mathbf{A})_z = \frac{1}{r}\frac{\partial}{\partial r}(rA_\varphi) - \frac{1}{r}\frac{\partial A_r}{\partial \varphi}$$

Laplacian

$$\nabla^2 f = \frac{1}{r}\frac{\partial}{\partial r}\left(r\frac{\partial f}{\partial r}\right) + \frac{1}{r^2}\frac{\partial^2 f}{\partial \varphi^2} + \frac{\partial^2 f}{\partial z^2} \tag{A1.46}$$

Laplacian of a Vector

$$(\nabla^2 \mathbf{A})_r = \nabla^2 A_r - \frac{2}{r^2}\frac{\partial A_\varphi}{\partial \varphi} - \frac{A_r}{r^2} \tag{A1.47}$$

$$(\nabla^2 \mathbf{A})_\varphi = \nabla^2 A_\varphi + \frac{2}{r^2}\frac{\partial A_r}{\partial \varphi} - \frac{A_\varphi}{r^2}$$

$$(\nabla^2 \mathbf{A})_z = \nabla^2 A_z$$

Components of $\mathbf{A} \cdot \nabla \mathbf{B}$

$$(\mathbf{A} \cdot \nabla \mathbf{B})_r = A_r\frac{\partial B_r}{\partial r} + \frac{A_\varphi}{r}\frac{\partial B_r}{\partial \varphi} + A_z\frac{\partial B_r}{\partial z} - \frac{A_\varphi B_\varphi}{r} \tag{A1.48}$$

$$(\mathbf{A} \cdot \nabla \mathbf{B})_\varphi = A_r\frac{\partial B_\varphi}{\partial r} + \frac{A_\varphi}{r}\frac{\partial B_\varphi}{\partial \varphi} + A_z\frac{\partial B_\varphi}{\partial z} + \frac{A_\varphi B_r}{r}$$

$$(\mathbf{A} \cdot \nabla \mathbf{B})_z = A_r\frac{\partial B_z}{\partial r} + \frac{A_\varphi}{r}\frac{\partial B_z}{\partial \varphi} + A_z\frac{\partial B_z}{\partial z}$$

Divergence of a Tensor

$$(\nabla \cdot \mathbf{T})_r = \frac{1}{r}\frac{\partial}{\partial r}(rT_{rr}) + \frac{1}{r}\frac{\partial}{\partial \varphi}(T_{\varphi r}) + \frac{\partial T_{zr}}{\partial z} - \frac{1}{r}T_{\varphi\varphi} \tag{A1.49}$$

$$(\nabla \cdot \mathbf{T})_\varphi = \frac{1}{r}\frac{\partial}{\partial r}(rT_{r\varphi}) + \frac{1}{r}\frac{\partial}{\partial \varphi}(T_{\varphi\varphi}) + \frac{\partial T_{z\varphi}}{\partial z} + \frac{1}{r}T_{\varphi r}$$

$$(\nabla \cdot \mathbf{T})_z = \frac{1}{r}\frac{\partial}{\partial r}(rT_{rz}) + \frac{1}{r}\frac{\partial}{\partial \varphi}(T_{\varphi z}) + \frac{\partial T_{zz}}{\partial z}$$

A.1.3.2 Spherical Coordinates

Divergence

$$\nabla \cdot \mathbf{A} = \frac{1}{r^2}\frac{\partial}{\partial r}(r^2 A_r) + \frac{1}{r\sin\theta}\frac{\partial}{\partial\theta}(A_\theta \sin\theta) + \frac{1}{r\sin\theta}\frac{\partial A_\varphi}{\partial\varphi} \qquad (A1.50)$$

Gradient

$$(\nabla f)_r = \frac{\partial f}{\partial r}, \qquad (\nabla f)_\theta = \frac{1}{r}\frac{\partial f}{\partial\theta}, \qquad (\nabla f)_\varphi = \frac{1}{r\sin\theta}\frac{\partial f}{\partial\varphi} \qquad (A1.51)$$

Curl

$$(\nabla \times \mathbf{A})_r = \frac{1}{r\sin\theta}\frac{\partial}{\partial\theta}(A_\varphi \sin\theta) - \frac{1}{r\sin\theta}\frac{\partial A_\theta}{\partial\varphi} \qquad (A1.52)$$

$$(\nabla \times \mathbf{A})_\theta = \frac{1}{r\sin\theta}\frac{\partial A_r}{\partial\varphi} - \frac{1}{r}\frac{\partial}{\partial r}(rA)$$

$$(\nabla \times \mathbf{A}) = \frac{1}{r}\frac{\partial}{\partial r}(rA_\theta) - \frac{1}{r}\frac{\partial A_r}{\partial\theta}$$

Laplacian

$$\nabla^2 f = \frac{1}{r^2}\frac{\partial}{\partial r}\left(r^2 \frac{\partial f}{\partial r}\right) + \frac{1}{r^2 \sin\theta}\frac{\partial}{\partial\theta}\left(\sin\theta \frac{\partial f}{\partial\theta}\right) + \frac{1}{r^2 \sin^2\theta}\frac{\partial^2 f}{\partial\varphi^2} \qquad (A1.53)$$

Laplacian of a Vector

$$(\nabla^2 \mathbf{A})_r = \nabla^2 A_r - \frac{2A_r}{r^2} - \frac{2}{r^2}\frac{\partial A_\theta}{\partial\theta} - \frac{2A_\theta \cot\theta}{r^2} - \frac{2}{r^2 \sin\theta}\frac{\partial A_\varphi}{\partial\varphi} \qquad (A1.54)$$

$$(\nabla^2 \mathbf{A})_\theta = \nabla^2 A_\theta + \frac{2}{r^2}\frac{\partial A_r}{\partial\theta} - \frac{A_\theta}{r^2 \sin^2\theta} - \frac{2\cos\theta}{r^2 \sin^2\theta}\frac{\partial A_\varphi}{\partial\varphi}$$

$$(\nabla^2 \mathbf{A})_\varphi = \nabla^2 A_\varphi - \frac{A_\varphi}{r^2 \sin^2\theta} + \frac{2}{r^2 \sin\theta}\frac{\partial A_r}{\partial\varphi} + \frac{2\cos\theta}{r^2 \sin^2\theta}\frac{\partial A_\theta}{\partial\varphi}$$

Components of $\mathbf{A}\cdot\nabla\mathbf{B}$

$$(\mathbf{A}\cdot\nabla\mathbf{B})_r = A_r\frac{\partial B_r}{\partial r} + \frac{A_\theta}{r}\frac{\partial B_r}{\partial\theta} + \frac{A_\varphi}{r\sin\theta}\frac{\partial B_r}{\partial\varphi} - \frac{A_\theta B_\theta + A_\varphi B_\varphi}{r} \qquad (A1.55)$$

$$(\mathbf{A}\cdot\nabla\mathbf{B})_\theta = A_r\frac{\partial B_\theta}{\partial r} + \frac{A_\theta}{r}\frac{\partial B_\theta}{\partial\theta} + \frac{A_\varphi}{r\sin\theta}\frac{\partial B_\theta}{\partial\varphi} + \frac{A_\theta B_r}{r} - \frac{A_\varphi B_\varphi \cot\theta}{r}$$

$$(\mathbf{A}\cdot\nabla\mathbf{B})_\varphi = A_r\frac{\partial B_\varphi}{\partial r} + \frac{A_\theta}{r}\frac{\partial B_\varphi}{\partial\theta} + \frac{A_\varphi}{r\sin\theta}\frac{\partial B_\varphi}{\partial\varphi} + \frac{A_\varphi B_r}{r} + \frac{A_\varphi B_\theta \cot\theta}{r}$$

Divergence of a Tensor

$$(\nabla \cdot \mathbf{T})_r = \frac{1}{r^2}\frac{\partial}{\partial r}(r^2 T_{rr}) + \frac{1}{r \sin \theta}\frac{\partial}{\partial \theta}(T_{\theta r}\sin \theta) + \frac{1}{r \sin \theta}\frac{\partial T_{\varphi r}}{\partial \varphi} - \frac{1}{r}(T_{\theta \theta} + T_{\varphi \varphi}) \quad (A1.56)$$

$$(\nabla \cdot \mathbf{T})_\theta = \frac{1}{r^2}\frac{\partial}{\partial r}(r^2 T_{r\theta}) + \frac{1}{r \sin \theta}\frac{\partial}{\partial \theta}(T_{\theta \theta}\sin \theta) + \frac{1}{r \sin \theta}\frac{\partial T_{\varphi \theta}}{\partial \varphi} + \frac{T_{\theta r}}{r} - \frac{\cot \theta}{r}T_{\varphi \varphi}$$

$$(\nabla \cdot \mathbf{T})_\varphi = \frac{1}{r^2}\frac{\partial}{\partial r}(r^2 T_{r\varphi}) + \frac{1}{r \sin \theta}\frac{\partial}{\partial \theta}(T_{\theta \varphi}\sin \theta) + \frac{1}{r \sin \theta}\frac{\partial T_{\varphi \varphi}}{\partial \varphi} + \frac{T_{\varphi r}}{r} + \frac{\cot \theta}{r}T_{\varphi \theta}$$

A.1.4 SELECTED INTEGRALS

$$\int_{-\infty}^{\infty} dx \, e^{-ax^2} = \left(\frac{\pi}{a}\right)^{\frac{1}{2}}$$

$$\int_{-\infty}^{\infty} dx \, x^{2n} e^{-ax^2} = \frac{2n!}{2^{2n}n!}\left(\frac{\pi}{a^{2n+1}}\right)^{\frac{1}{2}}$$

Appendix 2 FUNDAMENTAL CONSTANTS AND PLASMA PARAMETERS OF SPACE PHYSICS

A.2.1 FUNDAMENTAL CONSTANTS

Mass of a proton	1.6726×10^{-27} kg
Mass of an electron	9.1095×10^{-31} kg
Electron-to-proton mass ratio	1,836.2
Speed of light in vacuum	2.9979×10^{8} m·s^{-1}
Gravitational constant	6.672×10^{-11} N·m^2·kg^{-2}
Stefan-Boltzmann constant	5.6703×10^{-8} J·m^{-2}·s^{-1}·K^{-4}
Boltzmann constant	1.3807×10^{-23} J·K^{-1}
Electron volt	1.6022×10^{-19} J
Electronic charge	1.6022×10^{-19} C
Temperature of 1-eV particle	1.1605×10^{4} K
Permittivity of free space, ϵ_0	8.8542×10^{-12} F·m^{-1}
Permeability of free space, μ_0	$4\pi \times 10^{-7}$ H·m^{-1}

A.2.2 FUNDAMENTAL PLASMA PARAMETERS IN PRACTICAL UNITS

Electron gyrofrequency [Hz]	$28\, B\,[\mathrm{nT}]$
Proton gyrofrequency [Hz]	$0.01525\, B\,[\mathrm{nT}]$
Electron plasma frequency [Hz]	$8{,}980\, n_e^{\frac{1}{2}}\,[\mathrm{cm}^{-3}]$
	$8.98\, n_e^{\frac{1}{2}}\,[\mathrm{m}^{-3}]$
Proton plasma frequency [Hz]	$210\, n_p^{\frac{1}{2}}\,[\mathrm{cm}^{-3}]$
	$0.21\, n_p^{\frac{1}{2}}\,[\mathrm{m}^{-3}]$
Electron thermal speed [km·s^{-1}]	$5.50\, T_e^{\frac{1}{2}}\,[\mathrm{K}]$
Proton thermal[1] speed [km·s^{-1}]	$0.129\, T_p^{\frac{1}{2}}\,[\mathrm{K}]$
Sound speed[2] [km·s^{-1}]	$0.117\, T_e^{\frac{1}{2}}\,[\mathrm{K}]$

Electron gyroradius [km]	$0.0221\, T_e^{\frac{1}{2}}\,[K] \cdot B^{-1}\,[nT]$
Proton gyroradius [km]	$0.947\, T_i^{\frac{1}{2}}\,[K] \cdot B^{-1}\,[nT]$
Electron inertial length [km]	$5.31\, n_e^{-\frac{1}{2}}\,[cm^{-3}]$
Proton inertial length [km]	$228\, n_i^{-\frac{1}{2}}\,[cm^{-3}]$
Debye length [cm]	$6.90\, T_e^{\frac{1}{2}}\,[K] \cdot n^{-\frac{1}{2}}\,[cm^{-3}]$
Particles in Debye cube	$329\, T_e^{\frac{3}{2}}\,[K] \cdot n^{-\frac{1}{2}}\,[cm^{-3}]$
Alfvén velocity [km·s^{-1}]	$21.8\, B\,[nT] \cdot n_p^{-\frac{1}{2}}\,[(m_i/m_p)^{-\frac{1}{2}} \cdot cm^{-3}]$
Beta	$3.47 \times 10^{-5}\, n\,[cm^{-3}] \cdot T\,[K] \cdot B^{-2}\,[nT]$
Neutral–charged-particle collision frequency[3] [s^{-1}]	$4 \times 10^{-5}\, n_0\,[cm^{-3}] \cdot T_q^{\frac{1}{2}}\,[K] \cdot (m_q^{-\frac{1}{2}} \cdot m_p)$
Bremsstrahlung from hydrogenlike plasma [W·m^{-3}]	$1.46 \times 10^{-30}\, n_e\,[cm^{-3}] \cdot T_e^{\frac{1}{2}}\,[K] \cdot \Sigma[Z^2 N(Z)]$
Cyclotron radiation [W·m^{-2}]	$5.41 \times 10^{-31}\, B^2\,[nT] \cdot n_e\,[cm^{-3}] \cdot T_e\,[K]$
Cyclotron radiation, [W·m^{-2}] $T_e = T_i$; $\beta = 1$	$3.79 \times 10^{-35}\, n_e^2\,[cm^{-3}] \cdot T_e^2\,[K]$
Magnetic field from an infinite wire, B_θ [nT]	$0.200\, I\,[A] \cdot r^{-1}\,[km]$

[1] Most probable speed [see equation (2.22)].
[2] Assumes $T_e \gg T_i$.
[3] Approximate; mass of charged particle q expressed in proton masses.

Appendix 3 GEOPHYSICAL COORDINATE TRANSFORMATIONS

A.3.1 INTRODUCTION

MANY DIFFERENT COORDINATE SYSTEMS are used in experimental and theoretical work on solar-terrestrial relationships. These coordinate systems are used to display satellite trajectories, boundary locations, and vector-field measurements. The need for more than one coordinate system arises from the fact that often various physical processes are better understood, experimental data more ordered, or calculations more easily performed in one or another of the various systems. Frequently it is necessary to transform from one to another of these systems. It is possible to derive the transformation from one coordinate system to another in terms of trigonometric relations between angles measured in each system by means of the formulas of spherical trigonometry. However, the use of this technique can be very tricky and can result in rather complex relationships.

Another technique is to find the required Euler rotation angles and construct the associated rotation matrices. Then these rotation matrices can be multiplied to give a single transformation matrix. The vector-matrix formalism is attractive because it permits a shorthand representation of the transformation and because it permits multiple transformations to be performed readily by matrix multiplication and the inverse transformation to be derived readily.

The matrices required for coordinate transformations need not be derived from Euler rotation angles. This appendix explains another approach and describes the most common coordinate systems in use in the field of solar-terrestrial relationships.

A.3.2 GENERAL REMARKS

In defining a coordinate system, in general, we choose two quantities: the direction of one of the axes and the orientation of the other two axes in the plane perpendicular to this direction. This latter orientation is often specified by requiring one of the two remaining axes to be perpendicular to some direction. A fortunate feature of rotation matrices (the matrix that transforms a vector from one system to another) is that the inverse is simply its transpose. Thus, if the matrix A transforms the vector $\mathbf{V}^a$ measured in system a to $\mathbf{V}^b$ measured in system b, then the matrix that transforms $\mathbf{V}^b$ into $\mathbf{V}^a$ is A^T. Thus we can write

$$A \cdot \mathbf{V}^a = \mathbf{V}^b$$
$$A^T \cdot \mathbf{V}^b = \mathbf{V}^a$$

The simplest way to obtain the transformation matrix A is to find the directions of the three new coordinate axes for system b in the old system (system a). If the direction cosines of the new x-direction expressed in the old system are (x_1, x_2, x_3), those of the new y-direction are (y_1, y_2, y_3), and those of the new z-direction are (z_1, z_2, z_3), then the rotation matrix is formed by these three vectors as rows:

$$\begin{pmatrix} x_1 & x_2 & x_3 \\ y_1 & y_2 & y_3 \\ z_1 & z_2 & z_3 \end{pmatrix} \cdot \begin{pmatrix} V_x^a \\ V_y^a \\ V_z^a \end{pmatrix} = \begin{pmatrix} V_x^b \\ V_y^b \\ V_z^b \end{pmatrix}$$

Similarly, the transformation from system b to a is

$$\begin{pmatrix} x_1 & y_1 & z_1 \\ x_2 & y_2 & z_2 \\ x_3 & y_3 & z_3 \end{pmatrix} \cdot \begin{pmatrix} V_x^b \\ V_y^b \\ V_z^b \end{pmatrix} = \begin{pmatrix} V_x^a \\ V_y^a \\ V_z^a \end{pmatrix}$$

The following properties of rotation matrices are useful for error checking: (1) Each row and column is a unit vector. (2) The dot products of any two rows or any two columns is zero. (3) The cross product of any two rows or columns equals the third row or column or its negative (row 1 cross row 2 equals row 3; row 2 cross row 1 equals minus row 3).

A.3.3 COORDINATE SYSTEMS

A.3.3.1 Geocentric Equatorial Inertial System

A.3.3.1.1 DEFINITION The geocentric equatorial inertial (GEI) coordinate system has its x-axis pointing from the earth toward the first point of Aries (the position of the sun at the vernal equinox). This direction is the intersection of the earth's equatorial plane and the ecliptic plane, and thus the x-axis lies in both planes. The z-axis is parallel to the rotation axis of the earth, and y completes the right-handed orthogonal set ($\mathbf{y} = \mathbf{z} \times \mathbf{x}$).

A.3.3.1.2 USES This is the system commonly used in astronomy and satellite orbit calculations. The angles, right ascension and declination, are measured in this system. If (V_x, V_y, V_z) is a vector in GEI with magnitude V, then its right ascension, α, is $\tan^{-1}(V_y/V_x)$, $0° \leq \alpha \leq 180°$ if $V_y \geq 0$, and $180° \leq \alpha \leq 360°$ if $V_y \leq 0$. Its declination, θ, is $\sin^{-1}(V_z/V)$, $-90° \leq \theta \leq 90°$.

A.3.3.2 Geographic Coordinates

A.3.3.2.1 DEFINITION The geographic coordinate system (GEO) is defined so that its x-axis is in the earth's equatorial plane, but is fixed with the rotation of the earth, so that it passes through the Greenwich meridian (0° longitude). Its z-axis is parallel to the rotation axis of the earth, and its y-axis completes a right-handed orthogonal set ($\mathbf{y} = \mathbf{z} \times \mathbf{x}$).

A.3.3.2.2 USES This system is used for defining the positions of ground observatories and transmitting and receiving stations. Longitude and latitude in this system are defined in the same way as right ascension and declination in GEI. Longitude is measured positively moving eastward. Universal time (UT) is defined as 12 h minus the longitude of the sun converted from degrees to hours by dividing by 15. Local time is the universal time plus the geographic longitude of the observer converted to hours. Universal time is the local time of the Greenwich meridian.

A.3.3.2.3 TRANSFORMATIONS Because the GEO and GEI coordinate systems have the z-axis in common, we need only to know the position of the first point in Aries (the x-axis of GEI) relative to the Greenwich meridian to determine the required transformation. If we let the angle between the Greenwich meridian and the first point of Aries measured eastward from the first point of Aries in the earth's equator be θ, then the first point of Aries is at $(\cos\theta, -\sin\theta, 0)$ in the GEO system, and the transformation from GEO to GEI is

$$\begin{pmatrix} \cos\theta & -\sin\theta & 0 \\ \sin\theta & \cos\theta & 0 \\ 0 & 0 & 1 \end{pmatrix} \cdot \begin{pmatrix} V_x \\ V_y \\ V_z \end{pmatrix}_{GEO} = \begin{pmatrix} V_x \\ V_y \\ V_z \end{pmatrix}_{GEI}$$

and the inverse transformation is

$$\begin{pmatrix} \cos\theta & \sin\theta & 0 \\ -\sin\theta & \cos\theta & 0 \\ 0 & 0 & 1 \end{pmatrix} \cdot \begin{pmatrix} V_x \\ V_y \\ V_z \end{pmatrix}_{GEI} = \begin{pmatrix} V_x \\ V_y \\ V_z \end{pmatrix}_{GEO}$$

The angle θ is, of course, a function of the time of day and the time of year, since the earth spins 366.25 times per year around its axis in inertial space, rather than 365.25 times. Thus, the duration of a day, relative to inertial space (a sidereal day), is less than 24 h. The angle θ is called Greenwich mean sidereal time and can be calculated by means of the formulas given in Section A.3.5.

A.3.3.3 Geomagnetic Coordinates

A.3.3.3.1 DEFINITION The geomagnetic coordinate system (MAG) is defined so that its z-axis is parallel to the magnetic-dipole axis.

The geographic coordinates of the dipole axis from the International Geomagnetic Reference Field 1985 (IGRF) are 11.018° colatitude and −70.905° east longitude. Thus the z-axis is (0.06252, −0.18060, 0.98157) in geographic coordinates. The y-axis of this system is perpendicular to the geographic poles, such that if **D** is the dipole position and **P** is the South Pole, $\mathbf{y} = \mathbf{D} \times \mathbf{P} / |\mathbf{D} \times \mathbf{P}|$.

A.3.3.3.2 USES This system is often used for defining the positions of magnetic observatories. Also, it is a convenient system in which to do field-line tracing when current systems, in addition to the earth's internal field, are being considered. The magnetic longitude is measured eastward from the x-axis, and magnetic latitude is measured from the equator in magnetic meridians, positive northward and negative southward. Thus, if (V_x, V_y, V_z) is a vector in the MAG system with magnitude V, then its magnetic longitude, λ, is $\tan(V_y/V_x)$, $0° \leq \lambda \leq 180°$ if $V_y \geq 0$, and $180° \leq \lambda \leq 360°$ if $V_y \leq 0$. Its magnetic latitude, θ, is $\sin^{-1} V_z/V$, $-90° \leq \theta \leq 90°$. Except near the poles, magnetic longitude generally will be about 70° greater than geographic longitude. The magnetic local time is defined in this system as the magnetic longitude of the observer minus the magnetic longitude of the sun expressed in hours plus 12 h.

A.3.3.3.3 TRANSFORMATIONS This system is fixed in the rotating earth, and thus the transformation from the geographic coordinate system to the geomagnetic system is constant. From the foregoing definitions, we obtain

$$\begin{pmatrix} 0.32110 & -0.92756 & -0.19112 \\ 0.94498 & 0.32713 & 0 \\ 0.06252 & -0.18060 & 0.98157 \end{pmatrix} \cdot \begin{pmatrix} V_x \\ V_y \\ V_z \end{pmatrix}_{\text{GEO}} = \begin{pmatrix} V_x \\ V_y \\ V_z \end{pmatrix}_{\text{MAG}}$$

A.3.3.4 Geocentric Solar Ecliptic System

A.3.3.4.1 DEFINITION The geocentric solar ecliptic (GSE) system has its x-axis pointing from the earth toward the sun, and its y-axis is chosen to be in the ecliptic plane pointing toward dusk (thus opposing planetary motion). Its z-axis is parallel to the ecliptic pole. Relative to an inertial system, this system has a yearly rotation.

A.3.3.4.2 USES This system is used to display satellite trajectories, interplanetary magnetic-field observations, and data on solar-wind velocity. The system is useful for the latter display because the aberration of the solar wind caused by the earth's motion can easily be removed in this system. The velocity of the earth is approximately 30 km·s^{-1} in the −y-direction. Because the only important effect of the earth's orbital motion in solar-terrestrial relationships is to cause the aberration, other choices for the orientation of the y- and z-axes about the x-axis have been used. These are to be discussed later. Longitude, as with the geographic system, is measured in the x–y plane from the x-axis toward the y-axis, and latitude is the angle out of the x–y plane, positive for positive z-components.

A.3.3.4.3 TRANSFORMATION The most commonly required transformation into the GSE system from those discussed thus far is that from the GEI system. The direction of the ecliptic pole $(0, -0.398, 0.917)$ is constant in the GEI system. The x-axis, the direction of the sun, can be obtained in GEI from the subroutine given in Section A.3.5. If this direction is (S_1, S_2, S_3), then the y-axis in GEI (y_1, y_2, y_3) is

$$(0, -0.3978, 0.9175) \times (S_1, S_2, S_3)$$

and the transformation is

$$\begin{pmatrix} S_1 & S_2 & S_3 \\ y_1 & y_2 & y_3 \\ 0 & -0.3978 & 0.9175 \end{pmatrix} \cdot \begin{pmatrix} V_x \\ V_y \\ V_z \end{pmatrix}_{GEI} = \begin{pmatrix} V_x \\ V_y \\ V_z \end{pmatrix}_{GSE}$$

A.3.3.5 Geocentric Solar Equatorial System

A.3.3.5.1 DEFINITION The geocentric solar equatorial (GSEQ) system, like the GSE system, has its x-axis pointing toward the sun from the earth. However, instead of having its y-axis in the ecliptic plane, the GSEQ y-axis is parallel to the sun's equatorial plane, which is inclined to the ecliptic. We note that because the x-axis is in the ecliptic plane and therefore is not necessarily in the sun's equatorial plane, the z-axis of this system will not necessarily be parallel to the sun's axis of rotation. However, the sun's axis of rotation must lie in the x–z plane. The z-axis is chosen to be in the same sense as the ecliptic pole (i.e., northward).

A.3.3.5.2 USES This system has been used extensively to display interplanetary magnetic-field data. We note that this system is useful for ordering data controlled by the sun and therefore offers an improvement over the use of the GSE system for studying the interplanetary magnetic field and the solar wind. However, for studying the interaction of the interplanetary magnetic field with the earth, a third system is more relevant.

A.3.3.5.3 TRANSFORMATIONS The rotation axis of the sun, **R**, has a right ascension of $-74.0°$ and a declination of $63.8°$. Thus **R** is $(0.1217, -0.424, 0.897)$ in GEI. To transform from GEI to GSEQ, we must know the position of the sun (S_1, S_2, S_3) in GEI (see Section A.3.5). Then the y-axis in GEI (y_1, y_2, y_3) is parallel to $\mathbf{R} \times \mathbf{S}$. Note that because the cross product of two unit vectors is not a unit vector unless they are perpendicular to each other, this cross product must be normalized. Finally the x-axis in GEI $(z_1, z_2, z_3) = \mathbf{S} \times \mathbf{y}$. Then

$$\begin{pmatrix} S_1 & S_2 & S_3 \\ y_1 & y_2 & y_3 \\ z_1 & z_2 & z_3 \end{pmatrix} \cdot \begin{pmatrix} V_x \\ V_y \\ V_z \end{pmatrix}_{\text{GEI}} = \begin{pmatrix} V_x \\ V_y \\ V_z \end{pmatrix}_{\text{GSEQ}}$$

Because both the GSE and GSEQ coordinate systems have their x-axes directed toward the sun, they differ only by a rotation about the x-axis. Thus the transformation matrix from GSE to GSEQ must be of the form

$$\begin{pmatrix} 1 & 0 & 0 \\ 0 & \cos\theta & -\sin\theta \\ 0 & \sin\theta & \cos\theta \end{pmatrix} \cdot \begin{pmatrix} V_x \\ V_y \\ V_z \end{pmatrix}_{\text{GSE}} = \begin{pmatrix} V_x \\ V_y \\ V_z \end{pmatrix}_{\text{GSEQ}}$$

If the transformations from GEI to GSE and GEI to GSEQ are both known, then the angle θ can be determined by examining the angle between the y-axes in the two systems, or the z-axes (i.e., the angle between the vectors formed by the second row of each matrix, or the third row). If these transformation matrices are not available, θ can be calculated from the following formula:

$$\sin\theta = \frac{\mathbf{S} \cdot (-0.032, -0.112, -0.048)}{|(0.1217, -0.424, 0.897) \times \mathbf{S}|}$$

where $\mathbf{S}$ is the direction to the sun in GEI and can be calculated from the formulas in Section A.3.5. Because the sun's spin axis is inclined 7.25° to the ecliptic, θ ranges from $-7.25°$ (on approximately 5 December) to 7.25° (on 5 June) each year. The sun's spin axis is directed most nearly toward the earth on approximately 5 September, at which time the earth reaches its most northerly heliographic latitude. At this time, θ equals zero.

A.3.3.6 Geocentric Solar Magnetospheric System

A.3.3.6.1 DEFINITION The geocentric solar magnetospheric (GSM) system, like both the GSE and GSEQ systems, has its x-axis from the earth to the sun. The y-axis is defined to be perpendicular to the earth's magnetic dipole, so that the x–z plane contains the dipole axis. The positive z-axis is chosen to be in the same sense as the northern magnetic pole. The difference between the GSM system and the GSE and GSEQ systems is simply a rotation about the x-axis.

A.3.3.6.2 USES This system is useful for displaying magnetopause and shock-boundary positions, magnetosheath and magnetotail magnetic fields, and magnetosheath solar-wind velocities, because the orientation of the magnetic-dipole axis alters the otherwise cylindrical symmetry of the solar-wind flow. It also is used in models of magnetopause currents. It reduces the three-dimensional motion of the earth's dipole in GEI, GSE, and so forth, to motion in a plane (the x–z plane). The angle of the north magnetic pole to the GSM z-axis is called the dipole tilt angle and is positive when the north magnetic pole is tilted toward the sun. In addition to a yearly period due to the motion of the earth about the sun, this coordinate system rocks about the solar direction with a 24-h period. We note that because the y-axis is perpendicular to the dipole axis, the y-axis is always in the magnetic equator, and because it is perpendicular to the earth–sun line, it is in the dawn–dusk meridian (pointing toward dusk). GSM longitude is measured in the x–y plane from x toward y, and latitude is the angle northward from the x–y plane. However, another set of spherical polar angles is sometimes used. Here, the angle between the vector and the x-axis, called the solar-zenith angle (SZA), is the polar angle, and the angle of the projected vector in the y–z plane is the azimuthal angle. It is measured from the positive y-axis toward the positive z-axis. If the interplanetary magnetic field is being described, these two angles often are called the cone and clock angles, respectively.

A.3.3.6.3 TRANSFORMATIONS To transform from GEI to GSM, we need to know both the direction to the sun in GEI and the orientation of the earth's dipole axis. The direction of the sun (S_1, S_2, S_3) can be obtained from Section A.3.5. The direction of the dipole **D** must be obtained by transforming from geographic coordinates (see Section A.3.3.2). In geographic coordinates, the dipole is at 11.018° colatitude and −70.905° east longitude (IGRF epoch 1985.0). Thus, **D** in geographic coordinates is (0.06252, −0.18060, 0.98157). If **D'** is **D** transformed into GEI, the y-axis is

$$\frac{\mathbf{D'} \times \mathbf{S}}{|\mathbf{D'} \times \mathbf{S}|}$$

We note that the normalizing factor occurs because **D'** and **S** are not necessarily perpendicular. Finally, **z** is **S** × **y**, and the transformation becomes

$$\begin{pmatrix} S_1 & S_2 & S_3 \\ y_1 & y_2 & y_3 \\ z_1 & z_2 & z_3 \end{pmatrix} \cdot \begin{pmatrix} V_x \\ V_y \\ V_z \end{pmatrix}_{GEI} = \begin{pmatrix} V_x \\ V_y \\ V_z \end{pmatrix}_{GSM}$$

The transformation matrix between GSM and GSE or GSEQ is of the form

$$\begin{pmatrix} 1 & 0 & 0 \\ 0 & \cos\theta & -\sin\theta \\ 0 & \sin\theta & \cos\theta \end{pmatrix}$$

Because θ changes with both time of day and time of year, it is not derivable from a simple equation. If the transformation matrix from GEI to GSE, A_{GSE}, and that from GEI to GSM, A_{GSM}, are both known, then the transformation from GSM to GSE is simple: $A_{\text{GSE}} \cdot A_{\text{GSM}}^T$, where A_{GSM}^T is the transpose of A_{GSM}. An analogous formula holds for the transformation from GSM to GSEQ. We note that the amplitude of the diurnal variation of θ is $\pm 11.0°$, which is added to an annual variation of $\pm 23.5°$.

A.3.3.7 Solar Magnetic Coordinates

A.3.3.7.1 DEFINITION In solar magnetic (SM) coordinates, the z-axis is chosen parallel to the north magnetic pole, and the y-axis perpendicular to the earth–sun line toward dusk. The difference between this system and the GSM system is a rotation about the y-axis. The amount of rotation is simply the dipole tilt angle, as defined in the preceding section. We note that in this system the x-axis does not point directly at the sun. Like the GSM system, the SM system rotates with both a yearly period and a daily period with respect to inertial coordinates.

A.3.3.7.2 USES The SM system is useful for ordering data controlled more strongly by the earth's dipole field than by the solar wind. It has been used for magnetopause cross sections and magnetospheric magnetic fields. We note that because the dipole axis and the z-axis of this system are parallel, the cartesian components of the dipole magnetic field are particularly simple in this system (see Chapter 6).

A.3.3.7.3 TRANSFORMATIONS As for GSM, the transformation from GEI to SM requires a knowledge of the earth–sun direction **S** and the dipole direction **D** in GEI. Having obtained these as in Section A.3.5, we find $\mathbf{y} = (\mathbf{D} \times \mathbf{S})/(|\mathbf{D} \times \mathbf{S}|)$ and $\mathbf{x} = \mathbf{y} \times \mathbf{D}$. Then the transformation becomes

$$\begin{pmatrix} x_1 & x_2 & x_3 \\ y_1 & y_2 & y_3 \\ D_1 & D_2 & D_3 \end{pmatrix} \cdot \begin{pmatrix} V_x \\ V_y \\ V_z \end{pmatrix}_{\text{GEI}} = \begin{pmatrix} V_x \\ V_y \\ V_z \end{pmatrix}_{\text{SM}}$$

The transformation from GSM to SM is simply a rotation about the y-axis by the dipole tilt angle μ. Thus,

$$\begin{pmatrix} \cos \mu & 0 & -\sin \mu \\ 0 & 1 & 0 \\ \sin \mu & 0 & \cos \mu \end{pmatrix} \cdot \begin{pmatrix} V_x \\ V_y \\ V_z \end{pmatrix}_{\text{GSM}} = \begin{pmatrix} V_x \\ V_y \\ V_z \end{pmatrix}_{\text{SM}}$$

A.3.3.8 Other Planets

The coordinate systems discussed here are all designed for use on or around the earth. However, they can be generalized for the other planets. The generalization of the GSE coordinate system becomes the Venus solar orbital, and the Mars solar orbital at Mars. Instead of the ecliptic plane in which the earth orbits, we use the orbital plane of the planet to define the x–y plane. Likewise, at the magnetized planets, the GSM system can be adapted by keeping the y-axis perpendicular to the plane containing the magnetic-dipole axis and the solar direction. In studying unmagnetized planets, comets, and phenomena upstream from the bow shocks of even the magnetized planets it is useful to use a coordinate system whose orientation is ordered by the magnetic field. The most common choice is to orient the coordinate system so that x points to the sun, and the projection of the magnetic field in the y–z plane lies along the y-axis. Because the solar wind flows approximately in the $-x$-direction, the electric field of the solar wind points in the $+z$-direction. Any newly created ions at a comet or near a planet will initially be accelerated in the z-direction. It is suggested that such a coordinate system be called the Venus solar interplanetary (VSI) for Venus, and similarly for other planets.

A.3.4 LOCAL COORDINATE SYSTEMS

Often it is advantageous to define a coordinate system oriented with reference to the observing site. For example, when one is on the surface of the earth, a coordinate system oriented with respect to the local vertical is useful for many purposes. In this section we describe several such coordinate systems.

A.3.4.1 Dipole Meridian System

A.3.4.1.1 DEFINITION As with the SM system, the z-axis of the dipole meridian (DM) system is chosen along the north magnetic-dipole axis. However, the y-axis is chosen to be perpendicular to a radius vector to the point of observation, rather than the sun. The positive y-direction is chosen to be eastward, so that the x-axis is directed outward from the dipole. This is a local coordinate system in that it varies with position; however, because the x–z plane contains the dipole magnetic field, it is quite useful.

A.3.4.1.2 USES The DM system is used to order data controlled by the dipole magnetic field where the influence of the solar-wind interaction with the magnetosphere is weak. It has been used extensively to describe the distortions of the magnetospheric field in terms of the two angles, declination and inclination, that can be easily derived from measurements in this system. The inclination I is simply the angle that the field makes with the radius vector minus 90°. Thus, if $\mathbf{R}$ is the unit vector from the center of the earth to the point of observation in the DM system (we note that in this system, $R_y = 0$), and $\mathbf{b}$ is the direction of the magnetic field in the DM system, then $I = \cos^{-1}(R_x b_x + R_z b_z) - 90°$. The declination, D, is measured about the radius vector, with $D = 0°$ in the x–z plane, and positive D angles for positive b_y. Thus, $D = \tan^{-1}[b_y/(R_x b_z - R_z b_x)]$, $0° \leqslant D \leqslant 180°$ for $0° \leqslant b_y \leqslant 1$, and $0° \geqslant D \geqslant 180°$ for $0 \geqslant b_y \geqslant -1$. As in the SM system, the cartesian components of the dipole field can be expressed very simply in this system. In particular, $B_y = 0$, by definition.

A.3.4.1.3 TRANSFORMATIONS To transform from any system to the DM system, we must know the dipole axis $\mathbf{D}$ in this system and the unit-position vector of the point of observation relative to the center of the earth. Because $\mathbf{y}$ is perpendicular to $\mathbf{R}$ and $\mathbf{D}$, then $\mathbf{y} = (\mathbf{D} \times \mathbf{R})/(|\mathbf{D} \times \mathbf{R}|)$, and $\mathbf{x} = \mathbf{D} \times \mathbf{y}$. Thus,

$$\begin{pmatrix} x_1 & x_2 & x_3 \\ y_1 & y_2 & y_3 \\ D_1 & D_2 & D_3 \end{pmatrix} \cdot \begin{pmatrix} V_x \\ V_y \\ V_z \end{pmatrix} = \begin{pmatrix} V_x \\ V_y \\ V_z \end{pmatrix}_{DM}$$

We note that this transformation usually is particularly straightforward from geographic coordinates, because the geographic latitude and longitude of a point of observation often are known, and the dipole is fixed in geographic coordinates. From geomagnetic coordinates, it is a simple rotation about the z-axis by the angle between the projections of the sun and the local radius vector in the magnetic equator.

A.3.4.2 Surface Magnetic Measurements

Two local coordinate systems that are used for surface magnetic data differ slightly from the geographic (GEO) and geomagnetic (MAG) systems in the sense that they use the local vertical for their z-direction, but order their x- and y-axes according to geographic and geomagnetic directions, respectively. The first is called simply the XYZ system and is oriented so that the z-direction is downward and the x-direction points to geographic north. The y-direction points to geographic east. The second is the HDZ system, in which the z-direction is vertically downward, and the H-direction points to the north magnetic pole, with D roughly eastward orthogonal to H. Caution is urged in the use of this coordinate system, because in some, but not all, instances researchers express H and D as a magnitude and an angle, rather than as two components.

A.3.4.3 Boundary-Normal Coordinate Systems

Across an infinitesimally thin boundary, the normal component of the magnetic field is continuous or nearly so, because the magnetic field is divergenceless, and thus it is often useful to express solar-terrestrial data in a boundary-normal coordinate system. Situations in which this approach is helpful include studies of the bow shock, magnetopause, and plane waves. The key to using this approach is determining an accurate boundary normal.

A.3.4.3.1 SHOCK-NORMAL COORDINATES The normal direction to the bow shock may be satisfactorily defined geometrically under most circumstances. With interplanetary shocks and the occasional bow shock crossing, one may wish to use the so-called coplanarity theorem to derive the shock-normal direction, as discussed in Chapter 5. This technique makes use of the fact that the direction of the change in magnetic field across the shock is perpendicular to the normal to the shock, because the component along the shock normal is constant. Further, the upstream magnetic field (ahead of the shock structure), the shock normal, and the downstream magnetic field (downstream of the shock structure) are in the same plane. Hence the cross product of the upstream and downstream fields must be perpendicular to the shock normal. Thus, the triple cross product of the upstream field times the downstream field and then times the difference of the two fields, when properly normalized, is along the normal direction. This does not work for parallel or perpendicular shocks when the fields upstream and down are parallel. More sophisticated techniques are possible when plasma-velocity data are available or when there are observations from multiple spacecraft. For example, timing data from four spacecraft are sufficient to obtain the orientation and velocity of a planar boundary.

A.3.4.3.2 MAGNETOPAUSE-NORMAL COORDINATES Under most circumstances, the normal to the magnetopause, like the normal to the bow shock, can be obtained satisfactorily using a geometric model. The normal can also be obtained from magnetic-field observations if the field behaves according to certain assumptions. For example, often the magnetopause appears to be a tangential discontinuity, with no connection of magnetic field across the surface of the magnetopause. Under such conditions the magnetic fields on either side of the boundary are tangential to the boundary, and the direction of the normal can be obtained from the cross product of these two directions. Even when the magnetic field connects across the magnetopause, it is possible to determine a normal to the boundary by finding the direction in which the magnetic field remains most nearly constant or has a minimum variance. This direction should be the normal to the magnetopause, because the magnetic field is divergenceless, and across a thin discontinuity it should not change in the direction perpendicular to the surface. If the change in magnetic field across the boundary occurs along a single direction, however, this technique becomes indeterminant, because there are two orthogonal directions in which the field change is very small. The true normal may be along either direction or anywhere in the plane defined by these two directions.

A.3.4.3.3 PRINCIPAL-AXES COORDINATE SYSTEM By solving an eigenvalue problem, it is possible to calculate the rotation matrix that rotates a vector time series into a coordinate system in which the direction of the coordinate axes are those of the maximum variance, the minimum variance, and an intermediate variance. The eigenvectors are the rows of the transformation matrix, and the eigenvalues are the variances along each direction. These directions are called the principal axes. These can often be used at the magnetopause, but not at shocks. The change in the magnetic field at shocks, which is mandated by the Rankine-Hugoniot equations, is along a single direction in the plane of the upstream magnetic field and the normal. Variation of the field perpendicular to this plane frequently occurs because of waves that do not necessarily propagate along the shock normal.

Principal-axes coordinates are most useful for waves that are circularly or elliptically polarized, such as whistler-mode waves or ion-cyclotron waves. At very low frequencies, well below the ion-cyclotron frequency, waves tend to be linearly polarized and not amenable to this analysis.

A.3.5 CALCULATIONS OF THE POSITION OF THE SUN

In this section we present a simple subroutine to calculate the position of the sun in GEI coordinates. It is accurate for the years 1901 through 2099, to within 0.006°. The inputs are the year, the day of the year, and the seconds of the day in UT. The outputs are Greenwich mean sidereal time in degrees, the ecliptic longitude, and the apparent right ascension and declination of the sun in degrees. The listing of this program in FORTRAN follows. We note that the cartesian coordinates of the vector from the earth to the sun are

$$X = \cos(\text{SRASN})\cos(\text{SDEC})$$
$$Y = \sin(\text{SRASN})\cos(\text{SDEC})$$
$$Z = \sin(\text{SDEC})$$

A.3.5 POSITION OF THE SUN

```
      SUBROUTINE SUN (IYR, IDAY, SECS, GST, SLONG, SRASN, SDEC)
C     PROGRAM TO CALCULATE SIDEREAL, TIME AND POSITION OF THE SUN
C     GOOD FOR YEARS 1901 THROUGH 2099. ACCURACY 0.006 DEGREE
C     INPUT IS IYR, IDAY (INTEGERS), AND SECS, DEFINING UNIVERSAL TIME
C     OUTPUT IS GREENWICH MEAN SIDEREAL TIME (GST) IN DEGREES,
C     LONGITUDE ALONG ECLIPTIC (SLONG), AND APPARENT RIGHT ASCENSION
C     AND DECLINATION (SRASN, SDEC) OF THE SUN, ALL IN DEGREES.
      DATA RAD /57.29578/
      DOUBLE PRECISION DJ, FDAY
      IF (IYR.LT.1901.OR.IYR.GT.2099) RETURN
      FDAY = SECS/86400.
      DJ = 365*(IYR − 1900) + (IYR − 1901)/4 + IDAY + FDAY − 0.5D0
      T = DJ/36525.
      VL = DMOD(279.696678 + 0.9856473354*DJ, 360.D0)
      GST = DMOD(279.690983 + 0.9856473354* + 360.*FDAY + 180., 360.D0)
      G = DMOD(358.475845 + 0.985600267*DJ, 360.D0)/RAD
      SLONG = VL + (1.91946 − 0.004789*T)*SIN(G) + 0.020094* SIN(2.*G)
      OBLIQ = (23.45229 − 0.0130125*T)/RAD
      SLP = (SLONG − 0.005686)/RAD
      SIND = SIN(OBLIQ)*SIN(SLP)
      COSD = SQRT(1. − SIND**2)
      SDEC = RAD*ATAN(SIND/COSD)
      SRASN = 180. − RAD*ATAN2(COTAN(OBLIQ)*SIND/COSD, − COS(SLP)/COSD)
      RETURN
      END
```

REFERENCES

Akasofu, S.-I. 1964. The development of the auroral substorm. *Planet. Space Sci.* 12(4):273–82.

Akasofu, S.-I. 1968. *Polar and Magnetospheric Substorms.* Dordrecht: Reidel.

Akasofu, S.-I. 1977. *Physics of Magnetospheric Substorms.* Dordrecht: Reidel.

Akasofu, S.-I. 1979. Interplanetary energy flux associated with magnetospheric substorms. *Planet. Space Sci.* 27:425–31.

Akasofu, S.-I. 1980. The solar wind–magnetosphere energy coupling and magnetospheric disturbances. *Planet. Space Sci.* 28:495–509.

Akasofu, S.-I. 1981. Energy coupling between the solar wind and the magnetosphere. *Space Sci. Rev.* 28:121–90.

Akasofu, S.-I., and S. Chapman. 1972. *Solar-Terrestrial Physics.* Oxford: Clarendon Press.

Akasofu, S.-I., S. Chapman, and C.-I. Meng. 1965. The polar electrojet. *J. Atmos. Terr. Phys.* 27(11/12):1275–305.

Akasofu, S.-I., and C.-I. Meng. 1969. A study of polar magnetic substorms. *J. Geophys. Res.* 74(1):293–313.

Alfvén, H. 1942. Existence of electromagnetic-hydrodynamic waves. *Nature* 150:405.

Alfvén, H. 1957. On the theory of comet tails. *Tellus* 9:92–6.

Alfvén, H. 1968. Some properties of magnetospheric neutral surfaces. *J. Geophys. Res.* 73:4379.

Anders, E., and M. Ebihara. 1982. Solar-system abundance of the elements. *Geochim. Cosmochim. Acta* 46:2363.

Arnoldy, R. L. 1971. Signature in the interplanetary medium for substorms. *J. Geophys. Res.* 76(22):5189–201.

Atkinson, G. 1967. The current system of geomagnetic bays. *J. Geophys. Res.* 72(23):6063–7.

Aubry, M. P., and R. L. McPherron. 1971. Magnetotail changes in relation to the solar wind magnetic field and magnetospheric substorms. *J. Geophys. Res.* 76(19):4381–401.

Aubry, M. P., C. T. Russell, and M. G. Kivelson. 1970. Inward motion of the magnetopause before a substorm. *J. Geophys. Res.* 75(34):7018–31.

Axford, W. I. 1984. Magnetic field reconnection. In *Magnetic Reconnection in Space and Laboratory Plasmas,* Geophysical Monograph 30 (p. 1). Washington, DC: American Geophysical Union.

Baker, D. N., S.-I. Akasofu, W. Baumjohann, J. W. Bieber, D. H. Fairfield, E. W. Hones, Jr., B. Mauk, R. L. McPherron, and T. E. Moore. 1984. Substorms in the magnetosphere. In *Solar Terrestrial Physics: Present and Future* (pp. 8-1–8-55), NASA reference publication 1120.

Baker, D. N., R. C. Anderson, R. D. Zwickl, and J. A. Slavin. 1987. Average plasma and magnetic field variations in the distant magnetotail associated with near-earth substorm effects. *J. Geophys. Res.* 92(A1):71–81.

Baker, D. N., and R. L. McPherron. 1990. Extreme energy particle decreases near geostationary orbit: a manifestation of current diversion within the inner plasma sheet. *J. Geophys. Res.* 95(A5):6591–9.

Ballester, J. L., and E. R. Priest. 1987. A 2-D model for a solar prominence. *Solar Phys.* 109:335–50.

Banks, P. M., and T. E. Holzer. 1969. High-latitude plasma transport. *J. Geophys. Res.* 74:6317.

Banks, P. M., and G. Kockarts. 1973. *Aeronomy* (p. 785). New York: Academic Press.

Bargatze, L. F., D. N. Baker, R. L. McPherron, and E. W. Hones. 1985. Magnetospheric impulse response for many levels of geomagnetic activity. *J. Geophys. Res.* 90:6387–94.

Barker, F. S., D. R. Barraclough, V. P. Golovkov, P. J. Hood, F. J. Lowes, W. Mundt, N. W. Peddie, G.-z. Qi, S. P. Srivastava, R. Whitworth, D. E. Winch, T. Yukutake, and D. P. Zidarov. 1986. International Geomagnetic Reference Field Revision 1985. *EOS, Trans. Am. Geophys. Union* 67:523–4.

Basu, S., S. Basu, E. MacKenzie, W. R. Coley, J. R. Sharber, and W. R. Hoegy. 1990. Plasma structuring by the gradient drift instability at high latitudes and comparison with velocity shear driven processes. *J. Geophys. Res.* 95:7799–818.

Bateman, G. 1978. *MHD Instabilities*. Cambridge, MA: MIT Press.

Baumjohann, W., and K.-H. Glassmeier. 1986. The transient response mechanism and Pi 2 pulsations at substorm onset – review and outlook. *Planet. Space Sci.* 32:1361–70.

Baumjohann, W., G. Paschmann, and C. A. Cattell. 1989. Average plasma properties in the central plasma sheet. *J. Geophys. Res.* 94:6597.

Berchem, J., and C. T. Russell. 1984. Flux transfer events on the magnetopause: spatial distribution and controlling factors. *J. Geophys. Res.* 89:6689.

Bieber, J. W., E. C. Stone, E. W. Hones, Jr., D. N. Baker, and S. J. Bame. 1982. Plasma behavior during energetic electron streaming events: further evidence for substorm-associated magnetic reconnection. *Geophys. Res. Lett.* 9(6):664–7.

Biermann, L. 1951. Kometschwerfe und solare Korpuskularstrahlung. *Z. Astrophys.* 29:274.

Billings, D. G. 1966. *A Guide to the Solar Corona*. New York: Academic Press.

Birkeland, K. 1908. *The Norwegian Aurora Polaris Expedition, 1902–1903*, vol. 1, 1st sec. (pp. 1–315). Christiania (Oslo): H. Aschehoug & Co.

Birkeland, K. 1913. *The Norwegian Aurora Polaris Expedition, 1902–1903*, vol. 1, 2nd sec. (pp. 319–801). Christiana (Oslo): H. Aschehoug & Co.

Birn, J., and M. Hesse. 1991. The substorm current wedge and field-aligned currents in MHD simulations of magnetotail reconnection. *J. Geophys. Res.* 96:1611–18.

Birn, J., M. Hesse, and K. Schindler. 1989. Filamentary structure of a three-dimensional plasmoids. *J. Geophys. Res.* 94:241.

Bochsler, P., and J. Geiss. 1989. Composition of the solar wind. In *Solar System Plasma,* ed. J. H. Waite, Jr., J. L. Burch, and R. L. Moore (pp. 133–41). Washington, DC: American Geophysical Union.

Bostrom, R. 1964. A model of the auroral electrojets. *J. Geophys. Res.* 69:4963–99.

Bougher, S. W., R. E. Dickenson, E. C. Ridley, R. G. Robel, A. F. Nagy, and T. E. Cravens. 1986. Venus mesosphere and thermosphere: II. Global circulation, temperatures, and density variations. *Icarus* 68:285.

Brandt, J. C. 1970. *Introduction to the Solar Wind*. San Francisco: Freeman.

Brekke, A., and A. Egeland. 1994. *The Northern Lights, Their Heritage and Science,* Dreyer: Grøndahl.

Breneman, H. H., and E. C. Stone. 1985. Solar coronal and photospheric abundances from solar energetic particle measurements. *Astro. J.* 299:L57.

Browning, P. K., and E. R. Priest. 1984. The magnetic nonequilibrium of buoyant flux tubes in the solar corona. *Solar Phys.* 92:173–88.

Browning, P. K., and E. R. Priest. 1985. Heating of coronal arcades by magnetic tearing turbulence. *Astron. Astrophys.* 159:129–41.

Budden, K. E. 1985. *The Propagation of Radio Waves*. Cambridge University Press.
Burch, J. L. 1972. Preconditions for the triggering of polar magnetic substorms by storm sudden commencements. *J. Geophys. Res.* 77(28):5629–32.
Burton, R. K., R. L. McPherron, and C. T. Russell. 1975. An empirical relationship between interplanetary conditions and Dst. *J. Geophys. Res.* 80(31):4204–14.
Caan, M. N., R. L. McPherron, and C. T. Russell. 1975. Substorm and interplanetary magnetic field effects on the geomagnetic tail lobes. *J. Geophys. Res.* 80(1):191–4.
Caan, M. N., R. L. McPherron, and C. T. Russell. 1977. Characteristics of the association between the interplanetary magnetic field and substorms. *J. Geophys. Res.* 82(29):4837–42.
Cahill, L. J., and V. L. Patel. 1967. The boundary of the geomagnetic field, August to November 1961. *Planet. Space Sci.* 15:997–1033.
Calvert, W. 1981. The stimulation of auroral kilometric radiation by type III solar radio bursts. *Geophys. Res. Lett.* 8:1091–4.
Carlson, H. C. 1990. Dynamics of the polar cap. *J. Geomag. Geoelectr.* 42:697–710.
Carlson, H. C., R. A. Heelis, E. J. Weber, and J. R. Sharber. 1988. Coherent mesoscale convection patterns during northward interplanetary magnetic field. *J. Geophys. Res.* 93:14501–14.
Carpenter, D. L. 1970. Whistler evidence of the dynamic behavior of the duskside bulge in the plasmasphere. *J. Geophys. Res.* 75:3837.
Carpenter, D. L., and C. G. Park. 1973. On what ionospheric workers should know about the plasmapause-plasmasphere. *Rev. Geophys. Space Phys.* 11:133.
Carr, T. D., M. D. Desch, and J. K. Alexander. 1983. Phenomenology of magnetospheric radio emissions. In *Physics of the Jovian Magnetosphere*, ed. A. J. Dessler (pp. 226–316). Cambridge University Press.
Chamberlain, J. W. 1961. *Physics of the Aurora and Airglow*. New York: Academic Press.
Chapman, S., and J. Bartels. 1940. *Geomagnetism*. Oxford University Press.
Chapman, S., and J. Bartels. 1962. *Geomagnetism*, vol. 1. Oxford: Clarendon Press.
Chapman, S., and V. C. A. Ferraro. 1930. A new theory of magnetic storms. *Nature* 126:129.
Chapman, S., and V. C. A. Ferraro. 1931. A new theory of magnetic storms. *Terr. Magn. Atmosph. Elec.* 36:171–86.
Chapman, S., and V. C. A. Ferraro. 1932. A new theory of magnetic storms. *Terr. Magn. Atmosph. Elec.* 37:147–56.
Chappell, C. R. 1972. Recent satellite measurements of the morphology and dynamics of the plasmasphere. *Rev. Geophys. Space Phys.* 10:951–72.
Chen, L., and A. Hasegawa. 1974. A theory of long-period magnetic pulsations. 1. Steady state excitation of field line resonance. *J. Geophys. Res.* 79:1024.
Clauer, C. R., and Y. Kamide. 1985. DP 1 and DP 2 current systems for the March 22, 1979 substorms. *J. Geophys. Res.* 90(A2):1343–54.
Clauer, C. R., and R. L. McPherron. 1974. Mapping the local time–universal time development of magnetospheric substorms using mid-latitude magnetic observations. *J. Geophys. Res.* 79(19):2811–20.
Coroniti, F. V. 1985. Explosive tail reconnection: the growth and expansion phases of magnetospheric substorms. *J. Geophys. Res.* 90(A8):7427–47.
Coroniti, F. V., and C. F. Kennel. 1972. Changes in magnetospheric configuration during the substorm growth phase. *J. Geophys. Res.* 77(19)3361–70.
Coroniti, F. V., R. L. McPherron, and G. K. Parks. 1968. Studies of the magnetospheric substorm. 3. Concept of the magnetospheric substorm and its relation to electron precipitation and micropulsations. *J. Geophys. Res.* 73(5):1715–22.

Cowley, S. W. H. 1980. Plasma populations in a simple open model magnetosphere. *Space Sci. Rev.* 25:217.

Cowley, S. W. H. 1981. Magnetic asymmetries associated with the *y*-component of the IMF. *Planet. Space Sci.* 29:79.

Cowley, S. W. H. 1982. The causes of convection in the earth's magnetosphere: a review of developments during the IMS. *Rev. Geophys. Space Phys.* 20:531.

Cowley, S. W. H. 1984. Solar wind control of magnetospheric convection. In *Proc. Conf. Achievements IMS,* ESA SP-217 (p. 483) Noordwijk: ESA.

Cowley, S. W. H. 1986. Magnetic reconnection. In *Solar System Magnetic Fields,* ed. E. R. Priest (p. 121). Dordrecht: Reidel.

Cowley, S. W. H. 1991. The structure and length of tail-associated phenomena in the solar wind downstream from the earth. *Planet. Space Sci.* 39:1039.

Cowley, S. W. H., and Z. V. Lewis. 1990. Magnetic trapping of energetic particles on open dayside boundary layer flux tubes. *Planet. Space Sci.* 38:1343.

Cowley, S. W. H., and M. Lockwood. 1992. Excitation and decay of solar wind-driven flow in the magnetosphere-ionosphere system. *Ann. Geophys.* 10:103–15.

Cowling, T. G. 1957. *Magnetohydrodynamics.* New York: Interscience.

Cravens, T. E., H. Shinagawa, and A. F. Nagy. 1984. The evolution of large-scale magnetic fields in the ionosphere of Venus. *Geophys. Res. Lett.* 11:267.

Crook, W. R., E. C. Stone, and R. E. Vogt. 1984. Elemental composition of solar energetic particles. *Astrophys. J.* 279:827.

Crooker, N. U. 1977. The magnetospheric boundary layers: a geometrically explicit model. *J. Geophys. Res.* 82:3629.

Crooker, N. U. 1979. Dayside merging and cusp geometry. *J. Geophys. Res.* 84:951.

Crooker, N. U., J. Feynman, and J. T. Gosling. 1977. On the high correlation between long-term averages of solar wind speed and geomagnetic activity. *J. Geophys. Res.* 82(13):1933–7.

Davis, T. N., and M. Sugiura. 1966. Auroral electrojet activity index AE and its universal time variations. *J. Geophys. Res.* 71(3):785–801.

Demoulin, P., and E. R. Priest. 1988. Instability of a prominence supported in a linear force-free field. *Astron. Astrophys.* 206:336–47.

Dessler, A. J., and E. N. Parker. 1959. Hydromagnetic theory of magnetic storms. *J. Geophys. Res.* 64(12):2239–59.

Dungey, J. W. 1954a. Electrodynamics of the outer atmosphere. Pennsylvania State University Ionosphere Research Laboratory Report 69.

Dungey, J. W. 1954b. The propagation of Alfvén waves through the ionosphere. Pennsylvania State University Ionosphere Research Laboratory Science Report 57.

Dungey, J. W. 1961. Interplanetary magnetic field and the auroral zones. *Phys. Rev. Lett.* 6:47.

Dungey, J. W. 1963a. The structure of the exosphere or adventures in velocity space. In *Geophysics, The Earth's Environment,* ed. C. Dewitt, J. Hieblot, and A. Lebeau (p. 503). New York: Gordon & Breach.

Dungey, J. W. 1963b. Hydromagnetic waves and the ionosphere. In *Proc. Int. Conf. Ionosphere,* Institute of Physics, London (p. 230).

Dungey, J. W. 1965. The length of the magnetospheric tail. *J. Geophys. Res.* 70:1753.

Dungey, J. W. 1967. Hydromagnetic waves. In *Physics of Geomagnetic Phenomena,* ed. S. Matsushita and W. H. Campbell (p. 913). New York: Academic Press.

Eastman, T. E., L. A. Frank, and C. Y. Huang. 1985. The boundary layers as the primary transport regions of the earth's magnetotail. *J. Geophys. Res.* 90(A10):9541–60.

Eastman, T. E., L. A. Frank, W. K. Peterson, and W. Lennartsson. 1984. The plasma sheet boundary layer. *J. Geophys. Res.* 89:1553.

Eather, R. H. 1980. *Majestic Lights: The Aurora in Science, History, and the Arts.* Washington, DC: American Geophysical Union.

Egeland, A., O. Holter, and A. Omholt. 1973. *Cosmical Geophysics.* Oslo: Universiteisforlaget.

Einaudi, G., and G. Van Hoven. 1983. *Solar Phys.* 88:163–78.

Elphic, R. C. 1990. Observations of flux transfer events: Are FTE's flux ropes, islands or surface waves? In *Physics of Magnetic Flux Ropes,* Geophysical Monograph 58, ed. C. T. Russell, E. R. Priest, and L. C. Lee (p. 455). Washington, DC: American Geophysical Union.

Elphic, R. C., and S. P. Gary. 1990. ISEE observations of low frequency waves and ion distribution function evolution in the plasma sheet boundary layers. *Geophys. Res. Lett.* 17:2023.

Elphic, R. C., C. T. Russell, J. A. Slavin, and L. H. Brace. 1980. Observations of the dayside ionopause and ionosphere of Venus. *J. Geophys. Res.* 85:7679.

Elphinstone, R. D., J. S. Murphree, L. L. Cogger, D. Hearn, and M. G. Henderson. 1991. Observations of changes to the auroral distributions prior to substorm onset. In *Magnetospheric Substorms,* Geophysical Monograph 64. Washington, DC: American Geophysical Union.

Engle, I. M. 1991. Idealized *Voyager* Jovian magnetosphere shape and field. *J. Geophys. Res.* 96:7793–802.

Erickson, G. M. 1984. On the cause of x-line formation in the near-earth plasma sheet: results of adiabatic convection of plasma-sheet plasma. In *Magnetic Reconnection in Space and Laboratory Plasmas,* Geophysical Monograph 30, ed. E. W. Hones (p. 269). Washington, DC: American Geophysical Union.

Erickson, G. M., R. W. Spiro, and R. A. Wolf. 1991. The physics of the Harang discontinuity. *J. Geophys. Res.* 96:1633.

Erickson, G. M., and R. A. Wolf. 1980. Is steady convection possible in the earth's magnetotail? *Geophys. Res. Lett.* 7:897.

Fairfield, D. H. 1987. Structure of the geomagnetic tail. In *Magnetotail Physics,* ed. A. T. Y. Lui (pp. 23–33). Baltimore: Johns Hopkins University Press.

Fairfield, D. H., and L. J. Cahill, Jr. 1966. Transition region magnetic field and polar magnetic disturbances. *J. Geophys. Res.* 71:155.

Fairfield, D. H., and G. D. Mead. 1975. Magnetospheric mapping with a quantitative geomagnetic field model. *J. Geophys. Res.* 80(4):535–42.

Farrugia, C. J., M. P. Freeman, S. W. H. Cowley, D. J. Southwood, M. Lockwood, and A. Etemadi. 1989. Pressure driven magnetopause motions and attendant response on the ground. *Planet. Space Sci.* 37:589.

Fedder, J. A., J. G. Lyon, and J. L. Giuliani, Jr. 1986. Numerical simulations of comets: predictions for comet Giacobini-Zinner. *EOS, Trans. Am. Geophys. Union* 67:17.

Feldman, W. C., J. R. Asbridge, S. J. Bame, and J. T. Gosling. 1977. Plasma and magnetic fields from the sun. In *The Solar Output and Its Variations,* ed. O. R. White (pp. 351–82). Boulder: Colorado Associated University Press.

Feldstein, Y. I., and G. V. Starkov. 1967. Dynamics of auroral belt and polar geomagnetic disturbances. *Planet. Space Sci.* 15:209.

Forbes, T. G., and E. R. Priest. 1983. On reconnection and plasmoids in the geomagnetic tail. *J. Geophys. Res.* 88:863.

Forbes, T. G., and E. R. Priest. 1987. A comparison of analytical and numerical models for steadily driven magnetic reconnection. *Rev. Geophys.* 25:1583.

Frank, L. A., and J. D. Craven. 1988. Imaging results from Dynamics Explorer 1. *Rev. Geophys. Space Phys.* 26:249–83.

Frank L. A., J. D. Craven, D. A. Gurnett, S. D. Shawhan, D. R. Weimer, J. L. Burch, J. D. Winningham, C. R. Chappell, J. H. Waite, R. A. Heeles, N. C. Maynard, M. Sugiura, W. K. Peterson, and E. G. Shelly. 1986. The theta aurora. *J. Geophys. Res.* 91:3177–224.

Garrett, H. B., D. C. Schwank, and S. E. DeForest. 1981a. A statistical analysis of the low-energy geosynchronous plasma environment. I. Electrons. *Planet. Space Sci.* 29:1021–44.

Garrett, H. B., D. C. Schwank, and S. E. DeForest. 1981b. A statistical analysis of the low-energy geosynchronous plasma environment. II. Ions. *Planet. Space Sci.* 29:1045–60.

Gilbert, W. 1893. *De Magnete,* trans. P. Fleury Mottelay. Reprinted 1958, New York: Dover.

Goertz, C. K., and W. Baumjohann. 1991. On the thermodynamics of the plasma sheet. *J. Geophys. Res.* 96:20991.

Goertz, C. K., and R. A. Smith. 1989. Thermal catastrophe model of substorms. *J. Geophys. Res.* 94:6581.

Gosling, J. T., S. J. Bame, E. J. Smith, and M. E. Burton. 1988. Forward–reverse shock pairs associated with transient disturbances in the solar wind at 1 AU. *J. Geophys. Res.* 93:8741.

Gosling, J. T., M. F. Thomsen, S. J. Bame, R. C. Elphic, and C. T. Russell. 1990a. Plasma flow reversals at the dayside magnetopause and the origin of asymmetric polar cap convection. *J. Geophys. Res.* 95:8073.

Gosling, J. T., M. F. Thomsen, S. J. Bame, T. G. Onsager, and C. T. Russell. 1990b. The electron edge of the low latitude boundary layer during accelerated flow events. *Geophys. Res. Lett.* 17:1833.

Grebowsky, J. M. 1970. Model study of plasmapause motion. *J. Geophys. Res.* 75:4329–33.

Grebowsky, J. M., Y. K. Tulanay, and A. J. Chen. 1974. Temporal variations in the dawn and dusk midlatitude trough and plasmapause position. *Planet. Space Sci.* 22:1089.

Green, J. L., and J. L. Horwitz. 1986. Destiny of earthward streaming plasma in the plasma sheet boundary layer. *Geophys. Res. Lett.* 13:76.

Gurnett, D. A., R. R. Shaw, R. R. Anderson, and W. S. Kurth. 1979. Whistlers observed by *Voyager 1:* detection of lightning on Jupiter. *Geophys. Res. Lett.* 6:511–14.

Haerendel, G., G. Paschmann, N. Sckopke, H. Rosenbauer, and P. C. Hedgecock. 1978. The frontside boundary layer of the magnetosphere and the problem of reconnection. *J. Geophys. Res.* 83:3195.

Hanson, W. B., and G. P. Mantas. 1988. *Viking* electron temperature measurements: evidence for a magnetic field in the Martian ionosphere. *J. Geophys. Res.* 93:7538.

Hanson, W. B., S. Sanatani, and D. R. Zuccaro. 1977. The Martian ionosphere as observed by the *Viking* retarding potential analyzers. *J. Geophys. Res.* 82:4351.

Harel, M., R. A. Wolf, P. H. Reiff, R. W. Spiro, W. J. Burke, F. J. Rich, and M. Smiddy. 1981. Quantitative simulation of a magnetospheric substorm. 1. Model logic and overview. *J. Geophys. Res.* 86:2217–41.

Harris, E. G. 1962. On a plasma sheet separating regions of oppositely directed magnetic field. *Nuovo Cim.* 23:115.

Hartle, R. E., H. A. Taylor, Jr., S. J. Bauer, L. H. Brace, C. T. Russell, and R. E. Daniell, Jr. 1980. Dynamical response of the dayside ionosphere of Venus to the solar wind. *J. Geophys. Res.* 85:7739.

Hartz, T. R. 1971. Particle precipitation patterns. In *The Radiating Atmosphere,* ed. B. M. McCormac (pp. 225–38). Dordrecht: Reidel.

Heelis, R. A. 1988. Studies of ionospheric plasma and electrodynamics and their application to ionosphere-magnetosphere coupling. *Rev. Geophys. Space Phys.* 26:317–28.

Helliwell, R. A. 1988. VLF wave-injection experiments from Siple Station, Antarctica. *Adv. Space Res.* 8:279.

Heppner, J. P., and N. C. Maynard. 1987. Empirical high latitude electric field models. *J. Geophys. Res.* 92:4467.

Heyvaerts, J., and E. R. Priest. 1983. Coronal heating by phase-mixed Alfvén waves. *Astron. Astrophys.* 117:220–34.

Heyvaerts, J., and E. R. Priest. 1984. Coronal heating by reconnection in DC current systems. *Astron. Astrophys.* 137:63–78.

Hill, T. W. 1975. Magnetic merging in a collisionless plasma. *J. Geophys. Res.* 80:4689.

Hirshberg, J., and D. S. Colburn. 1969. Interplanetary field and geomagnetic variations – a unified view. *Planet. Space Sci.* 17:1183–206.

Hollweg, J. V. 1984. Resonances of coronal loops. *Astrophys. J.* 277:392–403.

Holzer, T. E. 1989. Interaction between the solar wind and the interstellar medium. *Ann. Rev. Astron. Astrophys.* 27:199.

Holzworth, R. H., and C.-I. Meng. 1975. Mathematical representation of the auroral oval. *Geophys. Res. Lett.* 2(9):377–80.

Hones, E. W., Jr. 1976. The magnetotail: its generation and dissipation. In *Physics of Solar Planetary Environments,* ed. D. J. Williams (p. 558). Washington, DC: American Geophysical Union.

Hones, E. W., Jr. 1977. Substorm processes in the magnetotail: comments on "On hot tenuous plasmas, fireballs, and boundary layers in the earth's magnetotail" by L. A. Frank, K. L. Ackerson, and R. P. Lepping. *J. Geophys. Res.* 82(35):5633–40.

Hones, E. W., Jr. 1979. Transient phenomena in the magnetotail and their relation to substorms. *Space Sci. Rev.* 23:393.

Hones, E. W., Jr., C. D. Anger, J. Birn, J. S. Murphree, and L. L. Cogger. 1987. A study of a magnetospheric substorm recorded by the *Viking* auroral imager. *Geophys. Res. Lett.* 14(4):411–14.

Hood, A. W., and E. R. Priest. 1979. Kink instability of solar coronal loops as the cause of small flares. *Solar Phys.* 64:303–21.

Hood, A. W., and E. R. Priest. 1980. Magnetic instability of coronal arcades as the origin of two-ribbon solar flares. *Solar Phys.* 66:113–34.

Huang, C. Y., and L. A. Frank. 1986. A statistical study of the central plasma sheet. *Geophys. Res. Lett.* 13:652–5.

Hughes, W. J. 1983. Hydromagnetic waves in the magnetosphere. In *Solar Terrestrial Physics,* ed. R. L. Carovillano and J. M. Forbes (p. 453). Dordrecht: Reidel.

Hughes, W. J., and D. G. Sibeck. 1987. On the three dimensional structure of plasmoids. *Geophys. Res. Lett.* 14:636.

Hundhausen, A. J. 1972. *Coronal Expansion and Solar Wind.* Berlin: Springer-Verlag.

Hundhausen, A. J. 1977. An interplanetary view of coronal holes. In *Coronal Holes and High Speed Wind Streams,* ed. J. B. Zirker (pp. 225–319). Boulder: Colorado Associated University Press.

Iijima, T., and T. Nagata. 1972. Signatures for substorm development of the growth phase and expansion area. *Planet. Space Sci.* 20:1095–112.

Iijima, T., and T. A. Potemra. 1978. Large-scale characteristics of field-aligned currents associated with substorms. *J. Geophys. Res.* 83:599.

Ip, W. H., and W. I. Axford. 1982. Theories of physical processes in the cometary comae and ion tails. In *Comets,* ed. L. L. Wilkening (p. 588). Tucson: University of Arizona Press.

Jackson, J. D. 1962. *Classical Electrodynamics.* New York: Wiley.

Jackson, J. D. 1975. *Classical Electrodynamics,* 2nd ed. New York: Wiley.

Jackson, D. J., and D. B. Beard. 1977. The magnetic field of Mercury. *J. Geophys. Res.* 82:2828–36.

Jacobs, J. A., Y. Kato, S. Matsushita, and V. A. Troitskaya. 1964. Classification of geomagnetic micropulsations. *J. Geophys. Res.* 69:180.

Jacquey, C., J. A. Sauvaud, and J. Dandouras. 1991. Location and propagation of the magnetotail current disruption during substorm expansion: analysis and simulation of an ISEE multi-onset event. *Geophys. Res. Lett.* 18(3):389–92.

Johnson, C. Y., 1969. Ion and neutral composition of the ionosphere. *Annals of the IQSY.* 5:197–213.

Kamide, Y. 1988. *Electrodynamic Processes in the Earth's Ionosphere and Magnetosphere.* Kyoto: Kyoto Sangyo University Press.

Kan, J. R. 1990. Developing a global model of magnetospheric substorms. *EOS, Trans. Am. Geophys. Union* 71(38):1086–7.

Kan, J. R., and S.-I. Akasofu. 1982. Dynamo process governing solar wind–magnetosphere energy coupling. *Planet. Space Sci.* 30(4):367–70.

Kan, J. R., and S.-I. Akasofu. 1988. A theory of substorms: onset and subsidence. *J. Geophys. Res.* 93:5624–40.

Kaufmann, R. I. 1987. Substorm currents: growth phase and onset. *J. Geophys. Res.* 92(A7):7471–86.

Kavanagh, L. D., Jr., J. W. Freeman, Jr., and A. J. Chen. 1968. Plasma flow in the magnetosphere. *J. Geophys. Res.* 73:5511.

Khurana, K. K., and M. G. Kivelson. 1989. Ultralow frequency MHD waves in Jupiter's middle magnetosphere. *J. Geophys. Res.* 94:5241.

Kidd, S. R., and G. Rostoker. 1991. Distribution of auroral surges in the evening sector. *J. Geophys. Res.* 96(A4):5697–706.

Kippenhahn, R., and A. Schlüter. 1957. Eine Theorie der solaren Filamente. *Z. Astrophys.* 3:36–62.

Kivelson, M. G., and W. J. Hughes. 1990. On the threshold for triggering substorms. *Planet. Space Sci.* 38:211.

Kivelson, M. G., J. A. Slavin, and D. J. Southwood. 1979. Magnetospheres of the Galilean satellites. *Science* 205:491.

Kivelson, M. G., and D. J. Southwood. 1986. Coupling of global magnetospheric MHD eigenmodes to field line resonances. *J. Geophys. Res.* 91:4345.

Kivelson, M. G., and H. E. Spence. 1988. On the possibility of quasi-static convection in the quiet magnetotail. *Geophys. Res. Lett.* 15:1541.

Knudsen, W. C., K. Spenner, K. L. Miller, and V. Novak. 1980. Transport of ionospheric O^+ ions across the Venus terminator and implications. *J. Geophys. Res.* 85:7803.

Kokubun, S. 1972. Relationship of interplanetary magnetic field structure with development of substorm and storm main phase. *Planet. Space Sci.* 20(7):1033–49.

Kokubun, S., R. L. McPherron, and C. T. Russell. 1976. Ogo 5 observations of Pc 5 waves: ground–magnetosphere correlations. *J. Geophys. Res.* 81:5141.

Kokubun, S., R. L. McPherron, and C. T. Russell. 1977. Triggering substorms by solar wind discontinuities. *J. Geophys. Res.* 82(1):74–86.

Kokubun, S., and T. Nagata. 1965. Geomagnetic pulsation Pc 5 in and near the auroral zones. *Ion Space Res. Japan* 19:158.

Kozyra, J. U., C. E. Valladares, H. C. Carlson, M. J. Buonsanto, and D. W. Slater. 1990. A theoretical study of the seasonal and solar cycle variations of stable auroral red arcs. *J. Geophys. Res.* 95:12219.

Kuperus, M., and M. A. Raadu. 1974. The support of prominences formed in neutral sheets. *Astron. Astrophys.* 31:189–93.

Kurth, W. S. 1991. *Voyager* plasma wave observations near the outer planets. *Adv. Space Res.* 11:(9)59–68.

Kurth, W. S., and D. A. Gurnett. 1991. Plasma waves in planetary magnetospheres. *J. Geophys. Res.* 96:18977–91.

Landau, L. D., and E. M. Lifshitz. 1960. *Mechanics*. Reading, MA: Addison-Wesley.

Lee, L. C. 1991. The magnetopause, a tutorial review. In *Physics of Space Plasmas (1990)*, ed. T. Chang, G. B. Crew, and J. R. Jasperse (p. 33). Cambridge, MA: Scientific Publishers.

Lee, L. C., and Z. F. Fu. 1985. A theory of magnetic flux transfer at the earth's magnetopause. *Geophys. Res. Lett.* 12:105.

Lemaire, J., M. J. Rycroft, and M. Roth. 1979. Control of impulsive penetration of solar wind irregularities into the magnetosphere by the interplanetary magnetic field direction. *Planet. Space Sci.* 27:47.

Lennartsson, W., and R. D. Sharp. 1982. A comparison of 0.1–17 keV/e ion composition in the near equatorial magnetosphere between quiet and disturbed conditions. *J. Geophys. Res.* 87:6109.

Leroy, J. L. 1989. Observation of prominence magnetic fields. In *Dynamics and Structure of Quiescent Solar Prominences*, ed. E. R. Priest (pp. 77–114). Dordrecht: Kluwer.

Lincoln, J. V. 1967. Geomagnetic indices. In *Physics of Geomagnetic Phenomena*, vol. 1, ed. S. Matsushita and W. H. Campbell (pp. 67–100). New York: Academic Press.

Linker, J. A., M. G. Kivelson, and R. J. Walker. 1988. An MHD simulation of plasma flow past Io, Alfvén and slow mode perturbations. *Geophys. Res. Lett.* 15:1311.

Lopez, R. E., D. N. Baker, A. Y. T. Lui, D. G. Sibeck, R. D. Belian, R. W. McEntire, T. A. Potemra, and S. M. Krimigis. 1988a. The radial and longitudinal propagation characteristics of substorm injections. *Adv. Space Res.* 8(9–10):(9)91–5.

Lopez, R. E., and A. T. Y. Lui. 1990. A multi-satellite case study of the expansion of a substorm current wedge in the near-earth magnetotail. *J. Geophys. Res.* 95(A6):8009–17.

Lopez, R. E., A. T. Y. Lui, D. G. Sibeck, R. W. McEntire, L. J. Zanetti, T. A. Potemra, and S. M. Krimigis. 1988b. The longitudinal and radial distribution of magnetic reconfigurations in the near-earth magnetotail as observed by AMPTE/CCE. *J. Geophys. Res.* 93(A2):997–1001.

Lopez, R. E., D. G. Sibeck, A. T. Y. Lui, K. Takahashi, R. W. McEntire, and T. A. Potemra. 1988c. Substorm variations in the magnitude of the magnetic field: AMPTE/CCE observations. *J. Geophys. Res.* 93(A12):14444–52.

Luhmann, J. G. 1977. Auroral bremsstrahlung spectra in the atmosphere. *J. Atmos. Terr. Phys.* 39:595.

Luhmann, J. G. 1986. The solar wind interaction with Venus. *Space Sci. Rev.* 44:241–306.

Luhmann, J. G., 1991. The solar wind interaction with Venus and Mars, Cometary analogies and contrasts. In *Cometary Plasma Processes, Geophysical Monograph 61*, ed. A. D. Johnstone (p. 7). Washington, DC: American Geophysical Union.

Luhmann, J. G., and L. H. Brace, 1991. Near-Mars Space. *Rev. Geophys.* 29:121–40.

Luhmann, J. G., and R. R. Elphic, 1985. On the dynamo generation of flux ropes in the Venus ionosphere. *J. Geophys. Res.* 90:12047–56.

Luhmann, J. G., C. T. Russell, F. L. Scarf, L. H. Brace, and W. C. Knudsen, 1987. Characteristics of the Marslike limit of the Venus-solar wind interaction. *J. Geophys. Res.* 92:8545–57.

Lui, A. T. Y. 1991a. A synthesis of magnetospheric substorm models. *J. Geophys. Res.* 96(A2):1849–56.

Lui, A. T. Y. 1991b. Extended consideration of a synthesis model for magnetospheric substorms. In *Magnetospheric Substorms,* Geophysical Monograph 64, ed. J. Kan, T. A. Potemra, S. Kokubun, and T. Iijima (pp. 43–60). Washington, DC: American Geophysical Union.

Lui, A. T. Y., R. E. Lopez, S. M. Krimigis, R. W. McEntire, L. J. Zanetti, and T. A. Potemra. 1988. A case study of magnetotail current sheet disruption and diversion. *Geophys. Res. Lett.* 15(7):721–4.

Lui, A. T. Y., A. Mankofsky, C.-L. Chang, K. Papadopolous, and C. S. Wu. 1990. A current disruption mechanism in the neutral sheet: a possible trigger for substorm expansions. *Geophys. Res. Lett.* 17(6):745–8.

McPherron, R. L. 1970. Growth phase of magnetospheric substorms. *J. Geophys. Res.* 75(28):5592–9.

McPherron, R. L. 1979. Magnetospheric substorms. *Rev. Geophys. Space Phys.* 17(4):657–81.

McPherron, R. L. 1991. Physical processes producing magnetospheric substorms and magnetic storms. In *Geomagnetism,* vol. 4, ed. J. Jacobs (pp. 593–739). London: Academic Press.

McPherron, R. L., D. N. Baker, L. F. Bargatze, C. R. Clauer, and R. E. Holzer. 1988. IMF control of geomagnetic activity. *Adv. Space Res.* 8(9):71–86.

McPherron, R. L., and R. H. Manka. 1985. Dynamics of the 1054 UT, March 22, 1979 substorm event: CDAW-6. *J. Geophys. Res.* 90(A2):1175–90.

McPherron, R. L., A. Nishida, and C. T. Russell. 1987. Is near-earth current sheet thinning the cause of auroral substorm expansion onset? In *Quantitative Modeling of Magnetosphere-Ionosphere Coupling Processes,* ed. Y. Kamide and R. A. Wolf (pp. 252–7). Kyoto: Kyoto Sangyo University Press.

McPherron, R. L., C. T. Russell, and M. Aubry. 1973. Satellite studies of magnetospheric substorms on August 15, 1978. 9. Phenomenological model for substorms. *J. Geophys. Res.* 78(16):3131–49.

Maezawa, K., and T. Murayama. 1986. Solar wind velocity effects on the auroral zone magnetic disturbances. In *Solar Wind Magnetosphere Coupling,* ed. Y. Kamide and J. Slavin (pp. 59–83). Tokyo: Terra Scientific Publishing.

Malherbe, J. M., and E. R. Priest. 1983. Current sheet models for solar prominences. *Astron. Astrophys.* 123:80–8.

Manka, R. H., and F. C. Michel. 1973. Lunar ion energy spectra and surface potential. In *Proc. Fourth Lunar Science Conf.,* vol. 3 (p. 2897). Cambridge, MA: MIT Press.

Martin, S. F. 1986. Recent observations of the formation of filaments. In *Coronal and Prominence Plasmas,* ed. A. Poland (pp. 73–80), NASA CP-2422.

Maxwell, J. C. 1873. *Treatise on Electricity and Magnetism.* Edinburgh.

Mayaud, P. N. 1980. *Derivation, Meaning and Use of Geomagnetic Indices,* Geophysical Monograph 22. Washington, DC: American Geophysical Union.

Mead, G. D., and D. B. Beard. 1964. Shape of the geomagnetic field–solar wind boundary. *J. Geophys. Res.* 69:1169.

Melrose, D. B. 1986. *Instabilities in Space and Laboratory Plasmas.* Cambridge University Press.

Menvielle, M., and A. Berthelier. 1991. The K-derived planetary indices: description and availability. *Rev. Geophys.* 29:415–32.

Midgely, J. E., and L. Davis, Jr. 1963. Calculation by a moment technique of the perturbation of the geomagnetic field by the solar wind. *J. Geophys. Res.* 68:5111.

Mikic, Z., D. C. Barnes, and D. D. Schnack. 1988. Dynamic evolution of a solar coronal magnetic field arcade. *Astrophys. J.* 328:830–47.

Mikic, Z., D. D. Schnack, and G. Van Hoven. 1989. Creation of current filaments in the solar corona. *Astrophys. J.* 338:1148.

Milne, A., E. R. Priest, and B. Roberts. 1979. A model for quiescent prominences. *Astrophys. J.* 232:304–17.

Miura, A., and T. Sato. 1980. Numerical simulation of global formation of auroral arcs. *J. Geophys. Res.* 85:73–91.

Moldwin, M. B., and W. J. Hughes. 1992. On the formation and evolution and plasmoids: a survey of the ISEE 3 geotail data. *J. Geophys. Res.* 97:19259.

Murayama, T., and K. Hakamada. 1975. Effects of solar wind parameters on the development of magnetospheric substorms. *Planet. Space Sci.* 23:75–91.

Murphree, J. S., and L. L. Cogger. 1992. Observations of substorm onset. In *Proceedings of the First International Conference on Substorms,* ESA SP-335. Noordwijk: European Space Agency.

Murphree, J. S., R. D. Elphinstone, L. L. Cogger, and D. Hearn. 1991. *Viking* optical substorm signatures. In *Magnetospheric Substorms,* Geophysical Monograph 64, ed. J. Kan, T. A. Potemra, S. Kokubun, and T. Iijima (pp. 241–55). Washington, DC: American Geophysical Union.

Nagai, T., J. H. Waite, Jr., J. L. Green, C. T. Chappell, R. C. Olsen, and R. H. Comfort. 1984. First measurement of supersonic polar wind in the polar magnetosphere. *Geophys. Res. Lett.* 11:669.

Nagy, A. F., and T. E. Cravens. 1988. Hot oxygen atoms in the upper atmospheres of Venus and Mars. *Geophys. Res. Lett.* 15:433.

Ness, N. F. 1987. Magnetotail research: the early years. In *Magnetotail Physics,* ed. A. T. Y. Lui (p. 11). Baltimore: Johns Hopkins University Press.

Ness, N. F., M. H. Acuna, K. W. Behannon, and F. M. Neubauer. 1982. The induced magnetosphere of Titan. *J. Geophys. Res.* 87:1369.

Neugebauer, M., and C. W. Snyder. 1966. *Mariner 2* observations of the solar wind. 1. Average Properties. *J. Geophys. Res.* 71:4469.

Newkirk, G., Jr. 1967. Structure of the solar corona. *Ann. Rev. Astron. Astrophys.* 5:213.

Noyes, R. W. 1982. *The Sun, Our Star.* Harvard University Press.

Ogino, T., R. J. Walker, and M. Ashour-Abdalla. 1992. Global magnetohydrodynamic simulation of the magnetosheath and magnetosphere when the interplanetary magnetic field is northward. *IEEE Trans. Plasma Sci.* 20:1.

Ogino, T., R. J. Walker, M. Ashour-Abdalla, and J. M. Dawson. 1986. An MHD simulation of the effects of the interplanetary magnetic field by component on the interaction of the solar wind with the earth's magnetosphere during southward IMF. *J. Geophys. Res.* 91:10029.

Ohtani, S., S. Kokubun, and C. T. Russell. 1992. Radial expansion of the tail current disruption during substorms: a new approach to the substorm onset region. *J. Geophys. Res.* 97:3129–36.

Olsen, R. C., S. D. Shawhan, D. L. Gallagher, J. L. Green, C. R. Chappell, and R. R. Anderson. 1987. Plasma observations at the earth's magnetic equator. *J. Geophys. Res.* 92:2385–407.

Omholt, A. 1971. *The Optical Aurora.* Berlin: Springer-Verlag.

Onsager, T. G., M. F. Thomsen, R. C. Elphic, and J. T. Gosling. 1991. Model of electron and ion distributions in the plasma sheet boundary layer. *J. Geophys. Res.* 96:20999.

Opgenoorth, H. J., J. Oksman, K. U. Kaila, E. Nielsen, and W. Baumjohann. 1983. Characteristics of eastward drifting omega bands in the morning sector of the auroral oval. *J. Geophys. Res.* 88:9171–85.

Orlowski, D. S., C. T. Russell, and R. P. Lepping. 1992. Wave phenomena in the upstream region of Saturn. *J. Geophys. Res.* 97:19187–99.

Parker, E. N. 1957. Sweet's mechanism for merging magnetic fields in conducting fluids. *J. Geophys. Res.* 62:509.

Parker, E. N. 1958. Dynamics of the interplanetary gas and magnetic fields. *Astrophys. J.* 128:664–76.

Parker, E. N. 1963. *Interplanetary Dynamical Processes.* New York: Wiley-Interscience.

Parker, E. N. 1972. Topological dissipation and small-scale fields in turbulent gases. *Astrophys. J.* 174:499–510.

Parkinson, W. D. 1983. *Introduction to Geomagnetism.* Amsterdam: Elsevier.

Paschmann, G., I. Papamastorakis, W. Baumjohann, N. Sckopke, C. W. Carlson, B. U. O. Sonnerup, and H. Luhr. 1986. The magnetopause for large magnetic shear: AMPTE/IRM observations. *J. Geophys. Res.* 91:11099.

Paschmann, G., B. U. O. Sonnerup, I. Papamastorakis, N. Sckopke, G. Haerendel, J. R. Ashbridge, S. J. Bame, J. T. Gosling, and C. T. Russell. 1982. Plasma and magnetic field characteristics of magnetic flux transfer events. *J. Geophys. Res.* 87:2159.

Paulikas, G. A. 1974. Tracing of high-latitude magnetic field lines by solar particles. *Rev. Geophys. Space Phys.* 12:117–28.

Perraut, S., R. Gendrin, P. Robert, A. Roux, and C. de Villedary. 1978. ULF waves observed with magnetic and electric sensors on *GEOS-1*. *Space Sci. Rev.* 22: 347.

Perreault, P., and S. -I. Akasofu. 1978. A study of geomagnetic storms. *Geophys. J. R. Astr. Soc.* 54:547–73.

Petschek, H. E. 1964. Magnetic field annihilation. In *The Physics of Solar Flares,* ed. W. N. Hess, NASA SP-50 (p. 425). Washington, DC: NASA.

Phillips, J. L., J. G. Luhmann, and C. T. Russell. 1984. Growth and maintenance of large-scale magnetic fields in the dayside Venus ionosphere. *J. Geophys. Res.* 89:10676.

Pilipp, W. G., and G. Morfill. 1978. The formation of the plasma sheet resulting from plasma mantle dynamics. *J. Geophys. Res.* 83:5670.

Pizzo, V. J. 1985. Interplanetary shocks on the large scale: a retrospective on the last decade's theoretical efforts. In *Collisionless Shocks in the Heliosphere: Reviews of Current Research,* ed. B. T. Tsurutani and R. G. Stone (pp. 51–68). Washington, DC: American Geophysical Union.

Pizzo, V. J., T. Holzer, and D. G. Sime (eds.). 1987. *Proceedings of the Sixth International Solar Wind Conference,* NCAR technical note NCAR/TN-3064. Boulder: National Center for Atmospheric Research.

Pneuman, G. W., and R. A. Kopp. 1971. Gas–magnetic field interactions in the solar corona. *Solar Phys.* 18:258.

Podgorny, I. M. 1976. Laboratory experiments: intrusion into the magnetic field. In *Physics of Solar Planetary Environment,* ed. D. J. Williams (pp. 241–54). Washington, DC: American Geophysical Union.

Poedts, S., and M. Goossens. 1987. The continuous spectrum of MHD waves in 2D solar loops and arcades. *Solar Phys.* 109:265–86.

Poland, A. (ed.). 1986. *Coronal and Prominence Plasmas,* NASA CP-2422. Washington, DC: NASA.

Pontius, D. H., Jr., and R. A. Wolf. 1990. Transient flux tubes in the terrestrial magnetosphere. *Geophys. Res. Lett.* 17:49.

Priest, E. R. 1982. *Solar MHD.* Dordrecht: Reidel.

Priest, E. R. 1985. *Solar System Magnetic Fields.* Dordrecht: Reidel.

Priest, E. R. 1989. *Dynamics and Structure of Quiescent Solar Prominences.* Dordrecht: Kluwer.

Priest, E. R., and T. G. Forbes. 1986. New models for fast steady state magnetic reconnection. *J. Geophys. Res.* 91:5579.

Priest, E. R., A. Hood, and U. Anzer. 1989. A twisted flux tube model for solar prominences. *Astrophys. J.* 344:1010–25.

Quest, K. B. 1988. Theory and simulation of collisionless parallel shocks. *J. Geophys. Res.* 93:9649–80.

Quest, K. B., and F. V. Coroniti. 1981. Tearing at the dayside magnetopause. *J. Geophys. Res.* 86:3289.

Ratcliffe, J. A. 1972. *An Introduction to the Ionosphere and Magnetosphere*. Cambridge University Press.
Rees, M. H. 1989. *Physics and Chemistry of the Upper Atmosphere*. Cambridge University Press.
Rees, M. H., and R. G. Roble. 1986. Excitation of O(^{1}S) atoms in aurora and emission of the (OI) 6300A line. *Can. J. Phys.* 64:1608–13.
Reiff, P. H., and J. G. Luhmann. 1986. Solar wind control of the polar cap voltage. In *Solar-Wind Magnetosphere Coupling*, ed. Y. Kamide and J. A. Slavin (pp. 453–76). Tokyo: Terra Scientific Publishing.
Rishbeth, H., and O. K. Garriott. 1969. *Introduction to Ionospheric Physics*, Int. Geophys. Ser., vol. 14 (pp. 1–330). New York: Academic Press.
Roederer, J. G. 1967. On the adiabatic motion of energetic particles in a model magnetosphere. *J. Geophys. Res.* 72:981–92.
Roederer, J. G. 1970. *Dynamics of Geomagnetically Trapped Radiation*. Berlin: Springer-Verlag.
Rosenberg, T. J., D. L. Detrick, D. Venkatesan, and G. van Bauel. 1991. A comparative study of imaging and broad-beam riometer measurements: the effect of spatial structure on the frequency dependence of auroral absorption. *J. Geophys. Res.* 96:17793–803.
Rossi, B., and S. Olbert, 1970. *Introduction to Space Physics*. New York: McGraw-Hill.
Rostoker, G. 1972. Geomagnetic indices. *Rev. Geophys. Space Phys.* 10:935–50.
Rostoker, G. 1983. Triggering of expansive phase intensifications of magnetospheric substorms by northward turnings of the interplanetary magnetic field. *J. Geophys. Res.* 88:6981–93.
Rostoker, G., S.-I. Akasofu, W. Baumjohann, Y. Kamide, and R. L. McPherron. 1987a. The roles of direct input of energy from the solar wind and unloading of stored magnetotail energy in driving magnetospheric substorms. *Space Sci. Rev.* 46:93–111.
Rostoker, G., S.-I. Akasofu, J. Foster, R. A. Greenwald, Y. Kamide, K. Kawasaki, A. T. Y. Lui, R. L. McPherron, and C. T. Russell. 1980. Magnetospheric substorms – definition and signatures. *J. Geophys. Res.* 85(A4):1663–8.
Rostoker, G., and T. Eastman. 1987. A boundary layer model for magnetospheric substorms. *J. Geophys. Res.* 92:12187–201.
Rostoker, G., A. Vallance Jones, R. L. Gattinger, C. D. Anger, and J. S. Murphree. 1987b. The development of the substorm expansive phase: the "eye" of the substorm. *Geophys. Res. Lett.* 14(4):399–402.
Rothwell, P. L., L. P. Block, M. B. Silevitch, and C.-G. Falthammar. 1988. A new model for substorm onsets: the pre-breakup and triggering regimes. *Geophys. Res. Lett.* 15(11):1279–82.
Rothwell, P. L., L. P. Block, M. B. Silevitch, and C.-G. Falthammar. 1989. A new model for auroral breakup during substorms. *IEEE Trans. Plasma Sci.* 17(2):150–7.
Rothwell, P. L., M. B. Silevitch, and L. P. Block. 1984. A model for the propagation of the westward traveling surge. *J. Geophys. Res.* 89(A10):8941–8.
Russell, C. T. 1972. The configuration of the magnetosphere. In *Critical Problems of Magnetospheric Physics*, ed. E. R. Dyer (p. 1). Washington, DC: IUCSTP, National Academy of Sciences.
Russell, C. T. 1990. The magnetopause. In *Physics of Magnetic Flux Ropes, Geophysical Monograph 58*, ed. C. T. Russell, E. R. Priest, and L. C. Lee (p. 439). Washington, DC: American Geophysical Union.
Russell, C. T., D. N. Baker, and J. A. Slavin. 1988. The magnetosphere of Mercury. In *Mercury*, ed. F. Vilas, C. R. Chapman, and M. S. Matthews (pp. 514–61). Tucson: University of Arizona Press.

Russell, C. T., and R. C. Elphic. 1978. Initial ISEE magnetometer results: magnetopause observations. *Space Sci. Rev.* 22:681.

Russell, C. T., and R. C. Elphic. 1979. Observations of flux ropes in the Venus ionosphere. *Nature* 279:616.

Russell, C. T., and M. M. Hoppe. 1983. Upstream waves and particles. *Space Sci. Rev.* 34:155–72.

Russell, C. T., R. P. Lepping, and C. W. Smith. 1990. Upstream waves at Uranus. *J. Geophys. Res.* 95:2273–9.

Russell, C. T., and B. R. Lichtenstein. 1975. On the source of lunar limb compressions. *J. Geophys. Res.* 80:4700.

Russell, C. T., and R. L. McPherron. 1973. The magnetotail and substorms. *Space Sci. Rev.* 11:111–22.

Russell, C. T., R. L. McPherron, and P. J. Coleman, Jr. 1972. Fluctuating magnetic fields in the magnetosphere. 1. ELF and VLF fluctuations. *Space Sci. Rev.* 12:810.

Russell, C. T., E. J. Smith, B. T. Tsurutani, J. T. Gosling, and S. J. Bame. 1983. Multiple spacecraft observations of interplanetary shocks: characteristics of the upstream ULF turbulence. In *Solar Wind Five,* NASA CP 2280, ed. M. Neugebauer (pp. 385–400). Washington, DC: NASA.

Russell, C. T., P. Song, and R. P. Lepping. 1989. The Uranian magnetopause: lessons from earth. *Geophys. Res. Lett.* 16:1485–8.

Russell, C. T., and O. Vaisberg. 1983. The interaction of the solar wind with Venus. In *Venus,* ed. D. M. Hunten, L. Colin, T. M. Donahue, and V. I. Moroz (p. 873). Tucson: University of Arizona Press.

Samson, J. C., and G. Rostoker. 1972. Latitude-dependent characteristics of high latitude Pc 4 and Pc 5 micropulsations. *J. Geophys. Res.* 77:6133.

Sandholt, P. E., and A. Egeland (eds.). 1989. *Electromagnetic Coupling in the Polar Clefts and Caps.* Dordrecht: Kluwer.

Sato, T. 1978. A theory of quiet auroral arcs. *J. Geophys. Res.* 83(A3):1042–8.

Saunders, M. A., and C. T. Russell. 1986. Average dimension and magnetic structure of the distant Venus magnetotail. *J. Geophys. Res.* 91:5589.

Scarf, F. L., D. A. Gurnett, and W. S. Kurth. 1979. Jupiter plasma wave observations: an initial *Voyager 1* overview. *Science* 204:991–5.

Scarf, F. L., D. A. Gurnett, and W. S. Kurth. 1981. Measurements of plasma wave spectra in Jupiter's magnetosphere. *J. Geophys. Res.* 86:8181–98.

Scarf, F. L., D. A. Gurnett, W. S. Kurth, and R. L. Poynter. 1979b. Plasma wave turbulence at Jupiter's bow shocks. *Nature* 280:796–8.

Schatten, K. H., and J. M. Wilcox. 1967. Response of the geomagnetic activity index Kp to the interplanetary magnetic field. *J. Geophys. Res.* 72(21):5185–91.

Schindler, K. 1974. A theory of the substorm mechanism. *J. Geophys. Res.* 79:2803.

Scholer, M. 1988. Magnetic flux transfer at the magnetopause based on single x-line bursty reconnection. *Geophys. Res. Lett.* 15:291.

Schulz, M., and L. J. Lanzerotti. 1974. *Particle Diffusion in the Radiation Belts.* Berlin: Springer-Verlag.

Schunk, R. W. 1983. The terrestrial ionosphere. In *Solar Terrestrial Physics,* ed. R. L. Carovillano and J. M. Forbes (pp. 609–76). Dordrecht: Reidel.

Schunk, R. W., and A. F. Nagy. 1980. Ionospheres of the terrestrial planets. *Rev. Geophys. Space Phys.* 18:813.

Sckopke, N. 1966. A general relation between the energy of trapped particles and the disturbance field near the earth. *J. Geophys. Res.* 71:3125.

Sckopke, N., G. Paschmann, S. J. Bame, J. T. Gosling, and C. T. Russell. 1983. Evolution of ion distributions across the nearly perpendicular bow shock: specularly and non-specularly reflected-gyrating ions. *J. Geophys. Res.* 88:6121–36.

Shinagawa, H., and T. E. Cravens. 1988. A one-dimensional multi-species magnetohydrodynamic model of the dayside ionosphere of Venus. *J. Geophys. Res.* 93:11263.

Shinagawa, H., and T. E. Cravens. 1989. A one-dimensional multispecies magnetohydrodynamic model of the dayside ionosphere of Mars. *J. Geophys. Res.* 94:6506.

Shinagawa, H., T. E. Cravens, and A. F. Nagy. 1987. A one-dimensional time-dependent model of the magnetized ionosphere of Venus. *J. Geophys. Res.* 92:7317.

Siscoe, G. L. 1987. The magnetospheric boundary. In *Physics of Space Plasmas (1987),* ed. T. Chang, G. B. Crew, and J. R. Jasperse (p. 3). Cambridge, MA: Scientific Publishers.

Slavin, J. A., D. N. Baker, J. D. Craven, R. C. Elphic, D. H. Fairfield, L. A. Frank, A. B. Galvin, W. J. Hughes, R. H. Manka, D. G. Mitchell, I. G. Richardson, T. R. Sanderson, D. J. Sibeck, E. J. Smith, and R. D. Zwickl. 1989. CDAW 8 observations of plasmoid signatures in the geomagnetic tail: an assessment. *J. Geophys. Res.* 94(A11):15153–75.

Slavin, J. A., and R. E. Holzer. 1982. The solar wind interaction with Mars revisited. *J. Geophys. Res.* 87:10285.

Slavin, J. A., E. J. Smith, D. G. Sibeck, D. N. Baker, R. D. Zwickl, and S.-I. Akasofu. 1985. An ISEE 3 study of average and substorm conditions in the distant magnetotail. *J. Geophys. Res.* 90:10875.

Smith, E. J., and B. T. Tsurutani. 1983. Saturn's magnetosphere: observations of ion cyclotron waves near the Dione L shell. *J. Geophys. Res.* 88:7831–6.

Smith, M. F., and M. Lockwood. 1990. The pulsating cusp. *Geophys. Res. Lett.* 17:1069.

Smith, P. H., and R. A. Hoffman. 1973. Ring current particle distributions during the magnetic storms of December 16–18, 1971. *J. Geophys. Res.* 78:4731.

Smith, R. A., C. K. Goertz, and W. Grossman. 1986. Thermal catastrophe in the plasma sheet boundary layer. *J. Geophys. Res.* 13(13):1380–3.

Snyder, C. W., M. Neugebauer, and U. R. Rao. 1963. The solar wind velocity and its correlation with cosmic-ray variations and with solar and geomagnetic activity. *J. Geophys. Res.* 68(24):6361–70.

Sonnerup, B. U. O. 1970. Magnetic field reconnection in a highly conducting incompressible fluid. *J. Plasma Phys.* 4:161.

Sonnerup, B. U. O. 1984. Magnetic field reconnection at the magnetopause: an overview. In *Magnetic Reconnection in Space and Laboratory Plasmas, Geophysical Monograph 30,* ed. E. W. Hones, Jr. (p. 92). Washington, DC: American Geophysical Union.

Sonnerup, B. U. O., G. Paschmann, I. Papamastorakis, N. Sckopke, G. Haerendel, S. J. Bame, J. R. Ashbridge, J. T. Gosling, and C. T. Russell. 1981. Evidence for magnetic field reconnection at the earth's magnetopause. *J. Geophys. Res.* 86:10049.

Southwood, D. J. 1974. Some features of field line resonances in the magnetosphere. *Planet. Space Sci.* 22:483.

Southwood, D. J., C. J. Farrugia, and M. A. Saunders. 1988. What are flux transfer events? *Planet. Space Sci.* 36:503.

Southwood, D. J., and M. G. Kivelson. 1990. The magnetohydrodynamic response of the magnetospheric cavity to changes in solar wind pressure. *J. Geophys. Res.* 95:2301.

Southwood, D. J., M. G. Kivelson, R. J. Walker, and J. A. Slavin. 1980. Io and its plasma environment. *J. Geophys. Res.* 85:5959–68.

Speiser, T. W. 1965. Particle trajectories in model current sheets. 1. Analytical solutions. *J. Geophys. Res.* 70:4219.

Spence, H. E., and M. G. Kivelson. 1990. The variation of the plasma sheet polytropic index along the midnight meridian for a finite width magnetotail. *Geophys. Res. Lett.* 17:591–4.

Spiro, R. W., R. A. Wolf, and B. G. Fejer. 1988. Penetration of high-latitude-electric-field effects to low latitudes during SUNDIAL 1984. *Ann. Geophys.* 6:39–50.

Spjeldvik, W. N., and P. L. Rothwell. 1983. The earth's radiation belts. In *Handbook of Geophysics and the Space Environment,* AFGL-TR-88-0240, ed. A. S. Jursa. Air Force Geophysics Laboratory, Air Force Systems Command.

Spreiter, J. R., M. C. Marsh, and A. L. Summers. 1970. Hydromagnetic aspects of solar wind flow past the moon. *Cosmic Electrodyn.* 1:5.

Spreiter, J. R., and S. S. Stahara. 1980. A new predictive model for determining solar wind–terrestrial planet interactions. *J. Geophys. Res.* 85:6769–77.

Spreiter, J. R., A. L. Summers, and A. Y. Alksne. 1966. Hydromagnetic flow around the magnetosphere. *Planet. Space Sci.* 14:223–53.

Stewart, B. 1861. On the great magnetic disturbance which extended from August 28 to September 7, 1859, as recorded by photography at the Kew Observatory. *Philos. Trans. R. Soc. London* 423.

Størmer, C. 1955. *The Polar Aurora.* Oxford: Clarendon Press.

Sugiura, M., and D. J. Poros. 1973. A magnetospheric magnetic field model incorporating the *OGO 3* and *5* magnetic field observations. *Planet. Space Sci.* 21:1763.

Sweet, P. A. 1958. The neutral point theory of solar flares. In *Electromagnetic Phenomena in Cosmical Physics,* ed. B. Lehnert. Cambridge University Press.

Takahashi, K., R. L. McPherron. 1982. Harmonic structure of Pc 3-4 pulsations. *J. Geophys. Res.* 87:1504.

Takahashi, K., and R. L. McPherron, and W. J. Hughes. 1984. Multispacecraft observations of the harmonic structure of Pc 3-4 magnetic pulsations. *J. Geophys. Res.* 89:6758.

Takahashi, K., L. J. Zanetti, R. E. Lopez, R. W. McEntire, T. A. Potemra, and K. Yumoto. 1987. Disruption of the magnetotail current sheet observed by AMPTE/CCE. *Geophys. Res. Lett.* 14(10):1019–22.

Tandberg-Hanssen, E. (ed.). 1990. *Dynamics of Quiescent Prominences,* IAU Colloquium 117. Berlin: Springer-Verlag.

Taylor, J. B. 1974. Relaxation of toroidal plasma and generation of reverse magnetic fields. *Phys. Rev. Lett.* 33:1139.

Tinsley, B. A. 1976. Evidence that the recovery phase ring current consists of helium ions. *J. Geophys. Res.* 81:6193–6.

Tromholt, S. 1885. *Under the Rays of the Aurora Borealis: In the Land of the Lapps and Kvaens,* ed. C. Siewers. Boston: Houghton Mifflin.

Troshichev, O. A., V. G. Andrezen, S. Vennerstrom, and E. Friss-Christensen. 1988. Relationship between the polar cap activity index PC and the auroral zone indices AU, AL, AE. *Planet. Space Sci.* 36:1095.

Tsurutani, B. T., and R. G. Stone (eds.). 1985. *Collisionless Shocks in the Heliosphere: Reviews of Current Research.* Washington, DC: American Geophysical Union.

Tsyganenko, N. A., and A. V. Usmanov. 1982. Determination of the magnetospheric current system parameters and development of experimental geomagnetic field models based on data from IMP and HEOS satellites. *Planet. Space Sci.* 30:985–98.

Unti, T., and G. Atkinson. 1968. Two-dimensional Chapman-Ferraro problem with neutral sheet. 1. The boundary. *J. Geophys. Res.* 73:2319.

Valladares, C. E., and H. C. Carlson. 1991. The electrodynamics, thermal and energetic character of intense sun-aligned arcs in the polar cap. *J. Geophys. Res.* 96:1379–400.

Vallance Jones, A. 1974. *Aurora*. Dordrecht: Reidel.

Van Ballegooijen, A., and P. Martens. 1989. Formation and eruption of solar prominences. *Astrophys. J.*, 343:971–84.

Vasyliunas, V. M. 1968a. A survey of low-energy electrons in the evening sector of the magnetosphere with *OGO 1* and *OGO 3*. *J. Geophys. Res.* 73:2839–84.

Vasyliunas, V. M. 1968b. Low-energy electrons in the magnetosphere as observed by *OGO-1* and *OGO-3*. In *Physics of the Magnetosphere*, ed. R. L. Carovillano (p. 622). Dordrecht: Reidel.

Vasyliunas, V. M. 1975. Theoretical models of magnetic field line merging. *Rev. Geophys. Space Phys.* 13:303–36.

Vasyliunas, V. M. 1979. Interaction between the magnetospheric boundary layers and the ionosphere. In *Proceedings of the Magnetospheric Boundary Layer Symposium*, ESA SP-148, ed. B. Battrick (p. 387). Noordwijk: ESA.

Vasyliunas, V. M. 1983. Plasma distribution and flow. In *Physics of the Jovian Magnetosphere*, ed. A. J. Dessler (pp. 395–453). Cambridge University Press.

Vasyliunas, V. M., J. R. Kan, G. L. Siscoe, and S. -I. Akasofu. 1982. Scaling relations governing magnetospheric energy transfer. *Planet. Space Sci.* 30(4):359–65.

Vegard, L. 1939. Hydrogen showers in the auroral region. *Nature* 144:1089.

Vennerstrom, S., and E. Friis-Christensen. 1987. On the role of IMF B-y in generating the electric field responsible for the flow across the polar cap. *J. Geophys. Res.* 92(A1):195–202.

Voigt, G. -H. 1981. A mathematical magnetospheric model with independent physical parameters. *Planet. Space Sci.* 29:1–20.

Walker, R. J., K. N. Erickson, R. L. Swanson, and J. R. Winckler. 1976. Substorm-associated particle boundary motion at synchronous orbit. *J. Geophys. Res.* 81(31):5541–50.

Walker, R. J., K. N. Erickson, and J. R. Winckler. 1978. Pitch angle dispersions of drifting energetic protons at synchronous orbit. *J. Geophys. Res.* 83(A4):1595–600.

Wallis, D. D., J. R. Burrows, T. J. Hughes, and M. D. Wilson. 1982. Eccentric dipole coordinates for MAGSAT data presentation and analysis of external current effects. *Geophys. Res. Lett.* 9(4):353–6.

Wentzel, D. G. 1989. *The Restless Sun*. Washington, DC: Smithsonian Institution Press.

Wilcox, J., and N. F. Ness. 1965. Quasi-stationary corotating structure in the interplanetary medium. *J. Geophys. Res.* 70(23):5793–805.

Williams, D. J. 1980. Ring current composition and sources. In *Dynamics of the Magnetosphere*, ed. S. -I. Akasofu (p. 407). Dordrecht: Reidel.

Williams, D. J. 1987. Ring current and radiation belts. *Rev. Geophys.* 25:570–8.

Winckler, J. R. 1980. The application of artificial electron beams to magnetospheric research. *Rev. Geophys. Space Phys.* 18:659–82.

Wolf, R. A. 1983. The quasi-static (slow-flow) region of the magnetosphere. In *Solar Terrestrial Physics*, ed. R. L. Carovillano and J. M. Forbes (pp. 303–68). Dordrecht: Reidel.

Wolf, R. A., S.-I. Akasofu, S. W. H. Cowley, R. L. McPherron, G. Rostoker, G. L. Siscoe, and B. V. O. Sonnerup. 1986. Coupling between the solar wind and the earth's magnetosphere: summary comments. In *Solar Wind–Magnetosphere Coupling*, ed. Y. Kamide and J. Slavin (pp. 769–807). Tokyo: Terra Scientific.

Wolff, R. S., B. E. Goldstein, and C. M. Yeates. 1980. The onset and development of Kelvin-Helmholtz instability at the Venus ionopause. *J. Geophys. Res.* 85:7697.

Woltjer, L. 1958. A theorem on force-free fields. *Proc. Natl. Acad. Sci. USA* 44:489.

Wu, C. S., D. Winske, Y. M. Zhou, S. T. Tsai, P. Rodriguez, M. Tanaka,

K. Papadopoulos, K. Akimoto, C. S. Lin, M. M. Leroy, and C. C. Goodrich. 1984. Microinstabilities associated with a high Mach-number, perpendicular shock. *Space Sci. Rev.* 37:65.

Zhu, X. M., and M. G. Kivelson. 1989. Global mode ULF pulsations in a magnetosphere with a nonmonotonic Alfvén velocity profile. *J. Geophys. Res.* 94:1479.

Zwingmann, W. 1986. Quasistationare Entwicklung und einsatz eruptive Prozesse von plasma Strukturen in Magnetfeld mit Andwendurg auf die Sonnenatmosphäre. Ph.D. thesis, Bochum University.

INDEX

active region, solar, 61, 79
adiabatic invariants
   bounce, 167
   first, 33, 308–9
   second, 33, 167, 308–9
   third, 33–4, 50
Alfvén, Hannes, 14–15
Alfvén current sheet, 253
Alfvén layer, 314–20
Alfvén velocity, 51, 67, 337
Alfvén waves, 94, 390–1
Alfvén wings, 220–1, 513
alpha effect, 74
ambipolar diffusion, 195
Appleton, E. V., 10
Appleton-Hartree dispersion, 386–90
Arnoldy, Roger, 19, 415
attachment recombination, 192
aurora australis, 6, 459
aurora borealis, 2, 6, 459
auroral break-up, 486
auroral clutter, 493
auroral emissions, 10, 467
auroral excitation
   impact excitation, 467
   ionization excitation, 468
   thermal excitation, 473, 475
auroral forms
   arc, 474, 476
   curls, 476
   rays, 474, 476
   spirals, 476
auroral height, 6, 477
auroral intensities, 471, 475–6
auroral oval, 420, 478
auroral photography, 8–9, 476
auroral precipitation, 326, 463–7
auroral scintillations, 493
auroral spectrum, 463, 467–75
auroral zone, 7, 477
auroras
   cleft, 466, 481–3
   cusp, 466, 481–3
   diffuse, 420–1, 476
   discrete, 420–1
   early references, 1–3, 6–9
   polar cap, 483–5

Babcock, H., 11
Barkhausen, H., 12–13
Barnett, M. A. F., 10
Beard, D., 172

Bernouilli's equation, 170
beta, plasma, 17, 50, 505
Biermann, L., 14, 228
bi-Maxwellian distribution, 37
Birkeland, K., 7–8, 463
Boltzmann equation, collisionless, 392–3
bounce period, 54, 306–7
boundary layer, 259–65
   entry layer, 260–3
   high latitude, 261–3
   low latitude, 260–5, 433
   plasma mantle, 261–3
   plasma sheet, 276–83, 291, 432
bow shocks, 17–18, 135–6, 159–60
   location, 178, 504, 509–10
   overshoot, 505
Breit, G., 10
bemsstrahlung, 183, 191, 466

Carpenter, D. L., 298
Carrington, Richard, 6
cascading, 468
Cassini spacecraft, 25, 519
Cavendish, Henry, 6
Celsius, Anders, 5
Chapman, Sydney, 11–12
Chapman and Ferraro model, 14, 164, 168, 228, 410
Chapman production function, 187, 193
Chapman theory, ionosphere, 184
charge exchange, 42, 205, 325–6, 473
chemiluminescence, 468
chromosphere, 59–60
comet tails, 14, 223–4
cometopause, 222
compass, 1, 3–4, 443
conductivity
   Hall, 201
   ionospheric electrical, 6, 48, 201
   Pedersen, 201
conservation equations (*see also* continuity equation), 41–3, 138–9
   of energy, 47, 103, 139, 175
   of mass, 42, 97, 112, 138, 175
   of momentum, 42, 48, 67, 97, 101, 111, 138, 175, 333, 358; ionosphere, 194
contact surface, 222
continuity equation, 41, 68, 100, 138, 175, 332, 358
continuum radiation, 388
convection
   ionospheric, 424, 490

convection (*cont.*)
  magnetospheric, 243, 300–4, 315–18
  polar cap, 424
  separatrix, 316
convection zone, solar, 60, 74
Cook, Captain, 6
coordinate systems, 532–42
  geomagnetic, 403, 533–4
coordinates, GSM, 415, 536–8
coplanarity theorem, 53, 142–3, 162
corona, magnetic structure, 111–18
corona, solar, 59–61, 94–6
coronal features
  bright points, 65
  helmet streamers, 117, 120, 126
  holes, 65, 79, 118, 120
  loops, 65
  mass ejections, 63, 126–7
corotation velocity, 315, 510
Cowling diffusion time, *see* diffusion time
critical radius, 101
Crookes, Sir William, 7
current continuity, 45
current sheets
  Alfvén, 253
  Harris, 250–2
  interplanetary, 116
currents
  auroral zone, 490–2
  Birkeland, 289, 302–3, 321–3, 491
  Chapman-Ferraro, 229–31, 288
  DP-1, 426
  DP-2, 424–6
  field-aligned, 7
  Hall, 491
  ionospheric, 6, 9, 323–4
  partial-ring, 289, 409
  Pedersen, 302, 491
  ring, 12, 289, 297, 310–12, 407, 429
  Sq, 404
  tail, 233–4, 240–2, 251–4, 289, 400
curvature, radius of, 31, 307
cut-off frequency, 382–3
cyclotron radius, 29, 54, 306

Davis–Williamson protons, 296
de Castro, Joao, 3
de Hoffman–Teller frame, 147, 156–8, 275
de Mairan, Jean, 2
Debye length, 39, 41, 364, 377
Debye sphere, 39, 41
decametric emissions, 518
decimetric emissions, 518
declination, 3–4, 403, 532, 540
degrees of freedom, 36
density
  charge, 42, 356
  current, 42, 65, 356
  electron, 11, 13, 41, 192–6, 214, 295, 489
  enthalpy, 47
  mass, 34
  number, 34
  phase space, 34
  solar mean, 58
Descartes, René, 2
Dessler-Parker-Sckopke relation, 312

differential directional flux, 38
diffusion, radial, 319
diffusion/convection equation, 215
diffusion equation, ionosphere, 195
diffusion time, 66, 20
dip angle, 403
dipole magnetic field, 165–6
dipole moment
  magnetic, 12, 166–7
  tilt, 168
distribution functions
  bi-maxwellian, 37
  maxwellian, 35
  phase space, 34
Doppler shift, 367, 395, 473
drift
  curvature, 32, 307, 309–10
  ExB, 31, 307
  gradient, 32, 307, 309–10
  period, 307
  shell-splitting, 312–14
  westward, 168
Dungey, James, 19
Dungey model
  pulsations, 332
  solar wind interaction, 19–20, 243–5, 415
dynamo equation, *see* induction equation

Eckersley, T. L., 13
electrojet, 424, 491–2
  eastward, 424
  westward, 424
electron Volt, 29
energy
  density, internal, 47
  deposition, particles, 190
  kinetic, 35
entropy conservation, 333
Explorer spacecraft
  Explorer 45, 296
  Explorers 10 and 12, 15–16
  Explorers 33 and 35, 204
extraordinary mode wave, *see* waves, X-mode

field line, magnetic, 166–7
filament, solar, 61, 74
flux
  differential directional, 38
  frozen-in, 48–9, 168, 204, 208, 236
  heat, 47
  magnetic, 44
  Poynting, 341–2, 494
flux ropes, 77, 218
Flux Transfer Events (FTEs), 272
flux tube, 49, 63
  model, twisted, 77
force
  curvature, 50
  ion drag, 497
  magnetic, 50
force-free field, 69
foreshock, 155, 158–61
  geometry, 505, 507
Franklin, Captain John, 7
frequencies
  cut-off, 382–3
  cyclotron, 29, 305

gyrofrequency, 29, 40, 305
lower hybrid, 378
plasma, 40, 362
resonant, 345, 383
ultralow, 331
upper hybrid, 376, 383
frozen-in field, *see* flux, frozen-in

Galileo Galilei, 2, 5
Galileo spacecraft, 25, 519
Gassendi, Pierre, 2
Gauss, C. F., 6, 164
geomagnetic cavity, 12, 168–74
geomagnetic disturbances, 6, 228
geomagnetic field, early measurements of, 3–4
geomagnetic pulsations, 7, 330–2, 408
geomagnetic storm, 12–13, 19, 296, 406–8
geomagnetic tail, 232–6, 239
geomagnetism, early studies of, 3–4
geosynchronous orbit plasma, 293–5
Gilbert, William, 3–4
Giotto spacecraft, 23
Goldstein, E., 7
Graham, George, 5
green line, auroral, 11, 469
Gringauz, K. I., 15, 298
group velocity, MHD, *see* velocity, group
gyro resonance, 395–7
gyroradius, 29, 53–4, 306

Halley, Edmund, 2, 4
Harris current sheet, 250–2
Hartman, Georg, 3
heating
    coronal, 79–81
    ionospheric, 495
    Joule, 495–6
    thermospheric, 496–7
Heaviside, O., 10
helio seismology, 62
heliopause, 110
high-speed streams, 122–4
Hiorter, O., 5–6
Hirshberg, Joan, 19
Hoffmeister, C., 14

ICE spacecraft, 23
ideal gas law, 36, 43
impact ionization, 183, 188–91
inclination, 3, 403, 540
induction equation, 66, 215
instabilities
    convective, 70
    interchange modes, 70, 322, 351, 513
    Kelvin-Helmholtz, 70, 349, 493, 510
    magnetic buoyancy, 70
    MHD, 70, 351
    mirror, 351
    radiative, 69–70
    Rayleigh-Taylor, 70
    two-stream, 367–72
International Geophysical Year, 15, 421
Interplanetary Monitoring Platform, 17
invariant latitude, 166
Io, 220, 322, 512–13
    plasma torus, 322, 513

ion drag, 497
ion loss processes, 192–3
ion reflection, 152–4
ionization processes, 489–90
    photoionization, 183–91
ionopause, 209, 214
ionospheres, 9–13, 183–99
    auroral, 489–92
ionospheric layers
    D region, 10, 197–8; absorption, 493
    E region, 10, 197–8
    F region, 10, 197–8
ionospheric outflow, 199–200
ISEE spacecraft
    ISEE 1, 20
    ISEE 2, 20, 266, 271–3, 280–1
    ISEE 3, 235

kappa distribution, 38
Kelvin, Lord, 8–9
Kelvin-Helmholtz instability, 70, 349, 493, 510
Kennelly, A. E., 10
kinetic wave theory, 393

L value, 166–7, 296–300
Landau resonance, 393–5
Larmor radius, 29, 54, 306
Lindemann, Frederick, 12
Liouville theorem, 393
Loomis, Elias, 7
Lorentz force law, 28–9, 357
lunar core, 205
lunar–solar wind interaction, 204
lunar wake, 204

Mach number, 17, 132, 148
    Alfvén, 132, 144
    fast, 132
    slow, 132
    sonic, 132
McIlwain, C. E., 166
McLennan, J., 11
magnetic activity, 402–30
magnetic barrier, 213
magnetic buoyancy, 73
magnetic diffusivity, 66
magnetic field, interplanetary (IMF), 17, 104–7, 118–20
magnetic indices, 408–9
    AE, AL, AU, 455–6
    Ap, 453–4
    Ci, 451–3
    C9, 451–3
    Dst, 312, 407, 456–7
    Kp, 453–4
    PC, 454–5
magnetic induction, 28, 402
magnetic moment, 33, 165–8, 308
    dipole tilt, 168, 508–9
    planetary, 507–8
magnetic reconnection, *see* reconnection, magnetic
magnetic Reynolds number, 66, 237
magnetic storms, 6, 8–9, 12–13, 406–8, 420, 429
magnetohydrodynamics, 41–53, 65–70, 136–45

magnetometers, 1, 5–6
 fluxgate, 448–9
 liquid, 446–8
 vapor, 446–8
magnetopause, 15–16, 228–32
 boundary layers, see boundary layer
 location, 171
 radius, 508
 shape, 171
 tangential stress, 172–3
magnetosheath, 135, 177–9, 209
magnetosphere, 12, 18–21, 68–76, 288–304
 box model of, 343, 347
 planetary, 507–11
 pressure exerted by, 171–2
 shape, 172–4
 sizes, 168–74, 507–11
magnetosphere-ionosphere coupling, 320–3, 436–7
magnetospheric pulsations, see geomagnetic pulsations
magnetospheric substorm, 232, 402, 405–6, 420, 426–9
magnetotail, 232–6
 flaring angle, 234
 length, 245
 lobe, 232–6, 291
 magnetic flux, 173–4, 234
 width, 173–4, 234
Marconi, G., 10
Mariner spacecraft
 Mariner 2, 15, 22
 Mariner 4, 22
 Mariner 5, 22
 Mariner 10, 23–4
mass-loading, 221
Maxwellian distribution, 35–8
Maxwell's equations, 29, 44–50, 65, 175, 333, 356
 Ampere's law, 29, 44, 46, 65, 175, 333
 divergence of the magnetic field, 29, 44, 46, 65, 333
 Faraday's law, 29, 32, 44, 46, 65, 175, 333
 Lorentz force law, 28–9, 357
 Poisson's equation, 29, 44, 46, 65
MHD (see also Maxwell's equations), equations of, 41, 65
MHD discontinuities, 140–1
 contact, 141
 fast mode shock, 53, 140–2
 intermediate shock, 140–1
 oblique shocks, 141, 147
 parallel shock, 141, 143
 perpendicular shock, 141, 143–5
 rotational, 141
 slow mode shock, 53, 140–2
 tangential, 141–2
MHD shocks, see MHD discontinuities
MHD simulation, 175–6
 comet, 222–4
 gasdynamic, 177–80
 hybrid, 176
MHD waves, 51–3, 69–70, 331
 Alfvén, 94, 390–1
 Alfvén, shear, 51, 337, 342
 fast mode, 52, 339–40
 intermediate mode, 52
 slow mode, 52, 334, 342
mirror force, 33
mirror point, 33
moon, 203
moons, interactions, 512–14
multipole expansion, 167–8

Neekan, Alexander, 3
Neugebauer, Marcia, 15
Norman, Robert, 3
northern light, see aurora borealis

O-line, 428
observatories, magnetic, 408, 444, 450–1
Ohm's law, 45, 47–8, 65, 333
omega bands, 421
optical depth, 185
Orbiting Geophysical Observatory, 17
ordinary mode wave, see waves, O-mode

Parker, E. W., 14, 81, 100
particle loss, 325–6
particles, upstream, 17–19, 158–61, 505–6
penumbra, 71
phase space, 34
phase velocity, 51
Phobos spacecraft, 22
photochemical equilibrium, 193
photoionization, 183–91
photosphere, 59, 61
pickup ion, 206
Pioneer 10 and 11 spacecraft, 24
Pioneer Venus Orbiter, 22, 213
pitch angle, 306
 scattering, 325–6
plasma, two-fluid, 356
plasma beta, 17, 50, 505
plasma frequency, 40, 362
plasma parameter, 39
plasma sheet, central (see also boundary layer, plasma sheet), 276–7, 291
plasmapause, 298–300, 314–20
plasmasphere, 13, 298–300
Podgorny, Igor, 20–1
polar cap (see also auroras, polar cap), 232–3, 420, 424, 430
 arcs, 482, 498–500
 convection, 424
 potential, 302
 potential drop, 498
polar wind, 200, 261, 299
polytropic
 index, 47, 333
 law, 358
positive bays, 423
potential
 electrostatic, 38
 magnetic scalar, 164
 shielded, 38–9
potential drop, field-aligned, 304, 324–5
pressure
 magnetic, 50
 partial, 35
prominences, solar, 61, 74–8, 117
proton precession, 447

pulsations
  continuous, 331
  irregular, 331

quenching, 470

radial diffusion, 319
radiation belts, 8, 15, 20, 295–8
  planetary, 514–15
  Van Allen belts, 297
radio emissions
  planetary, 517–19
  type III solar, 370
radius of curvature, 31, 307
range-energy relation, 188
Rankine-Hugoniot relations, 137–40, 170
ratio of specific heat, 47
Rayleigh, 475
recombination
  attachment, 192
  coefficient, 192
  dissociative, 192
  radiative, 192
reconnection, magnetic, 20, 81–8, 227, 236–42, 415, 438–9
  fluid theory, 246–50
  magnetopause, 265–75
  particle theory, 250–7
  Petschek, 247–50
  planetary, 511–12
  Sweet-Parker, 247
red line, auroral, 11, 469
resonance
  gyro, 395–7
  upper hybrid, 383
resonant frequency, 345, 383

Sabine, Edward, 6
Sakigake, 23
scalar potential, magnetic, 164
scale height, 67, 99, 184
scavenging solar wind, 213
Schuster, A., 9
Schwabe, Heinrich, 6
sector structure, interplanetary, 119–20
secular variation, 4, 168, 457
separatrix, convection, 316
shock dissipation, 130–3, 150
  collective dissipation mechanism, 150
  dissipation mechanism, 137
shock jump conditions, 137–42
shock normal, 17, 161–2
shock structure
  foot, 148, 153–4
  overshoot, 149, 153–4
  ramp, 148, 152
shocks
  bow shock, *see* bow shocks
  collisional, 129
  collisionless, 7, 124, 129–30, 134–6
  interplanetary, 7, 125–6, 136
  magnetohydrodynamic, *see* MHD discontinuities
  planetary, 506
Shon-Kau, 3

Snyder, Conway, 15
solar cycle variations, 410
solar flares, 6, 61, 82–8, 127
solar granulation, 61
solar neutrinos, 60
solar seismology, 62
solar wind, 13–15, 91–127
  dynamic pressure, 169
  properties, 92–6
  radial variation, 504–6
  scavenging, 213
  termination, 107–10
  theory, 96–104
solar wind–comet interaction, 23, 221–4
solar wind–ionosphere interaction, 22, 210
solar wind–magnetosphere coupling, 401
South Atlantic anomaly, 312–14
speed of sound, 47, 367
Spreiter, J. R., 204
sputtering, 205, 512
Stefan-Boltzmann law, 59
Stewart, Balfour, 7, 9
stopband, 383
Storey, L. R. O., 13
Størmer, Carl, 8
substorm, 19
  auroral, 421–3, 486–9
  boundary layer model, 434–5
  coupling model, 436–7
  current disruption, 439–41
  current wedge, 409, 424–6
  driven model, 434
  expansion phase, 421, 428
  growth phase, 427
  magnetospheric, 232, 402, 405–6, 420, 426–9
  near-earth neutral line, 437–9
  polar, 420–6
  recovery phase, 423, 428, 486
  thermal catastrophes, 435–6
  triggering, 418–20
sudden impulse, 229
sudden storm commencement, 229
Suisei spacecraft, 23
sunspot cycle (*see also* variations, solar cycle), 6, 478
sunspots, 5, 71
supergranulation, 63
Sweet-Parker rate, 83–5
Sweet-Parker solution, 246

telescope, 5
temperature
  parallel, 36
  perpendicular, 36
tension, magnetic, 50
thermal speed, 36
Thomson, J. J., 7
Titan, 219, 513–14
transition, forbidden, 470
transition probability, 470
trough, 298
Tuve, M. A., 10

umbra, 71

Van Allen, James, 15
Van Allen belts, 297
variations
   annual, 168, 410–12
   semiannual, 168, 411
   solar cycle, 410
   27 day, 119, 412–14, 478
variometers, 443–6
VEGA 1 and 2 spacecraft, 23
Vegard, Lars, 10, 471
VELA spacecraft, 17
velocity
   Alfvén, 51, 67, 337
   bulk, 35
   drift, 31, 304–7
   flow, 35
   group, 51, 339, 341, 363
   phase, 51
Venera 9 and 10 spacecraft, 22
Vlasov equation, 392–3
VLF emissions, 13
Voyager spacecraft
   Voyager 1, 24, 367, 513, 519
   Voyager 2, 24, 352, 511, 517, 519

waves (*see also* MHD waves)
   compressional, 339
   electromagnetic, 372, 375
   electromagnetic, magnetized, 380–90
   electromagnetic, unmagnetized, 372–9
   electrostatic, 375
   electrostatic ion, 377–9
   ion-acoustic, 366
   ion-plasma, 365–7
   L-O mode, 388
   L-X mode, 388
   Langmuir, 360–4
   longitudinal, 375
   magnetized plasma, 375
   magnetoacoustic, 340, 391
   magnetosonic, 340, 391
   O-mode, 381–2
   Pc, 331
   Pi, 331
   R-X mode, 388
   standing, 330, 343–4
   transverse, 375
   ULF, 331, 408; planetary, 516
   upper hybrid, 376–7
   upstream, 17, 19, 158–61
   VLF, planetary, 516
   whistlers, 12, 298, 385–6
   X-mode, 382
   Z-mode, 388
westward traveling surge, 421
whistler, *see* waves, whistler
Wilcke, J. C., 6

X-line, 428–9
Xenophanes, 1–2

Zeeman effect, 447